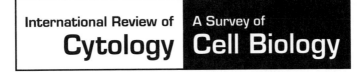

International Review of Cytology A Survey of Cell Biology

VOLUME 157

International Review of

Cytology

A Survey of

Cell Biology

Edited by

Kwang W. Jeon
Department of Zoology
University of Tennessee
Knoxville, Tennessee

Jonathan Jarvik
Department of Biological Sciences
Carnegie Mellon University
Pittsburgh, Pennsylvania

VOLUME 157

ACADEMIC PRESS
San Diego   New York   Boston   London   Sydney   Tokyo   Toronto

Academic Press, Inc.
A Division of Harcourt Brace & Company
525 B Street, Suite 1900, San Diego, California 92101-4495

United Kingdom Edition published by
Academic Press Limited
24-28 Oval Road, London NW1 7DX

International Standard Serial Number: 0074-7696

International Standard Book Number: 0-12-364560-3

PRINTED IN THE UNITED STATES OF AMERICA
95 96 97 98 99 00 BB 9 8 7 6 5 4 3 2 1

CONTENTS

Centrifuge Microscope as a Tool in the Study of Cell Motility

Yukio Hiramoto and Eiji Kamitsubo

Murine B Cell Development: Commitment and Progression from Multipotential Progenitors to Mature B Lymphocytes

Barbara L. Kee and Christopher J. Paige

Signal Transduction by the Antigen Receptors of B and T Lymphocytes

Michael R. Gold and Linda Matsuuchi

Neuroendocrine Epithelial Cell System in Respiratory Organs of Air-Breathing and Teleost Fishes

Giacomo Zaccone, Salvatore Fasulo, and Luigi Ainis

Structure–Function Relationships in Gap Junctions

Hartwig Wolburg and Astrid Rohlmann

CONTRIBUTORS

Numbers in parentheses indicate the pages on which the authors' contributions begin.

Luigi Ainis (277), *Department of Animal Biology and Marine Ecology, Faculty of Science, University of Messina, I-98166 Messina, Italy*

Salvatore Fasulo (277), *Department of Animal Biology, Faculty of Science, University of Catania, I-95124 Catania, Italy*

Michael R. Gold (181), *Department of Microbiology and Immunology, University of British Columbia, Vancouver, British Columbia, Canada V6T 1Z3*

Yukio Hiramoto (99), *Biological Laboratory, The University of the Air, Chiba 261, Japan*

Eiji Kamitsubo (99), *Biological Laboratory, Hitotsubashi University, Tokyo 186, Japan*

Barbara L. Kee (129), *The Wellesley Hospital Research Institute and Department of Immunology, University of Toronto, Toronto, Ontario, Canada M4Y 1J3*

Linda Matsuuchi (181), *Department of Zoology, University of British Columbia, Vancouver, British Columbia, Canada V6T 1Z3*

Christopher J. Paige (129), *The Wellesley Hospital Research Institute and Department of Immunology, University of Toronto, Toronto, Ontario, Canada M4Y 1J3*

Astrid Rohlmann (315), *Department of Anatomy, Developmental Neurobiology Unit, University of Göttingen, D-37075 Göttingen, Germany*

Elisabeth Strömberg (1), *Department of Zoophysiology, University of Göteborg, S-413 90 Göteborg, Sweden*

Sumio Takahashi (33), *Department of Biology, Faculty of Science, Okayama University, Tushima, Okayama 700, Japan*

Margareta Wallin (1), *Department of Zoophysiology, University of Göteborg, S-413 90 Göteborg, Sweden*

Hartwig Wolburg (315), *Institute of Pathology, University of Tübingen, D-72076 Tübingen, Germany*

Giacomo Zaccone (277), *Department of Animal Biology and Marine Ecology, Faculty of Science, University of Messina, I-98166 Messina, Italy*

Cold-Stable and Cold-Adapted Microtubules

Margareta Wallin and Elisabeth Strömberg
Department of Zoophysiology, University of Göteborg, S-413 90 Göteborg,
Sweden

Most mammalian microtubules disassemble at low temperatures, but some are
cold stable. This probably has little to do with a need for cold-stable microtubules,
but reflects that certain populations of microtubules must be stabilized for specific
functions. There are several routes by which to achieve cold stability. Factors that
interact with microtubules, such as microtubule-associated proteins, STOPs (stable
tubule only polypeptides), histones, and possibly capping factors, are involved.
Specific tubulin isotypes and posttranslational modifications might also be of
importance. More permanent stable microtubules can be achieved by bundling
factors, associations to membranes, as well as by assembly of microtubule doublets
and triplets. This is, however, not the explanation for cold adaptation of
microtubules from poikilothermic animals, that is, animals that must have all their
microtubules adapted to low temperatures. All evidence so far suggests that cold
adaptation is intrinsic to the tubulins, but it is unknown whether it depends on
different amino acid sequences or posttranslational modifications.
KEY WORDS: Microtubules, Tubulin, Microtubule-associated proteins, Cold
adaptation, Cold stability.

I. Introduction

Many animals are adapted to low temperatures, and must have microtubules that can fulfill their tasks at such temperatures. Cold adaptation
is, of course, only one specific example of adaptations of organisms to
environmental differences (for a discussion of cold adaptation in general,
see Clarke, 1991). Most studies of microtubules have been made on homeotherm mammalian material and many of the properties found are
thought of as original. Life was created in the sea and the first eukaryotic

1

cells developed at temperatures well beneath 37°C, a normal body temperature for the homeotherm mammal. Therefore, some of the properties (e.g., the fact that mammalian microtubules disassemble when the temperature is lowered to 20°C) might be due to secondary changes of microtubules in these animals. Stability to the temperature of the pre-Cambrian sea must have been a prerequisite for the creation and survival of the first eukaryotic organism.

What characteristics at the molecular level cause microtubules to become cold adapted? Even if subpopulations of stable microtubules that do not disassemble at low temperatures are found in mammals, organisms living at low temperatures must adapt all their microtubules. It is not necessary to demonstrate that cold-adapted animals have cold-adapted microtubules, but to understand the mechanisms by which they achieve this; this is the topic of this chapter. We do not restrict it to cold-adapted microtubules, but also include cold-stable microtubules found in other organisms. The latter does not necessarily reflect cold adaptation, but we deem it valuable to describe and understand different mechanisms that render microtubules cold stable. Rather than describing all presently known cold-stable microtubules, we try to give as many different examples as possible. Our chapter starts with a short general description of microtubules, based mainly on experiments on the well-studied mammalian microtubules, in order to make it easier for the reader to follow the different arguments given for the reason(s) for cold adaptation and stability.

II. Microtubule Structure and Functions

Microtubules are organelles found in all eukaryotic cells. They participate in a wide variety of cellular functions, many of which are transport processes. The movement of flagella and cilia renders a whole cell motile, or moves liquid over a cell surface. The mitotic spindle is composed of microtubules, the role of which is to segregate chromosomes during meiosis and mitosis. Microtubules are also involved in intracellular transport processes such as axonal transport and transport of pigment granulae in chromatophores, as well as in cell motility, secretion, determination, and maintenance of cell shape (for reviews, see Dustin, 1984; Bershadsky and Vasiliev, 1988; Amos and Amos, 1991; Bray, 1992).

The structure of microtubules is similar in all eukaryotic cells. They are long, tubelike structures composed of tubulin with an outer and inner diameter of about 25 and 15 nm, respectively. They differ in length, but can be up to 100 μm long. Tubulin is a heterodimer consisting of two related (but not identical) tubulins, α- and β-tubulin, with a size of 4 $\times$

8 nm. The tubulin dimers form the backbone of microtubules, and align longitudinally in rows called protofilaments. Generally 13 protofilaments form a microtubule. A heterogeneous group of proteins, the microtubule-associated proteins (MAPs), bind to and decorate the outer surface of microtubules.

Microtubules are dynamic polymers that continually assemble and disassemble in cells and *in vitro,* a property termed dynamic instability (Mitchison and Kirschner, 1984; Saxton *et al.,* 1984; Wadsworth and Salmon, 1986; Cassimeris *et al.,* 1987). Individual microtubules increase or decrease in length, with frequent transitions between the two phases (Fig. 1). The rates of growth and shortening differ markedly; lengthening is much slower than shortening. During incorporation of tubulin dimers into a microtubule, the GTP bound to the tubulin dimers is hydrolyzed to GDP. Mitchison and Kirschner (1984) proposed that dynamic instability depends on the presence of a "cap" of tubulin dimers with unhydrolyzed GTP at the plus end, or growing end, of the microtubule. This means that hydrolyzis of GTP lags slightly behind assembly. The assembly continues as long as the cap remains, but if it is lost and the microtubule end contains only tubulin dimers with GDP, the microtubule rapidly disassembles. In many cases populations of microtubules are selectively stabilized, for example, during morphogenesis (Kirschner and Mitchison, 1986), and during mitosis when kinetochore microtubules are more stable than the other mitotic microtubules (Brinkley and Cartwright, 1975). Because the overall assembly of microtubules is the net result of growth and shortening, the cell can by a so far unknown mechanism modify the stability of the microtubules by modulating rates of assembly, disassembly, as well as the frequency of catastrophes and rescues. Kirschner and Mitchison (1986) have suggested that one way to stabilize microtubules is to cap their ends against disassembly. Capping could be caused either by a blocking of the microtubule end directly, or by a lateral interaction near the end. Prevention of GTP hydrolysis could also stabilize a microtubule end. This stabilization may thereafter gradually be converted to a more permanent stabilization.

Microtubules assemble into more complex and stable arrangements. In centrioles, one finds nine triplets of microtubules. A triplet consist of one complete microtubule and two partial microtubules, the first of which binds to the complete microtubule, and the other to the partial microtubule. In cilia and flagella, microtubules are arranged in nine doublets with two single microtubules in the middle of the axoneme. Microtubules assemble into other complex arrangements. Several unicellular organisms capture food, move, or fix themselves to a surface by means of long, slender extensions. Such axopodia contain up to more than 100 microtubules linked to each other in a specific way.

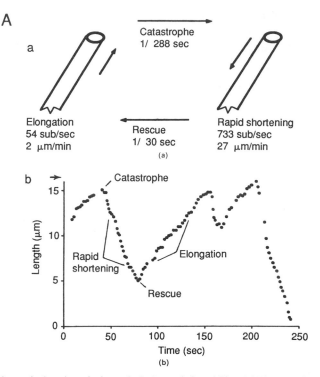

FIG. 1   (A) Schematic drawing of microtubule dynamic instability. (a) The two phases (elongation and rapid shortening) and the two transitions (catastrophe and rescue) of dynamic
instability. The numerical values are rates observed for the plus end of free tubulin at a
concentration of 11 μM, the estimated steady state concentration. sub, subunit. (b) Life
history of a microtubule undergoing dynamic instability inside a living cell. The rates of
growth and rapid shortening are constant for extended periods, and the transitions are abrupt
and stochastic. [Reproduced with permission, from Erickson, H. P., and O'Brien, E. T.
(1992). *Annu. Rev. Biophys. Biomol. Struct.* **21,** 145–166, © 1992, by Annual Reviews, Inc.]
(B) Growth and shortening of microtubules in dynamic instability. A solution of tubulin
subunits is placed between coverslips to which short pieces of flagellar axonemes have been
allowed to adhere. The axonemes serve as nucleating structures. Microtubule assembly is
initiated by raising the temperature from 0 to 37°C. The resulting spontaneous growth and
shortening microtubules can be seen in this series of photographs, taken from videotaped
data. In each frame two microtubules are visible, extending rightward from the thicker-
appearing axoneme (a third, whose end is not visible, extends leftward). Changes in the
lengths of these two microtubules can be observed between one frame and the next. Notice
that each microtubule grows and shortens apparently independent of the other. The time,
in seconds, is shown by the number in each frame. The distance across each photograph
is approximately 20 μm. Conditions: 13 μM bovine tubulin in 0.1 M PIPES buffer (pH 6.9),
1 mM MgSO$_4$, 1 mM dithioerythritol, 1 mM EGTA, 1 mM GTP. [R. C. Williams, Jr.,
unpublished observations; further information can be found in Williams (1992) and in Gildersleeve *et al.* (1992).]

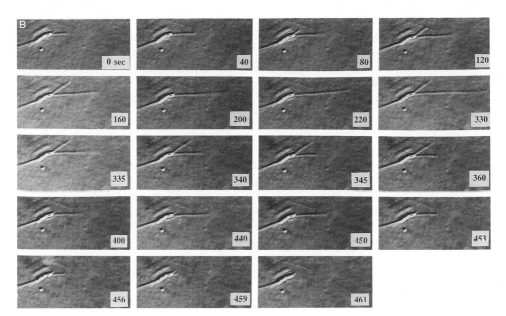

III. Properties of Microtubules

A. Tubulin Isotypes and Posttranslational Modifications

How can microtubules, composed of tubulin and microtubule-associated proteins (MAPs), take part in so many different functions? Several possibilities exist. Different tubulin forms can be generated either by the expression of distinct tubulin genes, or by posttranslational modifications. There are several different isotypes of α- and β-tubulins (see Luduena, 1993, for a review). In mammals there are up to six α-tubulin genes and six β-tubulin genes that can give rise to many different $\alpha\beta$-tubulin heterodimers, but in some protists there is only one α-tubulin and one β-tubulin. A new tubulin has also unexpectedly been found and named γ-tubulin (Oakley and Oakley, 1989). This tubulin does not seem to be incorporated in cytoplasmic microtubules, but is localized to the centrosomes in the cell. It represents only a minor fraction of about 1% of the cellular tubulin. In spite of an extensive effort to relate different isotypes to different functions of microtubules, it seems that most microtubules are a mixture of all tubulin isotypes, and many of these have redundant functions. However, several studies provide evidence that certain isotypes have unique functions (for a review, see Moritz, 1993).

The cell can modify the different tubulins selectively after synthesis, by posttranslational modifications of the tubulins. These modifications have the advantage that they can be rapid and reversible. Several different modifications are known, of which the most studied are acetylation (L'Hernault and Rosenbaum, 1985) and detyrosination (Barra *et al.*, 1973) of α-tubulin, phosphorylation of β-tubulin (Eipper, 1974), and polyglutamylation (Eddé *et al.*, 1990) of both α- and β-tubulin. Although acetylated and detyrosinated microtubules seem to colocalize with more stable microtubules in the cell, all evidence so far suggests that these posttranslational modifications are not the cause of stability, but an event following stabilization (for a review, see Bulinski and Gundersen, 1991).

B. Microtubule-Associated Proteins

Microtubule function can be influenced by the MAPs, which vary in composition both between cells, and within a cell. The expression of several MAPs changes during development by a mechanism where alternative splicing of the primary gene transcript is involved. This gives rise to a size, and most probably a functional, difference between microtubules during the lifetime of an organism (for a review, see Matus, 1991). Microtubule-associated proteins can be divided into two groups: those that possess microtubule-activated ATPase or GTPase activities (kinesin, dynein, and dynamin), and those that are not sensitive to nucleotides (e.g., MAP1, MAP2, MAP4, and tau) (for a review, see Chapin and Bulinski, 1992). Several of the nucleotide-independent MAPs are known to stimulate assembly of microtubules and to stabilize the microtubules, and mammalian microtubule dynamics is modulated by MAPs (Horio and Hotani, 1986; Pryer *et al.*, 1992; Drechsel *et al.*, 1992; Kowalski and Williams, 1993). The MAPs can also attach to preassembled microtubules, thereby reducing the rate of cold-induced disassembly (Sloboda and Rosenbaum, 1979).

The binding between MAPs and tubulins is ionic, and MAPs predominantly bind to the acidic carboxy terminal of tubulin. It is unclear how the MAPs exert their effects on microtubules, but it may be a combination of cross-linking of tubulin subunits by repetetive tubulin-binding sites on MAPs, and a supression of the highly acidic C terminals of tubulin. Padilla *et al.* (1993) have found an intramolecular interaction between the MAP-binding carboxy terminus and the β-tubulin region containing the GTP-binding site. It is possible that such an interaction can affect microtubule dynamics and explain how MAPs can facilitate assembly.

An interesting feature of several MAPs is that they, similar to tubulin, can be posttranslationally modified by phosphorylation. Evidence exist that phosphorylation of some MAPs reduces their ability to stimulate the assembly of microtubules, and destabilizes microtubules (for a review, see Lee, 1993). One can furthermore not exclude the possibility that certain tubulin isotypes specifically bind certain MAPs to the microtubules, thereby influencing the properties of the microtubules. These interactions must be studied carefully in order to determine functional distinctions between different kinds of microtubules.

IV. Cold-Adapted Microtubules in Poikilotherms

A. Assembly of Microtubules at Low Temperatures

Most of our present knowledge of cold-adapted microtubules comes from studies of fishes living in two completely different habitats. Microtubules have been isolated from the Atlantic cod (*Gadus morhua*), which preferably lives at temperatures between 5 and 15°C, and from Antarctic rockcods (mainly *Notothenia gibberifrons* and *Notothenia coriiceps neglecta*), which have body temperatures as low as −1.8°C. These fishes appear to have diverged evolutionarily a long time ago, and their cold-adapted phenotypes have probably developed independently. One might therefore be able to determine whether cold adaptation has arisen only once during evolution, or if there has been convergent development, possibly using different mechanisms achieving similar results.

Isolated microtubules from the Atlantic cod assemble within a broad temperature interval, between 11 and 30°C (Fig. 2), with a critical concentration for assembly of 0.8 mg/ml (Wallin *et al.*, 1993). When seeded with axonemes, pure cod tubulins assemble at 8°C (Billger *et al.*, 1994). Microtubules isolate mainly from the Antarctic rockcods also assemble within a broad temperature interval, and the assembly is strongly entropy driven (Detrich *et al.*, 1989). The temperature interval is, however, different for the different fishes. Antarctic fish microtubules are more assembly competent at temperatures below 8°C (Williams *et al.*, 1985) than are Atlantic cod microtubules, which readily disassemble at 0°C (Strömberg

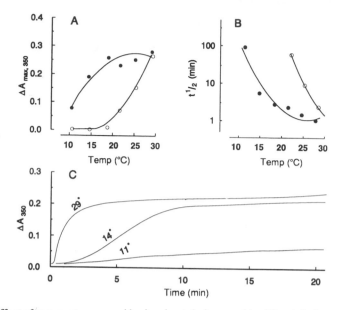

FIG. 2   Effect of temperature on cod brain microtubule assembly. Microtubules were assembled from cod tubulin in the absence and presence of cod MAPs at different temperatures at 2.5 mg/ml (total concentration). The assembly was followed spectrophotometrically at 350 nm in a thermostatted cuvette. (A) The steady state value for the assembly is plotted against the assembly temperature for cod tubulin and cod MAPs (●) and pure cod tubulin (○). (B) The rate of assembly is expressed as $t_{1/2}$ (the time needed to reach half the steady state) and plotted against the different assembly temperatures for cod tubulin and cod MAPs (●) and cod tubulin alone (○). (C) The assembly curves are shown for cod tubulin and cod MAPs at the indicated temperatures. [Reproduced from Wallin et al., 1993.]

et al., 1986), an adaptation that probably reflects the different habitat temperatures. The assembly of pure Antarctic fish tubulin has a very low critical concentration for assembly, 0.87 mg/ml at 0°C and 0.02 mg/ml at 18°C, which is lower than for isolated cod microtubules, which in turn is lower than for cow microtubules assembly.

Does cold adaptation mean that such microtubules have the same dynamic instability at low temperatures as mammalian microtubules at 37°C? To answer this question, the dynamic instability of Atlantic cod and Antarctic fish microtubules assembled from pure tubulin was studied at low temperatures in a collaborative study. Isolated cod microtubules, as well as Antarctic fish microtubules, showed dynamic instability at low temperatures, but both assembled and disassembled with lower rates, and had a lower frequency of catastrophes during in vitro experiments, than did mammalian microtubules (Billger *et al.*, 1994). This was, however, not a phenomenon caused by low temperatures, because cod tubulin showed the same kind of dynamic instability at 30°C, indicating that it is an intrinsic property of the tubulin.

Microtubules isolated from sea urchin eggs are less dynamic than mammalian brain microtubules (Simon *et al.*, 1992), but dynamic instability has not been investigated for other cold-adapted species. Microtubules from a number of other poikilothermic species have, however, been found to assemble within a broad temperature interval, including higher temperatures than their normal habitat temperatures (Detrich *et al.*, 1985; Suprenant and Marsh, 1987; Maccioni and Mellado, 1981; Langford, 1978; Wallin and Jönsson, 1983). Isolated catfish (*Heteropneustes fossilis*) tubulins assemble optimally between 18 and 37°C with the same critical concentration (Chaudhuri, 1986). Brain microtubules isolated from carp (*Cyprinus carpio*), acclimated to summer temperatures (16–20°C), assemble betwen 15 and 37°C (Maccioni and Mellado, 1981). It therefore seems that temperature optima found *in vitro* do not necessarily reflect habitat temperature. This is true not only for microtubules; several neuronal enzymes have been shown to have temperature optima much higher than the body temperature of the species from which they are isolated (for examples, see Abrahamsson, 1979, and Jönsson and Nilsson, 1979). Cells from poikilothermic animals can furthermore adapt to different temperatures. A cell line from rainbow trout has been cultured at 24 or 4–5°C for more than 3 years, and microtubules in the warm temperature-cultured cells disassemble markedly after exposure to 0°C, whereas the microtubule network remains in those cells that were cultured at 4–5°C (Tsugawa and Takahashi, 1987). The biochemical difference in the microtubules from these cells has, however, not yet been determined.

B. Role of Tubulin Isotypes in Assembly of
    Cold-Adapted Microtubules

On the basis of data from two-dimensional gel electrophoresis it can be seen that the Antarctic fish α-tubulin is more basic than both mammalian and Atlantic cod α-tubulin, as well as α-tubulin from the temperate channel catfish *Ictalurus punctatus* (Detrich and Overton, 1986; Detrich *et al.*, 1987; Billger *et al.*, 1994). Detrich and co-workers have suggested that the greater basicity of Antarctic fish α-tubulins is significant for cold adaptation, and that hydrophobic interactions make a larger contribution to the assembly of these tubulins than for mammalian tubulins. If it is involved in cold adaptation, this suggests that there are at least two different ways to achieve cold-adapted microtubules, or that such an alteration

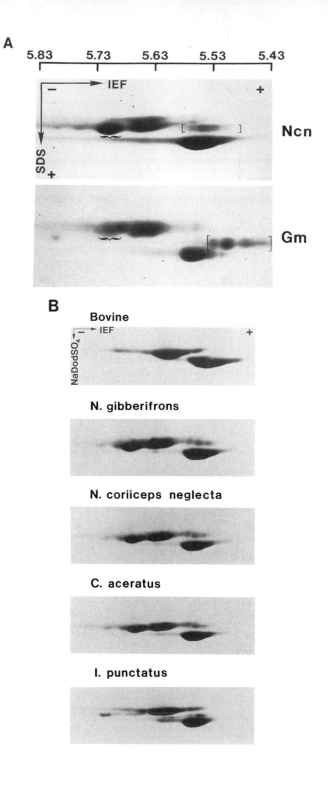

is important for the adaption to the extreme low temperatures at the Antarctic.

Atlantic cod β-tubulins exhibit large amounts of several highly acidic β-tubulin isoforms, in contrast to Antarctic fish (Billger et al., 1994), catfish, and cow tubulins as shown in Fig. 3 (Detrich and Overton, 1986). Whether this reflects different β-tubulin isotypes or posttranslational modifications such as polyglutamylation remains to be elucidated. Because more acidic β-tubulins were specific for cod tubulin, it is not known whether this actually is involved in cold adaptation. One β-tubulin has been sequenced from Antarctic fish DNA (Detrich and Parker, 1993). It is closely related to mammalian neural β-tubulin chains, but there are several unique amino acid substitutions, and an unusual carboxy-terminal residue insertion, that might be of importance for cold adaptation.

Is the brain tubulin of Antarctic fish representative of microtubules in all its cells? Studies on egg tubulin from the same species showed that egg and brain tubulin differed in their assembly capacities; the capacity of egg tubulin to assemble was much higher than that of neural tubulin (Detrich et al., 1992). The critical concentration for assembly of egg tubulin was extremely low, 0.057 mg/ml at 3°C and 0.0022 mg/ml at 19°C. Detrich (1994) has suggested that these differences provide evidence for the multi-tubulin hypothesis, originally proposed by Stephens (1975) and Fulton and Simpson (1976), which claims that chemically distinct tubulins differ in their assembly properties, or form microtubules with different characteristics. The greater assembly efficiency of egg tubulin was proposed to

---

FIG. 3 (A) Two-dimensional electrophoretic analysis of brain tubulins from an Antarctic rockcod and the Atlantic cod. Samples: Ncn, tubulin from the rockcod *N. coriiceps neglecta*, 20 µg; Gm, tubulin from the cod *G. morhua*, approximately 10 µg. Only the central regions of the second-dimension slab gels are shown, and the focusing (IEF) and electrophoretic (SDS) dimensions are indicated. The gels are aligned with respect to a common pH gradient (basic end on the left), a portion of which is shown on the horizontal axis (top). Basic α-tubulin isoforms discussed in text are denoted by curly brackets, and acidic β-tubulin isoforms are enclosed by square brackets. From Billger et al., 1994, *Cell Motil. Cytoskel.*, **21.** Copyright © 1994. Reprinted by permission of Wiley-Liss, a division of John Wiley and Sons. (B) Two-dimensional electrophoretic analysis of brain tubulin from cold-adapted and temperate fishes and from a mammal. Tubulins isolated from the three Antarctic fishes *N. gibberifrons*, *N. coriiceps neglecta*, and *Chaenocephalus aceratus* were compared with tubulins from the temperate channel fish *I. punctatus* and cow tubulins. Aliquots of about 20 µg of tubulin were focused in the first dimension, then electrophoresed in the second dimension. The focusing (IEF) and electrophoretic (SDS) dimensions are indicated. The gels are aligned horizontally with respect to a common pH gradient (basic end on left), and electrophoretic migration was from top to bottom. [Reproduced from Detrich, H. W., III, and Overton, S. A. (1986). *J. Biol. Chem.* **261,** 10922–10930, with the permission of the American Society for Biochemistry and Molecular Biology.]

result from a reduced charge–charge repulsion, because the egg α-tubulin subunits lacked several acidic peptides present in brain α-tubulins.

C. Role of Posttranslational Modifications of Tubulin in Cold Adaptation

Isolated cod microtubules are highly acetylated (Billger *et al.*, 1991), a posttranslational modification that usually correlates with stable microtubules (Fig. 4). Isolated cod brain microtubules are not only cold adapted but are also more stable to colchicine than are mammalian microtubules (Billger *et al.*, 1991). The effect of colchicine could be modulated by either cod or cow MAPs. Cod tubulin was more sensitive to colchicine in the absence of MAPs. However, cow MAPs had little effect on the stability of cow microtubules, indicating that the amount of stability depends on intrinsic properties of cod tubulins and not on cod MAPs. Cod brain microtubule stability did not seem to be a function of posttranslational modifications of cod tubulin per se, but intrinsic to cod tubulin and modu-

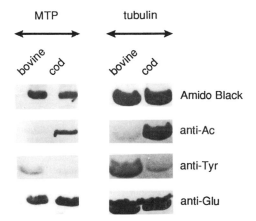

FIG. 4   Immunoblotting of microtubule proteins. Microtubule proteins, consisting of either tubulin and MAPs (MTP) (left lanes) or phosphocellulose-purified tubulin (right lanes) from bovine and cod brain, were electrophoresed on 12% PhastGels, transferred to nitrocellulose membranes, and immunoblotted with the 6-11B-1 antibody against acetylated α-tubulin (anti-Ac) and antibodies against tyrosinated (anti-Tyr) and detyrosinated (anti-Glu) α-tubulin, as indicated. A control membrane was stained with Amido Black to detect tubulin, as indicated. Each lane was loaded with 8 μg of microtubule proteins (Billger *et al.*, 1991). [Reproduced from Billger, M., Strömberg, E., and Wallin, M. (1991). *J. Cell Biol.* **113**, 331–338, by copyright permission of the Rockefeller University Press.]

lated by MAPs. Colchicine has been found to induce unfolding of a small region in the carboxyl-terminal region of rat brain β-tubulin, an unfolding that probably prevents tubulin dimer contacts necessary for assembly (Sackett and Varma, 1993). One could speculate that small sequence differences may affect the three-dimensional structure of the tubulin dimer in such a way that binding of colchicine does not induce an unfolding, or induces a slightly different unfolding with no such marked effect on the assembly properties of tubulin.

Ca^{2+} is a well-known inhibitor of mammalian brain microtubules, but cod brain microtubules are more stable to Ca^{2+} than mammalian microtubules (Strömberg et al., 1989; Strömberg and Wallin, 1994). Ca^{2+} induces coiled ribbons instead of being inhibitory, showing that the assembly process can be affected differently, either by a complete inhibition of assembly or by induction of morphologically altered polymers, depending on the tubulins. Coiled ribbons are not unique for cod microtubules; they have previously been found in dogfish microtubules (Langford, 1978; Wallin and Jönsson, 1983). It was found that the effect of Ca^{2+} is temperature dependent (Strömberg and Wallin, 1994), the reason for which is unknown.

To address the question of whether acetylation is involved in cold adaptation of cod microtubules, cells have been cultured from cod brain and skin. The results showed that only one of three different cultured brain cell types contained acetylated microtubules, but that in spite of that all cells were able to grow at 4°C with intact microtubules. Skin cells could also grow at such low temperatures, and did not contain any acetylated microtubules. The microtubules of all three types of brain cells were, however, more stable to microtubule inhibitors than the skin cells, indicating that there is no correlation between acetylation and cold and drug resistance of microtubules in cells cultured from cod skin and brain (Rutberg et al., 1994). This conclusion is supported by the lack of correlation between colchicine sensitivity, inability of Ca^{2+} to inhibit assembly, and acetylation of microtubules isolated from three other cold-temperate fishes: the rainbow trout (Oncorhynchus mykiss), ballan wrasse (Labrus berggylta), and viviparous eelpout (Zoarces viviparus) (Modig et al., 1994). Microtubules from all three species were highly acetylated, but ballan wrasse microtubules were inhibited by colchicine and Ca^{2+} at much lower concentrations than were microtubules from the other species. However, acetylation does not seem to be a general property of cold-adapted microtubules, as Antarctic fish microtubules are not acetylated to a higher degree than mammalian tubulin (Skoufias et al., 1992), and the assembly of these microtubules is inhibited by Ca^{2+} and colchicine.

Isolated cod brain microtubules are a mixture of tyrosinated and detyrosinated microtubules, and Antarctic fish tubulin is more tyrosinated than detyrosinated (Skoufias *et al.*, 1992), indicating that neither posttranslational modification is involved in cold adaptation. It is, however, not yet known whether the tubulins are modified by other posttranslational modifications of importance.

D. Role of Microtubule-Associated Proteins in Cold Adaptation of Microtubules

Atlantic cod and Antarctic fish MAPs differ in composition from mammalian MAPs. A heat-labile 400-kDa MAP is found in cod MAPs (Fig. 5), a protein that is not present in microtubule preparations from higher vertebrates (Strömberg *et al.*, 1989; Billger *et al.*, 1991). A similar protein was also found in Antarctic fish MAPs (Detrich *et al.*, 1990), and in MAP fractions from rainbow trout, ballan wrasse, and viviparous eelpout (Modig *et al.*, 1994). Little is known about this protein so far, but it is excluded from cod microtubules by cow MAPs (Fridén *et al.*, 1992), indicating that the 400-kDa MAP has a tubulin-binding site similar to that of one or several of the cow MAPs. MAP1 and tau have not been found in cod MAPs, but a heat-stable MAP2 was present (Modig *et al.*, 1994). In Antarctic fish microtubules, MAPs in the molecular weight ranges of 220,000–270,000, 140,000–155,000, 85,000–95,000, 40,000–45,000, and 32,000–34,000 have been found (Detrich *et al.*, 1990), but it has not been determined whether these are homologs of MAP1, MAP2, or tau.

Are these MAPs of importance for cold adaptation? Cod MAPs stimulate the rate and extent of assembly of cod microtubules at low temperatures, but the assembly was independent of MAPs at higher temperatures *in vitro* (Wallin *et al.*, 1993). At 30°C, MAPs bound to the microtubules, but did not stimulate assembly. This was also shown with the help of a relative new microtubule inhibitor, estramustine phosphate, which binds to MAPs and thereby inhibits assembly of mammalian microtubules and disassembles preformed microtubules (Fridén *et al.*, 1992). However, the ability of MAPs to stimulate assembly at low temperatures was not intrinsic to the unusual cod MAPs, because they could be replaced by cow MAPs. The molecular mechanism of action for different characteristics at low and high temperatures is so far unknown, but fluorescence and circular dichroism spectra differ for cod tubulin between 4 and 30°C, indicating that there may be a small conformational change in the tubulin between the different temperatures (Wallin *et al.*, 1993), a change that may be modulated by MAPs. During seasonal temperature variations in the water mass the Atlantic cod avoids temperature stress by migrating between

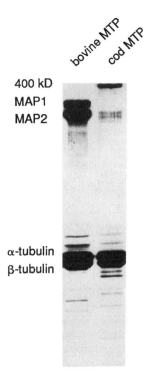

FIG. 5 Analysis of protein contents in preparations of microtubule proteins isolated from bovine (left lane) and cod (right lane) brains. Proteins were separated under identical conditions at 5–12% SDS-polyacrylamide vertical slab gels and stained with Coomassie blue. Identified proteins were denoted as indicated. Each lane was loaded with 50 μg of microtubule proteins (Billger et al., 1991). [Reproduced from Billger, M., Strömberg, E., and Wallin, M. (1991). J. Cell Biol. **113,** 331–338, by copyright permission of the Rockefeller University Press.]

different habitats of similar characteristics. It therefore seems unlikely that the cod will alter its microtubule characteristics in response to a change of body temperature. The ability to change its dependency on MAPs, may, however, be used by the cells to regulate microtubule stability and dynamics in a so far unknown way.

Antarctic fish microtubules have a higher propensity to assemble at low temperatures than do cod microtubules (Williams et al., 1985; Wallin et al., 1993), but MAPs stimulated the assembly at low temperatures, and could be replaced by cow MAPs (Detrich et al., 1990). These characteristics also show similarities with sea urchin egg tubulin, which assembled in a broad temperature interval with a slow assembly at low temperatures, an assembly that was stimulated by cow MAPs (Suprenant and Rebhun, 1983). Thus, the tubulin-binding motifs and assembly-stimulating properties of MAPs seem to be rather conservative, even if little is known about sequence similarities between MAPs from different organisms.

Tubulin and MAPs seem, furthermore, to have different temperature optimum intervals. Mammalian tubulin is not able to assemble at low temperatures, whereas mammalian MAPs are fully competent to bind to tubulin and stimulate assembly of cold-adapted tubulins at low temperatures. The cold-adapted cod MAPs can stimulate assembly of cow tubulin at 37°C (Fridén et al., 1992), but cod tubulin seems to denature at 37°C (Strömberg et al., 1989). Cold adaptation must therefore be caused predominantly by modifications of the tubulins, either in the form of sequence differences and/or posttranslational modifications. These changes are probably small, but might alter subunit charge and hydrophobicity.

V. Cold-Stable Microtubules in Homeotherms

A population of cold-stable microtubules is present in mammalian cells. What is the function of such microtubules in homeotherms? These microtubules may perhaps only reflect the need of a cell to stabilize certain microtubules for specific microtubule functions. However, some homeotherms must at least partially adapt to cold, because they have a temperature gradient in their body. Arctic animals such as Eskimo dogs, caribou, and sea gulls have temperatures at their extremities that are far lower than their internal body temperature (Irving, 1966). The temperature of some extremities even approaches the outside temperature. Hibernating animals such as bears and hedgehogs must be able to cope with seasonal body temperature changes. To our knowledge, there is no information available about molecular mechanisms for cold adaptations of such microtubules. We have therefore chosen to present a survey of what is known about cold-stable microtubules in mammals, to discuss possible functions and causes of microtubule stability. Further studies will yield valuable information as to whether this is involved in cold adaptation of microtubules in extremities, or seasonal cold adaptations, as well as to why cold-stable microtubules exist in cells with a constant temperature.

A. Stable Microtubules in Axons

It has been proposed that stable regions of microtubules in axons play an important role in cytoskeletal organization (Brady et al., 1984; Morris and

Lasek, 1982), and that they might represent nucleating elements required for microtubule assembly (Fig. 6) (Heidemann *et al.*, 1984; Ahmad *et al.*, 1993). Axonal microtubules, in contrast to cytoplasmic microtubules, do not emanate from the microtubule-organizing center located close to the cellular nucleus. Neurons have dendrites and long axons, which are of importance for the ability of the nerve cell to receive and transmit information over long distances. These extensions are not found embryonically, but form during maturation. It is therefore important to stabilize microtubules in order for the neurons to be able to fulfill their tasks, and it is well known that axonal microtubules are less dynamic and more stable to drugs and cold than are interphase microtubules. Brown and co-workers (1993) have shown that axonal microtubules are composite; they are composed of one stable and one labile region. The stable region is detyrosinated and highly acetylated, less dynamic, and more resistant to nocodazole than the labile region, which is found in the growing end of the microtubule. The latter is tyrosinated, acetylated very little, and rapidly broken down by nocodazole. The authors suggested that the posttranslationally modified microtubules represent the more long-lived or older microtubule ends. Additional evidence for a difference in tyrosination has been presented by Ahmad *et al.* (1993). Brady *et al.* found in 1984 three different fractions of axonal microtubules; one cold labile, one cold stable, but extractable with Ca^{2+}, and one that is both cold and Ca^{2+} insoluble. No differences were found in the β-tubulins, whereas α-tubulin was modified in the cold-stable fractions in such a way that it did not even enter the two-dimensional

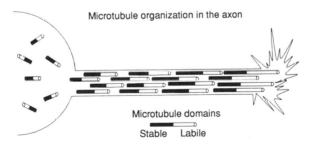

FIG. 6 Schematic drawing ilustrating the organization of axonal microtubules, based on information derived from electron microscopic analyses. The plus ends (growing ends) of the microtubules are directed away from the cell body. The microtubules consist of two domains, stable and labile, with the labile domain situated at the plus end of the stable domain. Both stable and labile domains are present all along the length of the axon, except in the distal region of the axon, into which extend only the labile domains of microtubules originating in the main shaft of the axon (Ahmad *et al.*, 1993). [Reproduced with permission from Ahmad, F. J., Pienkowski, T. P., and Baas, B. W. (1993). *J. Neurosci.* **13**(2), 856–866.]

gel. Tubulin sequence differences, posttranslational modifications, interactions with neurofilaments, or interactions with different MAPs have been implicated to be involved in the stability. A modified α-tubulin has been found by Tashiro and Komiya (1983) in guinea pig vagal nerve, and a specific α-tubulin associated with cold stability was suggested by Binet et al. (1987), even if the authors later discussed the possibility that other microtubule components might be involved (Binet and Meininger, 1988).

One can, of course, not exclude the possibility that β-tubulin characteristics may be of importance. A different utilization of β-tubulin isotypes has been found during differentiation of neurites from cultured PC-12 cells (Joshi and Cleveland, 1989), indicating that functional differences between different isotypes may exist and it is known that altered proportions of different isotypes within a microtubule can change its dynamics (Joshi et al., 1987). Arai and Matsumoto (1988) demonstrated a regional distribution of MAPs and β-tubulin isotypes, which was suggested to explain the differences in stability among axonal microtubules.

B. Other Examples of Cold-Stable Microtubules in Cells

There are other examples in which one finds a difference in stability of microtubules within a single cell. A population of cold-stable microtubules was found in the cytoplasm of mouse embryo fibroblasts by Bershadsky et al. (1979). In chick anterior latissimuris dorsi muscle, cold-labile microtubules are detected throughout the cytoplasm, whereas a subset of cold-stable and acetylated microtubules is located underneath the motor endplates (Jasmin et al., 1990). These subsynaptic cold-stable microtubules are probably involved in the transport of the nicotinic acetylcholine receptor, which is highly accumulated at the motor endplate. The reason for a localized stabilization of microtubules can therefore be the same as for the axons: a need for a stable track for intracellular transport of vesicles and organelles.

Primary cultures of chick heart fibroblasts contain one cold-stable and one cold-labile subset of microtubules (Brown and Warn, 1993). The stable microtubules with slower turnover rates are generally curly and perinuclear in distribution. In spite of two subsets, all microtubules were found to be highly posttranslationally modified, both by acetylation and detyrosination, further suggesting that these modifications are not the cause of cold stability. All cellular microtubules were, however, sensitive to colcemid, indicating that cold and drug stability are two separate characteristics. Another example is found in human monocytes. Two populations of microtubules are found in these cells; 70% of the microtubules are labile, whereas 30% are more stable to both nocodazole and 3°C (Cassimeris *et al.*, 1986). The rate-limiting step for disassembly seemed to be initiation of disassembly; once it was initiated depolymerization continually progressed rapidly. The reason for the differences in stability was unclear, but "capping" of microtubules, a cooperative binding of certain stabilizing MAPs, or a modification of MAPs was suggested.

C. Bundling of Microtubules

Nucleated erythrocytes have a stable marginal band of microtubules adjacent to the plasma membrane. Chicken erythrocytes have been studied thoroughly, and these cells contain a unique β-tubulin (class VI), which is expressed only in blood cells (for a review, see Murphy, 1991). The erythrocyte β-tubulin has relatively increased hydrophobic properties and erythrocyte microtubules are more stable and less dynamic than chicken brain microtubules. Frinczek *et al.* (1993) found that the dynamic instability of erythrocyte microtubules can be controlled by the tubulin isotype, independently of MAPs, because these microtubules were much less dynamic than brain microtubules. This means that the individual tubulin isotypes can differ in assembly properties *in vitro,* which can be of importance for different functions of microtubules. Transfection of the erythrocyte β-tubulin into monkey kidney cells showed that it can coassemble randomly with all cellular microtubules (Joshi *et al.*, 1987). However, although microtubules containing a mixture of all isotypes were disassembled by cold, some microtubules had less erythrocyte β-tubulin and were less stable to temperature-induced disassembly than those enriched in erythrocyte β-tubulin, showing that microtubule characteristics can differ depending on the proportions of different isotypes within a microtubule.

Factors leading to lateral interactions or bundles of microtubules have been suggested to be involved in microtubule stability in erythrocytes, as well as in the stability of other cellular microtubules (for a review, see MacRae, 1992), thus representing an alternative way to stabilize microtubules. Bundles of laterally associated microtubules increase during the development of erythrocytes in correlation with an increased drug stability of the microtubules. Several bundling proteins have been isolated both from erythrocytes and other cells. One interesting example is in epidermal tendon cells of the river crab, *Potamon dehaani* (Nakazawa *et al.*, 1992), in which large amounts of cold-resistant microtubule bundles are present. Within these bundles, fine filamentous structures, the nature of which is unknown, cross-linked microtubules. MAP2, MAP2C, and tau all induce bundling of microtubules *in vivo* (for a review, see Hirokawa, 1994), suggesting that MAPs probably are of importance for this phenomenon. Dynamin is a 100-kDa protein that forms regular arrays of periodic cross-bridges between microtubules, thereby arranging them in bundles (Shpetner and Vallee, 1989). Extensively cross-linked microtubules are further present in the parasitic hemoflagellate *Trypanosoma brucei* (Hemphill *et al.*, 1992 and references therein). Two unique MAPs with highly repetitive sequences of 38 amino acids and with potentially high microtubule-binding ability have been suggested as microtubule stabilizers in these parasites.

D. Acquired Cold Stability

Cold-stable microtubules have been found in mammalian crude brain extract (Webb and Wilson, 1990). This cold stability was, however, in contrast to the stability of microtubules from cold-adapted animals, acquired during assembly of cold-labile microtubules. A protein factor designated STOP (stable tubule only polypeptide) has been found to induce microtubule stability at concentrations that are substoichiometric to tubulin (Job *et al.*, 1982; Margolis *et al.*, 1986). Its stabilizing activity was lost *in vitro* in the presence of ATP (Job *et al.*, 1983), calmodulin (Job *et al.*, 1981), or by shearing (Job *et al.*, 1982). The STOP protein was first suggested to be a 64-kDa protein (Margolis and Rauch, 1981), but has later been identified as a 145-kDa protein with an intrinsic capacity to slide on microtubules (Pirollet *et al.*, 1989). Disassembly was prevented by substoichiometric blocks randomly distributed along microtubules (Job *et al.*, 1982). Bovine myelin basic protein is another protein found to be associated with cold-stable microtubules, but this protein appeared to induce cross-linking (Pirollet *et al.*, 1992).

Biochemical changes of microtubules occur with age in rat brains (Leterrier and Eyer, 1992). The composition of MAPs is not identical between cold-labile and cold-stable microtubules, and the degree of phosphorylation differs. STOP proteins may be involved, but the authors could not exclude the possibility that interactions with other cytoskeletal elements could be of importance.

During development of dendrites in cerebellar Purkinje cells microtubules are first cold labile, whereafter cold-stable microtubules appear during a period of extensive growth of dendrites (Faivre *et al.*, 1985). STOP proteins were already present before induction of cold-stable microtubules (Faivre-Sarrailh and Rabié, 1988). The authors hypothesize that there is a strong inhibition of STOP activity during early development, which decreases during later stages and correlates with the appearance of cold-stable microtubules.

Kinetochore microtubules are more stable than other microtubules in the mitotic spindle. Calmodulin colocalizes with cold-stable and nocodazole-stable kinetochore microtubules in living PtK$_1$ cells (Sweet and Welsh, 1988). Larson *et al.* (1985) found a calmodulin-dependent kinase activity enriched in the cytosol of cold-stable microtubules. It phosphorylates α- and β-tubulin, as well as MAP2, an 80-kDa doublet, and several other minor proteins. Job *et al.* (1981) have previously reported that cold-stable microtubules disassemble when STOP proteins are phosphorylated by a calmodulin-dependent protein kinase. How these results correlate with each other is unclear.

E. Histones as Microtubule Stabilizers

Microtubules in the complex 9 + 2 structures in eukaryotic cilia and flagella are very stable. These microtubules are not single but exist in doublets, which has been discussed as one reason for stability. Sea urchin flagellar microtubules were suggested to be stabilized by histone H1 (Multigner *et al.*, 1992). The authors found a protein identical to nuclear histone H1 in the axonemes. H1 induced stability to cold, freezing, and high concentrations of Ca^{2+} to preformed microtubules assembled from pure brain tubulin, but did not induce microtubule doublets. The mechanism of action is unclear, because addition of H1 initially during the assembly was inhibitory; only a few microtubules and mainly amorphous aggregates formed. The precice localization of H1 in axonemes is uncertain.

Eyer *et al.* (1990) have previously found that a bull sperm dynein induces cold stability to brain microtubules. The authors could not rule out the possibility that molecules other than dynein, which might contribute to the cold-stabilizing properties, were present in their preparation. In view of the results of Multigner *et al.* (1992), H1 might be one candidate. It was, however, not determined what kind of tubulin polymer this dynein preparation induced. Light scattering increased steadily to much higher values than can be expected from the amount of proteins that can assemble, and according to Gaskin *et al.* (1974) the light scattering is proportional to the amount of assembled microtubule proteins only if no abnormal polymers are induced.

F. Induction of Cold Stability of Microtubules Assembled from Pure Tubulin

Cold stability of microtubules can be induced in the absence of MAPs and STOPs. Bonne and Pantaloni (1982) have shown that microtubules newly assembled from pure tubulin at 37°C *in vitro* are more stable to disassembly induced by lowering the temperature to 20°C than are older microtubules, even if no change in average length or shape is seen. It is not known how this is coupled to GTP hydrolysis and dynamic instability, but microtubules that assemble in the presence of nonhydrolyzable analogs of GTP (Penningroth and Kirschner, 1977) appear to be less sensitive to cold than microtubules containing GDP at the E site. Burns and Surridge (1990) have, after analysis of β-tubulin sequences, suggested that there is a conformational change in the tubulin involved in dynamic instability that might affect the stability of microtubules. Further studies are needed to clarify this issue.

VI. Induction of Cold Stability by Taxol

Most drugs that inhibit microtubule functions induce depolymerization of microtubules, and are therefore valuable chemical tools in the studies of microtubules. Taxol, isolated from *Taxus brevifolia,* is an inhibitor of microtubule functions with another mechanism of action. It binds to tubulin and lowers the critical concentration for assembly, and makes the assembly MAP independent (Schiff *et al.,* 1979; Schiff and Horwitz, 1981). The formed microtubules are extremely stable; they are not depolymerized by cold or calcium. Even if no evidence exists so far, one cannot exclude the possibility that a molecule with characteristics similar to those of taxol exists in eukaryotic cells.

VII. Cold-Stable Microtubules in Plants

Some plants need cold-adapted microtubules, whereas others, probably just like animals, have microtubules that are cold stable without a specific need for cold stability, only a need for less dynamic microtubules. Within a single plant, the cellular microtubules respond differently to cold. It seems, however, that the ability of chilled roots to survive depends on the ability of certain tissues to have microtubules that can adapt to the cold, either by a preservation of microtubules or by a repolymerization of microtubules during the cold treatment (Baluska *et al.,* 1993). Microtubules in cells that were rapidly elongating were far more sensitive to cold than were cells in areas were elongation is slow, indicating that dynamic microtubules are less cold stable. How is the cold stability achieved? Plant microtubules seem to be associated to the membrane, and cross-linked to a higher degree than animal cells (Lloyd *et al.*, 1980). Cold- and nocodazole-stable acetylated cortical microtubules can also be found in *Paramecium,* in addition to plants, whereas the cytoplasmic microtubules are labile and not acetylated (Torres and Delgado, 1989).

Transmembrane proteins have been shown to be involved in the association of microtubules with the plasma membrane, and dissociation of these bindings rendered the microtubules more cold sensitive (Akashi and Shibaoka, 1991). When plant microtubules were isolated and studied *in vitro,* such microtubules were cold labile, further indicating that cross-linking to the membrane might be involved. In cells that have secondary thickened cell walls, microtubules were cold resistant. Akashi *et al.* (1990) have found a cell wall protein called extensin, the amount of which increased markedly in cold-resistant cells, suggesting that it might be of importance for microtubule stability. Transmembrane proteins seem to be involved in this stabilization, because digestion of the extracellular portion of transmembrane proteins dissociated cortical microtubules from the plasma membrane, and made them cold labile, and the organized arrays of these microtubules were often disturbed (Akashi and Shibaoka, 1991).

Different routes to achieve cold stability also seem to exist in plant cells. Chaudhuri and Biswas (1993) showed that microtubules from *Mimosa pudica* are cold stable, and that this stability was intrinsic to the tubulin. Addition of the mimosa tubulin to goat tubulin induced a cold-stability in these hybrid microtubules, again indicating that not all tubulin dimers must necessarily have this intrinsic ability to be cold stable in order to induce a cold stability in the whole microtubule structure. An altered proportion of isotypes with different degrees of stability can therefore affect the whole microtubule.

In suspension-cultured tobacco (*Nicotiana tabacum*) cells the amount of cold-stable microtubules was proposed to be regulated by the action of a phosphorylation/dephosphorylation system (Mizuno, 1992). In the presence of a protein kinase inhibitor, most cells acquired cold-stable microtubules. The target for phosphorylation is, however, so far unknown. Another example is an MAP isolated from suspension-cultured carrot cells, which bundles not only carrot microtubules but also brain microtubules, and induces cold stability (Cyr and Palevitz, 1989), further showing the conservative interaction between MAPs and tubulins from evolutionarily very distant species. A bundling effect is also caused by a 65-kDa MAP isolated from tobacco BY-2 cells (Chang-Jie and Sonobe, 1993).

VIII. Summary

It is obvious in view of the data presented in this chapter that there are several routes to achieve cold stability of microtubules (Fig. 7). Factors that can interact with microtubules, such as MAPs, STOPs, histones, and perhaps capping factors, can all be involved. The composition of isotypes in a single microtubule can affect its dynamics, and posttranslational modifications may also be important. More permanent stability seems to be achieved by bundling factors, associations to membranes, as well as assembly of microtubule doublets and triplets. Is this the explanation for cold adaptation of microtubules? Probably not. Cold-adapted microtubules must fulfill functions similar to those of microtubules in mammalian cells with a constant high temperature, but at lower temperatures. All evidence so far points to the fact that cold adaptation is intrinsic to the tubulins,

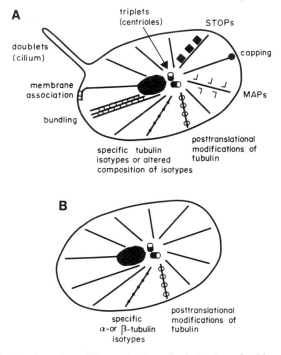

FIG. 7   Schematic drawing of possible mechanisms for induction of cold stability (top) and cold adaptation (bottom) of microtubules. In addition, histones have also been suggested to induce stability to flagella microtubules.

even if it is unclear whether it depends on different amino acid sequences or posttranslational modifications. All cold-stabilizing factors mentioned above can probably modulate the degree of stability of cold-adapted microtubules as well. On the basis of available data one cannot yet decide whether cold adaptation of microtubules has arisen only once during evolution.

It is unclear how some birds and mammals with a high body temperature adapt their extremities to low temperatures. Cold stability of most mammalian microtubules seems, however, to have little to do with a need for cold-adapted microtubules, but may be a way to stabilize certain populations of cellular microtubules for specific microtubule functions. It is therefore important to study both cold adaptation and cold stabilization in order to learn how microtubules can fulfill different functions, and at different temperatures.

Acknowledgments

This work was aided by grants from the Swedish Natural Science Research Council and the Swedish Cancer Society. We are indebted to Dr. H. W. Detrich III, for generously providing a review on Antarctic fish microtubules while in press, and for his collaboration and that of Prof. R. Williams, Jr., with our laboratory, making it possible to compare cold-adapted fish microtubules of evolutionary interest. We are also grateful to Prof. Jan Bergström for enjoyable discussions of evolution, and to the staff at the Kristineberg Marine Biological Station (Fiskebäckskil, Sweden), at which part of the studies on cod microtubules was conducted.

References

Abrahamsson, T. (1979). Axonal transport of adrenaline, noradrenaline and phenylethanol-amine-*N*-methyl transferase (PNMT) in sympathetic neurons of the cod, *Gadus morhua*. *Acta Physiol. Scand.* **105,** 316–325.

Ahmad, F. J., Pienkowski, T. P., and Baas, P. W. (1993). Regional differences in microtubule dynamics in the axon. *J. Neurosci.* **13,** 856–866.

Akashi, T., Kawasaki, S., and Shibaoka, H. (1990). Stabilization of cortical microtubules by the cell wall in cultured tobacco cells. Effects of extensin on the cold-stability of cortical microtubules. *Planta* **182,** 363–369.

Akashi, T., and Shibaoka, H. (1991). Involvement of transmembrane proteins in the association of cortical microtubules with the plasma membrane in tobacco BY-2 cells. *J. Cell Sci.* **98,** 169–174.

Amos, L. A., and Amos, W. B. (1991). "Molecules of the Cytoskeleton." Macmillan Education Ltd., London.

Arai, T., and Matsumoto, G. (1988). Subcellular localization of functionally differentiated microtubules in squid neurons: Regional distribution of microtubule-associated proteins and β-tubulin isotypes. *J. Neurochem.* **51,** 1825–1838.

Baluska, F., Parker, J. S., and Barlow, P. W. (1993). The microtubular cytoskeleton in cells of cold-treated roots of maize (*Zea mays* L.) shows tissue-specific responses. *Protoplasma* **172**, 84–96.

Barra, H. S., Rodriguez, A., Arce, C. A., and Caputto, R. (1973). A soluble preparation from rat brain that incorporates into its own proteins [14C]arginine by a ribonuclease-sensitive system and [14C]tyrosine by a ribonuclease-insensitive system. *J. Neurochem.* **20**, 97–108.

Bershadsky, A. D., and Vasiliev, J. M. (1988). "Cytoskeleton." Plenum, New York.

Bershadsky, A. D., Gelfand, V. I., Svitkina, T. M., and Tint, I. S. (1979). Cold-stable microtubules in the cytoplasm of mouse embryo fibroblasts. *Cell Biol. Int. Rep.* **3**, 45–50.

Billger, M., Strömberg, E., and Wallin, M. (1991). Microtubule-associated protein-dependent colchicine stability of acetylated cold-labile brain microtubules from the Atlantic cod, *Gadus morhua. J. Cell Biol.* **113**, 331–338.

Billger, M., Wallin, M., Williams, R. C., Jr., and Detrich, H. W. III (1994). Dynamic instability of microtubules from cold-living fishes. *Cell Motil. Cytoskel.* **28**, 327–332.

Binet, S., and Meininger, V. (1988). Biochemical basis of microtubule cold stability in the peripheral and central nervous systems. *Brain Res.* **450**, 231–236.

Binet, S., Cohen, E., and Meininger, V. (1987). Heterogeneity of cold-stable and cold-labile tubulin in axon- and soma-enriched portions of the adult mouse brain. *Neurosci. Lett.* **77**, 166–170.

Bonne, D., and Pantaloni, D. (1982). Mechanism of tubulin assembly: Guanosine 5′-triphosphate hydrolysis decreases the rate of microtubule depolymerization. *Biochemistry* **21**, 1075–1081.

Brady, S. T., Tytell, M., and Lasek, R. J. (1984). Axonal tubulin and axonal microtubules: Biochemical evidence for cold stability. *J. Cell Biol.* **99**, 1716–1724.

Bray, D. (1992). "Cell Movements." Garland, New York and London.

Brinkley, B. R., and Cartwright, J., Jr. (1975). Cold-labile and cold-stable microtubules in the mitotic spindle of mammalian cells. *Ann. N.Y. Acad. Sci.* **253**, 428–439.

Brown, A., Li, Y., Slaughter, T., and Black, M. M. (1993). Composite microtubules of the axon: Quantitative analysis of tyrosinated and acetylated-tubulin along individual axonal microtubules. *J. Cell Sci.* **104**, 339–352.

Brown, D. A., and Warn, R. M. (1993). Primary and secondary chick heart fibroblasts: Fast and slow-moving cells show no significant difference in microtubule dynamics. *Cell Motil. Cytoskel.* **24**, 233–244.

Bulinski, J. C., and Gundersen, G. G. (1991). Stabilization and posttranslational modification of microtubules during cellular morphogenesis. *BioEssays* **13**, 285–293.

Burns, R. G., and Surridge, C. (1990). Analysis of β-tubulin sequences reveals highly conserved, coordinated amino acid substitutions. Evidence that these "hot spots" are directly involved in the conformational change required for dynamic instability. *FEBS Lett.* **271**, 1–8.

Cassimeris, L. U., Wadsworth, P., and Salmon, E. D. (1986). Dynamics of microtubule depolymerization in monocytes. *J. Cell Biol.* **102**, 2023–2032.

Cassimeris, L U., Walker, R. A., Pryer, N. K., and Salmon, E. D. (1987). Dynamic instability of microtubules. *BioEssays* **7**, 149–154.

Chang-Jie, J., and Sonobe, S. (1993). Identification and preliminary characterization of a 65 kDa higher-plant microtubule-associated protein. *J. Cell Sci.* **105**, 891–901.

Chapin, S. J., and Bulinski, J. C. (1992). Microtubule stabilization of assembly-promoting microtubule-associated proteins: A repeat performance. *Cell Motil. Cytoskel.* **23**, 236–243.

Chaudhuri, A. (1986). Purification and characterization of tubulin from the catfish *Heteropneustes fossilis. J. Biosci.* **10**, 323–333.

Chaudhuri, A. R., and Biswas, S. (1993). Cold stability of microtubules of *Mimosa pudica*. *Biochem. Mol. Biol. Int.* **29**, 421–428.

Clarke, A. (1991). What is cold adaptation and how should we measure it? *Am. Zool.* **31**, 81–92.

Cyr, R. J., and Palevitz, B. A. (1989). Microtubule-binding proteins from carrot. I. Initial characterization and microtubule bundling. *Planta* **177**, 245–260.

Detrich, H. W. III (1994). Cold adaptation of microtubule assembly in Antarctic fishes. *In* "Advances in Molecular and Cell Biology: Thermobiology" (J. S. Willis, ed.). JAI Press, Greenwich, CT (in press).

Detrich, H. W. III, and Overton, S. A. (1986). Heterogeneity and structure of brain tubulins from cold-adapted Antarctic fishes: Comparison to brain tubulins from a temperate fish and a mammal. *J. Biol. Chem.* **261**, 10922–10930.

Detrich, H. W. III, and Parker, S. K. (1993). Divergent neural β tubulin from the Antarctic fish *N. coriiceps neglecta:* Potential sequence contributions to cold adaptation of microtubule assembly. *Cell Motil. Cytoskel.* **24**, 156–166.

Detrich, H. W. III, Jordan, M. A., Wilson, L., and Williams, R. C., Jr. (1985). Mechanism of microtubule assembly. Changes in polymer structure and organization during assembly of sea urchin egg tubulin. *J. Biol. Chem.* **260**, 9479–9490.

Detrich, H. W. III, Prasad, V., and Luduena, R. F. (1987). Cold-stable microtubules from Antarctic fishes contain unique α tubulins. *J. Biol. Chem.* **262**, 8360–8366.

Detrich, H. W. III, Johnson, K. A., and Marchese-Ragona, S. P. (1989). Polymerization of Antarctic fish tubulins at low temperatures: Energetic aspects. *Biochemistry* **28**, 10085–10093.

Detrich, H. W. III, Neighbors, B. W., Sloboda, R. D., and Williams, R. D., Jr. (1990). Microtubule-associated proteins from Antarctic fishes. *Cell Motil. Cytoskel.* **17**, 174–186.

Detrich, H. W. III, Fitzgerald, T. J., Dinsmore, J. H., and Marchese-Ragona, S. P. (1992). Brain and egg tubulins from Antarctic fishes are functionally and structurally distinct. *J. Biol. Chem.* **267**, 18766–18775.

Drechsel, D. N., Hyman, A. A., Cobb, M. H., and Kirschner, M. W. (1992). Modulation of the dynamic instability of tubulin assembly by the microtubule-associated protein tau. *Mol. Biol. Cell* **3**, 1141–1154.

Dustin, P. (1984) "Microtubules." Springer-Verlag, Berlin and New York.

Eddé, B., Rossier, J., Le Caer, J. P., Desbruyéres, E., Gros, F., and Denoulet, P. (1990). Posttranslational glutamylation of α-tubulin. *Science* **247**, 83–85.

Eipper, B. A. (1974). Properties of rat brain tubulin. *J. Biol. Chem.* **249**, 1407–1416.

Erickson, H. P., and O'Brien, E. T. (1992). Microtubule dynamic instability and GTP hydrolysis. *Annu. Rev. Biophys. Biomol. Struct.* **21**, 145–166.

Eyer, J., White, D., and Gagnon, C. (1990). Presence of a new microtubule cold-stabilizing factor in bull sperm dynein preparations. *Biochem. J.* **270**, 821–824.

Faivre, C., Legrand C. H., and Rabié, A. (1985). The microtubular apparatus of cerebellar Purkinje cell dendrites during postnatal development of the rat: The density and cold-stability of microtubules increase with age and are sensitive to thyroid hormone deficiency. *Int. J. Dev. Neurosci.* **3**, 559–565.

Faivre-Sarrailh, C., and Rabié, A. (1988). Developmental study of factors controlling micro-tubule in vitro cold-stability in rate cerebrum. *Dev. Brain Res.* **42**, 199–214.

Fridén, B., Strömberg, E., and Wallin, M. (1992). Different assembly properties of cod, bovine, and rat brain microtubules. *Cell Motil. Cytoskel.* **21**, 305–312.

Fulton, C., and Simpson, P. A. (1976). Selective synthesis and utilization of flagellar tubulin. The multitubulin hypothesis. *In* "Cell Motility" (R. Goldman, T. Pollard, and J. L. Rosenbaum, eds.), pp. 987–1005. Cold Spring Harbor Lab., Cold Spring Harbor, NY.

Gaskin, F., Cantor, C. R., and Shelanski, M. L. (1974). Turbidimetric studies of the in vitro assembly and disassembly of porcine brain neurotubules. *J. Mol. Biol.* **89,** 737–758.

Gildersleeve, R. F., Cross, A. R., Cullen, K. E., Fagen, A. P., and Williams, R. C., Jr. (1992). Microtubules grow and shorten at intrinsically variable rates. *J. Biol. Chem.* **267,** 7995–8006.

Heidemann, S. R., Hamborg, M. A., Thomas, S. J., Song, B., Lindley, S., and Chu, D. (1984). Spatial organization of axonal microtubules. *J. Cell Biol.* **99,** 1289–1295.

Hemphill, A., Affolter, M., and Seebeck, T. (1992). A novel microtubule-binding motif identified in a high molecular weight microtubule-associated protein from *Trypanosoma brucei. J. Cell Biol.* **117,** 95–103.

Hirokawa, N. (1994). Microtubule organization and dynamics dependent on microtubule-associated proteins. *Curr. Biol.* **6,** 74–81.

Horio, T., and Hotani, H. (1986). Visulization of the dynamic instability of individual microtubules by dark-field microscopy. *Nature (London)* **321,** 605–607.

Irving, L. (1966). Adaptations to cold. *Sci. Am.* **214,** 94–101.

Jasmin, B. J., Changeux, J.-P., and Cartaud, J. (1990). Compartmentalization of cold-stable and acetylated microtubules in the subsynaptic domain of chick skeletal muscle fibre. *Nature (London)* **344,** 673–675.

Job, D., Fischer, E. H., and Margolis, R. L. (1981). Rapid disassembly of cold-stable microtubules by calmodulin. *Proc. Natl. Acad. Sci. U.S.A.* **78,** 4679–4682.

Job, D., Rauch, C. T., Fischer, E. H., and Margolis, R. L. (1982). Recycling of cold-stable microtubules: Evidence that cold stability is due to substoichiometric polymer blocks. *Biochemistry* **21,** 509–515.

Job, D., Rauch, C. T., Fischer, E. H., and Margolis, R. L. (1983). Regulation of microtubule cold stability by calmodulin-dependent and -independent phosphorylation. *Proc. Natl. Acad. Sci. U.S.A.* **80,** 3894–3898.

Jönsson, A.-C., and Nilsson, S. (1979). Effects of pH, temperature and Cu^{2+} on the activity of dopamine-β-hydroxylase from the chromaffin tissue of the cod, *Gadus morhua. Comp. Biochem. Physiol. C* **62C,** 5–8.

Joshi, H. C., and Cleveland, D. W. (1989). Differential utilization of β-tubulin isotypes in differentiating neurites. *J. Cell Biol.* **109,** 663–673.

Joshi, H. C., Yen, T. J., and Cleveland, D. W. (1987). *In vivo* coassembly of a divergent β-tubulin subunit (cβ6) into microtubules of different function. *J. Cell Biol.* **105,** 2179–2190.

Kirschner, M., and Mitchinson, T. (1986). Beyond self-assembly: From microtubules to morphogenesis. *Cell (Cambridge, Mass.)* **45,** 329–342.

Kowalski, R. J., and Williams, R. C., Jr. (1993). Microtubule-associated protein 2 alters the dynamic properties of microtubule assembly and disassembly. *J. Biol. Chem.* **268,** 9847–9855.

Langford, G. M. (1978). *In vitro* assembly of dogfish brain tubulin and the induction of coiled ribbon polymers by calcium. *Exp. Cell Res.* **11,** 139–151.

Larson, R. E., Goldenring, J. R., Vallano, M. L., and DeLorenzo, R. J. (1985). Identification of endogenous calmodulin-dependent kinase and calmodulin-binding proteins in cold-stable microtubule preparations from rat brain. *J. Neurochem.* **44,** 1566–1574.

Lee, G. (1993). Non-motor microtubule-associated proteins. *Curr. Opin. Cell Biol.* **5,** 88–94.

Leterrier, J. F., and Eyer, J. (1992). Age-dependent changes in the ultrastructure and in the molecular composition of rat brain microtubules. *J. Neurochem.* **59,** 1126–1137.

L'Hernault, S. W., and Rosenbaum, J. L. (1985). *Chlamydomonas* α-tubulin is posttranslationally modified by acetylation on the ε-amino group of a lysine. *Biochemistry* **24,** 473–478.

Lloyd, C. W., Slabas, A. R., Powell, A. J., and Lowe, S. B. (1980). Microtubules, protoplasts and plant cell shape. An immunofluorescent study. *Planta* **147,** 500–506.

Luduena, R. F. (1993). Are tubulin isotypes functionally significant? *Mol. Biol. Cell.* **4,** 445–457.

Maccioni, R. B., and Mellado, W. (1981). Characteristics of the in vitro assembly of brain tubulin in *Cyprinos carpio. Comp. Biochem. Physiol. B* **70B,** 375–380.

MacRae, T. H. (1992). Microtubule organization by cross-linking and bundling proteins. *Biochim. Biophys. Acta* **1160,** 145–155.

Margolis, R. L., and Rauch, C. T. (1981). Characterization of rat brain crude extract microtubule assembly: Correlation of cold stability with the phosphorylation state of a microtubule-associated 64K protein. *Biochemistry* **20,** 4451–4458.

Margolis, R. L., Rauch, C. T., and Job, D. (1986). Purification and assay of cold-stable microtubules and STOP protein. (1986). *In* "Methods in Enzymology" (R. B. Vallee, ed.), Vol. 134, pp. 160–170. Academic Press, Orlando, FL.

Matus, A. (1991). Microtubule-associated proteins and neuronal morphogenesis. *J. Cell Sci., Suppl.* **15,** 61–67.

Mitchison, T., and Kirschner, M. W. (1984). Dynamic instability of microtubule growth. *Nature (London)* **312,** 237–242.

Mizuno, K. (1992). Induction of cold stability of microtubules in cultured tobacco cells. *Plant Physiol.* **100,** 740–748.

Modig, C., Strömberg, E., and Wallin, M. (1994). Different stability of posttranslationally modified brain microtubules isolated from cold-temperate fish. *Mol. Cell. Biochem.* **130,** 137–147.

Moritz, M. (1993). Watching the tube. *Curr. Biol.* **3,** 387–390.

Morris, J. R., and Lasek, R. J. (1982). Stable polymers of the axonal cytoskeleton: The axoplasmic ghost. *J. Cell Biol.* **92,** 192–198.

Multigner, L., Gagnon, J., Van Dorsselaer, A., and Job, D. (1992). Stabilization of sea urchin flagellar microtubules by histone H1. *Nature (London)* **360,** 33–39.

Murphy, D. B. (1991). Functions of tubulin isoforms. *Curr. Opin. Cell Biol.* **3,** 43–51.

Nakazawa, E., Katoh, K., and Ishikawa, H. (1992). The association of microtubules with the plasmalemma in epidermal tendon cells of the river crab. *Biol. Cell.* **75,** 111–119.

Oakley, C. E., and Oakley, B. R. (1989). Identification of γ-tubulin, a new member of the tubulin superfamily encoded by *mip*A gene of *Aspergillus nidulans. Nature (London)* **338,** 662–664.

Padilla, R., Otin, C. L., Serrano, L., and Avila, J. (1993). Role of the carboxy terminal region of β tubulin on microtubule dynamics through its interaction with the GTP phosphate binding region. *FEBS Lett.* **325,** 173–176.

Penningroth, S. M., and Kirschner, M. W. (1977). Nucleotide binding and phosphorylation in microtubule stability in vitro. *J. Mol. Biol.* **115,** 643–673.

Pirollet, F., Rauch, C. T., Job, D., and Margolis, R. L. (1989). Monoclonal antibody to microtubule-associated STOP proteins: Affinity purification of neuronal STOP activity and comparison of antigen with activity in neuronal and nonneuronal cell extracts. *Biochemistry* **28,** 835–842.

Pirollet, F., Derancourt, J., Haiech, J., Job, D., and Margolis, R. L. (1992). Ca²⁺-calmodulin regulated effectors of microtubule stability in bovine brain. *Biochemistry* **31,** 8849–8855.

Pryer, N. K., Walker, R. A., Skeen, V. P., Bourns, B. D., Soboeiro, M. F., and Salmon, E. D. (1992). Brain microtubule-associated proteins modulate microtubule dynamic instability in vitro. *J. Cell Sci.* **103,** 965–976.

Rutberg, M., Billger, M., and Wallin, M. (1994). To be published.

Sackett, D. L., and Varma, J. K. (1993). Molecular mechanism of colchicine action: Induced local unfolding of β-tubulin. *Biochemistry* **32,** 13560–13565.

Saxton, W. M., Stemple, D. L., Leslie, R. J., Salmon, E. D., Zavortink, M., and McIntosh, J. R. (1984). Tubulin dynamics in cultured mammalian cells. *J. Cell Biol.* **99,** 2175–2186.

Schiff, P. B., Fant, J., and Horwitz, S. B. (1979). Promotion of microtubule assembly in vitro by taxol. *Nature (London)* **277**, 665–667.

Schiff, S. B., and Horwitz, S. B. (1981). Taxol assembles tubulin in the absence of exogenous GTP or microtubule-associated proteins. *Biochemistry* **20**, 3247–3252.

Shpetner, H. S., and Vallee, R. B. (1989). Identification of dynamin, a novel mechanochemical enzyme that mediates interactions between microtubules. *Cell (Cambridge, Mass.)* **59**, 421–432.

Simon, J. R., Parsons, S. F., and Salmon, E. D. (1992). Buffer conditions and non-tubulin factors critically affect the microtubule dynamic instability of sea urchin egg tubulin. *Cell Motil. Cytoskel.* **21**, 1–14.

Skoufias, D. A., Wilson, L., and Detrich, H. W., III (1992). Colchicine-binding sites of brain tubulins from an Antarctic fish and from a mammal are functionally similar, but not identical—implications for microtubule assembly at low temperature. *Cell Motil. Cytoskel.* **21**, 272–280.

Sloboda, R. D., and Rosenbaum, J. L. (1979). Decoration and stabilization of intact smooth-walled microtubules with microtubule-associated proteins. *Biochemistry* **18**, 48–55.

Stephens, R. (1975). Structural chemistry of the axoneme: Evidence for chemically and functionally unique tubulin dimers in outer fibers. *In* "Molecules and Cell Movement" (S. Inoué and R. E. Stephens, eds.), pp. 181–206. Raven Press, New York.

Strömberg, E., and Wallin, M. (1994). Differences in the effect of Ca^{2+} on isolated microtubules from cod and cow brain. *Cell Motil. Cytoskel.* **28**, 59–68.

Strömberg, E., Jönsson, A.-C., and Wallin, M. (1986). Are microtubules cold-stable in the Atlantic cod, *Gadus morhua*? *FEBS Lett.* **204**, 111–116.

Strömberg, E., Serrano, L., Avila, J., and Wallin, M. (1989). Unusual properties of a cold-labile fraction of Atlantic cod (*Gadus morhua*) brain microtubules. *Biochem. Cell Biol.* **67**, 791–800.

Suprenant, K. A., and Marsh, J. C. (1987). Temperature and pH govern the self-assembly of microtubules from unfertilized sea-urchin egg extracts. *J. Cell Sci.* **87**, 71–84.

Suprenant, K. A., and Rebhun, L. I. (1983). Assembly of unfertilized sea urchin egg tubulin at physiological temperatures. *J. Biol. Chem.* **258**, 4518–4525.

Sweet, S. C., and Welsh, M. J. (1988). Calmodulin colocalization with cold-stable and nocodazole-stable microtubules in living PtK$_1$ cells. *Eur. J. Cell Biol.* **47**, 88–93.

Tashiro, T., and Komiya, Y. (1983). Subunit composition specific to axonally transported tubulin. *Neuroscience* **9**, 943–950.

Torres, A., and Delgado, P. (1989). Effects of cold and nocodazole treatments on the microtubular systems of *Paramecium* in interphase. *J. Protozool.* **36**, 113–119.

Trinczek, B., Marx, A., Mandelkow, E.-M., Murphy, D. B., and Mandelkow, E. (1993). Dynamics of microtubules from erythrocyte marginal bands. *Mol. Biol. Cell* **4**, 323–335.

Tsugawa, K., and Takahashi, K. P. (1987). Direct adaptation of cells to temperature: Cold-stable microtubule in rainbow trout cells cultured *in vitro* at low temperature. *Comp. Biochem. Physiol. A* **87A**, 745–748.

Wadsworth, P., and Salmon, E. D. (1986). Analysis of the treadmilling model during meta-phase of mitosis using fluorescence redistribution after photobleaching. *J. Cell Biol.* **102**, 1032–1038.

Wallin, M., and Jönsson, A. C. (1983). Characterization of microtubules isolated from dogfish (*Squalus acanthias* and *Scyliorhinus canicula*) brain in the absence of glycerol. *Comp. Biochem. Physiol. B* **75B**, 625–634.

Wallin, M., Billger, M., Strömberg, T., and Strömberg, E. (1993). Assembly of Atlantic cod (*Gadus morhua*) brain microtubules at different temperatures: Dependency of microtubule-associated proteins is relative to temperature. *Arch. Biochem. Biophys.* **307**, 200–215.

Webb, B. C., and Wilson, L. (1980). Cold-stable microtubules from brain. *Biochemistry* **19,** 1993–2001.

Williams, R. C., Jr. (1992). Analysis of microtubule dynamics *in vitro. In* "The Cytoskeleton: A Practical Approach" (K. L. Carraway, ed.), pp. 151–166. Oxford Univ. Press, Oxford.

Williams, R. C., Jr., Correia, J. J., and DeVries, A. L. (1985). Formation of microtubules at low temperatures by tubulin from Antarctic fish. *Biochemistry* **24,** 2790–2798.

Development and Heterogeneity of Prolactin Cells

Sumio Takahashi
Department of Biology, Faculty of Science, Okayama University,
Okayama 700, Japan

Prolactin (PRL) is synthesized in pituitary cells called mammotrophs (PRL cells).
Ample evidence demonstrates that the PRL cell population consists of structurally
and functionally heterogeneous PRL cells. Multiple variants of PRL molecules are
found in various species. Prolactin cells may be divided into various subtypes in the
rat and mouse. Secretory activities differ among the PRL cell population. These
heterogeneities may reflect various phases of the maturation process of PRL cells,
or the integrated outcome of various functional differences in PRL cells. To clarify
the significance of heterogeneities among PRL cells, we present updated reports on
the differentiation, proliferation, and development of PRL cells, and discuss factors
responsible for the functional differences in PRL cell population. The age-related
alteration in PRL secretion in the rat is summarized, because it is one of the most
important aspects of the developmental changes in PRL cells. A mammosomatotroph, which secretes growth hormone and PRL, is found in various species.
Prolactin cells and somatotrophs are derived from the same lineage. The possible
relationship among PRL cells, somatotrophs, and mammosomatotrophs is
discussed.
KEY WORDS: Pituitary, Prolactin, Mammotroph, Prolactin cell.

I. Introduction

Prolactin (PRL) is a protein hormone (about 200 amino acids) that is
synthesized in specific pituitary secretory cells called mammotrophs (PRL
cells) (Kurosumi, 1986; Tougard and Tixier-Vidal, 1988). Prolactin is well
known to have many diverse functions including reproduction, lactation,
osmoregulation, and immunomoduration (Nicoll, 1974; Nicoll *et al.*, 1986;

Wallis, 1988). Alterations of PRL secretion produce many disorders in body functions (Meites *et al.*, 1987; vom Saal and Finch, 1988; Meites, 1990).

Growth hormone (GH) is an anterior pituitary protein hormone (about 200 amino acids), and is synthesized in somatotrophs (GH cells). Growth hormone is mainly involved in growth regulation and protein metabolism. Dysfunction of the GH secretory mechanism results in dwarfism (Sonntag *et al.*, 1982, and senile decline in GH secretion causes disorders in many body functions (Sonntag *et al.*, 1985; Meites *et al.*, 1987; Takahashi and Meites, 1987). Molecular characteristics and physiological functions of GH are quite similar to PRL. These two hormones are considered to come from a common ancestor molecule (Cooke *et al.*, 1980, 1981). Molecular structures of these hormones and evolutionary relationship have been discussed by Seo (1985).

Studies of the ontogeny of pituitary cells showed that GH cells and PRL cells derived from common progenitor cells (Borrelli *et al.*, 1989). There is evidence that the GH cell population consists of morphologically and functionally heterogeneous cells. I reviewed the relationship between GH cells and PRL cells and extensively discussed the significance of GH cell heterogeneity (Takahashi, 1992a). Similarly, many studies have indicated that the PRL cell population consists of morphologically and functionally heterogeneous cells (Tougard and Tixier-Vidal, 1988; Kurosumi, 1991; Takahashi, 1992a). Moreover, variants of PRL molecules have been found (Wallis, 1988). It is highly probable that such heterogeneity observed in various aspects of PRL cells reflects functional differences among PRL cells and various phases of the maturation process of PRL cells and PRL molecules. Therefore, we hope that an analysis of the heterogeneity of the PRL cell population will make it possible for us to understand more clearly the differentiation, growth, development, and functions of PRL cells. This chapter discusses the development and heterogeneity of PRL cells primarily in the rat and mouse, because analyses of functional diversity of PRL and PRL cells seem to have advanced further in these animals.

It is not appropriate to disregard the relationship between GH cells and PRL cells. Therefore, we must pay attention to another type of pituitary cell. A novel pituitary cell, a mammosomatotroph (MS cell) or somatomammotroph, contains GH and PRL in the same cell and is found in several species. In this chapter the term *mammosomatotroph* is used to describe this secretory cell, instead of somatomammotroph or lactosomatotrophs.

There are several ideas about the developmental and functional significance of MS cells. The purpose of this chapter is to explore several views explaining the heterogeneity of PRL and PRL cells. As one of the most important aspects of the pituitary gland, age-related changes in PRL cells are discussed.

II. Identification of Prolactin Cells

Prolactin cells were first identified light microscopically as erythrosinophilic cells (Herlant, 1964). Prolactin cells were immunocytochemically identified in various species (rat: Baker *et al.*, 1969; Nakane, 1970; mouse: Baker and Gross, 1978; human: Halmi *et al.*, 1975; Baker and Yu, 1977). Prolactin cells were found sparsely in the anterior–ventral portion of the gland, and found in the areas near the intermediate lobe in the rat (Baker *et al.*, 1969; Nakane, 1970; Smets *et al.*, 1987). Regional distribution of PRL cells was reported in mice (Sasaki and Iwama, 1988a). The densities of PRL cells in the rostral and caudal pituitaries of mice were significantly greater than those of GH cells. The number and size of PRL cells differed significantly depending on the sex, age, and reproductive states (estrous cycle, pregnancy, lactation). Prolactin cells showed various shapes; polygonal, elongated, and cup-shaped. The cup-shaped PRL cells had long, slender cytoplasmic processes and these processes surrounded large, oval gonadotrophic cells (Sato, 1980; Nogami and Yoshimura, 1982).

III. Differentiation of Prolactin Cells

A. First Appearance of Prolactin Cells

Several laboratories reported the ontogenesis of PRL cells in rat pituitaries, and these reports are somewhat contradictory. Sétáló and Nakane (1972) and Nemeskéri *et al.* (1988) found PRL cells on day 16 of gestation. Chatelain *et al.* (1979) reported that PRL cells were detected on day 21 of gestation, and Watanabe and Daikoku (1979) reported that PRL cells were first detected postnatally. The reverse hemolytic plaque assay demonstrated that PRL cells were detectable by 4 days of age (Hoeffler *et al.*, 1985). In our laboratory, GH cells and PRL cells were both immunocytochemically detected on fetal day 18 in Wistar rats (Figs. 1 and 2), which is in agreement with the recent report by Watanabe and Haraguchi (1994). In the mouse pituitary, immunoreactive PRL cells were detected at birth, but PRL cells might have appeared in fetal pituitary glands (Harigaya and Hoshino, 1985). Ontogenic changes in the number of GH and PRL cells in the mice of the ICR strain were observed (Figs. 3 and 4). A two-

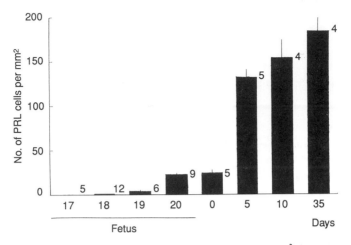

FIG. 1   The number of immunoreactive PRL cells per unit area (mm²) in male rats of the Wistar/Tw strain. Prolactin cells were identified using an anti-rat PRL serum. The number on each bar depicts the number of rats used. Occurrence of immunoreactive PRL in the fetal rats was as follows: day 17, 0; day 18, 1; day 19, 2; day 20, 9.

dimensional electrophoresis of pituitary extracts revealed that PRL synthesis in mice was first detected at 8 days of age (Slabaugh *et al.,* 1982). This disagreement on the date of first appearance of PRL cells is thought to be due to the difference in the strain and antisera used, and the difference

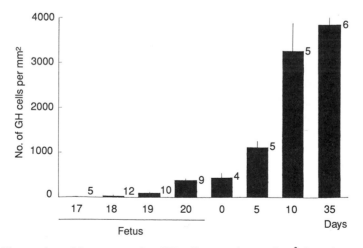

FIG. 2   The number of immunoreactive GH cells per unit area (mm²) in male rats of the Wistar/Tw strain. Growth hormone cells were identified using an anti-rat GH serum. The number on each bar depicts the number of rats used. Occurrence of immunoreactive GH in the fetal rats was as follows: day 17, 0; day 18, 3; day 19, 6; day 20, 9.

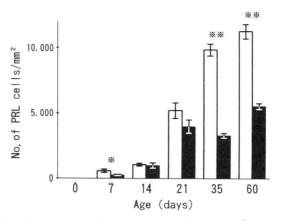

FIG. 3 The number of immunoreactive PRL cells per unit area (mm²) in male (■) and female (□) mice of the ICR strain. Prolactin cells were identified using an anti-mouse PRL serum. All age groups consisted of 10 mice. ※ $p < 0.02$, ※※ $p < 0.001$, significantly different between male and female rats.

in the sensitivity of assays used (immunocytochemistry, reverse hemolytic plaque assay, and two-dimensional electrophoresis). In human fetal pituitary glands the appearance of PRL cells was described by Baker and Jaffe (1975), Dubois and Hemming (1991), and Asa *et al.* (1988, 1991).

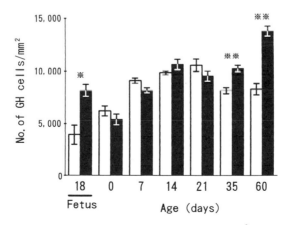

FIG. 4 The number of immunoreactive GH cells per unit area (mm²) in male (■) and female (□) mice of the ICR strain. Growth hormone cells were identified using an anti-rat GH serum. All age groups consisted of 10 mice. ※ $p < 0.02$, ※※ $p < 0.001$, significantly different between male and female rats.

B. Expression of Prolactin Gene

Prolactin gene expression has been studied during the perinatal period in the mouse and rat. Differentiation of PRL cells is first initiated by the gene expression of tissue-specific transcription factor Pit-1/GHF-1 (Dollé *et al.*, 1990; Simmons *et al.*, 1990). Pit-1 binds to a specific site to activate the PRL gene as well as the GH gene (Bodner *et al.*, 1988; Ingraham *et al.*, 1988; Nelson *et al.*, 1988). Expression of each gene (GHF-1, GH, and PRL) was viewed using the *in situ* hybridization technique. In the mouse pituitary GHF-1 transcript was first detected on fetal day 13.5. Growth hormone mRNA was first detected on fetal day 15.5. A few PRL mRNA-expressing cells were detected on the same day. The GH mRNA-expressing cells significantly increased on fetal days 16.5 and 17.5, but PRL-secreting cells decreased during the same period (Dollé *et al.*, 1990). Thus, PRL expression seems to be blocked during the fetal period.

In the rat pituitary, *in situ* hybridization demonstrated that PRL mRNA was first detected on gestation day 18–19 (Nogami *et al.*, 1989). On the other hand, Frawley and Miller (1989) reported that PRL genes had been expressed at least by 3 days of age, but the translation of the PRL message was, interestingly, blocked by the lack of association of the PRL message with ribosomes. Such blockage may possibly operate in the mouse pituitary. Release of this blockage may trigger PRL synthesis. Ontogenic analysis of human PRL cells with respect to PRL mRNA was reported by Suganuma *et al.* (1986).

C. Factors for Differentiation of Prolactin Cells

An inhibitory mechanism of PRL synthesis at the level of translation has been suggested, as stated above. We summarize here the factors for controlling PRL gene expression. The factors for PRL cell differentiation may not necessarily be specific substances, and spatial factors must be regarded.

1. Regional Difference in Differentiation of Prolactin Cells

Nemeskéri *et al.* (1988) reported the interesting finding of a regional difference in PRL cell appearance in the rat. They found that PRL cells were first detected in the pars tuberalis on fetal day 15, and in the pars distalis on fetal day 16, and also that pituitary cell differentiation in the pars distalis follows a rostrocaudal, ventrodorsal direction. The existence of the spatial and temporal difference in pituitary cell differentiation suggests the possibility that the committed cells may migrate in these directions (rostrocaudal and ventrodorsal). Another possibility is that inducing substance required for the cell differentiation may spread from the pars tuberalis side.

2. Factors for Differentiation of Prolactin Cells

The existence of a regional difference in PRL cell differentiation suggests that there are factors for the promotion of pituitary cell differentiation. Here, we discuss several possible factors, but brain–pituitary interaction is not discussed in detail in this chapter.

One possible factor is a maternal mile-borne substance. Normal differentiation of PRL cells in neonatal rats is stimulated by a maternal signal specific to early lactation (Porter et al., 1991a). This was identified as a peptide(s) whose molecular mass was 4–8 kDa (Porter and Frawley, 1991; Porter et al., 1993). This signal is thought to be transferred to the fetal circulation by ingestion of maternal milk early in lactation. This peptide may be the known peptide, or an unknown peptide. Possible candidates for this peptide are as follows: gonadotropin-releasing hormone (GnRH) (Baram et al., 1977; Amarant et al., 1982; Smith and Ojeda, 1984), thyrotropin-releasing hormone (TRH) (Strbak et al., 1980), neurotensin (Weström et al., 1987), parathyroid hormone-related peptide (Khosla et al., 1990), growth hormone-releasing hormone (GHRH) (Werner et al., 1986, 1988), somatostatin (Werner et al., 1985, 1988; Takeyama et al., 1990), a calcitonin-like peptide (Shah et al., 1989), epidermal growth factor (EGF) (Carpenter, 1980; Grueters et al., 1985), transforming growth factor β(TGF-β) (Eckert et al., 1985; Jin et al., 1991), insulin, and insulin-like growth factor I (IGF-I) (Baxter et al., 1984; Simmen et al., 1988; Donovan et al., 1991; Prosser et al., 1991). A mechanism of milk-borne peptide action needs to be clarified, and it is probable that this peptide may act in the expression of the PRL gene in collaboration with Pit-1/GHF-1.

Several reports are available to support the above-described hypothesis. The effects of TRH on the differentiation of pituitary cells from 11-day-old fetal rats were studied (Héritier and Dubois, 1993). The TRH treatment on day 1 of culture induced a differentiation of only a few immunoreactive PRL cells. Previously, GnRH (Bégeot et al., 1983) and the α subunit of glycoprotein hormones [luteinizing hormone (LH), follicle-stimulating hormone (FSH), and thyroid-stimulating hormone (TSH)] had been demonstrated to induce the differentiation of PRL cells (Bégeot et al., 1984).

Coculture of the rat posterior pituitary (the neurointermediate lobe) with GH_3 pituitary tumor cells significantly stimulated PRL gene expression, synthesis, and release (Corcia et al., 1993). They speculated that a subpopulation of the intermediate cells secreted a substance that is involved in PRL gene expression.

GH₃ pituitary tumor cells secrete GH and PRL *in vitro,* but when tumor cells were transplanted into host rats PRL gene expression was repressed (Day and Day, 1994). This repression is considered to be mediated by the alternatively spliced form of Pit-1 protein, because PRL promoter activity was inhibited by spliced small forms of Pit-1 (27 and 24 kDa), but not by 33- and 31-kDa forms of Pit-1. Further study is needed to understand the regulation of splicing of *pit-1* gene products and the interactions of the spliced form of Pit-1 with the PRL promoter. This finding indicates that alternative splicing can generate products that have opposite effects (namely, stimulatory or inhibitory for PRL gene expression).

IV. Morphological Heterogeneity of Prolactin Cells

A. Light Microscopic Observation

Prolactin cells showed various shapes ranging from oval to irregular with processes (Nakane, 1970). Sato (1980) immunocytochemically analyzed the postnatal development of PRL cells in the rat. Judging from the ontogenic changes in PRL cell shape, Sato suggested that the oval PRL cells were premature, the polygonal ones mature, and the cup-shaped ones particularly differentiated.

B. Electron Microscopic Observation

The ultrastructure of PRL cells has been extensively studied by immunoelectron microscopy and PRL cells have been divided into three subtypes, mainly based on the size of the secretory granules (Nogami and Yoshimura, 1980, 1982; Nogami, 1984; Kurosumi, 1986; Kurosumi *et al.,* 1987). Smets *et al.* (1987) subdivided rat PRL cells into two types, one containing large polymorphic granules and the other containing small round granules. Harigaya *et al.* (1983) also classified mouse PRL cells into three types on the basis of immunoelectron microscopy. In another electron microscopy study the PRL mRNA was localized in the rat pituitary to identify PRL cells, and two types of PRL-synthesizing cells were identified (Tong *et al.,* 1989). One type was characterized by large secretory granules, and the other by small secretory granules. Takahashi and Miyatake (1991) observed three subtypes of PRL cells in the rat, and classified them on the basis of Kurosumi's classification (Kurosumi, 1986; Kurosumi *et al.,* 1987). Kurosumi originally named three subtypes of PRL cell by numerals according to the order of discovery of each subtype of cell (type I, type II, and type III). On the other hand, Nogami and Yoshimura (1980) and Harigaya and Hoshino (1985) designated PRL cell subtypes according to the occurrence of cells. Kurosumi (1991) changed the nomenclature of PRL subtypes into the immature type, intermediate type, and mature type of PRL cells. Prolactin cells of the mature type (the former type I cell) contain irregularly shaped, large secretory granules with a diameter of 300–700 nm (Fig. 5), and appeared in adult animals. The mature type of PRL cell had been originally identified as mammotrophs. Prolactin cells of the intermediate type (the former type II cell) contain spherical granules with a diameter of 150–250 nm (Fig. 6), and appeared later. Prolactin cells of the immature type (the former type III cell) contain small, round granules with a diameter of 100 nm (Fig. 7), and appeared first during ontogeny.

V. Development of Prolactin Cells

A. Growth of Pituitary Cells

The pituitary gland is considered to be an "expanding organ" (Goss, 1978). Every pituitary cell has the capacity to divide during its life span, although the mitotic activity declines with aging (Takahashi and Kawashima, 1986). Differentiation of pituitary cells is not incompatible with mitosis, because mitotic figures of differentiated pituitary cells, which could be detected immunocytochemically, are encountered at any age in the rat and mouse. Pituitary cells divide into the same differentiated cells (simple duplication). Alternatively, daughter cells may turn into another type of cell. Although direct evidence for undifferentiated "stem" cells is not yet available, the "stem" cells may divide and one of the divided cells may differentiate into a specific type of pituitary cell. Agranular mitotic cells have been reported in electron microscopy studies (Kurosumi, 1971; Takahashi and Kawashima, 1982), and mitotic cells that are immunonegative with various pituitary hormone antisera have been observed in young adult rats (Shirasawa and Yoshimura, 1982). Such cells are candidate "stem" cells. At any rate, little is known about the mechanism of pituitary growth control or the mechanism of pituitary cell proliferation.

Dysfunction of the mechanism of pituitary cell proliferation results in pituitary tumorigenesis. In human pituitary GH-secreting tumors, mutant forms of the α subunit of the guanine nucleotide-binding protein G_s (the stimulatory regulator of adenylate cyclase) have been identified (Landis et al., 1989, 1990; Drews et al., 1992; Hosoi et al., 1993; Tian et al., 1994). Activating mutations in the adenylate cyclase system are closely

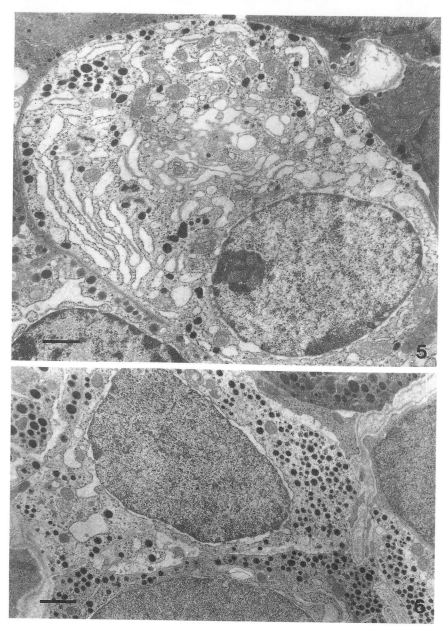

FIG. 5 Mature-type PRL cell in an adult female rat. The cell is elongated and a round nucleus is located slightly eccentrically. The rough endoplasmic reticulum and the Golgi apparatus are well developed. Large, round, irregularly shaped secretory granules (300–700 nm) are located in the peripheral cytoplasm. Bar: 1 μm. [Reprinted with permission from Takahashi, S., and Miyatake, M. (1991). *Zool. Sci.* **8,** 549–559.]

FIG. 6 Intermediate-type PRL cell in an adult female rat. The cell contains round secretory granules with a diameter of 150–250 nm. The number of secretory granules is larger than that in mature-type PRL cells. Bar: 1 μm. [Reprinted with permission from Takahashi, S., and Miyatake, M. (1991). *Zool. Sci.* **8,** 549–559.]

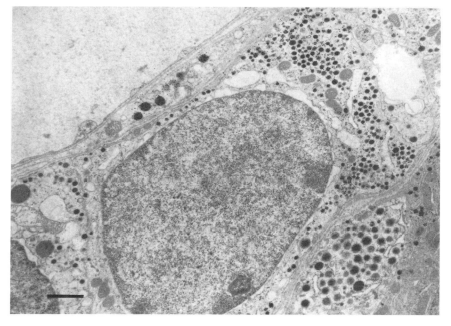

FIG. 7 Immature-type PRL cell in an adult female rat. The cell is characterized by the small amount of cytoplasm containing small secretory granules with a diameter of about 100 nm. Cell organelles are less developed. Bar: 1 nm. [Reprinted with permission from Takahashi, S., and Miyatake, M. (1991). *Zool. Sci.* **8,** 549–559.]

connected with pituitary tumorigenesis. This chapter briefly deals with the actions of estrogen, dopamine, epidermal growth factor (EGF), transforming growth factor α and β (TGF-α and-β), and other factors on this system in the following sections.

B. Development of Prolactin Cells

1. Ontogeny of Growth Hormone Cells and Prolactin Cells in Rat and Mouse

Ontogenic changes of PRL cells and GH cells in rats of the Wistar strain and mice of the ICR strain were studied (Figs. 1–4). The density of both cells was expressed as the number of immunoreactive PRL cells and GH cells per unit area. In the rat and mouse GH cells and PRL cells appeared at almost the same age (rat, fetal day 18 in our laboratory). However, the growth of the GH cell population was more active than that of the PRL cell population. Factors controlling the growth of each cell population, and the regulatory mechanism determining the proliferation of pituitary cells, must be clarified and are discussed below. In the rat, GH cells and PRL cells were first detected on fetal day 18. Growth hormone cells increased dramatically from fetal day 18; on the other hand, PRL cells increased significantly only after birth. In mouse pituitaries GH cells and PRL cells were already detected by fetal day 18. The developmental pattern of GH cells and PRL cells in the ICR mouse is shown in Figs. 3 and 4. Prolactin cells were rare by fetal day 18 and at birth, which is similar to the situation in rat pituitaries.

2. Proliferation of Prolactin Cells

a. Identification of Mitotic Cells   The proliferation of pituitary cells has been reported using various methods: the observation of colchicine-arrested mitotic figures (Pomerat, 1941; Hunt, 1943; Allanson *et al.*, 1969; Shirasawa and Yoshimura, 1982; Takahashi *et al.*, 1984), the autoradiographic observation of [³H]thymidine uptake by cells (Hunt and Hunt, 1966; Kalbermann *et al.*, 1979; Inoue and Kurosumi, 1981), the observation of bromodeoxyuridine (BrdU) uptake by cells (Wyndford-Thomas and Williams, 1986; Carbajo-Peréz *et al.*, 1989; Watanabe and Carbajo-Peréz, 1990), and the observation of proliferating cell nuclear antigen (PCNA, Miyachi *et al.*, 1978; Bravo *et al.*, 1987; Oishi *et al.*, 1993).

b. Changes in Mitotic Activity   The mitotic figures of pituitary cells are usually low, but during the pre- and peripubertal periods and at estrus the mitotic activity of pituitary cells is high (Hunt and Hunt, 1966; Allanson *et al.*, 1969; Shirasawa and Yoshimura, 1982; Takahashi and Kawashima, 1982; Takahashi *et al.*, 1984; Sakuma *et al.*, 1984; Carbajo-Peréz *et al.*, 1989; Carbajo-Peréz and Watanabe, 1990). Circadian variation of pituitary cells has been examined (Hunt, 1943; Nouët and Kujas, 1975; Carbajo-Peréz *et al.*, 1991), but there has been no consistent agreement concerning the timing of pituitary cell proliferation. However, flow cytometric analysis of nuclei stained with propidium iodine in the male rat pituitary cells demonstrated that the beginning of the light period may trigger a wave of cells to leave G_0/G_1 and enter into S phase (Carbajo-Peréz *et al.*, 1991).

c. Identification of Cell Types of Mitotic Pituitary Cells   The mitotic pituitary cells were identified by electron microscopy. The presence of mitotic PRL cells was clearly shown by Zambrano and Deis (1970), Kurosumi (1971, 1979), Smith and Keefer (1982), and Takahashi and Kawashima (1982). Kurosumi (1979) observed exocytotic figures in mitotic PRL cells, indicating that PRL was released even during mitotic division. Immunocytochemical identification of mitotic PRL cells as well as other pituitary cells was carried out by several laboratories (Shirasawa and Yoshimura, 1982; Sakuma *et al.*, 1984; Takahashi *et al.*, 1984; Carbajo-Peréz *et al.*, 1989).

d. Age-Related Changes in Mitotic Activity of Prolactin Cells in Rat and Mouse The ontogeny of the mitotic activity of pituitary cells, particularly PRL cells and GH cells, was studied (Shirasawa and Yoshimura, 1982; Takahashi *et al.*, 1984; Carbajo-Peréz *et al.*, 1989). During the prepubertal period, the mitotic activity of pituitary cells was relatively higher compared with that in adult animals. Shirasawa and Yoshimura (1982) reported that in male rat pituitaries the mitotic activity was highest at 30 days, during an observation period from 5 to 70 days of age. Growth hormone cells were the most actively proliferating cells, followed by PRL cells. However, Carbajo-Peréz *et al.* (1989) and Takahashi *et al.* (1984) found that the proliferating activity of pituitary cells was higher at 8 or 10 days of age than at any other age, up to 60 days of age (Fig. 8). Even at 1 or 2 years of age mitotic PRL cells were encountered, although the frequency was lower than at younger ages.

In adult female rats cyclic variation in the proliferation of pituitary cells, including PRL cells, had been observed. During the estrous cycle the mitotic activity of pituitary cells and PRL cells was highest at estrus, and was low and relatively constant at other stages of the estrous cycle (Takahashi *et al.*, 1984). This finding was confirmed by Oishi *et al.* (1993).

e. Factors for Controlling Proliferation of Prolactin Cells Proliferation of PRL cells is regulated by an estrogenic mechanism, a dopaminergic mechanism, hypothalamic hormones and other growth factors (including interleukins), and so on (Webster and Scanlon, 1991). GHF-1 is also required for pituitary cell proliferation (Castrillo *et al.*, 1991). In addition, most of these factors affect PRL secretion. For example, estrogen stimulates PRL synthesis and release (Chen and Meites, 1970; Maurer and Notides, 1987). Dopamine inhibits PRL synthesis and release (MacLeod and Lehmeyer, 1974; Maurer, 1980).

Pituitary cell proliferation is regulated by extrapituitary factors (estrogen, dopamine, other hypothalamic factors, etc.), and also by intrapituitary factors. A large number of studies exhibit lists of substances contained in pituitary glands. Intercellular communication within the pituitary glands may play a key role in pituitary growth (Schwartz ande Cherny, 1992). However, in this chapter we survey factors controlling pituitary PRL cell proliferation that fall into the classic category, such as estrogen, dopamine, and so on.

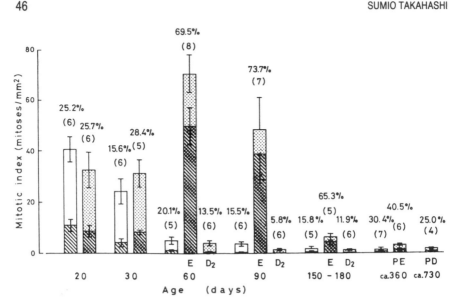

FIG. 8 Age-related changes in the mitotic indices of pituitary cells and PRL cells in male
(□) and female (□) rats. Colchicine-arrested mitotic cells and immunocytochemically identified PRL cells (positive cells, ▨), were observed. Female rats at estrus (E), second day of
diestrus (D₂), persistent estrus (PE), and persistent diestrus (PD) were used. The percentage
above the each column depicts the percentage of mitotic PRL cells in total mitotic pituitary
cells. The numbers in parentheses depict the number of rats, and bars depict the standard
errors of means. [Reprinted with permission from Takahashi, S. (1992a). *Zool. Sci.* **9**,
901–924.]

i. Estrogen A large body of evidence indicates that estrogen stimulates
pituitary growth, which results mainly from the hypertrophy and proliferation of certain pituitary cells. Most pituitary cells contain estrogen receptors (Ginsburg *et al.,* 1975; Menon and Gunaga, 1976; Sen and Menon,
1978; Keefer, 1980), and therefore are responsive to estrogen. Estrogen-
induced proliferation of PRL cells was clearly demonstrated using immunocytochemical techniques (Lloyd *et al.,* 1975; Burdman *et al.,* 1984;
Takahashi *et al.,* 1984; Pérez *et al.,* 1986; Motegi and Watanabe, 1990).

In vitro effects of estrogen on normal pituitary cells are still ambiguous,
although several reports indicated the mitogenic effect of estrogen
(Maurer, 1979). Estrogen stimulates GH₃ and GH₄ pituitary tumor cells
in vitro (Amara and Dannies, 1983; Shull, 1991). Importantly, population
density altered the responsiveness of GH₄ cells to estrogen, suggesting
the involvement of tumor cell-derived growth factors in cell proliferation
(Shull, 1991).

ii. Dopamine Hypothalamic tuberoinfundibular dopaminergic neurons release dopamine into the pituitary portal system, and dopamine acts on PRL cells as a PRL inhibitory factor (Ben-Jonathan, 1985; Weiner *et al.*, 1988). Dopamine D_2 receptor is present in the pituitary gland (Cronin *et al.*, 1978; Sibley *et al.*, 1982; Zabavnik *et al.*, 1993), and is also immunocytochemically demonstrated in the plasma membrane of PRL cells in the rat (Goldsmith *et al.*, 1979).

Bromocriptine, a dopamine agonist, is frequently used in the treatment of PRL-secreting pituitary tumors in humans (Thorner *et al.*, 1980; Archer *et al.*, 1982). In the rat bromocriptine inhibited DNA synthesis and PRL secretion in a spontaneous tumor (Prysor-Jones and Jenkins, 1981). Bromocriptine inhibited GH and PRL secretion, and decreased the GH mRNA and PRL mRNA content in GH_3 cells (Kamijo *et al.*, 1993). Vrontakis *et al.*, (1987) studied the effect of bromocriptine on the growth of a diethylstilbestrol (DES)-induced pituitary tumor. In their study, bromocriptine treatments after DES implants were removed significantly reduced the pituitary tumor weights and elevated the serum PRL levels. However, bromocriptine treatment together with DES was effective only in PRL synthesis and release, not in reducing pituitary tumor weights. These results suggest that estrogenic and dopaminergic mechanisms in PRL synthesis and release are independent from the mechanism for controlling the growth of PRL cells (the pituitary tumor).

Bromocriptine-induced regression of rat pituitary tumors was investigated by electron and light microscopy (Drewett *et al.*, 1993). Hyperplasia of PRL cells had been induced by chronic estrogen treatment. After estrogen withdrawal, apoptotic cells and cells with phagocytosed apoptotic bodies quickly increased in number. Administration of bromocriptine after estrogen withdrawal increased the number of apoptotic cells and cells with apoptotic bodies nearly twofold. Thus, bromocriptine accelerated apoptosis in the anterior pituitary gland.

Dopamine antagonists stimulate the proliferation of PRL cells (Kalbermann *et al.*, 1979; Jahn *et al.*, 1982) and PRL secretion (Takahashi *et al.*, 1979). On the other hand, dopamine and its agonists, which inhibit PRL secretion (Brooks and Welsch, 1974; Maurer, 1980), inhibit the proliferation of pituitary cells (Lloyd *et al.*, 1975; Burdman *et al.*, 1982). Takahashi and Kawashima (1987) demonstrated that bromocriptine inhibited the mitosis of PRL cells as well as PRL secretion in the rat. Thus, these findings suggest that dopamine agonists have an antimitogenic action in pituitary cells, and that the proliferation of PRL cells is closely correlated with PRL secretion.

Benzodiazepines (BZDs) are well known to be minor tranquillizers and inhibitors of estrogen-induced PRL-secreting pituitary tumors, and this inhibition seems to be calcium dependent (Kunert-Radek et al., 1994).

Benzodiazepine receptors have been reported in the rat and human pituitary (Brown and Martin, 1984; Voigt *et al.*, 1984). The molecular mechanism of the antiproliferative effects of BZD is not clear, but this effect may be potentially useful for understanding the mechanism of pituitary cell proliferation and the clinical therapy of pituitary tumors.

iii. Melanocyte-Stimulating Hormone and Endorphin Estrogen-induced acute secretion of PRL is dependent on the neurointermediate lobe *in vivo* (Murai and Ben-Jonathan, 1990) and *in vitro* (Ellerkmann *et al.*, 1991). Similarly, suckling-induced acute PRL secretion is mediated partly by α-melanocyte-stimulating hormone (α-MSH) from the intermediate lobe (Hill *et al.*, 1993). These results suggest that estrogen stimulated PRL secretion indirectly through the neurointermediate lobe. The active substances turned out to be acetylated forms of α-MSH and β-endorphin (β-END) (Ellerkman *et al.*, 1992a,b). α-Melanocyte-stimulating hormone and β-END increased both the numbers of PRL cells and the PRL secretion, but recruitment of PRL cells caused by these two hormones is restricted to the parts of the anterior pituitary gland proximal to the intermediate lobe (an inner part of the pituitary gland) (Porter and Frawley, 1992). The short portal vessel has been demonstrated to supply blood only to the region of the anterior pituitary gland proximal to the intermediate lobe (the region that is sensitive to α-MSH and β-END) (Daniel and Prichard, 1956). This short portal system may be involved in the recruitment of PRL cells induced by the two intermediate hormones.

iv. Hypothalamic Hormones Hypothalamic hormones that control pituitary hormone secretion affect the proliferation of pituitary cells (Billestrup *et al.*, 1986; McNicol *et al.*, 1988, 1990; Frawley and Hoeffler, 1988; Prysor-Jones *et al.*, 1989). Denef's group reported interesting findings that GnRH (LHRH) increased the number of PRL cells. The GnRH increased the number of PRL mRNA-expressing cells (PRL cells) in S phase. However, they stated that the GnRH-induced increase in the number of PRL cells may be based on differentiation of progenitor cells or immature cells into PRL cells rather than on the mitosis of preexisting PRL cells. The GnRH action is probably mediated by specific growth-promoting factors released from gonadotrophs in a paracrine manner (Tilemans *et al.*, 1992; Van Bael *et al.*, 1994).

Galanin is known to stimulate PRL secretion (Ottlecz *et al.*, 1988). Galanin, which is distributed in the brains of rats and humans, is also found in the anterior pituitary gland (Kaplan *et al.*, 1988; Steel *et al.*, 1989; Vronkatis *et al.*, 1989). Galanin is present in PRL cells, GH cells, and thyrotrophs. Colocalization of galanin and PRL within secretory granules has been imunocytochemically detected (Hyde *et al.*, 1991). Estrogen stimulated the expression of the galanin gene and the synthesis of galanin (Kaplan *et al.*, 1988; Vronkatis *et al.*, 1989). Bromocriptine inhibited estrogen-induced galanin gene expression (Leite *et al.*, 1993). Thus, galanin may participate in pituitary tumorigenesis.

v. Growth Factors   Human and rat pituitary glands contain various growth factors: fibroblast growth factor (FGF), epidermal growth factor (EGF), transforming growth factor α (TGF-α), TGF-β, insulin-like growth factor I (IGF-I), IGF-II, and TGFe (Kudlow and Kobrin, 1984; Murphy *et al.*, 1987; Bach and Bondy, 1992; Driman *et al.*, 1992; Halper *et al.*, 1992; Burns and Sarkar, 1993). Transforming growth factor α and β both inhibited GH$_4$ pituitary tumor cells, but differed in PRL production. Transforming growth factor α stimulated PRL secretion, but TGF-β did not (Ramsdell, 1991). The action of TGF-α on normal PRL cells is not clear, but Borgundvaag *et al.* (1992) presented the interesting finding that estrogen increased pituitary TGF-α mRNA preceding pituitary growth. Bromocriptine, a dopamine agonist, decreased TGF-α mRNA preceding regression of pituitary glands. These results show that the pituitary TGF-α level changes in accordance with pituitary growth. It is highly probable that released TGF-α stimulates the proliferation of PRL cells. Estrogen action on pituitary growth may be mediated by TGF-α. Epidermal growth factor and TGF-α were synthesized in the pituitary glands (Kudlow and Kobrin, 1984). Kudlow and Kobrin (1984) reported that EGF was localized in gonadotrophs and thyrotrophs, and TGF-α in somatotrophs and PRL cells in bovine pituitaries. Ren *et al.* (1994) reported that TGF-α was detected in PRL cells and glycoprotein-secreting cells (gonadotrophs and thyrotrophs) in the normal human pituitary. On the other hand, Finley *et al.* (1994) reported that TGF-α was localized in somatotrophs in an adult male human pituitary, and epidermal growth factor receptors were also identified in somatotrophs. The action of EGF is not known, but EGF and TGF α may act in an autocrine or paracrine manner within the pituitary glands. Estrogen action on cell proliferation is currently considered to be mediated through growth factors, and not directly (Sirbasku, 1978; Sutherland *et al.*, 1988). Several reports indicated that EGF and TGF-α act as mediators of estrogen action in the uterus (Tomooka *et al.*, 1986; Nelson *et al.*, 1991, 1992). The present data suggest that TGF-α stimulates the proliferation of normal pituitary cells, particularly PRL cells. Because the opposite effect had been reported in GH$_4$ pituitary tumor cells (Ramsdell, 1991), this discrepancy remains to be clarified.

Interesting data have been reported to the effect that TGF-β decreased the basal and estrogen-induced growth of pituitary cells, and concurrently decreased PRL secretion in a dose-dependent manner in the rat (Sarkar *et al.*, 1992). This result seems to be in agreement with Ramsdell's data examing GH$_4$ pituitary tumor cells (Ramsdell, 1991). An immunocytochemical study has demonstrated that about 60% of TGF-β-immunoreactive cells are PRL cells (Burns and Sarkar, 1993). They also found that estrogen decreased the pituitary TGF-β content. As TGF-β decreased PRL secretion and pituitary growth, it is highly probable that TGF-β directly inhibits the mitoses of PRL cells. Transforming growth factor β receptors are detected in GH_3 pituitary tumor cells (Cheifetz *et al.*, 1988), although evidence for TGF-β receptors in normal pituitary cells is not available. These findings suggest that pituitary TGF-β (probably generated from PRL cells) inhibits PRL secretion and the proliferation of PRL cells in an autocrine or paracrine fashion.

Insulin-like growth factor I (IGF-I) stimulated the proliferation of pituitary cells in our *in vitro* study (Oomizu *et al.*, 1993, 1994). Insulin-like growth factor I increased the number of PRL cells showing the uptake of BrdU into the nucleus (Fig. 9). We also found that insulin stimulated the proliferation of pituitary cells *in vivo* and *in vitro* (Fig. 10). Insulin-like growth factor I increased the number of DNA-synthesizing cells at a lower dose than insulin. High-dose insulin is known to act through the IGF-I receptor. Insulin adminstered *in vivo* may possibly stimulate the proliferation of pituitary cells through IGF-I receptors (Fig. 11). Insulin-like growth factor I is synthesized in pituitary glands (Murphy *et al.*, 1987; Bach and Bondy, 1992), and its synthesis may be altered by GH and estrogen (Michels *et al.*, 1993). Insulin-like growth factor I receptors were detected in

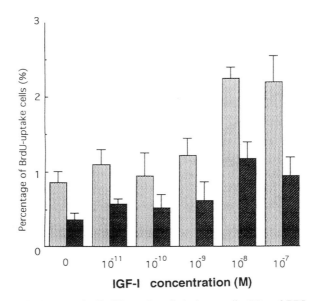

FIG. 9   Effects of IGF-I on the BrdU uptake of pituitary cells (□) and PRL cells (■) in a monolayer culture of male pituitary cells. IGF-I treatment for 5 days increased the percentage of BrdU uptake in pituitary cells and PRL cells.

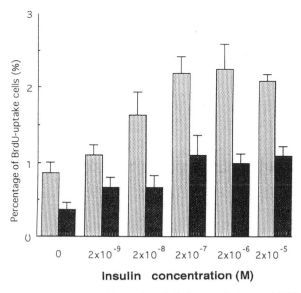

FIG. 10   Effect of insulin on the BrdU uptake of pituitary cells (☐) and PRL cells (■) in a monolayer culture of male pituitary cells. Insulin treatment for 5 days increased the percentage of BrdU uptake by pituitary cells and PRL cells.

the pituitary gland (Goodyer *et al.*, 1984). Thus, this evidence indicates that pituitary IGF-I is involved in pituitary cell proliferation. Pituitary IGF-I may act in an autocrine or paracrine manner to stimulate the proliferation of pituitary cells, at least PRL cells. The actual cell types of IGF-I-synthesizing cells other than GH cells are not known and remain to be identified. The original hypothesis concerning IGF-I function is that GH stimulates IGF-I synthesis in the liver. Insulin-like growth factor I released from the liver circulates and exerts most of its growth-promoting action at the peripheral tissues as a mediator of GH. At the pituitary level IGF-I acts on GH cells to adjust GH synthesis and secretion (negative feedback regulation of GH secretion) (Abe *et al.*, 1983; Tannenbaum *et al.*, 1983; Yamashita and Melmed, 1986a,b). Consequently, these findings suggest pituitary IGF-I exerts a growth-promoting action within the pituitary gland, and a modulating action on GH secretion.

Fibroblast growth factors (FGFs) are considered to be the most abundant growth factors in normal pituitary glands (Gospodarowicz, 1975; Amano *et al.*, 1993). In the rat anterior pituitary basic FGF immunoreactivity was found mostly in the folliculostellate cells, which were identified by the expression of S-100 protein (Amano *et al.*, 1993). Prysor-Jones *et al.*, (1989) found two FGFs in bovine pituitaries. One FGF stimulates, and the other inhibits, human pituitary tumors. In their system, EGF inhibited pituitary growth.

vi. Cytokines Cytokines are now known to exert important roles in the neuroendocrine system (Bateman *et al.,* 1989). The intrinsic pituitary production of interleukins (ILs) and the presence of their receptors (IL-1, IL-2, and IL-6) in the pituitary glands have been reported (Vankelecom *et al.,* 1989; Koenig *et al.,* 1990; Spangelo *et al.,* 1990; Jones *et al.,* 1991; Webster *et al.,* 1991; Ohmichi *et al.,* 1992; Arzt *et al.,* 1993). Arzt *et al.* (1993) studied the effects of IL-2 and IL-6 on the proliferation of GH_3 pituitary tumors cells and normal rat pituitary cells. In GH_3 cells IL-2 and IL-6 stimulated both DNA synthesis and cell proliferation, but on the other hand inhibited both parameters in normal pituitary cells. Interleukin 2 receptors were detected in GH_3 cells, PRL cells, GH cells, and adreno-corticotropic hormone (ACTH)-secreting cells (ACTH cells) by immuno-fluorescence detection. These findings indicate that IL-2 and IL-6 partici-pate in the control of pituitary cell proliferation. The primary function of IL-2 and IL-6 is considered to be the inhibition of cell proliferation in order to control pituitary cell numbers. Disruption of the growth inhibitory mechanism may result in uncontrolled cell growth (tumorigenesis). Their finding strongly suggest that pituitary cytokines are involved in pituitary growth control.

3. Sex Differences in Prolactin Cell Number

Sexually based differences in the number of PRL cells were light micro-scopically and electron microscopically observed in mice (Sano and Sa-saki, 1971; Sasaki, 1974; Sasaki and Sano, 1980; Sasaki and Iwama, 1988b) and in rats (Takahashi and Kawashima, 1982, 1983; Takahashi and Miya-take, 1991). Ontogenic changes in GH and PRL cells were demonstrated in Wistar rats and ICR mice (Figs. 3 and 4). On the other hand Dada *et al.* (1984) reported that sex differences were not detected in adult rats, although the volume of PRL cells was greater in females than in males. Using the reverse hemolytic plaque assay, postnatal development of PRL cells was studied, and it was clearly shown that the percentage of PRL-secreting cells did not differ between male and female rats at immature ages, but significantly increased in adult female rats as previously reported (Hoeffler *et al.,* 1985; Chen, 1987). These sex differences in the percentage and number of PRL cells were caused in part by the difference in ovarian estrogen level, because peripubertal overiectomy blocked the increase in the percentage and number of PRL cells (Takahashi and Kawashima, 1981, 1982, 1983; Sasaki and Sano, 1980, 1983).

The difference in the number of PRL cells may partly result from the difference in mitotic activity of PRL cells. The sexually based difference in the number of PRL cells is explained by the difference in mitotic activity of PRL cells. Actually, the mitotic activity of PRL cells in estrous female rats was significantly higher than that in male rats (Takahashi and Kawashima, 1982; Takahashi *et al.*, 1984). Oishi *et al.* (1993) provided similar results using various methods for the detection of DNA-synthesizing cells. The sharp increase in the mitotic activity of PRL cells in estrous female rats was closely correlated with the sharp increase in serum PRL levels in the afternoon of the proestrous day, as previously shown (Neill *et al.*, 1971; Smith *et al.*, 1975; Haisenleder *et al.*, 1989). Estrogen levels on the proestrous day are higher, and the portal dopamine levels are lower compared with those at other stages of the estrous cycle (Ben-Jonathan *et al.*, 1977). It is still not clear whether the PRL surge at proestrus could directly stimulate the mitosis of pituitary cells, particularly PRL cells, or whether the increase in estrogen secretion and the decrease in dopamine secretion could stimulate directly or via extrapituitary organs.

4. Involvement of Insulin or IGF-1 in Estrogen-Induced Proliferation of Prolactin Cells

In insulin-deficient rats and mice, the responsiveness to estrogen in terms of the proliferation of PRL cells was reduced, and insulin treatment recovered the responsiveness to estrogen (Fig. 11; Takahashi *et al.* 1994). This finding suggests the involvement of insulin in estrogen-induced proliferation of PRL cells. Our *in vitro* study confirmed the direct action of insulin on the proliferation of mouse pituitary PRL cells (Oomizu *et al.*, 1993; Oomizu and Takahashi, 1994). This *in vivo* effect of insulin may be mediated by IGF-I receptor (Straus, 1984), because IGF-I increased the number of PRL cells at a lower dose than insulin *in vitro*. In uterine tissue IGF-I is reported to phosphorylate estrogen receptors, resulting in the activation of estrogen receptors (Aronica and Katzenellenbogen, 1993). Recently, Newton *et al.* (1994) showed using the pituitary tumor cell line, GH_3 cells, that the growth stimulatory actions of insulin and IGF-I was inhibited by the steroidal antiestrogens, ICI 164384, and ICI 182780, under the serum-and estrogen-free culture condition. They also found that estrogen-mediated promoter activity was stimulated by insulin and IGF-I, and inhibited by antiestrogens. These indicate that estrogen receptors in the pituitary tumor cells are activated by peptide growth factors, insulin, and IGF-I, in the absence of estrogen and serum. Thus, in the pituitary gland insulin or IGF-I may activate estrogen receptor, resulting in the activation of the growth-promoting system.

Insulin, IGF-I, and EGF stimulated the proliferation of pituitary cells (Oomizu *et al.*, 1993; Oomizu and Takanashi, 1994). The receptors for these growth factors exhibit tyrosine kinase activity. Genistein, an inhibi-

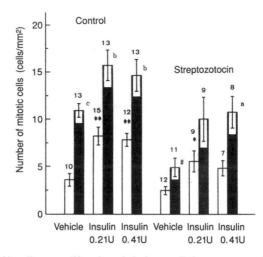

FIG. 11   Effects of insulin on proliferation of pituitary cells in streptozotocin-induced diabetic ovariectomized mice. Estradiol-17β (50 μg) was given to control and streptozotocin-treated mice. In control mice, estrogen increased the number of mitotic cells, but in streptozotocin-treated mice it did not. The ineffectiveness of estrogen was recovered by insulin treatment (0.2 or 0.4 IU twice for 3 days). The number above each column indicates the number of mice. *$p < 0.05$, **$p < 0.01$ vs respective vehicle group; a, $p < 0.05$, b, $p < 0.01$ c, $p < 0.001$ vs respective oil group; white bars, oil group; black bars, estradiol group. (From Takahashi *et al.* (1994, Fig. 1, p. 446.)

tor of tyrosine kinase (Akiyama *et al.*, 1987), inhibited the proliferation of pituitary cells *in vitro*. Therefore, these results strongly indicate that tyrosine kinase-associated signal transduction is involved in cell proliferation.

VI. Maturation of Prolactin Cells

A. Age-Related Changes in Percentage of Subtypes of Prolactin Cells

The percentage of PRL cell types changed during postnatal development in male and female rats (Fig. 12). At 10 days of age, most PRL cells were of the immature type, and mature-type cells were rarely encountered. At 30 days of age, about half of the PRL cells were intermediate type, and mature-type cells increased to about one-third of the PRL cell population. In adult male rats (60 days old), the relative proportion of each subtype

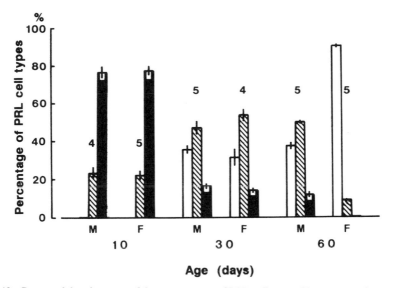

FIG. 12 Postnatal development of the percentages of PRL cell types (□, mature; ▨ intermediate; ■, immature) in male (M) and female (F) rats. The number above each group of columns depicts the number of rats. Bars depict the standard errors of the mean. There are significant differences ($p < 0.01$) between these groups: 10-day-old males vs 30-day-old males, 10-day-old females vs 30-day-old females, and 30-day-old females vs 60-day-old females. [Reprinted with permission from Takahashi, S., and Miyatake, M. (1991). *Zool. Sci.* **8**, 549–559.]

of PRL cells did not differ from that in 30-day-old male rats. In 60-day-old female rats most PRL cells were of the mature type, and immature-type cells were rarely observed. A similar age related change in PRL cell population was observed in the mouse pituitary (K. Nakatomi and S. Takahashi, unpublished data). These age-related changes in the relative proportion of PRL subtypes are explained by the changes in PRL secretion. Prolactin secretion is low during the first 3 weeks, and then starts to increase. Adult female rats at estrus showed higher PRL levels than did male rats (Negro-Vilar *et al.*, 1973; Döhler and Wuttke, 1975; Döhler *et al.*, 1977; Barkley, 1979; Takahashi and Kawashima, 1982). The increase in the number of mature-type PRL cells seems to occur concurrently with the elevation of blood PRL level.

B. Effect of Estrogen and Bromocriptine

Estrogen increased the percentage of mature-type PRL cells, and decreased the percentages of intermediate- and immature-type PRL cells.

On the other hand, ovariectomy and bromocriptine decreased the percentage of mature-type PRL cells and increased the percentages of the other two types (Figs. 13 and 14). Thus, the relative proportion of PRL cells changed in accordance with the change in PRL secretion. The immature type of PRL cell, containing small secretory granules, was predominantly present at immature ages, and was small in size. The mature type of PRL cell, containing large irregularly shaped secretory granules, constitutes most of the PRL cell population in adult female rats and is large in size. The PRL cell containing medium-sized granules is considered to be an intermediate cell between the immature and matrue PRL cells (Kurosumi et al., 1987; Takahashi and Miyatake, 1991). The change of secretory granules in size and shape is accounted for by the fusion and lysosomal degradation of preexisting secretory granules, which had previously been shown by Farquhar et al. (1978).

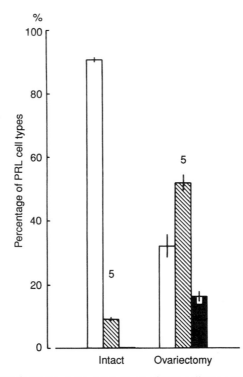

FIG. 13   Effects of ovariectomy on the percentages of PRL cell types (□, mature; ▨, intermediate; ■, immature) in adult female rats. Ten days after ovariectomy, the percentage of mature-type PRL cells decreased, and the other two types increased. The number above each group of columns depicts the number of rats. Bars depict the standard error of the mean. [Reprinted with permission from Takahashi, S., and Miyatake, M. (1991). *Zool. Sci.* **8**, 549–559.]

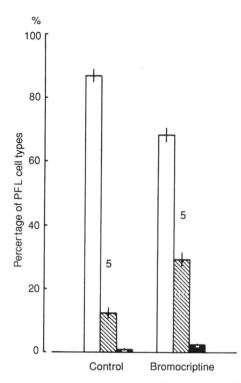

FIG. 14   Effect of bromocriptine on the percentage of PRL cell types (□, mature; ▨, interme-diate; ■, immature) in adult female rats. Five days of bromocriptine treatent (4 mg/kg) decreased the percentage of mature-type PRL cells, and increased that of the intermediate and immature types. The number above each group of columns depicts the number of rats. Bars depict the standard errors of the mean. [Reprinted with permission from Takahashi, S., and Miyatake, M. (1991). *Zool. Sci.* **8**, 549–559.]

C. Possible Hypothesis for Changes in Percentage of Prolactin Cell Subtypes

The changes in the percentages of subtypes of PRL cells observed during postnatal development and during estrogen treatment may be accounted for by the following possibilities (Fig. 15):

1. Each subtype of PRL cell may be independent of the others, and derived separately from the undifferentiated cells (stem cells). Each sub-type of PRL cell may divide into two daughter cells (simple duplication). It is probable that each subtype of PRL cell requires a specific signal for proliferation. For example, if estrogen is assumed to stimulate the

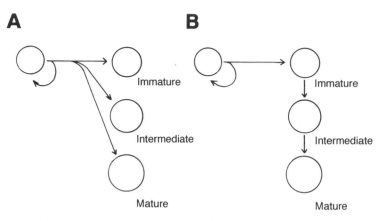

FIG. 15   Possible pathway explaining the changes in percentage of PRL subtypes. (A) Each subtype separately derives from "stem" cells (progenitor cells). (B) Each subtype represents one phase in the maturational process of PRL cells. Depending on the physiological demands, each cell changes its secretory activity, resulting in changes in the cell structure (cell size, cell shape, etc.). In both hypotheses, cell proliferation may be involved.

proliferation of mature-type PRL cells, and TRH stimulates immature-type PRL cells, then estrogen or TRH is able to increase the number of corresponding PRL cells only. It may also be possible that subtype-specific inhibitors inhibit the mitosis of PRL cells of the corresponding subtype. Luque *et al.* (1986) indicated that there were at least two populations responsive to dopamine on PRL release. Thus, several lines of evidence support this possibility. More detail is provided in Section VII. This finding supports the above possibility.

2. If each subtype of PRL cell corresponds to each stage in the maturational process of PRL cells, the transition from the immature type to the mature type of PRL cell through the intermediate form will occur. Estrogen, a stimulator of PRL secretion and cell proliferation, may accelerate the transition, and may lead to the increase in the number of mature-typed PRL cells. This possibility that subtypes of pituitary cells may be interconvertible depending on the secretory activity has already been suggested (Nogami 1984; Kurosumi *et al.,* 1986, 1987; Kurosumi and Tosaka, 1988; Takahashi, 1991, 1992a; Ozawa and Kurosumi, 1993). In this hypothesis, it is important to clarify whether or not such a transition requires mitosis. Currently we have no decisive evidence to determine which possibilities operate in the pituitary gland.

VII. Functional Heterogeneity of Prolactin Cells

A. Heterogeneity in Prolactin Synthesis and Release

Prolactin cell populations had been successfully dissociated into single cells by enzymatic digestion (Hopkins and Farquhar, 1973), and dissociated cells were separated using differences in unit gravity, depending on the difference in cell shape and secretory granule content (Hymer *et al.*, 1974). This method revealed that the intracellular content of PRL differed among the separated PRL cell fractions, and the amount of PRL released during the culture period of 14 days was positively correlated with the initial intracellular PRL content (Snyder *et al.*, 1976). A similar heterogeneity in pituitary cells was also demonstrated in GH cells (Synder *et al.*, 1977). Swearingen (1971) first described the heterogeneity in turnover of PRL in *in vivo* and *in vitro* studies. Walker and Farquhar (1980) further clarified by autoradiography the heterogeneity in PRL cells with respect to the PRL synthetic rate, which was visualized by the difference in the uptake of [^{3}H]leucine in PRL cells. They also found a subpopulation of PRL cells that preferentially secreted newly synthesized PRL. Velkeniers *et al.* (1988) separated PRL cell populations into high-density and low-density populations, using a discontinuous Percoll gradient, and found that low-density PRL cells have a high basal secretory activity and a higher PRL mRNA content, and that high-density PRL cells have a low basal secretory activity and a lower PRL mRNA content, but a higher responsiveness to vasoactive intestinal polypeptide. On the cortrary, in a long-term culture (7 days) of PRL cells separated on a Percoll density gradient, no major difference was detected among the cell fractions with respect to TRH, dopamine, and somatostatin repsonsiveness (Hofland *et al.*, 1990). The reason for this discrepancy in functional heterogeneity is not clear. Cell dissociation is considered to destroy intercellular communication (paracrine or juxtacrine control). Therefore, under these long-term culture conditions, particularly in monolayer culture, some of the properties of each cell that had been expressed *in situ* may be easily lost.

Functional heterogeneity of rat PRL cells had also been shown by the reverse hemolytic plaque assay (Boockfor *et al.*, 1986b; Luque *et al.*, 1986; Hymer and Motter, 1988). The bimodal distribution of plaque sizes indicated that the amount of hormones released from dissociated individual cells differed among PRL cells (Luque *et al.*, 1986; Horta and Cota, 1993). Prolactin cells were heterogeneous with respect to basal hormone secretion and responsiveness to TRH. Thus, there apparently seemed to be at least two subpopulations of PRL cells. One subpopulation of PRL cells secreted more PRL than the other subgroup under basal conditions and TRH stimulation. Luque *et al.* (1986) found that higher-secreting PRL cells were more responsive to TRH on PRL release than low-level secretors. Such a bimodal distribution of hormone secretion had also been demonstrated in rat GH cells (Frawley and Neill, 1984; Takahashi, 1992b), using the reverse hemolytic plaque assay.

Other evidence for the functional heterogeneity of PRL cell populations was reported by Arita *et al.* (1991, 1992), using the sequential cell immunoblot assay, which was originally developed by Kendall and Hymer (1987). A subpopulation of PRL cells (6% of total PRL cells) did not respond to dopamine of the highest dose (over 10^{-6} M). Furthermore, they divided PRL cells into three subpopulations with respect to plaque size and level of secretion (low, medium, and high). A subpopulation of small secretors of PRL cells contained a larger proportion of TRH-responsive cells than did a subpopulation of large secretors. Altogether, these studies indicated that there is a functional heterogeneity in PRL cell populations with respect to their responsiveness to dopamine and TRH.

B. Heterogeneity in Prolactin Cell Surface Antigen

Morphological heterogeneity in anti-PRL cell surface immunoreactivity had been previously shown in the rat pituitary (St. John *et al.*, 1986). Only half of all PRL cells from female rat pituitaries contained cell surface PRL immunoreactivity. This finding implied the presence of PRL receptors on the cell surface, or that some of the released PRL is retained on the surface of these cells. From this finding, PRL cell populations may also be divided into at least two subpopulations. However, it is not easy to correlate this heterogeneity of PRL cell populations with the PRL cell types described above.

Prolactin receptors had been identified in pituitary gland in the rat (Frantz *et al.*, 1975; Chiu *et al.*, 1992; Krown *et al.*, 1992). Morel *et al.* (1994) clearly demonstrated rat pituitary cells showing PRL receptor-like immunoreactivity. All types of pituitary cells showed PRL receptor-like immunoreactivity. The relative frequency of PRL binding was as follows: GH cells > PRL cells > thyrotrophs ≈ corticotrophs > gonadotrophs. Therefore, PRL secreted from PRL cells may possibly act in an autocrine or paracrine fashion to regulate pituitary functions in various cells (including PRL cell function). Several reports have already suggested the autoregulation of PRL secretion (Spies and Clegg, 1971; Herbert *et al.*, 1979; Kadowaki *et al.*, 1984; Ho *et al.*, 1989). A more recent finding favors the growth-promoting activity of PRL in pituitary cells (Krown *et al.*, 1992). They demonstrated that one of the PRL isoforms (24 kDa) had the autocrine growth activity in GH_3 tumor cells. The physiological significance of PRL isoforms is discussed in Section VIII,B.

C. Possible Mechanism of Functional Heterogeneity of Prolactin Cells

1. Difference in Prolactin Cell Location

A location dependent functional heterogeneity of PRL cells was shown by the reverse hemolytic plaque assay (Boockfor and Frawley, 1987), similar to the findings that have already been described in GH cells. In this study, PRL cells from the peripheral rim (outer zone) responded greatly to TRH, but only moderately to dopamine. Prolactin cells from the central region (inner zone) were affected slightly by TRH, but were markedly inhibited by dopamine. These regional differences in pituitary cells may be derived from the regional differences in the portal blood levels of hypothalamic releasing/inhibiting hormones. Another possibility is the paracrine effect on pituitary cells from the neighboring cells (Schwartz and Cherny, 1992).

As shown in Section V,B,2,e,iii, PRL cells that are located close to the intermediate lobe in the rat are more responsive to α-MSH and β-END in terms of proliferation (Porter and Frawley, 1992).

Three-dimensional reaggregated cell cultures from the rat anterior pituitary glands were analyzed to study whether juxtaposition of PRL cells and gonadotrophs is a selective or random phenomenon (Allaerts *et al.*, 1991), because several immunocytochemical observations had described cup-shaped PRL cells as frequently surrounding gonadotrophs (Nakane, 1970; Sato, 1980). Their conclusion was that there was no evidence for a selective adhesion between PRL cells and gonadotrophs. However, interestingly, the proportion of cup-shaped PRL cells juxtaposed at their concave side to gonadotrophs is markedly higher than the proportion of cup-shaped PRL cells associated with thyrotrophs or GH cells. From the statistical analysis, it was deduced that there is a selectivity of juxtaposition between cup-shaped PRL cells and gonadotrophs. This morphological correlate strongly suggests that there is a functional relationship between cup-shaped PRL cells and gonadotrophs in the form of a type of paracrine communication. In fact, this selective spatial association between cup-shaped PRL cells and gonadotrophs coincides well with the stimulatory effect of GnRH on the differentiation of PRL cells, as described above (Tilemans *et al.*, 1992).

2. Difference in Molecular Variants Secreted

It was demonstrated by gravitational sedimentation that diethylstilbestrol-induced prolactinomas consisted of three different subpopulations of PRL cells (Hymer and Motter, 1988). In their study by gravitational sedimentation, PRL cells were divided into large-, intermediate-, and small-sized PRL cells, which differed in their content and release of PRL. Large- and intermediate-sized PRL cells contained typical pleiomorphic secretory granules, but small-sized PRL cells were sparsely granulated or agranular. Small-sized PRL cell populations contained unique PRL variants, whose molecular masses ranged from 10 to 14 kDa. This study suggests that there may be a relationship between the molecular heterogeneity of PRL and the diversity of morphology and function of PRL cells. Molecular variants of PRL are discussed in Section VIII.

3. Difference in Electrophysical Properties of Prolactin Cells and Dopamine Receptors on Prolactin Cells

Prolactin cell populations were electrophysiologically divided into two subpopulations, which, in turn, correspond to two groups (light and heavy groups) separated by a bovine serum albumin (BSA) density gradient separation (Israel *et al.*, 1990). Most of the PRL cells of the light fraction showed a type 1 response, that is, dopamine induced a hyperpolarization from the resting potential. The other PRL cells, of the heavy fraction, mostly do not respond to dopamine (type 2 response), but when the membrane potential has been depolarized dopamine induces a repolarization. The expression of the two dopaminergic D_2 receptors, $D2_{415}$ and $D2_{444}$, was studied and found to be different in these two PRL cell populations (Kukstas *et al.*, 1991). The ratio of $D2_{415}$ to $D2_{444}$ was higher in the light fraction of PRL cells than in the heavy fraction. This result indicates that the two different responses to dopamine in PRL cells could be associated with the differential expression of two different D_2 receptors. Such differences may eventually result in a difference in PRL secretion, and/or even in the morphology of PRL cells.

Another difference in the electrophysiological property of rat PRL cells was also observed using the reverse hemolytic plaque assay (Horta and Cota, 1993). The exhibited bimodal distribution of plaque sizes, that is, small secretors and large secretors, which agreed well with Luque *et al.* (1986) and Arita *et al.* (1991). In addition, effects of φ-conotoxin and nifedipine, blockers of the high voltage-activated (HVA) Ca^{2+} channel, were studied in terms of PRL secretion from each cell type (Felix *et al.*, 1993). Their results indicated that blockers of the HVA Ca^{2+} channel suppress PRL release from large secretors, but not from small secretors.

They also analyzed two types of PRL cell with respect to a differential activity of two voltage-gated Ca^{2+} channels; low voltage-activated (LVA) and high voltage-activated (HVA) channels. High-level PRL secretors showed a greater activity of HVA channels than did low-level secretors. Thus, differences in electrophysiological properties may correspond to the difference in PRL secretion.

Ontogenic change in electrophysiological properties was studied (Felix *et al.*, 1993). Their finding was that neonatal male rat pituitaries (10 days old) contained only low-level secretors, whereas adult pituitaries contained both low- and high-level secretors. Electrophysiological properties of low-level secretors were not different from those of neonatal PRL cells (low-level secretors) and adult low-level secretors. High-level PRL secretors are considered to be the classic type of PRL cell. This result suggests that low-level PRL secretors, which are present in the neonatal period, persist at adult ages. Higher-level PRL secretors appear later during development. Low-level secretors may be mammosomatotrophs (GH- and PRL-secreting cells), which had already been shown to exist in the neonatal pituitary gland by the reverse hemolytic plaque assay (Frawley *et al.*, 1985; Hoeffler *et al.*, 1985), although Felix *et al.* (1993) did not present morphological evidence for mammosomatotrophs.

4. Differences in Intracellular Age of Prolactin

Mena *et al.* indicated that PRL secretion in the lactating rat required transformation of mature PRL from its prerelease form into a releasable form (Grosvenor and Mena 1980; Mena *et al.*, 1982, 1989a,b). Changes in solubility of PRL molecules were demonstrated in the lactating rat pituitary after suckling, and after *in vitro* incubation: PRL molecules became, over time, more insoluble (releasable form) than soluble (Mena *et al.*, 1982). In this transformation model of PRL molecules, the "intracellular age" of PRL molecules after synthesis may become an important factor in PRL secretion. The sequence of steps leading to PRL release after synthesis has been studied *in vivo* and *in vitro* (Mena *et al.*, 1989a,b). Under basal conditions (no stimulation and inhibition), a greater proportion of mature PRL molecules (synthesized 4–8 hr beforehand) was released than newly synthesized (~0.2–1 hr after synthesis) and old (~16–24 hr after synthesis) PRL molecules during the first 30–60 min of pituitary incubation (Mena *et al.*, 1989a,b). Dopamine had a significantly lower inhibitory effect on mature PRL (4–8 hr after synthesis) than on newly synthesized and older stored PRL. Thyrotropin-releasing hormone had a greater stimulatory effect on mature PRL (4–8 hr after synthesis), indicating that mature PRL molecules are more readily released than are newly synthesized or old stored PRL. Thus, the functional heterogeneity in PRL cell populations may be accounted for by the difference in intracellular age of PRL, which is determined by whether it is newly synthesized or old (stored). On the other hand, Chen *et al.* (1989) stated that only a fraction of GH cells or PRL cells depended on newly synthesized hormone for basal secretion, and the remainder appeared to secrete stored hormones even without stimulation. The ability to transform stored hormones into releasable hormones may differ among cells. Currently, we cannot describe the actual sequence of events leading from the synthesis of PRL molecules to their release. It seems important to know what changes occur in PRL molecules after synthesis.

VIII. Molecular Heterogeneity of Prolactin

Several laboratories have reported molecular variants of PRL in various species (human, rat, mouse, etc.) (Wallis, 1988). As an example, we will see variants of rat PRL molecules. The description of the molecular weights of each variant differs, depending on the laboratory. Molecular forms are as follows: the cleaved (16-kDa) form, the 23-kDa monomer (or 24 kDa), the polymer forms, the glycosylated 25-kDa (or 26-kDa) form, and the charge variants (Asawaroengchai *et al.*, 1978; Mittra, 1980; Sinha and Gilligan, 1984; Sinha *et al.*, 1985; Sinha and Jacobsen, 1988; Oetting and Walker, 1985, 1986; Ho *et al.*, 1989; Andries *et al.*, 1992; Anthony *et al.*, 1993; Bollengier *et al.*, 1993; Clapp *et al.*, 1993). Hymer and Motter (1988) found in diethylstilbestrol-induced prolactinomas that several vari-

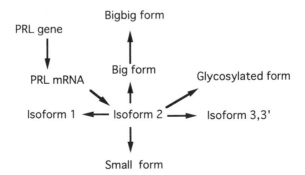

FIG. 16   Molecular heterogeneity of PRL generated from PRL mRNA. Isoforms 3 and 3′ are phosphorylated forms and isoform 1 is a dephosphorylated form. As molecular variants, small (cleaved) PRL and polymer forms are reported. Glycosylated PRL (26 kDa) is a glycosylated form of isoform 2.

ants of PRL molecules ranged from 12 to 64 kDa. Bollengier *et al.* (1989) also showed molecular heterogeneity of PRL, that is, 23-, doublet 25- to 26-, 40-, and 42-kDa forms. A variant of the 26-kDa form is considered to be glycosylated PRL. High molecular weight variants occur as a product of disulfide linkage between monomeric units. Cleaved PRLs may be generated by glandular kallikrein (a trypsin-like serine protease) (Powers and Hatala, 1990). However, there is a controversy concerning the genesis of cleaved forms of PRL (Casabiell *et al.*, 1989). In human pituitaries, the large polymer forms are known (big PRL, about 50 kDa; bigbig PRL, 150–170 kDa, etc.) (Bjøro *et al.*, 1993). The structures of the polymer forms of human PRL remain unclear (Garnier *et al.*, 1978; Tanaka *et al.*, 1989).

Two-dimensional electrophoresis of pituitary cell extracts and the sera revealed that there are charge isoforms of the 24-kDa PRL monomer in several species (Asawaroengchai *et al.*, 1978; Haro and Talamantes, 1985). In the rat (Fig. 16), four isoforms of 24-kDa PRL have been found, and described as isoforms 1, 2, 3, and 3′ (according to increasing acidity) (Oetting and Walker, 1985, 1986; Ho *et al.*, 1989, 1993). Isoform 2 is an unmodified mature PRL molecule. Other isoforms are modified posttranslationally. Isoforms 3 and 3′ are phosphorylated forms, and isoform 1 is a nonphosphorylated or dephosphorylated form generated by an unknown mechanism (Greenan *et al.*, 1989). Isoforms 2 and 3 are major types in normal rat pituitaries, and isoforms 2, 3, and 3′ are secreted into the medium. Isoform 1 is rarely secreted.

Physiological Significance of Molecular Heterogeneity

Frawley *et al.* (1986) indicated the possibility that each molecular variant of PRL differs in biological activity, and suggested that each molecular form may have specific target cells, and consequently, specific physiological roles. An example demonstrating the possible physiological significance of molecular variants of PRL was reported in ram pituitaries (Stroud *et al.*, 1992). The authors clearly showed that production of variant forms of PRL in the ram pituitary gland varied seasonably. In their study, the 23K form is a primary hormone, and the 25K form is a glycosylated form. High molecular weight forms (more than 25K), which are aggregated by a disulfide linkage between monomers, are significant in winter, and may be for storage. During the season when PRL secretion is active, high molecular weight forms disappeared. An explanation for this could be that synthesized hormones may be rapidly released into the circulation, and are not stored in the cell. On the other hand, during the season when PRL secretion is low or inhibited, synthesized hormones are more likely to aggregate and become the stored type. Thus, it is probable that changes in the molecular form of a hormone may parallel changes in the secretory activities of the hormone.

Prolactin variants of the 16-kDa form from the rat and humans inhibited the proliferation of vascular endothelial cells (Ferrara *et al.*, 1991; Clapp and Weiner, 1992; Clapp *et al.*, 1993). This antiangiogenic activity of 16-kDa PRL may be useful for the inhibition of pituitary tumor growth. Intact 23-kDa PRL had no effect on the growth of capillary endothelial cells (Ferrara *et al.*, 1991).

Krown *et al.* (1992) demonstrated that the growth of GH_3 tumor cells increased with increasing cell density, and was markedly inhibited by the addition of anti-rat PRL serum. Prolactin receptors were present in GH_3 cells, but in 40–50% of all cells. In GH_3 pituitary tumor cells, only isoform 2 had been detected, and secreted into the medium, and the other isoforms (1, 3, and 3') had not been detected (Ho *et al.*, 1989). From these findings they claimed that isoform 2 stimulated GH_3 cell proliferation as an autocrine growth factor.

GH_3 cells usually secrete PRL constitutively, without secretory granule formation. However, interestingly, administration of exogenous PRL derived from normal sources (containing isoforms 1, 2, and 3) was able to decrease PRL secretion (autoregulation), as previously shown by Kadowaki *et al.* (1984), and simultaneously secretory granules appeared in the cytoplasm. Ho *et al.* (1989) claimed that isoform 3 was an autocrine form for PRL secretion. Thus, initiation of granulogenesis coincided with the production of isoform 3. Therefore, they concluded that the autoregulatory defect of GH_3 tumor cells might be ascribed to the lack of production of isoform 3 rather than to an inability to produce isoform 3.

Prolactin stimulates the proliferation of Nb_2 lymphoma cells (Gout *et al.*, 1980). Nb_2 lymphoma cells possess a high number of high-affinity PRL receptors (Shiu *et al.*, 1983; Rao *et al.*, 1993). Using Nb_2 lymphoma cells, growth-promoting activities of various charge isoforms of PRL were assessed. Standard rat PRL (B-6, NIDDK) was partially dephosphorylated, and the dephosphorylated preparation showed significantly higher growth-promoting activity than did standard PRL (Wang and Walker, 1993).

Cyclic changes in 24-kDa PRL isoforms had been studied in the rat pituitary (Ho *et al.*, 1993). Isoform 1 was released only at estrus, and the lack of release of isoform 3' was evident on the afternoon of proestrus, accompanying a marked increase in isoform 2. As stated above, in GH_3 cells, isoform 2 may be an autocrine growth factor, and isoform 3 may be the active component for the regulation of PRL secretion. The dephosphorylated form, isoform 1, may also be the autocrine or paracrine growth factor in normal pituitary cells, because dephosphorylated PRL stimulated the growth of Nb_2 lymphoma cells as already discussed above.

These two results indicate that the transformation of PRL molecules by phosphorylation or dephosphorylation will finally determine the effect of PRL. In this case, alteration in the number of phosphate groups generated the opposite effect. Phosphorylation of PRLs from various species was examined. Chicken, turkey, ovine, and glycosylated turkey PRLs as well as rat PRL molecules were phosphorylated by protein kinase A (Arámburo et al., 1992). These findings imply that phosphorylation of PRL molecules is widely seen among various species. Further study is required to know how phosphorylation occurs in the cell.

A number of reports had showed serum patterns of various molecular forms of PRL in humans during pregnancy or in patients with hyperprolactinemia (Bjoro et al., 1993; Larrea et al., 1993). Because the data on variants of PRL molecules in various physiological states have become formidable, the physiological significance of polymer forms or glycosylated forms in the human is clarified below.

IX. Age-Related Changes in Prolactin Secretion

A number of physiological actions of PRL have been reported (Nicoll, 1974). The altered PRL secretion induces various diseases (e.g., Welsch and Nagasawa, 1977; Mori and Nagasawa, 1984; Meites et al., 1987; vom Saal and Finch, 1988; Meites, 1990; Ooka, 1993). In the rat, blood PRL levels increase with aging, and pituitary tumors frequently develop in female rats (Takahashi et al., 1980; Steger, 1981; Goya et al., 1990; Ooka and Shinkai, 1992). There is controversy concerning the changes in PRL secretion in men. Vekemans and Robyn (1975) reported that basal serum PRL levels increased with age, but the increase was modest. Hossdorf and Wagner (1980) found no change in basal serum PRL levels. Hyperprolactinemia (with prolactinoma) is well known to be closely associated with symptoms of amenorrhea and galactorrhea in women, and with impotence and decreased libido in men. It is valuable to study age-related changes in PRL secretion.

A. Changes in Morphology of Prolactin Cells with Age

Kawashima (1974) reported, in an electron microscopy sutdy, morphological changes in pituitary cells, particularly hypertrophy and hyperplasia of PRL cells in female rats, although an immunocytochemical identification

had not been done. Age-related changes in immunocytochemically identified PRL cells were reported by Takahashi and Kawashima (1983). The
percentage of PRL cells significantly increased in female rats with age
(Table I). The total number of PRL cells had not been measured, but it
had been estimated from the pituitary DNA content and the percentage
of PRL cells as described in GH cells (Takahashi *et al.,* 1990). Actually,
the DNA contents constituting the PRL cell population at 6, 12, and 18
months were as follows (the number of rats in parenthesis): in male rats,
3.7 ± 0.6 mg (7), 7.0 ± 0.8 mg (8), and 6.5 ± 0.8 mg (5), and in female
rats, 9.5 ± 1.4 mg (8), 25.8 ± 2.9 mg (8), and 40.3 ± 2.4 mg (7). The DNA
content of PRL cells increased in both sexes with age, but more markedly
in old female rats, suggesting a significant increase in the number of PRL
cells. Similar age-related changes in the number of PRL cells were reported
by Chuknyiska *et al.* (1986). The increase in the number of PRL cells with
aging was caused by the continuous elevated levels of ovarian estrogen (the
stimulatory factor) or the proliferation of PRL cells (Huang *et al.,* 1978;
Takahashi *et al.,* 1980; Kawashima and Takahashi, 1986), and the diminished secretion of hypothalamic dopamine (the PRL inhibitory factor)

TABLE I

Percentages of Prolactin Cells and Growth Hormone Cells in Male and
Female Rats

Sex	Age (months)	Percentage of cells (%)[a]	
		PRL cells	GH cells
Male	6	15.5 ± 1.4^b	68.3 ± 2.2^c
	12	22.1 ± 1.6	59.7 ± 4.1^d
	18	18.9 ± 3.0	$40.5 \pm 4.2^{c,d}$
Female	6	$34.0 \pm 4.4^{c,e}$	40.6 ± 2.0^c
	12	48.7 ± 2.9^e	34.8 ± 4.1^e
	18	52.1 ± 2.4^c	$23.4 \pm 2.2^{c,e}$

[a] In each age group five rats were used for the determination of
the percentages of PRL cells and GH cells. Statistical significance
was tested by ANOVA. When significant, the differences among age
groups of each sex were determined by Duncans's multiple range
test. In each sex, there is a significant difference between the values
for the age groups with the same superscripts.
[b] Means $\pm$ SE.
[c] $p < 0.01$.
[d] $p < 0.01$.
[e] $p < 0.05$.

(Simpkins *et al.*, 1977). Sarkar *et al.* (1982) demonstrated that estrogen-induced damage in hypothalamic dopaminergic neurons was closely associated with occurrence of pituitary tumors. Prepubertal ovariectomy prevented the increase in PRL cell number in old female rats (Kawashima and Takahashi, 1986). To elucidate the relationship between the arcuate nucleus (including dopaminergic neurons) and PRL secretion, neonatal rats were given monosodium glutamate (MSG) to destroy the arcuate nucleus (Takahashi *et al.*, 1982). This MSG-treated animal model, which had been originally reported by Olney (1969), could be utilized as a model system of aging animals.

Age-related changes in PRL cell mitosis in the rat were observed (Fig. 8). Even in 2-year-old female rats mitotic pituitary cells were encountered. Immunoelectron microscopical studies had been done in male rats (van Putten and Kiliaan, 1988). The relative proportion of each subtype changed with age in male rats. One type of PRL cell, containing small round secretory granules (immature type; originally type III cells in Kurosumi's classification), increased in percentage; on the other hand, another type of PRL cell with large, irregularly shaped secretory granules (mature type; originally type I cells in Kurosumi's classification) decreased in percentage in old male rats. These findings indicate that immature-type PRL cells increased with aging whereas mature-type PRL cells decreased.

B. Changes in Prolactin Synthesis and Secretion with Age

Prolactin secretion increased with age, and the enhanced secretion of PRL is partly due to the dysfunction of the hypothalamic dopaminergic mechanism (Simpkins *et al.*, 1977; Takahashi *et al.*, 1980; Meites *et al.*, 1987). A reverse hemolytic plaque assay revealed that the amount of PRL released per cell decreased in old rats (Larson and Wise, 1991). Pituitary PRL content significantly increased with age in female rats, but PRL concentration per PRL cell decreased (Takahashi *et al.*, 1990) (Fig. 17). Prolactin mRNA levels per PRL cell decreased with age in both sexes (Fig. 18). Prolactin synthesis in each PRL cell decreased at the transcription level with age. However, because PRL cells significantly increased in number in the pituitaries of old female Wistar/Tw rats (Takahashi and Kawashima, 1983), the total number of PRL cells significantly increased with age. Stewart *et al.* (1990) reported no significant change in PRL mRNA concentrations (per measured amount of pituitary DNA) with age in female rats, but did report a significant increase in serum PRL level. We did not find any significant difference in PRL mRNA concentrations (per microgram of pituitary cell DNA) in female rats, either (data not

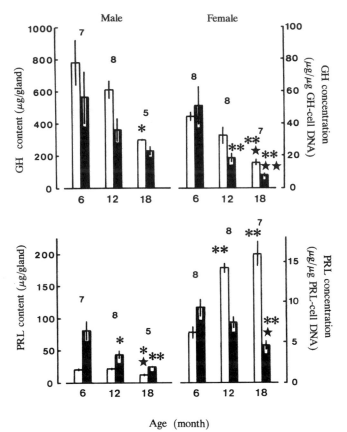

FIG. 17   Pituitary GH (top) and PRL (bottom) content (per gland) and concentrations (per microgram of GH cell DNA or PRL cell DNA) in male and female rats at 6, 12, and 18 months of age. Female rats in estrus or persistent estrus (middle-aged and old) were used; ★$p < 0.05$, ★★$p < 0.01$, compared with 6-month-old rats. White bars, GH/PRL content; black bars, GH/PRL concentrations. [Reprinted with permission from Takahashi, S., and Kawashima, S., Seo, H., and Matsui, N. (1990). *Endocrinol Jpn.* **37,** 827–840.]

shown). Crew *et al.* (1987) reported an age-related decrease in PRL mRNA in male mice. The decline in PRL synthesis at the transcription level in old rats and mice seems to coincide well with the morphological changes in PRL cells that indicate the relative increase in the number of immature-type PRL cells.

The age-related changes in GH secretion were different from those in PRL secretion. Growth hormone secretion declines with aging (Sonntag *et al.*, 1980; Ceda *et al.*, 1986; Takahashi *et al.*, 1987). Growth hormone

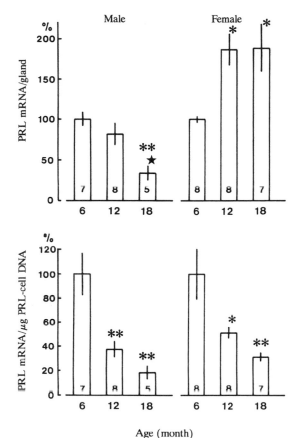

FIG. 18 Total pituitary PRL mRNA content (per gland) and concentration (per microgram of PRL cell DNA) in male and female rats at the ages of 6, 12, and 18 months. Female rats at estrus or persistent estrus (middle-aged and old) were used; $*p < 0.05$, $**p < 0.01$, compared with 6-month-old rats; ★$p < 0.05$, compared with 12-month-old rats. [Reprinted with permission from Takahashi, S., Kawashima, S., Seo, H., and Matsui, N. (1990). *Endocrinol Jpn.* **37**, 827–840.]

synthesis declined at a gene transcription level in old rats, and the amount of GH mRNA per GH cell was also lower in old rats than in young ones (Takahashi *et al.*, 1990). The GH secretion from each GH cell decreased with age in female rats (Takahashi, 1992b). The decline in GH synthesis per GH cell is similar to the change in PRL synthesis. A morphological study showed that the relative number of immature-type GH cells increased, and that of the mature-type GH cells decreased, with aging (Takahashi, 1991).

X. Mammosomatotrophs

A. Identification of Mammosomatotrophs

Mammosomatotrophs (MS cells), which contain both GH and PRL, were described in an immunoelectron microscopy study in intact adult rats (Nikitovitch-Winer *et al.*, 1987; Ishibashi and Shiino, 1989a; Losinski *et al.*, 1989). The MS cells were small in size and irregular in shape. Secretory granules, 50–150 nm in diameter, contained both hormones (Nikitovitch-Winer *et al.*, 1987).

B. Mammosomatotrophs in Various Animals

Mammosomatotrophs are rare in normal adult rats, but MS cells are usually encountered in lactating and pregnant females (Nikitovitch-Winer *et al.*, 1987; Ishibashi and Shiino, 1989a) and adenomatous rat pituitaries (Losinski *et al.*, 1989). Mammosomatotrophs have also been observed in mice (Sasaki and Iwama, 1988b, 1989), musk shrews (Ishibashi and Shiino, 1989a), bats (Ishibashi and Shiino, 1989b), cows (Fumagalli and Zanini, 1985), sheep (Thorpe *et al.*, 1990; Thorpe and Wallis, 1991), rhesus monkeys (Bethea and Freesh, 1991), and in human fetuses (Mulchahey and Jaffe, 1987; Heitz *et al.*, 1987; Asa *et al.*, 1988), normal adults (Lloyd *et al.*, 1988), and adenomatous adults (Kanie *et al.*, 1983; Bassetti *et al.*, 1986; Beckers *et al.*, 1988). Mammosomatotrophs in rat pituitary tumor lines are well known (Tashjian *et al.*, 1970; Chomczynski *et al.*, 1988; Kashio *et al.*, 1990). However, Shirasawa *et al.* (1990) could not detect any MS cells in fetal and male adult bovine pituitary glands, using three different immunocytochemical methods. Moreover, using the flip-flop (mirror) section technique Watanabe and Haraguchi (1994) demonstrated that PRL and GH were contained in different cells in the fetal rat pituitary. The difference between the report of Fumagalli and Zanini (1985) (nursing cows and virgin cows) and the reports of Shirawasa *et al.* (1990) (fetal and adult bulls) and Watanabe and Haraguchi (1994) is partly due to the age and sex of animals used. In mice, MS cells were further divided into two subtypes: the small, round, solid secretory granular type and the vesicular secretory granular type (Sasaki and Iwama, 1989).

Ishibashi and Shiino (1989a) found two types of colocalization of GH and PRL. In one type, GH and PRL are colocalized in the same secretory granules within a single cell, as described by Nikitovitch-Winer *et al.* (1987). In the other type, GH-secretory granules and PRL-secretory granules are separate and intermixed within closely aggregated and interdigitated cell clusters in pregnant rats and female musk shrews. This type is similar to the multinucleated mammosomatotrophs in cows reported by Fumagalli and Zanini (1985). This finding provides a possibility that the enhanced stimulation of hormone secretion induces the fusion of the secretory cells.

C. Development of Mammosomatotrophs

Using the reverse hemolytic plaque assay, MS cells were detected in neonatal and adult male and female rats (Frawley *et al.*, 1985), in bovine pituitaries (Kineman *et al.*, 1991), and also in human pituitaries (Mulchahey and Jaffe, 1987; Lloyd *et al.*, 1988). Mammosomatotrophs were 35.8% of all GH- and/or PRL-secreting cells of 5-day-old male rats (Hoeffler *et*

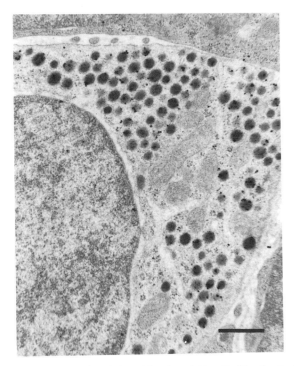

FIG. 19   A mammosomatotroph (in a 5-day-old male rat) identified by the double-immunocytochemical method, using antisera to GH and PRL. Growth hormone was labeled with small gold particles and PRL was labeled with large gold particles. Bar: 500 nm. [Reprinted with permission from Takahashi, S. (1992a). *Zool. Sci.* **9**, 901–924.]

al., 1985). In adult male rats about one-third of all GH- and/or PRL-secreting cells are MS cells (Frawley *et al.*, 1985). Leong *et al.* (1985) reported that about 5% of all pituitary cells were MS cells in adult male rats. Altogether, these findings provide evidence for the presence of cells that secrete both hormones at the same time from the neonatal period to the maturity.

In fetal mice at 15.5 days of gestation, a few pituitary cells colocalized GH and PRL mRNA, but the majority of cells containing PRL mRNA did not express GH mRNA (Dollé *et al.*, 1990). Chatelain *et al.* (1979) observed MS cells immunocytochemically in rats at 21 days of fetal age. In neonatal rats, MS cells were found and these cells resembled the immature-type GH or PRL cells (Fig. 19). The frequency of occurrence of MS cells during the neonatal period was not so high as reported by Hoeffler *et al.* (1985). We estimated the volume density of MS cells in prepubertal rats by the point-counting method. The volume density of MS cells in the Wistar rat at 3, 5, 10, and 30 days of age in male and female rats was as follows (number of rats in parenthesis): male rats, 1.6 ± 0.4% (4), 2.8 ± 0.4% (5), 3.2 ± 1.4% (5), and 3.3 ± 0.7% (4); female rats, 0.9 ± 0.8% (4), 2.9 ± 0.5% (4), and 3.1 ± 0.3% (4), respectively. Our immunoelectron microscopy observation showed that at 3 and 5 days of age, PRL cells were rare, but MS cells were detected in the percentages listed above.

D. Developmental and Physiological Significance of Mammosomatotrophs

Several possibilities were presented to explain the significance of MS cells (Fig. 20).

1. The MS cell is a transitional cell type in the interconversion of GH cells to PRL cells or PRL cells to GH cells.

2. The MS cell is a progenitor cell for GH and PRL cells (Mulchahey and Jaffe, 1987).

3. The MS cell is an independent type of cell, and may be terminally differentiated.

4. There is a interconversion between GH cells and PRL cells, but MS cells have no relation to either GH or PRL cells.

Analysis of the data which has been reported so far, and future analysis, particularly at perinatal period, may clarify the genesis and physiological roles of MS cells.

The following reports favor the transitional cell hypothesis. Growth hormone cells and PRL cells in rats appeared almost at the same age, but the growth of GH cells is more active than that of PRL cells (Nogami *et*

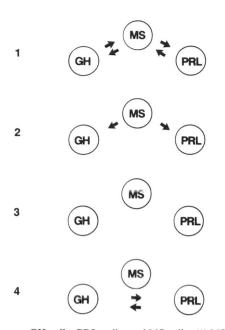

FIG. 20 Relationship among GH cells, PRL cells, and MS cells. (1) MS cells are transitional cells from GH cells to PRL cells, or PRL cells to GH cells. (2) MS cells are progenitor cells for both GH cells and PRL cells. (3) The three cell types are independent of each other. (4) Growth hormone cells and PRL cells are interconvertible, but MS cells are not involved in the transition. Other possibilities may be conceivable.

al., 1989; Baker and Jaffe, 1975). In fetal mice, GH synthesis preceded PRL synthesis (Slabaugh *et al.*, 1982). Stratmann *et al.* (1974) previously presented the possibility of the transition of GH cells to PRL cells by estrogen treatment, using both electron microscopy and autoradiographical analysis of [^{3}H]thymidine uptake in pituitary cells. They explained that some of the previously existing GH cells proliferated and were converted into PRL cells. Frawley's group had reported a large amount of evidence for the transition from GH cell to PRL cell, or from PRL cell to GH cell, using a hemolytic plaque assay (Hoeffler *et al.*, 1985; Boockfor *et al.*, 1986a; Porter *et al.*, 1990). Using transgenic mice, Borrelli *et al.* (1989) clearly demonstrated that some stem PRL cells were derived from some portion of the stem GH cells, stating that PRL cells originated from the GH cell lineage. One of the transcription factors for GH and PRL genes was the same, GHF-1 or Pit-1 (Bodner *et al.*, 1988; Ingraham *et al.*, 1988; Lira *et al.*, 1988; Nelson *et al.*, 1988; McCormick *et al.*, 1990; Castrillo *et al.*, 1991). Growth hormone and PRL molecules were considered to be derived from a common ancestor molecule (Cooke *et al.*, 1980, 1981). Lira *et al.* (1988) suggested that thyroid stimulating hormone-secreting cells (thyrotrophs) as well as GH and PRL cells are derived from a common lineage of pituitary cells.

Factors controlling the transition of pituitary cells must be clarified. Estrogen is essential for the genesis of stem PRL cells from stem GH cells in mice (Borrelli *et al.*, 1989). Insulin inhibited GH synthesis and secretion (Yamashita and Melmed, 1986b), and also reduced the number of fetal GH cells *in vitro* (Hemming *et al.*, 1984). On the contrary, insulin stimulated PRL synthesis through the activation of a PRL gene promoter (Keech and Gutierrez-Hartmann, 1991). Moreover, in GH_3 tumors cells, insulin and IGF-I stimulated PRL secretion and decreased GH secretion, although the numbers of GH- and PRL-secreting cells had not been studied (Prager *et al.*, 1988). Inoue and Sakai (1991) induced the transition from GH-secreting cells to PRL-secreting cells by insulin or insulin-like growth factor (IGF-I) in their newly established pituitary clonal cell line (Inoue *et al.*, 1990). Thus, insulin, and/or IGF-I, is closely associated with the development of GH and PRL cells, and probably MS cells. If stimulation of PRL synthesis and secretion can induce the transition from GH to PRL cells, which may occur through a transitional MS cell form, excessive stimulation of PRL secretion may enhance the occurrence of MS cells. Our preliminary study showed that estrogen treatment (50 μg for 3 days) increased the number of MS cells (Fig. 20) to about twice that in the neonatal male rats. Similarly, estradiol increased the proportion of MS cells in a monolayer culture of male pituitary cells (Boockfor *et al.*, 1986a).

The GH_3 cell line has been studied to understand the synthesis and release of GH and PRL (Tashjian *et al.*, 1968). The heterogeneity of GH_3 cells has been revealed by several reports (Boockfor *et al.*, 1985; Prysor-Jones *et al.*, 1985; Prager *et al.*, 1988). The number of GH-secreting cells and PRL-secreting cells changes with culture condition. Thyrotropin-releasing hormone and estradiol increase the percentage of PRL-secreting cells, and cortisol decreases the percentage of PRL-secreting cells (Boockfor *et al.*, 1985). Insulin and 17β-estradiol synergistically increased PRL secretion, but decreased the rate of cell proliferation of GH_4C_1 cells (Kiino and Dannies, 1981). Cooccurrence of the shift from GH-secreting to PRL-secreting cells in GH_3 and GH_4C_1 tumor cells induced by hormone treatments and the decrease in the growth activity of PRL-secreting cells suggests that PRL-secreting cells are a more differentiated cell type compared with GH-secreting cells, because GH-secreting cells have the higher activity of proliferation than PRL-secreting cells.

Constant stimulation of GHRH, using transgenic mice, caused a hyperplasia of MS cells (Stefaneanu *et al.*, 1989), although some of the MS cells in those transgenic mice were morphologically similar to those in adult mice; others, on the other hand, were morphologically different, and relatively close to those in adenomatous human pituitaries. Provided that the MS cell is the common progenitor cell, and that these progenitor cells (MS cells) exist even in adult pituitaries, MS cells in adenomatous tissues may be derived from unregulated proliferation of preexisting MS cells.

Although a large number of data have been accumulated so far, it is still difficult to determine conclusively the developmental and physiological significance of MS cells. In normal pituitary cells, PRL cells may transform into GH cells via a transitional MS cell stage, as shown in Porter *et al.* (1990, 1991b). If such bidirectional conversion between GH cells and PRL cells occurs in rat pituitaries, MS cells may be the common progenitor cells. As already explained, transplantation of GH_3 tumor cells into host animals blocked the PRL secretion. This indicates that the bidirectional conversion between GH-secreting cells and PRL-secreting cells is possible in GH_3 tumor cells.

XI. Conclusions

A large amount of evidence has shown that the PRL cell population consists of morphologically and functionally heterogeneous cells. A correlation between morphologically and functionally different subtypes has not been fully clarified. Further analysis at the single-cell level is required. Multiple molecular variants of PRL have been reported. Several reports clearly suggest that each molecular variant has specific physiological functions.

The functional heterogeneity of pituitary PRL cells may be the integrated outcome of the differences in responsiveness to hypothalamic hormones, steroid hormones, growth factors, cytokines and other substances, the mechanisms of intracellular signal transduction, the regulation of PRL gene expression, the PRL synthetic activity, the transformation of PRL molecules (the genesis of molecular variants of PRL), the storage of PRL hormones (intracellular age of hormones), the electrophysiological properties, and so on. Such wide differences may be regulated by intrinsic programs (genetically regulated), or by extracellular factors (regulation by cells other than PRL cells). Although factors that cause the heterogeneity of PRL cells are not clear, the paracrine control of cell activities is feasible. Ample evidence supports the existence of paracrine regulatory mechanisms within the pituitary gland. The microenvironment that surrounds the pituitary cells seems important for the development and function of pituitary cells. Although the interaction between pituitary cells and the extracellular matrix (ECM) has not been discussed in this chapter, a number of reports indicate the involvement of ECM in the morphogenesis, differentiation, growth, and functions of pituitary cells. The location of pituitary cells within the gland may be one of the essential factors for development of pituitary cells. The sum total of such differences may produce the morphologically different subtypes of PRL cells.

Subtypes of PRL cells seem to have completely different morphological characteristics (granule shape and size, cell shape). Each subtype may be separately derived from the progenitor cells. However, several observations including our reports suggest a transition among three types. Morphological heterogeneity of PRL cells may show the pathway by which immature cells become mature PRL cells.

Mammosomatotrophs are widely distributed among species. In particular, when PRL secretion is active (lactation and estrogen treatment), MS cells are more frequently observed. Therefore, MS cells may be involved in the recruitment of PRL cells. Mammosomatotrophs may be transitional cells between GH cells and PRL cells, or common progenitor cells of GH and PRL cells. Further study is necessary to clarify the developmental and phsiological significance of MS cells.

Previously, PRL cells had been considered to consist of one type (the classic type of PRL cell). Development of the immunocytochemical technique enabled us to find new PRL cell types. Similarly, separation of dissociated pituitary cells by velocity sedimentation at unit gravity revealed the presence of functional heterogeneity among PRL cells. The reverse hemolytic plaque assay greatly helped in the analysis of the function of pituitary cells at a single-cell level. Using such techniques, the heterogeneity of the PRL cell population is now well understood. The survey of heterogeneities among PRL cells implies that such a heterogeneous population of PRL cells, at various levels from the molecular to the pituitary level, could provide sufficient flexibility to respond to any demands for PRL secretion from the pituitary gland.

Acknowledgments

The author expresses cordial thanks to the late professor Yasuo Kobayashi (Faculty of Science, Okayama University) for his continuous hearty encouragement and helpful advice. The author also expresses thanks to Emeritus Professor Howard A. Bern (University of California at Berkeley) for his encouragement in writing the manuscript. The author is grateful to S. Oomizu, T. Osawa, and K. Nakatomi for their energetic help and discussion. This study was supported in part by Grants-in-Aid for Scientific Research from the Ministry of Education, Science and Culture, Japan, and by the Ryoubiteien Foundation.

References

Abe, H., Molitch, M. E., Van Wyk, J. J., and Underwood, L. E. (1983). Human growth hormone and somatomedin C suppress the spontaneous release of growth hormone in unanesthesized rats. *Endocrinology (Baltimore)* **113**, 1319–1324.

Akiyama, T., Ishida, J., Nakagawa, S., Ogawara, H., H., Watanabe, S., Itoh, N., Shibuya, M., and Fukami, Y. (1987). Genistein, a specific inhibitor of tyrosine-specific protein kinases. *J. Biol. Chem.* **262**, 5592–5595.

Allaerts, W., Mignon, A., and Denef, C. (1991). Selectivity of juxtaposition between cup-shaped lactotrophs and gonadotrophs from rat anterior pituitary in culture. *Cell Tissue Res.* **263**, 217–225.

Allanson, M., Foster, C. L., and Cameron, E. (1969). Mitotic activity in the adenohypophysis of pregnant and lactating rabbits. *J. Reprod. Fertil.* **19**, 121–131.

Amano, O., Yoshitake, Y., Nishikawa, K., and Iseki, S. I. (1993). Immunocytochemical localization of basic fibroblast growth factor in the rat pituitary gland. *Arch. Histol. Cytol.* **56**, 269–276.

Amara, J. F., and Dannies, P. S. (1983). 17β-estradiol has a biphasic effect on GH cell growth. *Endocrinology (Baltimore)* **112**, 1141–1143.

Amarant, T., Fridkin, M., and Koch, Y. (1982). Luteinizing hormone releasing hormone in human and bovine milk. *Eur. J. Biochem.* **127**, 647–650.

Andries, M., Tilemans, D., and Denef, C. (1992). Isolation of cleaved prolactin variants that stimulate DNA synthesis in specific cell types in rat pituitary cell aggregates in culture. *Biochem. J.* **281**, 393–400.

Anthony, P. K., Stoltz, R. A., Pucci, M. L., and Powers, C. A. (1993). The 22 K variant of rat prolactin: Evidence for identity to prolactin-(1–173), storage in secretory granules, and regulated release. *Endocrinology (Baltimore)* **132**, 806–814.

Arámburo, C., Montiel, J. L., Proudman, J. A., Berghman, L. R., and Scanes, C. G. (1992). Phosphorylation of prolactin and growth hormone. *J. Mol. Endocrinol.* **8**, 183–191.

Archer, D. F., Lattanzi, D. R., Moore E. E., Harper, H., and Hebert, D. L. (1982). Bromocriptine treatment of women with suspected pituitary prolactin-secreting microadenomas. *Am. J. Obstet. Gynecol.* **143**, 620–625.

Arita, J., Kojima, Y., and Kimura, F. (1991). Identification by the sequential cell immunoblot assay of a subpopulation of rat dopamine-unresponsive lactotrophs. *Endocrinology (Baltimore)* **128**, 1887–1894.

Arita, J., Kojima, Y., and Kimura, F. (1992). Lactotrophs secreting small amounts of prolactin reveal great responsiveness to thyrotropin-releasing hormone: Analysis by the sequential cell immunoblot assay. *Endocrinology (Baltimore)* **130**, 3167–3174.

Aronica, S. M., and Katzenellenbogen, B. S. (1993). Stimulation of estrogen receptor-mediated transcription and alteration in the phosphorylation state of the rat uterine estrogen receptor by estrogen, cyclic adenosine monophosphate, and insulin-like growth factor-I. *Mol. Endocrinol.* **7**, 743–752.

Arzt, E., Buric, R., Stelzer, G., Stalla, J., Sauer, J., Renner, U., and Stalla, G. K. (1993). Interleukin involvement in anterior pituitary cell growth regulation: Effect of IL-2 and IL-6. *Endocrinology (Baltimore)* **132**, 459–467.

Asa, S. L., Kovacs, K., Horvath, E., Losinski, N. E., Laszlo, F. A., Domokos, I., and Halliday, W. C. (1988). Human fetal adenohypophysis, electron microscopic and ultrastructural immunocytochemical analysis. *Neuroendocrinology* **48**, 423–431.

Asa, S. L., Kovacs, K., and Singer, W. (1991). Human fetal adenohypophysis: Morphologic and functional analysis *in vitro*. *Neuroendocrinology* **53**, 562–572.

Asawaroengchai, H., Russell, S. M., and Nicoll, C. S. (1978). Electrophoretically separable forms of rat prolactin with different bioassay and radioimmunoassay activities. *Endocrinology (Baltimore)* **102,** 407–414.

Bach, M. A., and Bondy, C. A. (1992). Anatomy of the pituitary insulin-like growth factor system. *Endocrinology (Baltimore)* **131,** 2588–2594.

Baker, B. L., and Gross, D. S. (1978). Cytology and distribution of secretory cell types in the mouse hypophysis as demonstrated with immunocytochemistry. *Am. J. Anat.* **153,** 193–215.

Baker, B. L., and Jaffe, R. B. (1975). The genesis of cell types in the adenohypophysis of the human fetus as observed with immnocytochemistry. *Am. J. Anat.* **143,** 137–162.

Baker, B. L., and Yu, Y. Y. (1977). An immunocytochemical study of human pituitary mammotropes from fetal life to old age. *Am. J. Anat.* **148,** 217–240.

Baker, B. L., Midgley, A. R. Jr., Gersten, B. E., and Yu, Y. Y. (1969). Differentiation of growth hormone and prolactin containing acidophils with peroxidase-labeled antibody. *Anat. Rec.* **164,** 163–171.

Baram, T., Koch, Y., Hazum, E., and Fridkin, M. (1977). Gonadotropin-releasing hormone in milk. *Science* **198,** 300–302.

Barkley, M. S. (1979). Serum prolactin in the male mouse from birth to maturity. *J. Endocrinol.* **83,** 31–33.

Bassetti, M., Spada, A., Arosio, M., Vallar, L., Brina, M., and Giannattasio, G. (1986). Morphological studies on mixed growth hormone (GH)- and prolactin (PRL)-secreting human pituitary adenomas. Coexistence of GH and PRL in the same secretory granule. *J. Clin. Endocrinol. Metab.* **62,** 1093–1100.

Bateman, A., Singh, A., Kral, T., and Solomon, S. (1989). The immune-hypothalamic-pituitary-adrenal axis. *Endocr. Rev.* **10,** 92–112.

Baxter, R. C., Zaltsman, Z., and Turtle, J. R. (1984). Immunoreactive somatomedin-C/insulin-like growth factor I and its binding protein in human milk. *J. Clin. Endocrinol. Metab.* **58,** 955–959.

Beckers, A., Courtoy, R., Stevenaert, A., Boniver, J., Closset, J., Frankenne, F., Reznik, M., and Hennen, G. (1988). Mammosomatotropes in human pituitary adenomas as revealed by electron microscopic double gold immunostaining method. *Acta Endocrinol. (Copenhagen)* **118,** 503–512.

Bégeot, M., Hemming, F. J., Martinat, N., Dubois, M. P., and Dubois, P. M. (1983). Gonadotropin releasing hormone (GnRH) stimulates immunoreactive lactotrope differentiation. *Endocrinology (Baltimore)* **112,** 2224–2226.

Bégeot, M., Hemming, F. J., Combarnous, Y., Dubois, M. P., Aubert, M. L., and Dubois, P. M. (1984). Lactotrope differentiation induced by LHα subunit. *Science* **226,** 566–568.

Ben-Jonathan, N. (1985). Dopamine: A prolactin inhibitory hormone. *Endocr. Rev.* **6,** 564–589.

Ben-Jonathan, N., Oliver, C., Weiner, H. J., Mical, R. S., and Porter, J. C. (1977). Dopamine in hypophysial portal plasma of the rat during the estrous cycle and throughout pregnancy. *Endocrinology (Baltimore)* **100,** 452–458.

Bethea, C. L., and Freesh, F. (1991). Estrogen action on growth hormone in pituitary cell cultures from adult and juvenile macaques. *Endocrinology (Baltimore)* **114,** 2110–2118.

Billestrup, N., Swanson, L. W., and Vale, W. (1986). Growth hormone-releasing factor stimulates proliferation of somatotrophs in vitro. *Proc. Natl. Acad. Sci. U.S.A.* **83,** 6854–6857.

Bjøro, T., Johansen, E., Frey, H. H., Turter, A., and Torjesen, P. A. (1993). Different responses in little and bigbig prolactin to metoclopromide in subjects with hyperprolactinemia due to 150–170 kD (bigbig) prolactin. *Acta Endocrinol. (Copenhagen)* **128,** 308–312.

Bodner, M., Castrillo, J.-L., Theill, L. E., Deerinck, T., Ellisman, M., and Karin, M. (1988).

The pituitary-specific transcription factor GHF-1 is a homeobox-containing protein. *Cell (Cambridge, Mass.)* **55,** 505–518.

Bollengier, F., Velkeniers, B., Hooghe-Peters, E., Mahler, A., and Vanhaelst, L. (1989). Multiple forms of rat prolactin and growth hormone in pituitary cell subpopulations separated using a Percoll gradient system: Disulphide-bridged dimers and glycosylated variants. *J. Endocrinol.* **120,** 201–206.

Bollengier, F., Geerts, A., Matton, A., Mahler, A., Velkeniers, B., Hooghe-Peters, E., and Vanhaelst, L. (1993). Identification and localization of 23,000 and glycosylated rat prolactin in subcellular fractions of rat anterior pituitary and purified secretory granules. *J. Neuroendocrinol.* **5,** 669–676.

Boockfor, F. R., and Frawley, L. S. (1987). Functional variations among prolactin cells from different pituitary regions. *Endocrinology (Baltimore)* **120,** 874–879.

Boockfor, F. R., Hoeffler, J. P., and Frawley, L. S. (1985). Cultures of GH$_3$ cells are functionally heterogeneous: Thyrotropin-releasing hormone, estradiol and cortisol cause reciprocal shifts in the proportions of growth hormone and prolactin secretors. *Endocrinology (Baltimore)* **117,** 418–420.

Boockfor, F. R., Hoeffler, J. P., and Frawley, L. S. (1986a). Estradiol induces a shift in cultured cells that release prolactin or growth hormone. *Am. J. Physiol.* **250,** E103–E105.

Boockfor, F. R., Hoeffler, J. P., and Frawley, L. S. (1986b). Analysis by plaque assays of GH and prolactin release from individual cells in cultures of male pituitaries. *Neuroendocrinology* **42,** 64–70.

Borgundvaag, B., Kudlow, J. E., Mueller, S. G., and George, S. R. (1992). Dopamine receptor activation inhibits estrogen-stimulated transforming growth factor-α gene expression and growth in anterior pituitary, but not in uterus. *Endocrinology (Baltimore)* **130,** 3453–3458.

Borrelli, E., Heyman, R. A., Arias, C., Sawchenko, P. E., and Evans, R. M. (1989). Transgenic mice with inducible dwarfism. *Nature (London)* **339,** 538–541.

Bravo, R., Frank, R., Blundell, P. A., and Macdonaldo-Bravo, H. (1987). Cyclin/PCNA is the auxiliary protein of DNA polymerase-δ. *Nature (London)* **326,** 515–517.

Brooks, C. L., and Welsch, C. W. (1974). Reduction of serum prolactin in rats by 2 ergot alkaloids and 2 ergoline derivatives: A comparison. *Proc. Soc. Exp. Biol. Med.* **146,** 863–867.

Brown, C., and Martin, I. L. (1984). Autoradiographic localization of benzodiazepine receptors in the rat pituitary gland. *Eur. J. Pharmacol.* **102,** 563–564.

Burdman, J. A., Calabrese, M. T., Harcus, C. T., and MacLeod, R. M. (1982). Lisuride, a dopamine agonist, inhibits DNA synthesis in the pituitary gland. *Neuroendocrinology* **35,** 282–286.

Burdman, J. A., Calabrese, M. T., Romano, M. I., Carricarte, V. C., and MacLeod, R. M. (1984). Effects of ovariectomy and clomiphene on the in-vitro incorporation of [^{3}H]thymidine into pituitary DNA and on prolactin synthesis and release in rats. *J. Endocrinol.* **101,** 197–201.

Burns, G., and Sarkar, D. K. (1993). Transforming growth factor-β_1-like immunoreactivity in the pituitary gland of the rat: Effect of estrogen. *Endocrinology (Baltimore)* **133,** 1444–1449.

Carbajo-Peréz, E., and Watanabe, Y. G. (1990). Cellular proliferation in the anterior pituitary of the rat during the postnatal period. *Cell Tissue Res.* **261,** 333–338.

Carbajo-Peréz, E., Motegi, M., and Watanabe, Y. G. (1989). Cell proliferation in the anterior pituitary of mice during growth. *Biomed. Res.* **10,** 275–281.

Carbajo-Peréz, E., Carbajo, S., Orfao, A., Vincente-Villardón, J. L., and Vázquez, R. (1991). Circadian variation in the distribution of cells throughout the different phases of the cell cycle in the anterior pituitary gland of adult male rats as analysed by flow cytometry. *J. Endocrinol.* **129,** 329–333.

Carpenter, G. (1980). Epidermal growth factor is a major growth-promoting agent in human milk. *Science* **210,** 198–199.

Casabiell, X., Robertson, M. C., Friesen, H. G., and Casanueva, F. F. (1989). Cleaved prolactin and its 16K fragment are generated by an acid protease. *Endocrinology (Baltimore)* **125,** 1967–1972.

Castrillo, J.-L., Theill, L. E., and Karin, M. (1991). Function of the homeodomain protein GHF-1 in pituitary cell proliferation. *Science* **253,** 197–199.

Ceda, G. P., Valenti, G., Butturini, U., and Hoffman, A. R. (1986). Diminished pituitary responsiveness to growth hormone-releasing factor in aging male rats. *Endocrinology (Baltimore)* **118,** 2109–2114.

Chatelain, A., Dupouy, J. P., and Dubois, M. P. (1979). Ontogenesis of cells producing polypeptide hormones (ACTH, MSH, LPH, GH, prolactin) in the fetal hypophysis of the rat: Influence of the hypothalamus. *Cell Tissue Res.* **196,** 409–427.

Cheifetz, S., Ling, N., Guillemin, R., and Massague, J. (1988). A surface component on GH₃ pituitary cells that recognizes transforming growth factor β, activin, and inhibin. *J. Biol. Chem.* **263,** 17225–17228.

Chen, C. L., and Meites, J. (1970). Effects of estrogen and progesterone on serum and pituitary prolactin levels in ovariectomized rats. *Endocrinology (Baltimore)* **86,** 503–510.

Chen, H. I. (1987). Postnatal development of pituitary lactotropes in the rat measured by reverse hemolytic plaque assay. *Endocrinology (Baltimore)* **120,** 247–253.

Chen, T. T., Kineman, R. D., Betts, J. G., Hill, J. B., and Frawley, L S. (1989). Relative importance of newly synthesized and stored hormone to basal secretion by growth hormone and prolactin cells. *Endocrinology (Baltimore)* **125,** 1904–1909.

Chiu, S., Koos, R. D., and Wise, P. M. (1992). Detection of prolactin receptor (PRL-R) mRNA in the rat hypothalamus and pituitary gland. *Endocrinology (Baltimore)* **130,** 1747–1749.

Chomczynski, P., Brar, A., and Frohman, L. A. (1988). Growth hormone synthesis and secretion by a somatomammotroph cell line derived from normal adult pituitary of the rat. *Endocrinology (Baltimore)* **123,** 2276–2283.

Chuknyiska, R. S., Blackman, M. R., Hymer, W. C., and Roth, G. S. (1986). Age-related alterations in the number and functions of pituitary lactotropic cells from intact and ovariectomized rats. *Endocrinology (Baltimore)* **118,** 1856–1862.

Clapp, C., and Weiner, R. I. (1992). A specific, high affinity, saturable binding site for the 16-kilodalton fragment of prolactin on capillary endothelial cells. *Endocrinology (Baltimore)* **130,** 1380–1386.

Clapp, C., Martial J. A., Guzman, R. C., Rentier-Dulrue, F., and Weiner, R. I. (1993). The 16-kilodalton N-terminal fragment of human prolactin is a potent inhibitor of angiogenesis. *Endocrinology (Baltimore)* **133,** 1292–1299.

Cooke, N. E., Coit, D., Weiner, R. I., Baxter, J. D., and Martial, J. A. (1980). Structure of cloned DNA complementary to rat prolactin messenger RNA. *J. Biol. Chem.* **255,** 6502–6510.

Cooke, N. E., Coit, D., Shine, J., Baxter, J. D., and Martial, J. A. (1981). Human prolactin cDNA structural analysis and evolutionary comparisons. *J. Biol. Chem.* **256,** 4007–4016.

Corcia, A., Steinmetz, R., Liu, J.-W., and Ben-Jonathan, N. (1993). Coculturing posterior pituitary and GH₃ cells: Dramatic stimulation of prolactin gene expression. *Endocrinology (Baltimore)* **132,** 80–85.

Crew, M. D., Spindler, S. R., Walford, R. L., and Koizumi, A. (1987). Age-related decrease of growth hormone and prolactin gene expression in the mouse pituitary. *Endocrinology (Baltimore)* **121,** 1251–1255.

Cronin, M. J., Roberts, J. M., and Weiner, R. I. (1978). Dopamine and dihydroergocriptine binding to the anterior pituitary and other brain areas of the rat and sheep. *Endocrinology (Baltimore)* **103**, 302–309.

Dada, M. O., Campbell, G. T., and Blake, C. A. (1984). Pars distalis cell quantification in normal adult male and female rats. *J. Endocrinol.* **101**, 87–94.

Daniel, P. M., and Prichard, M. M. L. (1956). Anterior pituitary necrosis. Infraction of the pars distalis produced experimentally in the rat. *Q. J. Exp. Physiol. Med. Sci. Cogn.* **41**, 215–229.

Day, R. N., and Day, K. H. (1994). An alternative spliced form of pit-1 represses prolactin gene expression. *Mol. Endocrinol.* **8**, 374–381.

Döhler, K. D., and Wuttke, W. (1975). Changes with age in levels of serum gonadotropins, prolactin, and gonadal steroids in prepubertal male and female rats. *Endocrinology (Baltimore)* **97**, 898–907.

Döhler, K. D., von zur Mühlen, A., and Döhler, U. (1977). Pituitary luteinizing hormone (LH), follicle stimulating hormone (FSH) and prolactin from birth to puberty in female and male rats. *Acta Encrinol. (Copenhagen)* **85**, 718–728.

Dollé, P., Castrillo, J.-L., Theill, L. E., Deerinck, T., Ellisman, M., and Karin, M. (1990). Expression of GHF-1 protein in mouse pituitaries correlates both temporally and spatially with the onset of growth hormone gene activity. *Cell (Cambridge, Mass.)* **60**, 809–820.

Donovan, S. M., Hintz, R. L., Wilson, D. M., and Rosenfeld, R. G. (1991). Insulin-like growth factors I and II and their binding proteins in rat milk. *Pediatr. Res.* **29**, 50–55.

Drewett, N., Jacobi, J. M., Willgoss, D. A., and Lloyd, H. M. (1993). Apoptosis in the anterior pituitary gland of the rat: Studies with estrogen and bromocriptine. *Neuroendocrinology* **57**, 89–95.

Drews, R. T., Gravel, R. A., and Collu, R. (1992). Identification of G protein α subunit mutations in human growth hormone (GH)- and GH/prolactin-secreting pituitary tumors by single-strand conformation polymorphism (SSCP) analysis. *Mole. Cell. Endocrinol.* **87**, 125–129.

Driman, D. K., Kobrin, M. S., Kudlow, J. E., and Asa, S. L. (1992). Transforming growth factor-α in normal and neoplastic human endocrine tissues. *Human Pathol.* **23**, 1360–1365.

Dubois, P. M., and Hemming, F. J. (1991). Fetal development and regulation of pituitary cell types. *J. Electron. Microsc. Tech.* **19**, 2–20.

Eckert, K., Lubbe, L., Schon, R., and Grosse, R., (1985). Demonstration of transforming grwoth factor activity in mammary epithelial tissues. *Biochem. Int.* **11**, 441–451.

Ellerkmann, E., Nagy, G. M., and Frawley, L. S. (1991). Rapid augmentation of prolactin cell number and secretory capacity by an estrogen-induced factor released from the neurointermediate lobe. *Endocrinology (Baltimore)* **129**, 838–842.

Ellerkmann, E., Nagy, G. M., and Frawley, L. S. (1992a) α-Melanocyte-stimulating hormone is a mammotrophic factor released by neurointermediate lobe cells after estrogen treatment. *Endocrinology (Baltimore)* **130**, 133–138.

Ellerkmann, E., Porter, T. E., Nagy, G. M., and Frawley, L. S. (1992b). N-Acetylation is required for the lactotrope recruitment activity of α-melanocyte-stimulating hormone and β-endorphin. *Endocrinology (Baltimore)* **131**, 566–570.

Farquhar, M. G., Reid, J. J., and Danicll, L. W.(1978). Intracellular transport and packaging of prolactin: Quantitative electron microscope autoradiographic study of mammotrophs dissociated from rat pituitaries. *Endocrinology (Baltimore)* **102**, 296–311.

Felix, R., Horta, J., and Cota, G. (1993). Comparison of lactotrope subtypes of neonatal and adult male rats: Plaque assays and patch-clamp studies. *Am. J. Physiol.* **265**, E121–E127.

Ferrara, N., Clapp, C., and Weiner, R. I. (1991). The 16K fragment of prolactin specifically inhibits basal or fibroblast growth factor stimulated growth of capillary endothelial cells. *Endocrinology (Baltimore)* **129**, 896–900.

Finley, E. L., King, J. S., and Ramsdell, J. S. (1994). Human pituitary somatotropes express transforming growth factor-α and its receptor. *J. Endocrinol.* **141**, 547–554.

Frantz, W. L., Payne, P., and Dombroske, O. (1975). Binding of ovine ^{125}I-PRL to cultured anterior pituitary tumor cells and normal cells. *Nature (London)* **255**, 636–638.

Frawley, L. S., and Hoeffler, J. P. (1988). Hypothalamic peptides affect the ratios of GH and PRL cells: Role of cell division. *Peptides (N.Y.)* **9**, 825–828.

Frawley, L. S., and Miller, H. A., III (1989). Ontogeny of prolactin secretion in the neonatal rats is regulated posttranscriptionally. *Endocrinology (Baltimore)* **124**, 3–6.

Frawley, L. S., and Neill, J. D. (1984). A reverse hemolytic plaque assay for microscopic visualization of growth hormone release from individual cells: Evidence for somatotrope heterogeneity.

Frawley, L. S., Boockfor, F. R., and Hoeffler, J. P. (1985). Identification by plaque assays of a pituitary cell type that secretes both growth hormone and prolactin. *Endocrinology (Baltimore)* **116**, 734–737.

Frawley, L.S., Clark, C L., Schoderbek, W. E., Hoeffler, J. P., and Boockfor, F. R. (1986). A novel bioassay for lactogenic activity: Demonstration that prolactin cells differ from one another in bio- and immunopotencies of secreted hormone. *Endocrinology (Baltimore)* **119**, 2867–2869.

Fumagalli, G., and Zanini, A. (1985). In cow anterior pituitary, growth hormone and prolactin can be packed in separate granules of the same cell. *J. Cell Biol.* **100**, 2019–2024.

Garnier, P. E., Aubert, M. I., Kaplan, S. I., and Grumbach, M. M. (1978). Heterogeneity of pituitary and plasma prolactin in man: Decreased affinity of "big" prolactin in a radioreceptor assay and evidence for its secretion. *J. Clin. Endocrinol. Metab.* **47**, 1273–1281.

Ginsburg, M., Maclusky, N. J., Morris, I. D., and Thomas, P. J. (1975). Physiological variations in abundance of oestrogen specific high-affinity binding sites in hypothalamus, pituitary and uterus of the rat. *J. Endocrinol.* **64**, 443–449.

Goldsmith, P. C., Cronin, M. J., and Weiner, R. I. (1979). Dopamine receptor sites in the anterior pituitary. *J. Histochem. Cytochem.* **27**, 1205–1207.

Goodyer, C. G., de Stephano, L., Wei Hsien Lai, Guyda, H. J., and Posner, B. I. (1984). Characterization of insulin-like growth factor receptors in rat anterior pituitary, hypothalamus, and brain. *Endocrinology (Baltimore)* **114**, 1187–1195.

Gospodarowicz, D. (1975). Purification of a fibroblast growth factor from bovine pituitary. *J. Biol. Chem.* **250**, 2515–2522.

Goss, R. J. (1978). "The Physiology of Growth," pp. 1–13. Academic Press, New York.

Gout, P. W., Beer, C. T., and Noble, R. L. (1980). Prolactin-stimulated growth of cell cultures established from malignant Nb rat lymphomas. *Cancer Res.* **40**, 2433–2436.

Goya, R. G., Quigley, K. L., Takahashi, S., Sosa, Y. E., and Meties, J. (1990). Changes in somatotropin and thyrotropin secretory patterns in aging rats. *Neurobiol Aging* **11**, 625–630.

Greenan, J. R., Balden, E., Ho, T. W. C., and Walker, A. M. (1989). Biosynthesis of the secreted 24 K isoforms of prolactin. *Endocrinology (Baltimore)* **125**, 2041–2048.

Grosvenor, C. E., and Mena, F. (1980). Evidence that thyrotropin-releasing hormone and hypothalamic prolactin-releasing factor may function in the release of prolactin in the lactating rat. *Endocrinology (Baltimore)* **107**, 863–868.

Grueters, A., Alm, J., Lakshamanan, J., and Fisher, D. A. (1985). Epidermal growth factor in mouse milk during early lactation: Lack of dependency on submandibular glands. *Pediatr. Res.* **19**, 853–856.

Haisenleder, D. J., Ortolano, G. A., Landefeld, T. D., Zmeili, S. M., and Marshall, J. C. (1989). Prolactin messenger ribonucleic acid concentrations in 4-day cycling rats and during the prolactin surge. *Endocrinology (Baltimore)* **124**, 2023–2028.

Halmi, N. S., Parsons, J. A., Erlandsen, S. L., and Duello, T. (1975). Prolactin and growth hormone cells in the human hypophysis: A study with immunoenzyme histochemistry and differential staining. *Cell Tissue Res.* **158**, 497–507.

Halper, J., Parnell, P. G., Carter, B. J., Ren, P., and Scheithauer, B. W. (1992). Presence of growth factos in human pituitary. *Lab. Invest.* **66**, 639–645.

Harigaya, T., and Hoshino, K. (1985). Immunohistochemical study of postnatal development of prolactin-producing cells in C57BL mice. *Acta Histochem. Cytochem.* **18**, 343–351.

Harigaya, T., Kohmoto, K., and Hoshino, K. (1983). Immunohistochemical identification of prolactin producing cells in the mouse adenohypophysis. *Acta Histochem. Cytochem.* **16**, 51–58.

Haro, L. S., and Talamantes, F. J. (1985). Secreted mouse prolactin (PRL) and stored ovine PRL. I. Biochemical characterization, isolation and purification of the electrophoretic isoforms. *Endocrinology (Baltimore)* **116**, 345–352.

Heitz, P. U. Landolt, A. M. Zenklusen, H.-R., Kasper, M., Reubi, J.-C. Oberholzer, M., and Roth, J. (1987). Immunocytochemistry of pituitary tumors. *J. Histochem. Cytochem.* **35**, 1005–1011.

Hemming, F J., Begeot, M., Dubois, M. P., and Dubois, P. M. (1984). Fetal rat somatotropes in vitro: Effects of insulin, cortisol, and growth hormone-releasing factor on their differentiation: A light and electron microscopic study. *Endocrinology (Baltimore)* **114**, 2107–2113.

Herbert, D. C., Ishikawa, H., and Rennels, E. G. (1979). Evidence for the autoregulation of hormone secretion by PRL. *Endocrinology (Baltimore)* **104**, 97–100.

Héritier, A. G., and Dubois, P. M. (1993). Influence of thyroliberin on the rat pituitary cell type differentiation: An *in vitro* study. *Endocrinology (Baltimore)* **132**, 634–639.

Herlant, M. (1964). The cells of the adenohypophysis and their functional significance. *Int. Rev. Cytol.* **17**, 299–382.

Hill, J. B., Lacy, E. R., Nagy, G. M., Gorcs, T. J., and Frawley, L. S. (1993). Dose α-melanocyte-stimulating hormone from the pars intermedia regulate suckling-induced prolactin release? Supportive evidence from morphological and functional studies. *Endocrinology (Baltimore)* **133**, 2991–2997.

Ho, T. W. C., Greenan, J. R., and Walker, A. M. (1989). Mammotroph autoregulation: The differential roles of the 24K isoforms of PRL. *Endocrinology (Baltimore)* **124**, 1507–1514.

Ho, T. W. C., Leong, F. S., Olaso, C. H., and Walker, A. M. (1993). Secretion of specific nonphosphorylated and phosphorylated rat prolactin isoforms at different stages of the estrous cycle. *Neuroendocrinlogy* **58**, 160–165.

Hoeffler, J. P., Boockfor, F. R., and Frawley, L. S. (1985). Ontogeny of prolactin cells in neonatal rats: Initial prolactin secretors also release growth hormone. *Endocrinology (Baltimore)* **117**, 187–195.

Hofland, L. J., van Koetsveld, P. M., Verleum, T. M., and Lamberts, S. W. J. (1990). Long-term culture of rat mammotrope and somatotrope subpopulations separated on continuous Percoll density gradients. Effects of dopamine, TRH, GHRH and somatostatin. *Acta Endocrinol. (Copenhagen)* **122**, 127–136.

Hopkins, C.. R., and Farquhar, M. G. (1973). Hormone secretion by cells dissociated from rat anterior pituitaries. *Endocrinology (Baltimore)* **59**, 276–303.

Horta, J., and Cota, G. (1993). Lactotrope subtypes are differentially responsive to calcium channel blockers. *Mol. Cell. Endocrinol.* **92**, 189–193.

Hosoi, E., Yokogoshi, Y., Hosoi, E., Horie, H., Sano, T., Yamada, S., and Saito, S. (1993). Analysis of the Gs α gene in growth hormone-secreting pituitary adenomas by the polymerase chain reaction-direct sequencing methods using paraffin-embedded tissues. *Acta Endocrinol. (Copenhagen)* **129**, 301–306.

Hossdorf, T., and Wagner, H. (1980). Secretion of prolactin in healthy men and women of different age. *Aktu. Gerontol.* **10**, 119–126.

Huang, H. H., Steger, R. W., Bruni, J. F., and Meites, J. (1978). Patterns of sex steroid and gonadotropin secretion in aging female rats. *Endocrinology (Baltimore)* **103**, 1855–1859..

Hunt, T. E. (1943). Mitotic activity in the anterior hypophysis of female rats of different age groups and at different periods of the day. *Endocrinology (Baltimore)* **32**, 334–339.

Hunt, T. E., and Hunt, E. A. (1966). A radioautographic study of the proliferative activity of adrenocortical and hypophyseal cells of the rat at different periods of the estrous cycle. *Anat. Rec.* **156**, 361–368.

Hyde, J., Engle, M. G., and Maley, B. E. (1991). Colocalization of galanin and prolactin within secretory granules of anterior pituitary cells in estrogen-treated Fischer 344 rats. *Endocrinology (Baltimore)* **129**, 270–276.

Hymer, W. C., and Motter, K. A. (1988). Heterogeneity in mammotrophs prepared from diethylstilbestrol-induced prolactinomas. *Endocrinology (Baltimore)* **122**, 2324–2338.

Hymer, W. C., Snyder, J., Wilfinger, W., Swanson, N., and Davis, J. A. (1974). Separation of pituitary mammotrophs from the female rat by velocity sedimentation at unit gravity. *Endocrinology (Baltimore)* **95**, 107–122.

Ingraham, H. A., Chen, R., Mangalam, H. J., Elsholtz, H. P., Flynn, S. E., Lin, C. R., Simmons, D. M., Swanson, L., and Rosenfeld, M. G. (1988). A tissue-specific transcription factor containing a homeodomain specifies a pituitary phenotype. *Cell (Cambridge, Mass.)* **55**, 519–529.

Inoue, K., and Kurosumi, K. (1981). Mode of proliferation of gonadotrophic cells of the anterior pituitary after castration-immunocytochemical and autoradiographic studies. *Arch. Histol. Jpn.* **44**, 71–85.

Inoue, K., and Sakai, T. (1991). Conversion of growth hormone-secreting cells into prolactin-secreting cells and its promotion by insulin and insulin-like growth factor-1 in vitro. *Exp. Cell Res.* **195**, 53–58.

Inoue, K., Hattori, M., Sakai, T., Inukai, S., Fujimoto, N., and Ito, A. (1990). Establishment of a series of pituitary clonal cell lines differing in morphlogy, hormone secretion, and response to estrogen. *Endocrinology (Baltimore)* **126**, 2313–2320.

Ishibashi, T., and Shiino, M. (1989a). Co-localization pattern of growth hormone (GH) and prolactin (PRL) within the anterior pituitary cells in the female rat and female musk shrew. *Anat. Rec.* **223**, 185–193.

Ishibashi, T., and Shiino, M. (1989b). Subcellular localization of prolactin in the anterior pituitary cells of the female Japanese house bat, *Pipistrellus abramus. Endocrinology (Baltimore)* **124**, 1056–1063.

Israel, J. M., Kuksts, L. A., and Vincent, J.-D. (1990). Plateau potentials recorded from lactating rat enriched lactotroph cells are triggered by thyrotropin releasing hormone and shortened by dopamine. *Neuroendocrinology* **51**, 113–122.

Jahn, G. A., Machiavelli, G. A., Kalbermann, L. E., Szijan, I., Alonso, G. E., and Burdman, J. A. (1982). Relationships among release of prolactin, synthesis of DNA and growth of the anterior pituitary gland of the rat: Effects of oestrogen and sulpiride. *J. Endocrinol.* **94**, 1–10.

Jin, Y., Cox, D. A., Knecht, R., Raschdorf, F., and Cerletti, N. (1991). Separation, purification, and identification of TGF-β1 and TGF-β2 from bovine milk. *J. Protein Chem.* **10**, 565–575.

Jones, T. H., Justice, S., Price, A., and Chapman, K. (1991). Interleukin-6 secreting human adenomas *in vitro. J. Clin. Endocrinol. Metab.* **73**, 207–209.

Kadowaki, J., Ku, N.., Oetting, W. S., and Walkers, A. M. (1984). Mammotroph autoregulation: Uptake of secreted PRL and inhibition of secretion. *Endocrinology (Baltimore)* **114**, 2060–2067.

Kalbermann, L. E., Szijan, I., Jahn, G. A., Krawiec, L., and Burdman, J. A. (1979). DNA synthesis in the pituitary gland of the rat. Effect of sulpiride and postnatal maturation. *Neuroendocrinology* **29,** 42–48.

Kamijo, K., Sato, M., Saito, T., Yabana, T., Yachi, A., Fujii, N., and Minase, T. (1993). Bromocriptine-induced reversible lysosomal change and reduction of prolactin and growth hormone messenger RNA in cultured GH₃ cell lines. *Endocr. Pathol.* **4,** 28–33.

Kanie, N., Kageyama, N., Kuwayama, A., Nakane, T., Watanabe, M., and Kawaoi, A. (1983). Pituitary adenomas in acromegalic patients: An immunohistochemical and endocrinological study with special reference to prolactin-secreting adenoma. *J. Clin. Endocrinol. Matab.* **57,** 1093–1101.

Kaplan, L. M., Gabriel, S. M., Koenig, J. I., Sunday, M. E., Spindel, E. R., Martin, J. B., and Chin, W. W. (1988). Galanin is an estrogen-inducible, secretory product of the rat anterior pituitary function. *Proc. Natl. Acad. Sci. U.S.A.* **85,** 7408–7412.

Kashio, Y., Chomczynski, P., Downs, T. R., and Frohman, L. A. (1990). Growth hormone and prolactin secretion in cultured somatomammotroph cells. *Endocrinology (Baltimore)* **127,** 1129–1135.

Kawashima, S. (1974). Morphology and function of prolactin cells in old female rats. *Gunma Symp. Endocrinol.* **11,** 129–141.

Kawashima, S., and Takahashi, S. (1986). Morphological and functional changes of prolactin cells during aging in the rat. *In* "Pars Distalis of the Pituitary Gland—Structure, Function and Regulation" (F. Yoshimura and A. Gorbman, eds.) pp. 51–56. Elsevier, Amsterdam.

Keech, C. A., and Gutierrez-Hartmann, A. (1991). Insulin activation of rat prolactin promoter activity. *Mol. Cell. Endocrinol.* **78,** 55–60.

Keefer, D. A. (1980). In vivo estogen uptake by individual cell types of the rat anterior pituitary after short-term castration-adrenalectomy. *Cell Tissue Res.* **209,** 167–175.

Kendall, M. E., and Hymer, W. C. (1987). Cell blotting: A new approach to quantify hormone secretion from individual rat pituitary cells. *Endocrinology (Baltimore)* **121,** 2260–2262.

Khosla, S., Johansen, K. L., Ory, S. J., O'Brien, P. C., and Kao, P. C. (1990). Parathyroid hormone-related peptide in lactation and in umbilical cord blood. *Mayo Clin. Proc.* **65,** 1408–1414.

Kiino, D., and Dannies, P. S. (1981). Insulin and 17β-estradiol increase the intercellular prolactin content of GH₄C₁ cells. *Endocrinology (Baltimore)* **109,** 1264–1269.

Kineman, R. D., Faught, W. J., and Frawley, L. S. (1991). Mammosomatotropes are abundant in bovine pituitaries: Influence of gonadal status. *Endocrinology (Baltimore)* **128,** 2229–2233.

Koenig, J. I., Snow, K., Clark, B. D., Toni, R., Cannon, J. G., Shaw, A. R., Dinarello, C. A., Reichlin, S., Lee, S. L., and Lechan, R. M. (1990). Intrinsic pituitary interleukin-1β is induced by bacterial lipopolysaccharide. *Endocrinology (Baltimore)* **126,** 3053–3058.

Krown, K. A., Wang, Y. F., Ho, T. W. C., Kelly, P. A., and Walker, A. M. (1992). Prolactin isoform 2: An autocrine growth factor for GH₃ cells. *Endocrinology (Baltimore)* **131,** 595–602.

Kudlow, J. E., and Kobrin, M. S. (1984). Secretion of epidermal growth factor-like mitogens by cultured cells from bovine anterior pituitary glands. *Endocrinology (Baltimore)* **115,** 911–917.

Kukstas, L. A., Domec, C., Bascles, L., Bonnet, J., Verrier, D., Israel, J.-M., and Vincent, J.-D. (1991). Different expression of the two dopaminergic D2 receptors, D2₄₂₅ and D2₄₄₄, in two types of lactotroph each characterised by their response to dopamine, and modification of expression by sex steroids. *Endocrinology (Baltimore)* **129,** 1101–1103.

Kunert-Radek, J., Stepien, H., and Pawlikowski, M. (1994). Inhibition of rat pituitary tumor cell proliferation by benzodiazepines in vitro. *Neuroendocrinology* **59,** 92–96.

Kurosumi, K. (1971). Mitosis of the rat anterior pituitary cells: An electron microscope study. *Arch. Histol. Jpn.* **33,** 145–160.

Kurosumi, K. (1979). Formation and release of secretory granules during mitosis in the anterior pituitary gland. *Arch. Histol. Jpn.* **42**, 481–486.

Kurosumi, K. (1986). Cell classification of the rat anterior pituitary by means of immunoelectron microscopy. *J. Clin. Electron Microsc.* **19**, 299–319.

Kurosumi, K. (1991). Ultrastructural immunocytochemistry of the adenohypophysis in the rat: A review. *J. Electron Microsc. Tech.* **19**, 42–56.

Kurosumi, K., and Tosaka, H. (1988). Prenatal development of growth hormone-producing cells in the rat anterior pituitary as studied by immunogold electron microscopy. *Arch. Histol. Cytol.* **51**, 139–204.

Kurosumi, K., Koyama, T., and Tosaka, H. (1986). Three types of growth hormone cells of the rat anterior pituitary as revealed by immunoelectron microscopy using a colloidal gold-antibody method. *Arch. Histol. Jpn.* **49**, 227–242.

Kurosumi, K., Tanaka, S., and Tosaka, H. (1987). Changing ultrastructures in the estrous cycle and postnatal development of prolactin cells in the rat anterior pituitary as studied by immunogold electron microscopy. *Arch. Histol Jpn.* **50**, 455–478.

Landis, C. A., Masters, S. B., Spada, A., Pace, A. M., Bourne, H. R., and Vallar, L. (1989). GTPase inhibiting mutations activate the α chain of Gs and stimulate adenylyl cyclase in human pituitary tumours. *Nature (London)* **340**, 692–696.

Landis, C. A., Harsh, G., Lyons, J., Davis, R. L., McCormick, F., and Bourne, H. R. (1990). Clinical characteristics of acromegalic patients whose pituitary tumors contain mutant Gs protein. *J. Clin. Endocrinol. Metab.* **71**, 1416–1420.

Larrea, F., Méndez, I., Parra, A., and de los Monteros, A. E. (1993). Serum pattern of different molecular forms of prolactin during normal human pregnancy. *Hum. Report.* **8**, 1617–1622.

Larson, G. H., and Wise, P. M. (1991). Age-related alterations in prolactin secretion by individual cells as assessed by the reverse hemolytic plaque assay. *Biol. Reprod.* **44**, 648–655.

Leite, V., Vrontakis, M. E., Kasper, S., and Friesen, H. G. (1993). Bromocriptine inhibits galanin gene expression in the rat pituitary gland. *Mol. Cell. Neurosci.* **4**, 418–423.

Leong, D. A., Lau, S. K., Sinha, Y. N., Kaiser, D. L., and Thorner, M. O. (1985). Enumeration of lactotropes and somatotropes among male and female pituitary cells in culture: Evidence in favor of a mammosomatotrope subpopulation in the rat. *Endocrinology (Baltimore)* **116**, 1371–1378.

Lira, S. A., Crenshaw E. B., III, Glass, C. K., Swanson, L. W., and Rosenfeld, M. G. (1988). Identification of rat growth hormone genomic sequences targeting pituitary expression in transgenic mice. *Proc. Natl. Acad. Sci. U.S.A.* **85**, 4755–4759.

Lloyd, H. M., Meares, J. D., and Jacobi, J. (1975). Effects of oestrogen and bromocriptine on in vivo secretion and mitosis in prolactin cells. *Nature (London)* **225**, 497–498.

Lloyd, R. V., Anagnostou, D., Cano, M., Barkan, A. L., and Chandler, W. F. (1988). Analysis of mammosomatotropic cells in normal and neoplastic human pituitary tissues by the reverse hemolytic plaque assay and immunocytochemistry. *J. Clin. Endocrinol. Metab.* **66**, 1103–1110.

Losinski, N. E., Horvath, E., and Kovacs, K. (1989). Double-labeling immunogold electron-microscopic study of hormonal colocalization in nontumorous and adenomatous rat pituitaries. *Am. J. Anat.* **185**, 236–243.

Luque, E. H., Munoz de Toro, M., Smith, P. F., and Neill, J. D. (1986). Subpopulations of lactotropes detected with the reverse hemolytic plaque assay show differential responsiveness to dopamine. *Endocrinology (Baltimore)* **118**, 2120–2124.

MacLeod, R. M., and Lehmeyer, J. E. (1974). Studies on the mechanism of the dopamine mediated inhibition of prolactin secretion. *Endocrinology (Baltimore)* **94**, 1077–1085.

Maurer, R. A. (1979). Estrogen-induced prolactin and DNA synthesis in immature female rat pituitaries. *Mol. Cell. Endocrinol.* **13**, 291–300.

Maurer, R. A. (1980). Dopaminergic inhibition of prolactin synthesis and prolactin messenger RNA accumulation in cultured pituitary cells. *J. Biol. Chem.* **255**, 8092–8097.

Maurer, R. A., and Notides, A. C. (1987). Identification of an estrogen-responsive element from the 5'-flanking region of the rat prolactin gene. *Mol. Cell. Endocrinol.* **7**, 4247–4254.

McCormick, A., Brady, H., Theill, L. E., and Karin, M. (1990). Regulation of the pituitary-specific homeobox gene GHF-1 by cell-autonomous and environmental cues. *Nature (London)* **345**, 829–832.

McNicol, A. M., Kubba, M. A. G., and McTeague, E. (1988). The mitogenic effects of corticotrophin-releasing factor on the anterior pituitary gland of the rat. *J. Endocrinol.* **118**, 237–241.

McNicol, A. M., Murray, J. E., and McMeekin, W. (1990). Vasopressin stimulation of cell proliferation in the rat pituitary gland *in vitro. J. Endocrinol.* **126**, 255 259.

Meites, J. (1990). Aging: Hypothalamic catecholamines, neuroendocrine-immune interactions, and dietary restriction. *Proc. Soc. Exp. Biol. Med.* **195**, 304–311.

Meites, J., Goya, R. G., and Takahashi, S. (1987). Why the neuroendocrine system is important in aging processes. *Exp. Gerontol.* **22**, 1–15.

Mena, F., Martinez-Escalera, G., Clapp, C., Aguayo, D., Forray, C., and Grosvenor, C. E. (1982). A solubility shift occurs during depletion-transformation of prolactin within the lactating rat pituitary. *Endocrinology (Baltimore)* **111**, 1086–1091.

Mena, F., Clapp, C., Aguayo, D., Morales, M. T., Grosvenor, C. E., and Martinez de la Escalera, G. (1989a). Regulation of prolactin secretion by dopamine and thyrotropin-releasing hormone in lactating rat adenohypophyses: Influence of intracellular age of the hormone. *Endocrinology (Baltimore)* **125**, 1814–1820.

Mena, F., Clapp, C., Aguayo, D., and Martinez-Escalera, G. (1989b). Differential effects of thyrotropin-releasing hormone on in vitro release of in vivo or in vitro newly synthesized and mature prolactin by lactating rat adenohypophyses. Further evidence for a sequential pattern of hormone release. *Neuroendocrinology* **49**, 207–214.

Menon, K. M. J., and Gunaga, K. P. (1976). Cytoplasmic and nuclear-estradiol complex in the hypothalamus and gonadotropin secretion in the rat. *Neuroendocrinology* **22**, 8–17.

Michels, K. M., Lee, W.-H., Seltzer, A., Saavedra, J. M., and Bondy, C. A. (1993). Up-regulation of pituitary [^{125}I]insulin-like growth factor-I (IGF-I) bunding and IGF binding protein-2 and IGF-I gene expression by estrogen. *Endocrinology (Baltimore)* **132**, 23–29.

Mittra, I. (1980). A novel "cleaved prolactin" in the rat pituitary. I. Biosynthesis, characterization and regulatory control. *Biochem. Biophys. Res. Commun.* **95**, 1750–1759.

Miyachi, K., Fritzler, M. J., and Tan, E. M. (1978). Autoantibody to a nuclear antigen in proliferating cells. *J. Immunol.* **121**, 2228–2234.

Morel, G., Ouhtit, A., and Kelly, P. A. (1994). Prolactin receptor immunoreactivity in rat anterior pituitary. *Neuroendocrinology* **59**, 78–84.

Mori, T., and Nagasawa, H. (1984). Alterations of the development of mammary hyperplastic alveolar nodules and uterine adenomyosis in SHN mice by different schedules of treatment with CB-154. *Acta Endocrinol. (Copenhagen)* **107**, 245–249.

Motegi, M., and Watanabe, Y. G. (1990). Effects of sex steroids on growth hormone and prolactin cells in immature mouse adenohypophysis. *Sci. Rep. Niigata Univ., Ser. D.* **27**, 1–10.

Mulchahey, J. J., and Jaffe, R. B. (1987). Detection of a progenitor cell in the human fetal pituitary that secretes both growth hormone and prolactin. *J. Clin. Endocrinol Metab.* **66**, 24–32.

Murai, I., and Ben-Jonathan, N. (1990). Acute stimulation of prolactin release by estradiol: Mediation by the posterior pituitary. *Endocrinology (Baltimore)* **126**, 3179–3183.

Murphy, L. J., Bell, G. I., and Friesen, H. G. (1987). Tissue distribution of insulin-like growth factor I and II messenger ribonucleic acid in the adult rat. *Endocrinology (Baltimore)* **120**, 1279–1282.

Nakane, P. K. (1970). Classifications of anterior pituitary cell types with immunoenzyme histochemistry. *J. Histochem. Cytochem.* **18**, 9–20.

Negro-Vilar, A., Krulich, L., and McCann, S. M. (1973). Changes in serum prolactin and gonadotropins during sexual development of the male rat. *Endocrinology (Baltimore)* **93**, 660–664.

Neill, J. D., Freeman, M. E., and Tillson, S. A. (1971). Control of the proestrous surge of prolactin and luteinizing hormone secretion by estrogens in the rat. *Endocrinology (Baltimore)* **89**, 1448–1453.

Nelson, C., Albert, V. R., Elsholtz, H. P., Lu, L. I.-W., and Rosenfeld, M. G. (1988). Activation of cell-specific expression of rat growth hormone and prolactin genes by a common transcription factor. *Science* **239**, 1400–1405.

Nelson, K. G., Takahashi, T., Bossert, N. L., Walmer, D. K., and McLachlan, J. A. (1991). Epidermal growth factor replaces estrogen in the stimulation of female genital tract growth and differentiation. *Proc. Natl. Acad. Sci. U.S.A.* **88**, 21–25.

Nelson, K. G., Takahashi, T., Lee, D. C., Luetteke, N. C., Bossert, N. L., Ross, K., Eitzman, B. E., and McLachlan, J. A. (1992). Transforming growth factor-α is a potential mediator of estrogen action in the mouse uterus. *Endocrinology (Baltimore)* **131**, 1657–1664.

Nemeskéri, A., Stétáló, G., and Halász, B. (1988). Ontogenesis of the three parts of the fetal rat adenohypophysis. A detailed immunohistochemical analysis. *Neuroendocrinology* **48**, 534–543.

Newton, C. J., Buric, R., Trapp, T., Brockmeier, S., Pagotto, U., and Stalla, G. K. The unliganded estrogen receptor (ER) transduces growth factor signals. *J. Steroid Biochem. Molec. Biol.* **48**, 481–486.

Nicoll, C. S. (1974). Physiological actions of prolactin. *In* "Handbook of Physiology" Sect 7, Vol. 4, (R. O. Greep and E. B. Astwood, eds.), Part 2, pp. 253–292. Williams & Wilkins, Baltimore, MD.

Nicoll, C. S., Mayer, G. L., and Russell, S. M. (1986). Structural features of prolactins and growth hormones that can be regulated to their biological properties. *Endocr. Rev.* **7**, 169–203.

Nikitovitch-Winer, M. B., Atkin, J., and Maley, B. E. (1987). Colocalization of prolatin and growth hormone within specific adenohypophyseal cells in male, female, and lactating rats. *Endocrinology (Baltimore)* **121**, 625–630.

Nogami, H. (1984). Fine-structural heterogeneity and morphologic changes in rat pituitary prolactin cells after estrogen and testosterone treatment. *Cell Tissue Res.* **237**, 195–202.

Nogami, H., and Yoshimura, F. (1980). Prolactin immunoreactivity of acidophils of the same granule type. *Cell Tissue Res.* **211**, 1–4.

Nogami, H., and Yoshimura, F. (1982). Fine structural criteria of prolactin cells identified immunohistochemically in the male rat. *Anat. Rec.* **202**, 261–274.

Nogami, H., Suzuki, K., Enomoto, H., and Ishikawa, H. (1989). Studies on the development of growth hormone and prolactin cells in the rat pituitary gland by *in situ* hybridization. *Cell Tissue Res.* **255**, 23–28.

Nouët, J. C., and Kujas, M. (1975). Variations of mitotic activity in the adenohypophysis of male and female rats during a 24-hour cycle. *Cell Tissue. Res.* **164**, 193–200.

Oetting, W. S., and Walker, A. M. (1985). Intracellular processing of prolactin. *Endocrinology (Baltimore)* **117**, 1565–1570.

Oetting, W. S., and Walker, A. M. (1986). Differential isoform distribution between stored and secreted prolactin. *Endocrinology (Baltimore)* **119**, 1377–1381.

Ohmichi, M., Hirota, K., Koike, K., Kurachi, H., Ohtsuka, S., Matsuzaki, N., Yamaguchi, M., Miyake, A., and Tanizawa, O. (1992). Binding sites for interleukin-6 in the anterior pituitary gland. *Neuroendocrinology* **55**, 199–203.

Oishi, Y., Okuda, M., Takahashi, H., Fujii, T., and Morii, S. (1993). Cellular proliferation in the anterior pituitary gland of normal adult rats: Influences of sex, estrous cycle, and circadian change. *Anat. Rec.* **235**, 111–120.

Olney, J. W. (1969). Brain lesions, obesity, and other disturbances in mice treated with monosodium glutamate. *Science* **164**, 719–721.

Ooka, H. (1993). Proliferation of anterior pituitary cells in relation to aging and longevity in the rat. *Zool. Sci.* **10**, 385–392.

Ooka, H., and Shinkai, T. (1992). Age-related changes in prolactin secretion and the population of mammotrophs in the rat: A longitudinal study. *Arch. Gerontol. Geriatr. Suppl.* **3**, 287–294.

Oomizu, S., Takahashi, S., Nakatomi, K., and Kobayashi, Y. (1993). Insulin stimulated the proliferation of pituitary cells in mice. *Zool. Sci.* **10**, Suppl., 146.

Oomizu, S., and Takahashi, S. (1994). Insulin and insulin-like growth factor I stimulate the proliferation of mouse pituitary cells. The Endocrine Society 76th Annual Meeting, *Program & Abstracts*, p. 489.

Ottlecz, A., Snyder, G. D., and McCann, S. M. (1988). Regulatory role of galanin in control of hypothalamic-anterior pituitary function. *Proc. Natl. Acad. Sci. U.S.A.* **85**, 9861–9865.

Ozawa, H., and Kurosumi, K. (1993). Morphofunctional study on prolactin-producing cells of the anterior pituitaries in adult male rats following thyroidectomy, thyroxine treatment and/or thyrotropin-releasing hormone treatment. *Cell Tissue Res.* **272**, 41–47.

Pérez, R. L., Machiavelli, G. A., Romano, M. I., and Burdman, J. A. (1986). Prolactin release, oestrogen and proliferation of prolactin-secreting cells in the anterior pituitary gland of adult male rats. *J. Endocrinol.* **108**, 399–403.

Pomerat, G. R. (1941). Mitotic activity in the pituitary of the white rat following castration. *Am. J. Anat.* **69**, 89–121.

Porter, T. E., and Frawley, L. S. (1991). Stimulation of prolactin cell differentiation *in vitro* by a milk-borne peptide. *Endocrinology (Baltimore)* **129**, 2707–2713.

Porter, T. E., and Frawley, L. S. (1992). Neurointermediate lobe peptides recruit prolactin-secreting cells exclusively within the central region of the adenohypophysis. *Endocrinology (Baltimore)* **131**, 2649–2652.

Porter, T. E., Hill, J. B., Wiles, C. D., and Frawley, L. S. (1990). Is the mammosomatotrope a transitional cell for the functional interconversion of growth hormone- and prolactin-secreting cells? Suggestive evidence from virgin, gestating, and lactating rats. *Endocrinology (Baltimore)* **127**, 2789–2794.

Porter, T. E., Chapman, L. E., Van Dolah, F. M., Frawley, L. S. (1991a). Normal differentiation of prolactin cells in neonatal rats requires a maternal signal specific to early lactation. *Endocrinology (Baltimore)* **128**, 792–796.

Porter, T. E., Wiles, C. D., and Frawley, L. S. (1991b). Evidence for bidirectional interconversion of mammotropes and somatotropes: Rapid reversion of acidophilic cell types to pregestational proportions after weaning. *Endocrinology (Baltimore)* **129**, 1215–1220.

Porter, T. E., Wiles, C. D., and Frawley, L. S. (1993). Lactotrope differentiation in rats is modulated by a milk-borne signal transferred to the neonatal circulation. *Endocrinology (Baltimore)* **133**, 1284–1291.

Powers, C. A., and Hatala, M. A. (1990). Prolactin proteolysis by glandular kallikrein: *In vitro* reaction requirements and cleaved sites, and detection of processed prolactin *in vivo*. *Endocrinology (Baltimore)* **127**, 1916–1927.

Prager, D., Yamashita, S., and Melmed, S. (1988). Insulin regulates prolactin secretion and messenger ribonucleic acid levels in pituitary cells. *Endocrinology (Baltimore)* **122**, 2946–2952.

Prosser, C. G., Fleet, I. R., Davis, A. J., and Heap, R. B. (1991). Mechanism of secretion of plasma insulin-like growth factor-I into milk of lactating goats. *J. Endocrinol.* **131,** 459–466.

Prysor-Jones, R. A., and Jenkins, J. S. (1981). Effect of bromocriptine on DNA synthesis, growth and hormone secretion of spontaneous pituitary tumors in the rat. *J. Endocrinol.* **88,** 463–469.

Prysor-Jones, R. A., Silverlight, J. J., and Jenkins, J. S. (1985). Differential effects of extracellular matrix on secretion of prolactin and growth hormone by rat pituitary tumour cells in vitro. *Acta Endocrinol. (Copenhagen)* **108,** 156–160.

Prysor-Jones, R. A., Silverlight, J. J., and Jenkins, J. S. (1989). Oestradiol, vasoactive intestinal peptide and fibroblast growth factor in the growth of human pituitary tumour cells *in vitro. J. Endocrinol.* **120,** 171–177.

Ramsdell, J. S. (1991). Transforming growth factor-α and -β are potent and effective inhibitors of GH$_4$ pituitary tumor cell proliferation. *Endocrinology (Baltimore)* **128,** 1981–1990.

Rao, Y., Olson, M. D., Buckley, D. J., and Buckley, A. R. (1993). Nuclear co-localization of prolactin and the prolactin receptor in rat Nb$_2$ node lymphoma cells. *Endocrinology (Baltimore)* **133,** 3062–3065.

Ren, P., Scheithauer, B. W., and Halper, J. (1994). Immunohistological localization of TGFα, EGF, IGF-I, and TGFβ in the normal human pituitary gland. *Endocr. Pathol.* **5,** 40–48.

Sakuma, S., Shirasawa, N., and Yoshimura, F. (1984). A histochemical study of immunohistochemically identified mitotic adenohypophysial cells in immature and mature castrated rats. *J. Endocrinol.* **100,** 323–328.

Sano, M., and Sasaki, F. (1971). Effects of prepubertal orchidectomy on the differentiation of prolactin cells in the mouse adenohypophysis. A quantitative study of electron microscopy. *J. Endocrinol.* **50,** 705–706.

Sarkar, D. K., Gottschall, P. E., and Meites, J. (1982). Damage to hypothalamic dopaminergic neurons is associated with development of prolactin-secreting pituitary tumors. *Science* **218,** 684–686.

Sarkar, D. K., Kim, K. H., and Minami, S. (1992). Transforming growth factor-β1 messenger RNA and protein expression in the pituitary gland: Its action on prolactin secretion and lactotropic growth. *Mol. Endocrinol.* **6,** 1825–1833.

Sasaki, F. (1974). Quantitative studies by electron microscopy on the sex-difference and the change during the oestrous cycle in the mouse anterior pituitary. *Arch. Histol. Jpn.* **37,** 41–57.

Sasaki, F., and Iwama, Y. (1988a). Correlation of spatial differences in concentrations of prolactin and growth hormone cells with vascular pattern in the female mouse adenohypophysis. *Endocrinology (Baltimore)* **122,** 1622–1630.

Sasaki, F., and Iwama, Y. (1988b). Sex difference in prolactin and growth hormone cells in mouse adenohypophysis: Stereological, morphometric, and immunohistochemical studies by light and electron microscopy. *Endocrinology (Baltimore)* **123,** 905–912.

Sasaki, F., and Iwama, Y. (1989). Two types of mammosomatotropes in mouse adenohypophysis. *Neuroendocrinology* **256,** 645–648.

Sasaki, F., and Sano, M. (1980). Role of the ovary in the sexual differentiation of prolactin and growth hormone cells in the mouse adenohypophysis during postnatal development: A stereological morphometric study by electron microscopy. *J. Endocrinol.* **85,** 283–289.

Sasaki, F., and Sano, M. (1983). Role of the ovary in sexual differentiation of lactotrophs and somatotrophs in the mouse adenohypophysis: A stereological morphometric study by electron microscopy. *J. Endocrinol.* **99,** 355–360.

Sato, S. (1980). Postnatal development, sexual difference and sexual cyclic variation of prolactin cells in the rats: Special reference to the topographic affinity to a gonadotroph. *Endocrinol. Jpn.* **27,** 573–583.

Schwartz, J., and Cherny, R. (1992). Intercellular communication within the anterior pituitary influencing the secretion of hypophysial hormones. *Endocr. Rev.* **13,** 453–475.

Sen, K. K., and Menon, K. M. J. (1978). Oestradiol receptors in the rat anterior pituitary gland during the oestrous cycle: Quantitation of receptor mediated luteinizing hormone release. *J. Endocrinol.* **76,** 211–218.

Seo, H. (1985). Growth hormone and prolactin: Chemistry, gene organization, biosynthesis, and regulation of gene expression. *In* "The Pituitary Gland" (H. Imura, ed.), pp. 57–82. Raven Press, New York.

Sétáló, G., and Nakane, P. K. (1972). Studies on the functional differentiation of cells in fetal anterior pituitary glands of rats with peroxidase labelled antibody method. *Anat. Rec.* **172,** 403–404.

Shah, G. V., Kacsoh, B., Seshadri, R., Grosvenor, C. E., and Crowley, W. R. (1989). Presence of calcitonin-like peptide in rat milk: Possible physiological role in regulation of neonatal prolactin secretion. *Endocrinology (Baltimore)* **125,** 61–67.

Shirasawa, N., and Yoshimura, F. (1982). Immunohistochemical and electron microscopical studies of mitotic adenohypophysial cells in different ages of rats. *Anat. Embryol.* **165,** 51–61.

Shirasawa, N., Hirano, M., and Ishikawa, H. (1990). Immunocytochemistry of mammotroph and somatotroph in fetal and adult bovine pituitary by three different methods. *Jikeikai Med. J.* **37,** 433–446.

Shiu, R. P. C., Elsholtz, H. P., Tanaka, T., Friesen, H. G., Gout, P. W., Beer, C. T., and Noble, R. L. (1983). Receptor-mediated mitogenic action of prolactin in a rat lymphoma cell line. *Endocrinology (Baltimore)* **113,** 159–165.

Shull, J. D. (1991). Population density alters the responsiveness of GH_4C_1 pituitary tumor cells to 17β-estradiol. *Endocrinology (Baltimore)* **129,** 1644–1652.

Sibley, D. R., De Lean, A., and Creese, I. (1982). Anterior pituitary dopamine receptors. Demonstration of interconvertible high and low affinity states of D-2 dopamine receptor. *J. Biol. Chem.* **257,** 6351–6357.

Simmen, F. A., Simmen, R. C. M., and Reinhart, G. (1988). Maternal and neonatal somatomedin C/insulin-like growth factor-I (IGF-I) and IGF binding proteins during early lactation in the pig. *Dev. Biol.* **130,** 16–27.

Simmons, D. M., Voss, J. W., Ingraham, H. A., Holloway, J. M., Broide, R. S., Rosenfeld, M. G., and Swanson, L. W. (1990). Pituitary cell phenotypes involve cell-specific Pit-1 mRNA translation and synergistic interactions with other classes of transcription factors. *Genes Dev.* **4,** 695–711.

Simpkins, J. W., Mueller, G. P., Huang, H. H., and Meites, J. (1977). Evidence for depressed catecholamine and enhanced serotonin metabolism in aging male rats: Possible relation to gonadotropin secretion. *Endocrinology (Baltimore)* **100,** 1672–1678.

Sinha, Y. N., and Gilligan, T. A. (1984). A cleaved form of prolactin in the mouse pituitary gland: Identification and comparison of *in vitro* synthesis and release in strains with high and low incidences of mammary tumors. *Endocrinology (Baltimore)* **114,** 2046–2053.

Sinha, Y. N., and Jacobsen, B. P. (1988). Three growth hormone- and two prolactin-related novel peptides of Mr 13,000–18,000 identified in the anterior pituitary. *Biochem. Biophys. Res. Commun.* **156,** 171–179.

Sinha, Y. N., Gilligan, T. A., Lee, D. W., Hollingsworth, D., and Markoff, E. (1985). Cleaved prolactin: Evidence for its occurrence in human pituitary gland and plasma. *J. Clin. Endocrinol. Metab.* **60,** 239–243.

Sirbasku, D. A. (1978). Estrogen induction of growth factors specific for hormone-responsive mammary, pituitary, and kidney tumor cells. *Proc. Natl. Acad. Sci. U.S.A.* **75,** 3786–3790.

Slabaugh, M. B., Lieberman, M. E., Rutledge, J. J., and Gorski, J. (1982). Ontogeny of growth hormone and prolactin gene expression in mice. *Endocrinology (Baltimore)* **110,** 1489–1497.

Smets, G., Velkeniers, B., Finne, E., Baldys, A., Gcⁿts, W., and Vanhaelst, L. (1987). Postnatal development of growth hormone and prolactin cells in male and female rat pituitary. An immunocytochemical light and electromicroscopic study. *J. Histochem. Cytochem.* **35,** 335–341.

Smith, P. F., and Keefer, D. A. (1982). Immunocytochemical and ultrastructural identification of mitotic cells in the pituitary gland of ovariectomized rats. *J. Reprod. Fertil.* **66,** 383–388.

Smith, M. S., Freeman, M. E., and Neill, J. D. (1975). The control of progesterone secretion during the estrous cycle and early pseudopregnancy in the rat: Prolactin, gonadotropin and steroid levels associated with rescue of the corpus luteum of pseudopregnancy. *Endocrinology (Baltimore)* **96,** 219–226.

Smith (White), S. S., and Ojeda, S. R. (1984). Maternal modulation of infantile ovarian development and available ovarian luteinizing hormone-releasing hormone (LHRH) receptors via milk LHRH. *Endocrinology (Baltimore)* **115,** 1973–1983.

Snyder, G., Hymer, W. C., and Snyder, J. (1977). Functional heterogeneity in somatotrophs isolated from the rat anterior pituitary. *Endocrinology (Baltimore)* **101,** 788–799.

Snyder, J. M., Wilfinger, W., and Hymer, W. C. (1976). Maintenance of separated rat pituitary mammotrophs in cell culture. *Endocrinology (Baltimore)* **98,** 25–32.

Sonntag, W. E., Steger, R. W., Forman, L. J., and Meites, J. (1980). Decreased pulsatile release of growth hormone in old male rats. *Endocrinology (Baltimore)* **107,** 1875–1879.

Sonntag, W. E., Lloyd, L. J., Miki, N., and Meites, J. (1982). Growth hormone and neuroendocrine regulation. *In* "CRC Handbook of Endocrinology" (G. H. Gass and H. M. Kaplan, eds.), pp. 35–59. CRC Press, Boca Raton, FL.

Sonntag, W. E., Hylka, V. W., and Meites, J. (1985). Growth hormone restores protein synthesis in skeletal muscle of old male rats. *J. Gerontol.* **40,** 689–694.

Spangelo, B. L., MacLeod, R. M., and Isakson, P. C. (1990). Production of interleukin-6 by anterior pituitary cells *in vitro. Endocrinology (Baltimore)* **126,** 582–586.

Spies, H. G., and Clegg, M. T. (1971). Pituitary as a possible site of PRL feedback in autoregulation. *Neuroendocrinology* **8,** 205–212.

St. John, P. A., Dufy-Barbe, L., and Baker, J. L. (1986). Anti-prolactin cell surface immunoreactivity identifies a subpopulation of lactotrophs from the rat anterior pituitary. *Endocrinology (Baltimore)* **119,** 2783–2795.

Steel, J. H., Gon, G., O'Halloran, D. J., Jones, P. M., Yanaihara, N., Ishikawa, H., Bloom, S. R., and Polak, J. M. (1989). Galanin and vasoactive intestinal polypeptide are colocalised with classical pituitary hormones and show plasticity of expression. *Histochemistry* **93,** 183–189.

Stefaneanu, L., Kovacs, K., Horvath, E., Asa, S. L., Losinski, N. E., Billestrup, N., Price, J., and Vale, W. (1989). Adenohypophysial changes in mice transgenic for human growth hormone-releasing factor: A histological, immunocytochemical, and electron microscopic investigation. *Endocrinology (Baltimore)* **125,** 2710–2718.

Steger, R. W. (1981). Age-related changes in the control of prolactin secretion in the female rat. *Neurobiol. Aging* **2,** 119–123.

Stewart, D. A., Blackman, M. R., Kowatch, M. A., Danner, D. B., and Roth, G. S. (1990). Discordant effects of aging on prolactin and luteinizing hormone-β messenger ribonucleic acid levels in the female rat. *Endocrinology (Baltimore)* **126,** 773–778.

Stratmann, I. E., Ezrin, C., and Sellers, E. A. (1974). Estrogen-induced transformation of somatotrophs into mammotrophs in the rat. *Cell Tissue Res.* **152,** 229–238.

Straus, D. S. (1984). Growth-stimulatory actions of insulin in vitro and in vivo. *Endocr. Rev.* **5,** 356–369.

Strbak, V., Alexandrova, M., Macho, L., and Ponec, J. (1980). Transport of ³H-TRH from plasma to rat milk: Accumulation and slow degradation in milk and presence of unaltered hormone in gastric content of pups. *Biol. Neonate* **37,** 313–321.

Stroud, C. M., Deaver, D. R., Peters, J. L., Loeper, D. C., Toth, B. E., Derr, J. A., and Hymer, W. C. (1992). Prolactin variants in ram adenohypophyses vary with season. *Endocrinology (Baltimore)* **130,** 811–818.

Suganuma, N., Seo, H., Yamamoto, N., Kikkawa, F., Narita, O., Tomoda, Y., and Matsui, N. (1986). Ontogenesis of pituitary prolactin in the human fetus. *J. Clin. Endocrinol. Metab.* **63,** 156–161.

Sutherland, R. L., Watts, C. K. W., and Clarke, C. L. (1988). Oestrogen action. In "Hormones and Their Actions" (B. A. Cooke, R. J. B. King, and H. J. van der Molen, eds.), Part I, pp. 197–215. Elsevier, Amsterdam.

Swearingen, K. C. (1971). Heterogeneous turnover of adenohypophysial prolactin. *Endocrinology (Baltimore)* **89,** 1380–1388.

Takahashi, S. (1991). Immunocytochemical and immunoelectron microscopical study of growth hormone cells in male and female rats of various ages. *Cell Tissue Res.* **266,** 275–284.

Takahashi, S. (1992a). Heterogeneity and development of somatotrophs and mammotrophs in the rat. *Zool. Sci.* **9,** 901–924.

Takahashi, S. (1992b). Growth hormone secretion in old female rats analyzed by the reverse hemolytic plaque assay. *Acta Endocrinol. (Copenhagen)* **127,** 531–535.

Takahashi, S., and Kawashima, S. (1981). Responsiveness to estrogen of pituitary glands and prolactin cells in gonadectomized male and female rats. *Annot. Zool. Jpn.* **54,** 73–84.

Takahashi, S., and Kawashima, S. (1982). Age-related changes in prolactin cell percentage and serum prolactin levels in intact and neonatally gonadectomized male and female rats. *Acta Anat.* **113,** 211–217.

Takahashi, S., and Kawashima, S. (1983). Age-related changes in prolactin cells in male and female rats of the Wistar/Tw strain. *J. Sci. Hiroshima Univ., Ser. B, Div. 1* **32,** 185–191.

Takahashi, S., and Kawashima, S. (1986). Mitotic potency of prolactin in rats. In "Pars Distalis of the Pituitary Gland: Structure, Function and Regulation" (F. Yoshimura and A. Gorbman, eds.), pp. 497–502. Elsevier, Amsterdam.

Takahashi, S., and Kawashima, S. (1987). Proliferation of prolactin cells in the rat: Effects of estrogen and bromocriptine. *Zool. Sci.* **4,** 855–860.

Takahashi, S., and Meites, J. (1987). GH binding to liver in young and old female rats: Relation to somatomedin-C secretion. *Proc. Soc. Exp. Biol. Med.* **186,** 229–233.

Takahashi, S., and Miyatake, M. (1991). Immuno-electron microscopical study of prolactin cells in the rat: Postnatal development and effects of estrogen and bromocriptine. *Zool. Sci.* **8,** 549–559.

Takahashi, S., Kawashima, S., and Wakabayashi, K. (1979). Effects of chlorpromazine and estradiol benzoate on prolactin secretion in gonadectomized male and female rats. *Endocrinol. Jpn.* **26,** 419–422.

Takahashi, S., Kawashima, S., and Wakabayashi, K. (1980). Effects of gonadectomy and chlorpromazine treatment on prolactin, LH, and FSH secretion in young and old rats of both sexes. *Exp. Gerontol.* **15,** 185–194.

Takahashi, S., Kawashima, S., and Wakabayashi, K. (1982). Serum and pituitary levels of prolactin and the responsiveness to pimozide on prolactin release in monosodium glutamate-treated rats. *J. Sci. Hiroshima Univ., Ser B, Div. 1* **30**, 221–228.

Takahashi, S., Okazaki, K., and Kawashima, S. (1984). Mitotic activity of prolactin cells in the pituitary glands of male and female rats of different ages. *Cell Tissue Res.* **235**, 497–502.

Takahashi, S., Gottschall, P. E., Quigley, K. L., Goya, R. G., and Meites, J. (1987). Growth hormone secretory patterns in young, middle-aged and old female rats. *Neuroendocrinology* **46**, 137–142.

Takahashi, S., Kawashima, S., Seo, H., and Matsui, N. (1990). Age-related changes in growth hormone and prolactin messenger RNA levels in the rat. *Endocrinol. Jpn.* **37**, 827–840.

Takahashi, S., Oomizu, S., and Kobayashi, Y. (1994). Proliferation of pituitary cells in streptozotocin-induced diabetic mice: Effect of insulin and estrogen. *Zool. Sci.* **11**, 445–449.

Takeyama, M., Yanaga, N., Yarimizu, K., Ono, J., Takai, R., Fujii, N., and Yajima, H. (1990). Enzyme immunoassay of somatostatin (SS)-like immunoreactive substance in bovine milk. *Chem. Pharm. Bull.* **38**, 456–459.

Tanaka, T., Yano, H., Umezawa, S., Shishiba, Y., Okada, K., Saito, T., and Hibi, I. (1989). Heterogeneity of bigbig hPRL in panhypopituitarism. *Horm. Metab. Res.* **21**, 84–88.

Tannenbaum, G. S., Guyda, H. J., and Posner, B. I. (1983). Insulin-like growth factors: A role in growth hormone negative feedback and body weight regulation via brain. *Science* **220**, 77–79.

Tashjian, A. H., Jr., Yasumura, Y., Levine, L., Sato, G. H., and Parkar, M. L. (1968). Establishment of clonal strains of rat pituitary tumor cells that secrete growth hormone. *Endocrinology (Baltimore)* **82**, 342–352.

Tashjian, A. H., Jr., Yasumura, Y., Levine, L., Sato, G. H., and Parker, M. L. (1970). Production of both prolactin and growth hormone by clonal strains of rat pituitary tumor cells. *J. Cell Biol.* **47**, 61–70.

Thorner, M. O., Martin, W. H., Rogol, A. D., Morris, J. L., Perryman, R. L., Conway, B. P., Howards, S. S., Wolfman, M. G., and MacLeod, R. M. (1980). Rapid regression of pituitary prolactinomas during bromocriptine treatment. *J. Clin. Endocrinol. Metab.* **51**, 438–445.

Thorpe, J. R., and Wallis, M. (1991). Immunocytochemical and morphometric studies of mammotrophs, somatotrophs, and somatomammotrophs in sheep pituitary cell cultures. *J. Endocrinol.* **129**, 417–422.

Thorpe, J. R., Ray, K. P., and Wallis, M. (1990). Occurrence of rare somatomammotrophs in ovine anterior pituitary tissue studied by immunogold labelling and electron microscopy. *J. Endocrinol.* **124**, 67–73.

Tian, J., Chen, J., and Bancroft, C. (1994). Expression of constitutively active Gs α-subunits in GH_3 pituitary cell stimulates prolactin promoter activity. *J. Biol. Chem.* **269**, 33–36.

Tilemans, D., Andries, M., and Denef, C. (1992). Luteinizing hormone-releasing hormone and neuropeptide Y influence deoxyribonucleic acid replication in three anterior pituitary cell types. Evidence for mediation by growth factors released from gonadotrophs. *Endocrinology (Baltimore)* **130**, 882–894.

Tomooka, Y., DiAugustine, R. P., and McLachlan, J. A. (1986). Proliferation of mouse uterine epithelial cells *in vitro*. *Endocrinology (Baltimore)* **118**, 1011–1018.

Tong, Y., Zhao, H. F., Simard, J., Labrie, F., and Pelletier, G. (1989). Electron microscopic autoradiographic localization of prolactin mRNA in rat pituitary. *J. Histochem. Cytochem.* **37**, 567–571.

Tougard, C., and Tixier-Vidal, A. (1988). Lactotropes and gonadotropes. *In* "The Physiology of Reproduction" (E. Knobil and J. Neill, eds.), pp. 1305–1333. Raven Press, New York.

Van Bael, A., Huygen, R., Himpens, B., and Denef, C. (1994). *In vitro* evidence that LHRH stimulates the recruitment of prolactin mRNA-expressing cells during the postnatal period in the rat. *J. Mol. Endocrinol.* **12**, 107–118.

Vankelecom, H., Carmeliet, P., Van Damme, J., Billiau, A., and Denef, C. (1989). Production of interleukin-6 by folliculo-stellate cells of the anterior pituitary gland in a histiotypic cell aggregate culture system. *Endocrinology (Baltimore)* **49**, 102–106.

van Putten, L. J. A., and Kiliaan, A. J. (1988). Immunoelectron-microscopic study of the prolactin cells in the pituitary gland of male Wistar rats during aging. *Cell Tisue Res.* **215**, 353–358.

Vekemans, M., and Robyn, C. (1975). Influence of age on serum prolactin levels in women and men. *Br. Med. J.* **4**, 738–739.

Velkeniers, B., Hooghe-Peters, E. L., Hooghe, R., Belayew, A., Smets, G., Claeys, A., Robberecht, P., and Vanhaelst, L. (1988). Prolactin cell subpopulations separated on discontinuous Percoll gradient: An immunocytochemical, biochemical, and physiological characterization. *Endocrinology (Baltimore)* **123**, 1619–1630.

Voigt, M. M., Davis, L. G., and Wyche, J. H. (1984). Benzodiazepine binding to cultured human pituitary cells. *J. Neurochem.* **43**, 1106–1113

vom Saal, F. S., and Finch, C. E. (1988). Reproductive senescence: Phenomena and mechanism in mammals and selected vertebrates. *In* "The Physiology of Reproduction" (E. Knobil and J. E. Neill *et al.*, eds.), pp. 2351–2413. Raven Press, New York.

Vrontakis, M. E., Thliveris, J. A., and Friesen, H. G. (1987). Influence of bromocriptine and oestrogen on prolactin synthesis, secretion and tumor growth *in vivo* in rats. *J. Endocrinol.* **113**, 383–388.

Vrontakis, M. E., Yamamoto, T., Schroedter, I. C., Nagy, J. I., and Friesen, H. G. (1989). Estrogen induction of galanin synthesis in the anterior pituitary gland demonstrated by *in situ* hybridization and immunohistochemistry. *Neurosci. Lett.* **100**, 59–64.

Walker, A. M., and Farquhar, M. G. (1980). Preferential release of newly synthesized prolactin granules is the result of functional heterogeneity among mammotrophs. *Endocrinology (Baltimore)* **107**, 1095–1104.

Wallis, M. (1988). Mechanism of action of prolactin. *In* "Hormones and Their Actions" (B. A. Cooke, R. J. B. King, and H. J. van der Molen, ed.), Part 2, pp. 295–319. Elsevier, Amsterdam.

Wang, Y.-F., and Walker, A. M. (1993). Dephosphorylation of standard prolactin produces a more biologically active molecule: Evidence for antagonism between nonphosphorylated and phosphorylated prolactin in the stimulation of Nb_2 cell proliferation. *Endocrinology (Baltimore)* **133**, 2156–2160.

Watanabe, Y. G., and Carbajo-Peréz, E. (1990). Cell proliferation in pituitary monolayers as revealed by the BrdU labelling method. *Biomed. Res.* **11**, 373–377.

Watanabe, Y. G., and Daikoku, S. (1979). An immunohistochemical study on the cytogenesis of adenohypophysial cells in fetal rats. *Dev. Biol.* **68**, 557–567.

Watanabe, Y. G., and Haraguchi, H. (1994). Immunohistochemical study of the cytogenesis of prolactin and growth hormone cells in the anterior pituitary gland of the fetal rat. *Arch. Histol. Cytol.* **57**, 161–166.

Webster, E. L., Tracey, D. E., and De Souza, E. B. (1991). Upregulation of interleukin-1 receptors in mouse AtT-20 pituitary tumor cells following treatment with corticotropin-releasing factor. *Endocrinology (Baltimore)* **129**, 796–2798.

Webster, J., and Scanlon, M. F. (1991). Growth factors and the anterior pituitary. *Bailliere's Clin. Endocrinol. Metab.* **5**, 699–726.

Weiner, R. I., Findell, P. R., and Kordon, C. (1988). Role of classic and peptide neuromediators in the neuroendocrine regulation of LH and prolactin. *In* "The Physiology of Reproduction" (E. Knobil and J. Neill *et al.*, eds.), pp. 1235–1281. Raven Press, New York.

Welsch, C. W., and Nagasawa, H. (1977). Prolactin and murine mammary tumorigenesis: A review. *Cancer Res.* **37,** 951–963.

Werner, H., Amarant, T., Millar, R. P., Fridkin, M., and Koch, Y. (1985). Immunoreactive and biologically active somatostatin in human and sheep milk. *Eur. J. Biochem.* **148,** 353–357.

Werner, H., Amarant, T., Fridkin, M., and Koch, Y. (1986). Growth hormone releasing factor-like immunoreactivity in human milk. *Biochem. Biophys. Res. Commun.* **135,** 1084–1089.

Werner, H., Katz, P., Fridkin, M., Koch, Y., and Levine, S. (1988). Growth hormone releasing factor and somatostatin concentrations in the milk of lactating women. *Eur. J. Pediatr.* **147,** 252–256.

Weström, B. R., Ekman, R., Svendsen, L., Svendsen, J., and Karlsson, B. W. (1987). Levels of immunoreactive insulin, neurotensin, and bombesin in porcine colostrum and milk. *J. Pediatr. Gastroenterol. Nutr.* **6,** 460–465.

Wyndford-Thomas, D., and Williams, E. D. (1986). Use of bromodeoxyuridine for cell kinetic studies in intact animals. *Cell Tissue Kinet.* **19,** 179–182.

Yamashita, S., and Melmed, S. (1986a). Insulin-like growth factor I action on rat anterior pituitary cells: Supression of growth hormone secretion and messenger ribonucleic acid levels. *Endocrinology (Baltimore)* **118,** 176–182.

Yamashita, S., and Melmed, S. (1986b). Effects of insulin on rat anterior pituitary cells. Inhibition of growth hormone secretion and mRNA levels. *Diabetes* **35,** 440–447.

Zabavnik, J., Wu, W.-X., Eidne, K. A., and McNeilly, A. S. (1993). Dopamine D_2 receptor mRNA in the pituitary during the oestrous cycle, pregnancy and lactation in the rat. *Mol. Cell. Endocrinol.* **95,** 121–128.

Zambrano, D., and Deis, R. P. (1970). The adenohypophysis of female rats after hypothalamic oestradiol implants: An electron microscopic study. *J. Endocrinol.* **47,** 101–110.

Centrifuge Microscope as a Tool in the Study of Cell Motility[1]

Yukio Hiramoto* and Eiji Kamitsubo†

*Biological Laboratory, The University of the Air, Chiba 261, Japan
†Biological Laboratory, Hitotsubashi University, Tokyo 186, Japan

The centrifuge microscope (CM) is composed of a centrifuge and a microscope optical system designed to observe minute objects, especially living cells, during the application of centrifugal acceleration. Structures and characteristics of various types of CM designed and constructed up to the present and studies done with the CM on cell biology, especially cell motility, are reviewed. These studies include observations of the behavior of cells and cell components in a centrifugal field, determination of the mechanical properties of the cell surface and cytoplasm, microsurgical operations on cells with centrifugal force, and determination of the magnitude and the site of generation of motive force for cell motility.

KEY WORDS: Centrifuge microscope, Cell motility, Microsurgery, Cytoplasmic streaming.

I. Introduction

Centrifugation is one way in which a force can be applied to a specific region of a body without direct contact. Therefore, centrifugation is used to separate specific substances and bodies in a mixture by the differences in their sizes and/or densities. In cell biology, centrifugation is one of the most important methods for separating cell organelles and substances from biological materials.

Components of living cells in a centrifugal field are subjected to various forces, depending on their sizes and densities. For example, the way

[1] This chapter is dedicated to Professor Noburo Kamiya, one of the pioneers who introduced the centrifuge microscope as a tool in cell biology studies, on the occasion of his eightieth birthday.

cytoplasmic particles move in a centrifugal field is governed by the difference in their densities from the cytoplasmic matrix surrounding them. In a centrifugal field, the consistency of the cytoplasm can be determined from the behavior of its protoplasmic constituents (e.g., nuclei, particles, and oil droplets), or by minute bodies introduced into the cell from outside (e.g., microinjected oil droplets and metal particles) (Heilbrunn, 1958).

When a cell suspended in a medium with a graded density is centrifuged, the cell moves to a position where the density of the ambient medium is the same as the mean density of the cell, and heavy components in the cell move in the centrifugal direction while light components move in the centripetal direction. In consequence, the cytoplasm is stratified. In cells without rigid envelopes, such as cell walls, the cell is stretched by the centrifugal movement of the heavy components and the centripetal movement of the light components. Finally, it is broken into heavy and light halves if the centrifugal force is sufficiently large. When a cell is centrifuged in a medium with a density different from that of the cell, while it is supported in position with an appropriate plane (e.g., the surface of a rigid plate or an interface between two kinds of liquid), the cell is compressed by the centrifugal force, if the force is sufficiently large. Thus, a cell can be either stretched or compressed by centrifugation. If a cell can be observed during centrifugation, it is possible to determine its mechanical properties from the relationship between its deformation and the centrifugal force applied to it (E. N. Harvey, 1954; Hiramoto, 1970, 1987). Plant cells are scarcely deformed by centrifugation because of their rigid cell walls, while the protoplasm is stratified.

The force behind cell motility can be determined by finding the size of the counterforce that stops the movement. Kamiya and Kuroda (1958) determined the motive force for cytoplasmic streaming in characean cells by applying centrifugal forces in the direction opposite to the streaming, and Kuroda and Kamiya (1989) determined the motive force for *Paramecium* locomotion by applying centrifugal forces in the direction opposite to the locomotion of the organism.

In these investigations, direct observation and measurement of the cell during application of the centrifugal force are required. The centrifuge microscope (CM), is an instrument that meets this requirement. This instrument is useful in studying the mechanisms of cell motility because quantitative information on the motive force and physical properties of the cell are of primary importance in such studies.

The ability to stratify protoplasmic components and break the cell without damaging it is useful in studying biological functions of cell components (e.g., the nucleus), because it is possible to obtain living cell fragments with different components (e.g., nucleated and enucleated fragments). The CM is a powerful instrument in this branch of cell biology.

II. Design and Construction of Centrifuge Microscope

In using a CM, it is necessary to keep the microscope image of the cell fixed, while the cell is moving at the speed of the rotating centrifuge. Several optical and mechanical systems were designed since the first report of a CM by E. N. Harvey and Loomis (1930). Centrifuge microscopes are classified into five types: HL type (E. N. Harvey and Loomis, 1930), B type (Beams, 1937; Brown, 1940), KK type (Kuroda and Kamiya, 1981, 1989, Kuroda *et al.*, 1986), K type (Kamitsubo *et al.*, 1988, 1989), and the H type designed by one of the authors of this chapter.

A. HL-Type Centrifuge Microscope

E. N. Harvey and Loomis (1930) designed an instrument with which microscopic objects could be observed while under the influence of a centrifugal force (the HL-type CM). Figure 1 shows their original design. A microscope objective lens (OB) is built into one end of a metal bar (B), which is mounted as a head on the centrifuge axis. The objective is reversed from the ordinary position on a microscope so that the light (L), the focusing screw (V), and the centrifuge chamber (CC) containing living cells will be easily accessible on top of the centrifuge.

After passing the objective lens, the light is reflected by two total reflecting prisms (P1 and P2) and passes along the axis of the centrifuge upward. The ocular lens (OC) is stationary and mounted above the axis of the centrifuge. The counterweight (W) is used for balancing. The centrifuge chamber is laid on a flat metal strip with holes for the passage of light, and its position in the direction of centrifugal acceleration can be adjusted by screws (H and V) in the direction of the centrifugal force and in the direction parallel to the centrifuge axis, respectively. Additional focusing during centrifugation is obtained by changing the vertical position of the ocular lens. The cells in the centrifuge chamber are illuminated by a mercury vapor lamp with a 2000- to 3000-V condenser discharge that is synchronized with the passage of the cells under the lamp.

With this CM, magnified images of cells under 450- to 1000-g centrifugal acceleration could be observed through a $62 \times$ objective lens. HL-type CMs were placed on the market by Bausch & Lomb Optical Co. (Rochester, NY) and H. Struers Chemiske Laboratorium (Copenhagen, Denmark) with some modifications of the original E. N. Harvey and Loomis design (1930). Figure 1 shows a photograph of a Bausch & Lomb CM in which the mercury vapor lamp is replaced with an ordinary tungsten lamp.

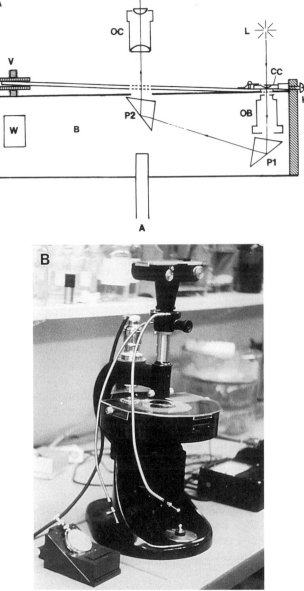

FIG. 1 HL-type CM. (a) Original design of E. N. Harvey and Loomis CM. A, Axis of the centrifuge; B, metal bar that revolves on the centrifuge axis, carrying the centrifuge chamber, objective lens, prisms, etc.; CC, centrifuge chamber; H, screw for adjusting horizontal position of CC; L, light source; OB, objective lens; OC, ocular lens; P1 and P2, total reflecting prisms; V, screw adjusting vertical position of CC; W, counterweight. (Modified from Fig. 1 of E. N. Harvey and Loomis, 1930). (b) HL-type CM made by Bausch & Lomb Optical Co., Rochester, NY. 1930.)

E. N. Harvey (1933b) described a CM for simultaneous observation of control and experimental material.

E. N. Harvey (1932, 1934) adapted the principle of the HL-type CM to an air turbine ultracentrifuge and succeeded in taking clear photographs of eggs centrifuged at 84,000 g, although the magnification was low because an objective lens with a long working distance was placed above the axis of the centrifuge in place of the objective lens that was set below the centrifuge chamber in the original HL-type design.

B. B-Type Centrifuge Microscope

Figure 2 shows a diagram of the B-type CM with an air turbine motor (Beams, 1937). The rotor (R) is a small, cone-shaped structure. A centrifuge chamber (CC) is attached at the top of the rotor. The lower cone-shaped part of the motor, which is grooved with a series of flutings, is

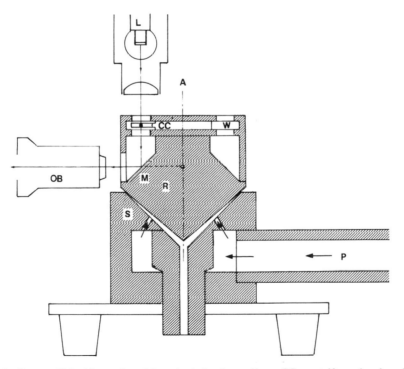

FIG. 2   B-type CM with an air turbine. A, Axis of centrifuge; CC, centrifuge chamber; L, light source; M, mirror; OB, objective lens; P, pipe connected to compressed air source; R, rotor; S, strator; W, counterweight. (Modified from Fig. 3 of Beams, 1937.)

mounted on a cup-shaped stator (S). Both the lower part of the rotor and the cup-shaped stator are of conical, but slightly different, angles so that when it is running the rotor cone touches the stator cup only at its top.

The rotor is started by applying compressed air through a pipe (P) connected to the stator. The air passes through eight diagonal holes into the stator cup, where it impinges on the flutings of the rotor. This flow causes the rotor to turn and then escapes between the surface of the cone and the stator cup. During rotation, the rotor floats on a cushion of air above the stator cup but does not leave the stator because of the reduced pressure in the space between the rotor and the vertex of the stator (by Bernoulli's principle). This air turbine motor is suitable for generating high centrifugal accelerations (10,000–200,000 g), because there is no mechanical contact between the rotor and the stator.

The optical system of the B-type CM is different from that of the HL-type CM. A mirror is placed at a 45° angle to the centrifuge chamber on the rotor so that the image of the specimen in the CC might be formed on the axis of the rotor. The light source is placed directly over the specimen. When the rotor is running fast, an almost motionless image of the specimen being centrifuged can be seen through the objective lens (OB) of the microscope. In this type of CM, large centrifugal accelerations can be generated, but a high-magnification (and therefore, high resolving power) objective lens cannot be used because the working distance of the objective lens cannot be larger than the effective radius of the rotor, that is, the distance of the object from the center of rotation as shown in Fig. 2. A similar CM was constructed by Brown (1940) to determine the cytoplasmic viscosity of *Paramecium*.

Some investigators (Hayashi, 1957, 1960, 1964; Kamitsubo, 1966a, 1972; Hiramoto and Yoneda, 1986) made CMs with the same optical system (B type) but with an electric motor. Figure 3 shows a B-type CM made by Kamitsubo. These can be made at a low cost by combining a handmade rotor, an off-the-shelf motor, and a microscope with a long-working-distance objective lens. They are suitable for preparing microscopic specimens affected by centrifugation at a desired degree, because the centrifugation can be stopped whenever the desired effect is recognized in the microscope.

C. KK-Type Centrifuge Microscope

Figure 4 shows the optical system of a KK-type CM (Kuroda and Kamiya, 1981, 1989; Kuroda *et al.*, 1986). The specimen in the centrifuge chamber (CC) set in the rotor (R) is continuously illuminated by a light source (L) placed on the axis of rotation, after being reflected by two total reflecting

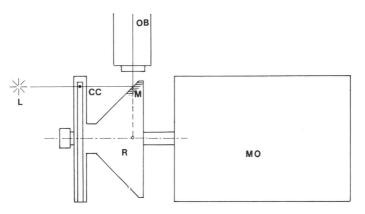

FIG. 3   B-type CM with an electric motor. CC, Centrifuge chamber; L, light source; M, mirror; MO, motor; OB, objective lens of microscope; R, rotor.

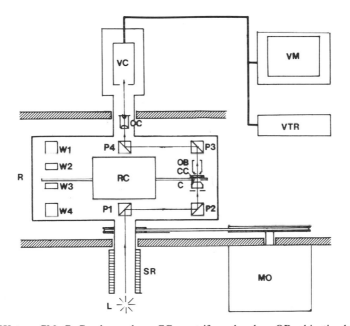

FIG. 4   KK-type CM. C, Condenser lens; CC, centrifuge chamber; OB, objective lens; OC, ocular lens; L, light source; MO, motor; P1, P2, P3, and P4, total reflecting prisms; R, rotor; RC, remote-control mechanism to move centrifuge chamber vertically and radially from the centrifuge axis; SR, slip rings leading electric current to remote-control mechanism; VC, video camera; VM, video monitor; VTR, videotape recorder; W1, W2, W3, and W4, counter-weights. (Modified from Kuroda *et al.,* 1986.)

prisms (P1 and P2) and passing through a condenser lens (C) set in the rotor. On the same principle as that of HL-type CM, the image of the specimen is formed at the axis of the rotor, where a charge-coupled device (CCD) camera is set (VC), after magnification with the objective lens (OB) and projection (ocular) lens (OC), and reflected by prisms (P3 and P4) set in the light path. The CCD camera moves with the rotor around its axis so that the camera can follow the rotation of the microscopic image formed on the image plane of the camera. In this way, the microscopic image of the object is continuously recorded with the camera. Therefore, the flickering light and insufficient brightness of an image that is inevitable in other types of CMs are completely eliminated, even at very low centrifugal speeds. The CC can be moved vertically and horizontally during centrifugation with a remote-control mechanism (RC). Ten- and $20\times$ objective lenses can be used with a $10\times$ ocular lens. The maximal centrifugal acceleration produced by this type of CM is about 2600 g. Kamiya made three CMs of this type that are slightly different from one another (cf. Kuroda and Kamiya, 1981, 1989; Kuroda et al., 1986) in cooperation with Sanki Engineering Co. (Kyoto, Japan).

D. K-Type Centrifuge Microscope

Kamitsubo, with collaborators, made a novel CM (K-type CM; Kamitsubo et al., 1988, 1989) in which a moving specimen on the rotor is directly observed with a microscope by illuminating the specimen with a flashing light triggered synchronously with the passage of the specimen across the optical axis of the microscope (Fig. 5). In this type, a special xenon flashbulb with a short flash time (Nano Pulse Light; Sugawara Kenkyusho, Tokyo, Japan) is used. Because of the short flash time (170 nsec), the movement of the specimen image during illumination is minimized. A signal generated by an electric circuit with a photocoupled switch (PS) set close to the edge of the rotor (R) triggers the flash. The timing of the flash can be adjusted by moving the position of the switch along the edge of the rotor. This makes it possible to scan the microscopic field in the direction in which the centrifuge rotor moves the specimen.

The specimen can be observed during centrifugation through the video monitor (VM). A record is made with a video system consisting of a video camera (VC), image processor (IP), video monitor, and videotape recorder (VTR). Noise bars caused by differences in the flash rate and the scanning rate of the video system are erased by an electric differentiation circuit of the image signal set in the video camera (C-2847; Hamamatsu Photonics, Hamamatsu, Japan). The microscope is mounted on a stage that can be moved in the direction of centrifugal acceleration with a micrometer screw.

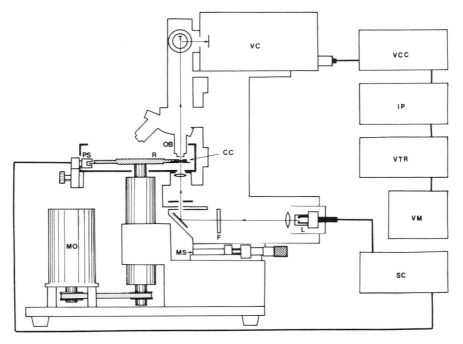

FIG. 5   K-type CM. CC, Centrifuge chamber; F, filter; IP, image processor; L, light source (Nano Pulse Light; Sugawara); MO, motor; MS, mechanical stage; OB, objective lens; PS, photocoupled switch; SC, stroboscope controller; R, rotor; T, telescope for direct observation; VC, video camera; VCC, video camera controller; VM, video monitor; VTR, videotape recorder. (Modified from Kamitsubo *et al.*, 1989, and Kaneda *et al.*, 1990.)

Thus, the microscopic field can be scanned by turning the micrometer screw and adjusting the position of the timing switch. The microscope is equipped with objective lenses up to $60 \times$. The maximum centrifugal acceleration of this CM is 1500 g. Kamitsubo made two CMs of this type—K type I (Kamitsubo *et al.*, 1988; Kaneda *et al.*, 1990) and K type II (improved K type I, Kamitsubo *et al.*, 1989)—in cooperation with Nikon Engineering Co. (Yokohama, Japan).

E. H-Type Centrifuge Microscope

Hiramoto designed a CM (H type) consisting of a centrifuge with a special rotor for setting a centrifuge chamber for observation, a Nano Pulse Light with a triggering system similar to that of the K-type CM for illumination, and an inverted microscope (TMD; Nikon, Tokyo, Japan) from which the stage had been removed. A CM following this design was made by Nikon Engineering Co. (Yokohama, Japan) in 1992.

Figure 6 shows a photograph of the H-type CM. The centrifuge rotor is driven by a motor through belts and pulleys at a speed of 0–60 rps, corresponding to up to 1000 g. Two thousand to 3000 g acceleration may be achieved by changing the motor. The position of the rotor can be adjusted in a horizontal plane with an $X-Y$ stage so that the optical axis of the inverted microscope can pass through the specimen to be observed.

Two light sources, a xenon flash lamp (Nano Pulse Light; Sugawara Kenkyusho) and a halogen lamp attached to the microscope body, can be exchanged by using a servomotor to move a mirror inserted in the light path. The halogen lamp is used to observe the specimen in order to find an object suitable for the experiment and then to bring it to the center of the microscopic field. Because a commercially available Nikon inverted microscope (TMD; Nikon) is used without altering the optical system, all the accessories for it (e.g., differential interference contrast equipment, phase-contrast equipment, and fluorescence microscope equipment) can be used. Therefore, it is possible to choose the optical system most suitable for particular observations. A single switch button starts centrifugation and changes the light source from the halogen lamp to the xenon flash lamp.

During centrifugation the specimen can be observed either through the binocular eyepiece of the microscope or by a video monitor. The

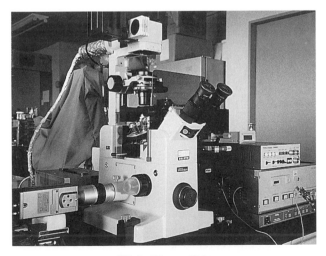

FIG. 6   H-type CM.

microscope image is recorded with a video system consisting of a CCD camera, an image processor, a video monitor, and a videotape recorder. The monitor displays the rotation speed of the centrifuge rotor, which is recorded by a tachometer. The intensity of the flashing light is sufficient to record differential interference contrast images with a 40× objective lens in real time (cf. Fig. 7).

In the H-type (and K-type) CM, the displacement of the specimen during a single flash illumination is about 4.5 μm when the rotation speed of the centrifuge is 60 rps (3600 rpm), corresponding to 1000 g. This is fairly large compared with the resolving limit of the microscope, but the resolving power in the direction of centrifugal acceleration is high. At lower centrifugal accelerations, the displacment of the image during a single flash illumination becomes smaller.

III. Brief History of the Centrifuge Microscope in Cell Biology Studies (1930–1980)

A. Behavior of Cells and Cell Components in a Centrifugal Field

E. N. Harvey (1931a) observed living *Amoeba, Paramecium,* and *Stentor;* and granule movement in unfertilized eggs of *Arbacia, Cumingia,* and *Chaetopterus.* At 1200 g, the amebas moved normally, adhering to the slide. The crystals in the cell were forced to the bottom but were redistributed rapidly by streaming movement after centrifuging. As the centrifugal

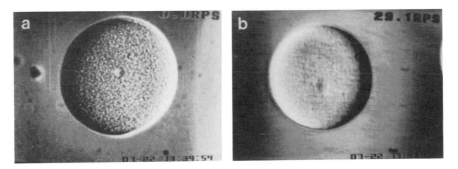

FIG. 7   Images recorded with an H-type CM. Differential interference contrast image of an unfertilized egg of the heart urchin *Clypeaster japonicus,* shown by the video monitor. (A) An egg under no centrifugal force. (B) Another egg during centrifugation at 29.1 rps. The direction of centrifugal acceleration is from left to right.

force was increased, the amebas suddenly let go of the substratum and rounded up at the bottom of the centrifuge chamber. Paramecia were forced to the bottom, forming a tightly packed mass. Individuals in the upper part of the mass swam back and forth by ciliary movement. Crystals and the nucleus were also forced down, but the contractile vacuoles remained in position. *Stentor* was forced down in such a position that the ciliated end was toward the axis of the centrifuge. In high centrifuge fields, the ciliated end was pinched off and the cilia kept beating.

   Arbacia and *Cumingia* eggs were stratified by centrifugation. The granules moved in a centrifugal direction and oil droplets in a centripetal direction. *Chaetopterus* eggs were pulled apart into fragments by the centripetal movement of a spherule formed from oil at one end of the cell. The spherule broke away, and the egg with a remaining stalk slowly rounded up. The granules in these cells did not always have a uniform velocity but moved and stopped, or moved slowly, then faster, then slowly, then faster, etc. Similar discontinuous movements of cytoplasmic inclusions in centrifugal fields were observed by E. N. Harvey and Marsland (1932) and Allen (1960) in amebas, which suggests that there are invisible structures in the cytoplasm.

   E. N. Harvey (1934) reported effects of high centrifugal accelerations (about 10^5 g) on some marine eggs, using an air turbine CM.

B. E. B. Harvey's Observations of Fertilized Sea
   Urchin Eggs

E. B. Harvey (1933) reported the results of observations of fertilized sea urchin (*Arbacia punctulata*) eggs with the HL-type CM. The cytoplasm was stratified by the centrifugal force, and the eggs elongated in a dumbbell fashion and finally broke into halves in the same way as unfertilized eggs, but the modes of stratification, elongation, and breaking were different from those in unfertilized eggs, and changed as development proceeded.

   Using fertilized eggs of *Parechinus microtuberculatus,* which are larger and clearer than *Arbacia* eggs, E. B. Harvey (1935) found that the mitotic apparatus is displaced either centripetally or centrifugally by centrifugation, depending on the batches of eggs, and that the cleavage furrow may be formed in either the plane correlated with the original position of the mitotic figure ("primary" plane) or that correlated with its final position ("induced" plane), or in both of these planes. From these facts E. B. Harvey concluded that cleavage may take place without relation to the final position of the asters. This conclusion was later confirmed by experiments on cell division without a mitotic apparatus (Swann and Mitchison,

1953; Hiramoto, 1956, 1965) and by experiments in which the mitotic apparatus was moved by micromanipulation (Rappaport, 1971, 1986). E. B. Harvey (1960) was able to demonstrate cell division while the nucleus was still intact.

On the basis of CM observations of fertilized sea urchin eggs and of eggs and egg fragments after centrifugation, E. B. Harvey (1934, 1935, 1938, 1940, 1946; reviewed by E. B. Harvey, 1951) carried out a series of analyses of the role of the nucleus in the development of sea urchins in fertilized and parthenogenically activated eggs and egg fragments (parthenogenic merogones, haplones, diplones, etc.). The results of these studies contributed greatly to our understanding of the role of the nucleus in development.

C. Mechanical Properties of the Cell

1. Tension at the Surface

Using an HL-type CM, E. N. Harvey determined the tension at the cell surface in various kinds of cells (Harvey and Danielli, 1938; E. N. Harvey, 1954; Hiramoto, 1970). In unfertilized *Arbacia* eggs, the protoplasm is stratified and the cell takes on a dumbbell shape and finally breaks into two halves (a heavy and light half) (E. N. Harvey, 1931c). The tension at the surface of the cell was calculated by assuming that the centrifugal and centripetal forces (calculated from the volumes of heavy and light halves, the difference in their specific gravities and that of the medium, and the magnitude of centrifugal acceleration) all balanced with the tension at the surface multiplied by the circumference of the cell at the point of the break.

In *Chaetopterus* eggs, oil droplets were accumulated at the centripetal end of the cell and formed a drop that finally separated from the cell, breaking the cell surface. The tension at the cell surface was calculated from the force applied to the oil drop, that is, it was the product of the volume of the drop and the difference in the density between the oil and the protoplasm, divided by the circumference at the amputated part of the cell (E. N. Harvey, 1931b). E. N. Harvey and Marsland (1932) determined the tension at the surface of amebas from the centrifugal force required to pull out a microinjected oil drop. In these experiments, values of 1 dyn/cm or smaller were obtained, which is relatively small compared with surface tensions at water–air, oil–air, and oil–water interfaces. With this and other methods, values of a similar order were obtained in many kinds of "naked" cells that are without definite membranes, for example, sea urchin eggs, molluscan eggs, and amphibian eggs (Cole, 1932; Cole and Michaelis, 1932; E. N. Harvey, 1933a; E. N. Harvey and Fankhauser, 1933; Hiramoto, 1963, 1967; Yoneda, 1964).

E. B. Harvey (1943) found that *Arbacia* eggs, when centrifuged with a force of 10,000 *g*, broke less readily in hypotonic sea water and more readily in hypertonic sea water than in normal sea water, suggesting that the tension at the cell surface increases when the surface is stretched and decreases when the surface shrinks.

With an HL-type CM, Hiramoto (1967) determined the tension at the surface of unfertilized sea urchin eggs from the degree of deformation of the cell during centrifugation, using a theoretical relation connecting the size of the cell, the density difference between the cell and the ambient medium, the magnitude of centrifugal acceleration, and the tension at the surface. Hiramoto (1967) found that the tension at the surface increased as the deformation increased, and that the deformation of the cell gradually increased when a constant centrifugal force was continuously applied to unfertilized sea urchin (*Arbacia* and *Lytechinus*) eggs. These facts indicate viscoelasticity of the cell.

Hiramoto (1967) also investigated the change in the stiffness of the cell during early development in *Arbacia* and *Lytechinus* eggs by using a CM to measure the degree of deformation of the fertilized egg under a constant centrifugal acceleration, and found that the stiffness peak exists before and during the first cleavage. This result is different from previous results on the mechanical properties of fertilized sea urchin eggs, in which only one stiffness peak was seen at the beginning of cleavage (Cole and Michaelis, 1932; Danielli, 1952). Later it was shown that the difference in the results was due to the difference in species of sea urchin used (cf. Hiramoto, 1981).

E. N. Harvey and Shapiro (1934) and E. N. Harvey and Schopfle (1939) determined the interfacial tension between the oil drop and the protoplasm from the degree of deformation of the drop in the cell when the cell is centrifuged, and obtained values of the interfacial tension similar to those of the tension at the cell surface in the "naked" cells mentioned earlier. This result was one of the important supports for the Davison-Danielli model of the cell membrane (cf. E. N. Harvey and Danielli, 1938).

2. Consistency of the Protoplasm

E. N. Harvey (1931a) determined protoplasmic viscosity in *Cumingia* eggs by using a CM to directly measure the speeds of granules in the protoplasm, using Stokes's equation. The values obtained (0.068 and 0.06 poise) were similar to those determined by Heilbrunn (1926) from the movement of granules across one-half of the egg diameter in the same materials, using an ordinary centrifuge. In *Chaetopterus* eggs, E. N. Harvey (1931a) obtained larger values (0.68, 1.62, and 1.54 poise) of the protoplasmic viscosity from the movement of the boundary of yolk granules in the protoplasm and the movement of individual yolk granules as determined with a CM.

Brown (1940) determined the protoplasmic viscosity of paramecia by measuring, with a B-type CM, the speed of a food vacuole containing iron particles; a value of 0.5 poise was obtained.

Hiramoto (1967) used an HL-type CM to determine the viscosity of the protoplasm in unfertilized sea urchin (*Lytechinus*) eggs from the speed of the nucleus in the centrifugal field, and obtained about 30 poise, a value much larger than the values of protoplasmic viscosity in sea urchin eggs obtained by previous investigators. The differences in the results may be due to the presence of network structures that support the nucleus in the egg cytoplasm.

D. Motive Force of Cytoplasmic Streaming

With a handmade B-type CM, Hayashi (1957) observed the effect of centrifugal force on cytoplasmic streaming in *Chara* internodal cells. The streaming was accelerated when a centrifugal force was applied in the direction of the streaming and was retarded when the force was applied in the opposite direction. The streaming could be brought to a standstill by applying a centrifugal force of a certain magnitude in the direction opposite to the streaming. Hayashi found that a 40 g centrifugal acceleration was necessary to stop the streaming. However, careful observation revealed that minute particles were still moving very near the cortical gel even when the bulk of the endoplasm stopped streaming. It was necessary to apply 200 g to balance this streaming near the cortical gel.

Using an HL-type CM, Kamiya and Kuroda (1958) determined the motive force of cytoplasmic streaming in *Nitella*. They assumed that the motive force located at the sol–gel interface of the cell (Kamiya and Kuroda, 1956) balances with the centrifugal forces applied to the endoplasm and the vacuole when the cytoplasmic streaming is stopped by centrifugal acceleration in the direction opposite to the cytoplasmic streaming. They used the equation

$$F = Ad(D_p - D_s)\alpha \qquad (1)$$

where F is the motive force generated at the sol–gel interface, α is the centrifugal acceleration required to stop the endoplasm flowing in the centripetal direction, d is the thickness of the endoplasm when it is brought to a standstill by centrifugal force, A is the area of the interface between the cortex and the endoplasm where active shearing takes place, and D_p and D_s are the density of the endoplasm and of the vacuolar sap, respectively. Irrespective of the thickness of the endoplasm, about 1.6 dyn/cm^2 of sol–gel interface was obtained for the motive force.

Using the same principle, Hayashi (1960) determined the motive force of cytoplasmic streaming of characean internodal cells at various temperatures, using a handmade B-type CM. The motive force was almost constant in the range of 5–20°C, and decreased slightly as the temperature increased.

Hayashi (1964) examined the effect of destruction of the cortical gel layer by centrifugation with a B-type CM. At accelerations from 400 to 1000 g, chloroplasts at the cortical gel layer were partially torn off the cortex and were accompanied by a substantial part of the cortical gel. In those cells whose cortical gel layer had been partially torn off, cytoplasmic streaming recovered normally after centrifugation at the area where chloroplasts normally remained, but the same cytoplasm immediately lost the ability to stream when it reached the cortical layer from which chloroplasts had been removed. Within several hours or several days after the partial destruction of the cortical gel layer by centrifugation, chloroplasts returned to and settled down at the injured area. Along with the restitution of chloroplasts, cytoplasmic streaming recovered at that area. These facts suggest that the cortical gel layer is of fundamental importance in the generation of cytoplasmic streaming.

E. Motile Phenomena in Centrifuged Characean Cells

1. Observation of Motile Protoplasmic Fibrils

The motile protoplasmic fibrils described by Jarosch (1956) and Kuroda (1964) in endoplasmic drops isolated *in vitro* from cells of Characeae seem to have a close bearing on cytoplasmic streaming. However, it is extremely difficult to observe or detect the fibrils *in vivo* because chloroplasts affixed densely to the cortical layer of the cell seriously disturb observation of the endoplasm. Furthermore, incessant active streaming of the endoplasm in this material does not permit precise and continuous observation of activity in the endoplasm.

To overcome these difficulties, Kamitsubo (1966a) centrifuged the cells with a handmade B-type CM, so that 10–30% of the chloroplasts were detached from the cell cortex. The merits of this procedure are, first, that it improves viewing conditions and, second, that it stops streaming of the endoplasm at the loci from which chloroplasts are detached (Hayashi,

1957). Use of the CM had a decided advantage for this kind of experiment, because it was possible to stop centrifugation at the time when an appropriate chloroplast-detached area (clear area) had formed in the cell cortex. The appropriate strength and duration of centrifugation for the above purposes were 600–1000 g for 3 min. The clear area was nearly transparent, and details of cytoplasmic movement in this area were observable with a high-magnification oil immersion, phase-contrast objective lens (Kamitsubo, 1966a,b, 1972).

Motile fibrils similar to those described in endoplasmic drops isolated *in vitro* from characean cells were also found *in vivo* in the endoplasm of centrifuged cells. The intracellular, freely moving fibrils appeared mostly in the form of polygons that rotated or undulated in the quiescent endoplasm at the clear area. The modes of movement of the fibrillar loops were classified into two categories: (1) undulatory and (2) rotatory. Generally the fibrils belonging to the undulatory type were associated with many cytoplasmic particles whereas the rotatory type were almost free of particles. The width of the naked fibril was estimated to be 0.2 μm or less. The speed of the rotatory fibrils was similar to the rate of normal cytoplasmic steaming (70–80 μm/sec at 23°C). However, the freely moving fibrils were not considered to be responsible for the cytoplasmic streaming in intact characean cells because they could not generate any organized mass streaming of endoplasm in the centrifuged cells.

Thin (ca. 0.2 μm), linear, stationary fibrillar structures were found in the rhizoid and etiolated and intact internodal cells of *Nitella flexilis,* and in normal internodal cells of both *Lamprothamnium succinctum* and *Chara australis*. The fibrils (subcortical fibrils) were also observed with electron microscopy by Nagai and Rebhum (1966). Later, the subcortical fibrils were identified as bundles of F-actin (Palevitz *et al.,* 1974; Williamson, 1974; Palevitz and Hepler, 1975).

2. Rotation of Cytoplasmic Protrusions in Centrifuged Characean Internodal Cells

Hemispherical cytoplasmic protrusions have been occasionally observed (Kamitsubo, 1979) in the centrifuged internodal cells of *Nitella*. The cytoplasmic protrusions were located in the cortical gel layer where chloroplasts had previously existed. The protrusion had a base 150–300 μm in diameter and protruded into vacuoles of 70–200 μm. The most remarkable characteristic of the protrusion was that the cytoplasmic mass rotated vigorously around the axis of the protrusion perpendicular to the cortex of the cell. This rotation was quite different from the normal rotational cytoplasmic streaming in intact internodal cells of *Netella* and also from any other patterns of cytoplasmic streaming in various plant cells.

The dynamic characteristics of the rotation of the cytoplasmic protrusion were studied cinematographically by means of "optical sectioning" (Allen *et al.,* 1969) with Nomarski optics (Kamitsubo, 1979). No velocity gradient was found within the rotating protrusion. The protrusion rotated as a whole with an angular velocity of ca. 40°/sec, sliding at its base against the cortex of the cell at a point where chloroplasts had been detached by centrifugation.

At the base of the protrusion, a homocentric arrangement of protoplasmic fibrils, identical with the subcortical fibrils in intact cells, was observed that seems to be the structure generating the motive force for rotation of the protrusion. The mechanism of rotation may be identical to that of normal cytoplasmic streaming in an intact *Nitella* cell, although the mode of movement is different.

IV. Centrifuge Microscopy Studies of Cell Motility

A. Behavior of Cells and Organelles under Centrifugation

1. *Paramecium* Behavior under Centrifugation

Kuroda *et al.* (1986) examined the behavior of *Paramecium* and *Tetrahymena* with a KK-type CM. Cells were suspended in a Ficoll–phosphate buffer density gradient medium. With a centrifugal acceleration of 300–400 *g,* paramecia moved to their longitudinal axis in parallel with the direction of the acceleration, pointing either upward (centripetal rotation) or downward (centrifugal). The paramecia pointing upward swam to an increasingly lighter area until they reached a level where they could no longer maintain their position and finally sank rapidly with some gyration to the level from which they had started swimming upward; then they started the upward swimming again. This process was repeated many times during centrifugation. The cycle was about 20 sec, with the amplitude of the rising and falling movement being 300–600 μm. The paramecia pointing downward displayed the opposite behavior. They dived downward into an increasingly heavy medium until they had to yield to their buoyancy and floated up again. This process was also repeated many times. The organisms pointing upward were usually found in an upper level and those pointing downward in a lower level in the graded density medium.

Kuroda *et al.* (1986) explained that this behavior of protozoa under centrifugal acceleration has a physical cause. In ellipsoidal or ovoidal cells, shifting of the center of gravity from the center of buoyancy gives rise to a torque that turns the longitudinal axis of the cell. Once a small shift occurs in one direction, this tendency is amplified by the centrifugal movement of heavy components and the centripetal movement of light components in the cell oriented in the direction of the acceleration. Thus, the organisms swim upward or downward against the negative or positive density gradient, respectively.

When *Tetrahymena* cells were subjected to the same experiment, they separated into two distinct layers of upward and downward-oriented organisms. This separation may be brought about by the active efforts of the organisms to move forward and not by density differences.

2. Light-Dependent Changes in Mechanical Properties of the Cytoplasmic Matrix in *Vallisneria*

Takagi *et al.* (1991) examined the effects of centrifugal acceleration on chloroplast behavior in epidermal cells of the aquatic monocot, *Vallisneria gigantea,* using a K-type CM II. The chloroplasts in the cytoplasmic layer that faces the outer periclinal wall (P side) were passively moved by a centrifugal acceleration of 25–40 g, and the movement was accelerated by irradiation with blue light (451 nm) at a high intensity (6.7 W/m^2). A similar acceleration was caused by applying ethylene glycol-bis (β-aminoethyl ether)-N,N,N',N'-tetraacetic acid (EGTA) in the dark to cells that had been treated with cytochalasin B. In cells preirradiated with red light (650 nm) at a low intensity (0.4 W/m^2) for 3–4 hr, the minimal centrifugal acceleration to induce the passive movement of chloroplasts increased to more than 50 g. In cells that had been preirradiated with red light in the presence of cytochalasin B, only the chloroplasts in the peripheral area of the P side could be moved by centrifugation after application of EGTA. These results suggest that some mechanical properties of the cytoplasmic matrix are affected by irradiation with light of different wavelengths at different intensities.

Takagi *et al.* (1992) also examined the effects of light irradiation on passive movement by centrifuging chloroplasts in the mesophyll cells of the same plant. Whereas passive movement was accelerated by irradiation with red light, blue light was ineffective. The effect of red light was negated by far-red light (746 nm). Red light was completely effective in accelerating passive movement even in the presence of 3- (3,4-dichlorophenyl)-1,1-dimethylurea, an inhibitor of photosynthesis. These results may indicate that light of different wavelengths affects some mechanical properties of the cytoplasmic matrix in different ways.

3. Movements of Cell Organelles in *Spirogyra* under Centrifugal Acceleration

Kuroda and Kamiya (1991) used a KK-type CM to observe the mode of dislocation of cell organelles in *Spirogyra* under centrifugal acceleration. At 40 g, chloroplasts, the nucleus, and most of the cytoplasm were shifted toward the centrifugal end of the cell and packed there in the course of a few minutes. The formation of chloroplast clusters surrounding the nucleus, which could not be observed owing to obstruction by chloroplasts during centrifugation, was considered to be in good conformity with the morphological data, that is, the existence of a nucleus-sustaining scaffold structure (Grolig, 1990) and the netlike cage of F-actin bundles surrounding the nucleus (Goto and Ueda, 1988).

When the nucleus, chloroplasts, and most of cytoplasm were packed at the centrifugal end of the cell, a large clear area that was more than 80% of total cell length appeared at the centripetal end of the cell, where only a thin cytoplasmic layer was left beneath the plasma membrane. In this area, active but complicated movement of tubular organelles was observed by Nomarski optics. Stained with 3,3' -dihexyloxacarbocyanine iodide, an endoplasmic reticulum (ER)-specific fluorescent dye, the organelles were shown to be ER. Bundles of F-actin running more or less longitudinally or transversely to the cell axis were also observed by fluorescent microscopy at the clear area when stained with rhodamine-conjugated phalloidin.

B. Physical Properties of the Cytoplasm

1. Apparent Viscosity of Endoplasm of Characean Internodal Cells

Kamitsubo et al. (1988) measured the viscosity of the endoplasm in internodal cells of *Nitella* with a K-type CM I. Prior to measurement of the viscosity, a sufficient amount of the endoplasm was collected at one end of the cell by using a B-type CM to apply centrifugal acceleration in the direction of the longitudinal axis of the cell. (Hayashi, 1957; Kamitsubo, 1966a, 1972). During this treatment, chloroplasts were detached from the cell cortex in the centrifugal part of the cell, resulting in an improvement in optical conditions for observing the movement of oil droplets through the accumulated endoplasm, while chloroplasts remained *in situ* in the centripetal part of the cell because the magnitude of centrifugal acceleration was smaller than that in the centrifugal part of the cell. The appropriate strength and duration of centrifugation for the above purpose were about 1000 g for about 5 min. Figure 8 schematically shows part of an internodal

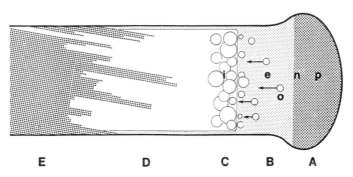

FIG. 8 Schematic view of a part of the centrifuged internodal cell of *Nitella*. The long arrow indicates the direction of centrifugal acceleration. At the centrifugal end of the cell (A), chloroplasts (p) detached from the cell cortex, nuclei (n), and endoplasm (e) were accumulated and stratified. Cytoplasmic streaming does not take place for several hours in the chloroplast-detached (subcortical fibril-free) area (B, C, and D). The viscosity of the endoplasm was determined by measuring the movement of oil droplets (o) in the accumulated endoplasm. i, Cell sap inclusion accumulated at the centrifugal end of the central vacuole; D (surface view), clear area where chloroplasts together with subcortical fibrils were detached. (Redrawn after Kamitsubo *et al.*, 1988.)

cell after the treatment. At the centrifugal end of the cell, chloroplasts (p) detached from the cell cortex, nuclei (n), and endoplasm (e) were accumulated and stratified. Cell sap inclusions (i) were accumulated at the centrifugal end of the central vacuole.

The internodal cells prepared as described above were observed with a K-type CM I. As a result of centifugation, oil droplets (ca. 10 μm in diameter, *o* in Fig. 8) left, one after the other, the accumulated chloroplasts and nuclei and moved in the centripetal direction through the accumulated endoplasm (Fig. 8). The velocity of the oil droplets was measured at various centrifugal accelerations from 100 to 500 *g*. From the values of the velocities and the diameter of the oil droplets, and the density difference between the oil and the endoplasm, the apparent viscosity of the endoplasm was calculated by Stokes's formula.

Figure 9A shows the relationship between shear stress and the rate of shear of the endoplasm at the surface of the oil droplet in internodes of *Nitella axilliformis*. As shown, the endoplasm exhibits a non-Newtonian character, having a yield value of about 5 dyn/cm^2. Figure 9B shows the relationship between the rate of shear and the apparent viscosity of the endoplasm. The apparent viscosity is high if the shear rate is low, and is low at a high shear rate. The above-described results on the mechanical properties of the endoplasm determined *in vivo* in an internodal cell of

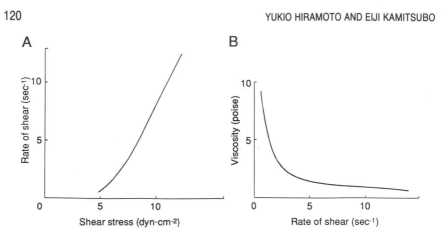

FIG. 9 (A) The relationship between the shear stress and the rate of shear in the endoplasm of *Nitella axilliformis* measured *in vivo* by the centrifuge method. (B) The relationship between the rate of shear and the apparent viscosity of the endoplasm of *N. axilliformis*. (Redrawn after Kamitsubo *et al.*, 1988.)

Nitella coincide well with those of the endoplasm isolated from internodal cells of *Nitella megacarpa* obtained *in vitro* by the agar capillary method (Kamiya and Kuroda, 1965).

2. Effects of Electric Stimulation on Cytoplasmic Streaming in Characean Internodal Cells under Centrifugal Accleration

It is well known that cytoplasmic streaming in characean internodal cells can be halted temporarily and reversibly by electrical stimulus (Tazawa and Kishimoto, 1968) and other physical and chemical treatments (Kamiya, 1981, 1986; Kuroda, 1990; Shimmen, 1992a,b). Streaming stops at the moment of excitation by electrical stimulation as a consequence of a transient increase in the Ca^{2+} concentration in the cytoplasm (Hayama and Tazawa, 1980; Kikuyama and Tazawa, 1982; Williamson and Ashley, 1982).

Using a K-type CM II, Kamitsubo *et al.* (1989) observed the behavior of the endoplasm in *Nitella* internodal cells under centrifugal acceleration before and after application of an electrical stimulus. The cell was electrically stimulated during centrifugation by a wireless method (Kamitsubo *et al.*, 1989).

Under centrifugal acceleration (ca. 50–100 *g*), the direction of cytoplasmic streaming in the internodal cell of *Nitella* is parallel to the orientation of the subcortical fibrils (Kamitsubo, 1966b, 1972) that were identified as bundles of F-actin (actin cables), generating the motive force that drives the endoplasm, as mentioned earlier. The speed of the peripheral endoplasm contiguous to the subcortical fibrils (ca. 80 μm/sec) is neither accelerated nor retarded by moderate (50–100 g) centrifugal acceleration. The endoplasmic flow, however, stops suddenly following electrical stimulus. The endoplasm contiguous to the subcortical fibrils is immobilized transiently at the time the electrical stimulus causes the streaming to stop. This suggests that transitory cross-bridges are formed between the immobilized endoplasm and the subcortical fibrils at the time streaming stops.

This immobilization of the endoplasm was further studied by Kamitsubo and Kikuyama (1992). Using a K-type CM II, they demonstrated that, when subcortical fibrils were previously removed locally from the cell cortex of *Nitella* internodes, the endoplasm passively flowing under centrifugal force did not stop at all on electrical stimulus. By contrast, the peripheral endoplasm contiguous to the subcortical fibrils in the normal cortex that was actively flowing in the centripetal direction promptly stopped, following the stimulus, and was immobilized for several seconds. These results indicate that, on electrical stimulus, the presence of subcortical fibrils is needed to immobilize the endoplasm flowing contiguous to them in a state that resists displacement by centrifugal force. The peripheral flow of endoplasm resumed gradually with nearly the same time course as that under gravity.

In characean internodal cells, the bulk endoplasm flows as a whole in a direction parallel to that of the subcortical fibrils. Its speed was neither accelerated nor retarded by moderate (50–100 g) centrifugal acceleration. Soon after stopping, the bulk endoplasm starts to flow passively in the direction of centrifugal acceleration (Fig. 10). Kamitsubo *et al.* (1989) suggest that the viscosity of the bulk endoplasm significantly decreases at the moment streaming ceases. Immediately after the instantaneous cessation of cytoplasmic streaming, oil droplets in the bulk endoplasm begin to move in the centripetal direction (Fig. 10). This movement supports the above inference that the viscosity of the bulk endoplasm decreases remarkably at the moment streaming stops.

After the internodal cell was centrifuged at about 250 g for about 30 min, most of the endoplasm was accumulated at the centrifugal end of the cell. When the cell thus treated was further centrifuged at about 900 g, the bulk endoplasmic flow became barely discernible, whereas the peripheral flow of endoplasm both in the centrifugal and centripetal directions was still observable. These authors could not measure the peripheral flow rate for endoplasm under a centrifugal acceleration of 900 g because the cytoplasmic particles, which serve as markers for the flow rate, were rare in this cell. However, Kamitsubo *et al.* (1989) could confirm the direction of peripheral flow of enoplasm, which was parallel

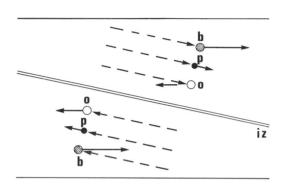

FIG. 10   Schematic illustration of cytoplasmic streaming in a *Nitella* internodal cell under moderate centrifugal acceleration before (broken arrows) and after (solid arrows) application of an electrical stimulus. The length of the arrow indicates relative speed. Thick arrow indicates the direction of centrifugation. o, Oil droplets; b, inclusions flowing in the bulk endoplasm; p, inclusions in the peripheral endoplasm contiguous to subcortical fibrils; iz, indifferentiation zone. For further explanation, see text. (Modified from Kamitsubo *et al.*, 1989.)

to that of the subcortical fibrils. The peripheral flow at 900 *g* stops instantaneously following an electrical stimulus. Complete recovery of the flow takes place in about 10 min under the centrifugal conditions.

Later, E. Kamitsubo and M. Kikuyama (unpublished) succeeded in measuring the peripheral flow rate for the endoplasm in question under high (1400 *g*) centrifugal acceleration. They used internodal cells of *N.axilliformis*, whose endoplasm contained abundant cytoplasmic particles and chloroplasts that were flat and thin enough for observation of the streaming with a 40× objective lens during centrifugation. The speed of peripheral flow was about 50 μm/sec under gravity (21°C). Centrifugal acceleration did not accelerate the peripheral flow of endoplasm in the centrifugal direction nor retard that in the centripetal one; both were in the range of 4–1400 *g*.

3. Stiffness Changes in Artificially Activated Sea Urchin Eggs

Using a KK-type CM, Yoshimoto *et al.* (1986) determined changes in stiffness in sea urchin eggs activated artificially by procaine. The stiffness was determined from the degree of deformation of the egg under constant centrifugal acceleration (Hiramoto, 1967). Following activation with 5–8 m*M* procaine, the stiffness fluctuated rhythmically. It attained three peaks (21 ± 3 min, 69 ± 7 min, and 118.8 min after activation). The second and third peaks almost coincided with two peaks of resistance to detergent lysis (60 and 120 min after activation) in procaine-activated sea urchin eggs reported previously, and with stiffness peaks at the first and the second cleavages in fertilized sea urchin eggs (70–80 and 110–125 min after fertilization, respectively).

4. Mechanical Structure of the Cytoplasm of the Echinoderm Egg

Kaneda *et al.* (1990) developed a method to investigate, using a K-type CM I, the mechanical structure of the cytoplasm of the echinoderm egg, based on the movement of a gold particle in the cytoplasm under centrifugal acceleration. Before experimentation, a spherical gold particle 2–5 μm in diameter was introduced into the oocytes or eggs in echinoderms (starfish and sand dollars). The movement of the particle deviated slightly from the direction of the centrifugal acceleration, and the speed flucutuated during the movement, suggesting the existence of a network structure in the cytoplasm. In fertilized eggs, movement of the gold particle by centrifugal acceleration was impeded by the sperm aster and the cleavage diaster in the cell, showing the high consistency of these structures. The apparent viscosity of the cytoplasm, calculated from the mean speed of the particle asssuming that Stokes's equation for Newtonian fluid can be applied, was on order of 10^2 poise, which changed as development proceeded.

C. Measurement of the Motive Force

1. Propulsive Force of Paramecia

Kuroda and Kamiya (1989) used their KK-type CM to determine the propulsive force generated by a single paramecium by finding the amount of centrifugal acceleration sufficient to stop the organism swimming against the centrifugal force. When paramecia in graded density medium were centrifuged, they oriented themselves with their longitudinal axis parallel to the direction of centrifugal acceleration, turning their anterior ends toward either the centripetal or centrifugal direction (Kuroda *et al.*, 1986). Because they were still swimming, it was possible to calculate their propulsive force from the difference between their density (1.04 g/cm³) and the density of the ambient medium (upper or lower layer) that they could reach, the volume of organism, and the magnitude of centrifugal acceleration. A value of 7×10^{-4} dyn was obtained for the propulsive force generated by a single paramecium. This value corresponds to 7×10^{-8} dyn contributed by single cilium, and to 3.5 dyn/cm² of propulsive force per unit area of the cell.

2. Force–Velocity Relation in the ATP-Dependent Sliding Movement of Myosin-Coated Beads on Actin Cables *in Vitro*

Using a K-type CM II, Oiwa *et al.* (1990) investigated the force–velocity relation in the ATP-dependent actin-myosin sliding system by measuring the movement of myosin-coated beads on actin cables in characean internodal cells *in vitro*. Tosyl-activated polyethylene beads (2.8 μm in diameter) coated with skeletal muscle myosin were placed on well-organized subcortical fibrils (actin cables) of the internodal cell. The movements of the particles along the cables in the presence of ATP were observed with the CM as centrifugal acceleration was applied in the direction in which the particle was moving, or in the opposite direction.

The velocity of the beads along the actin cables was affected to various degrees by the centrifugal force applied in the direction of centrifugal acceleration, depending on the magnitude of the applied force. Under constant centrifugal force opposite to their movement ("positive" loads), the beads continued to move with a constant velocity that decreased with increasing centrifugal force. The steady state force–velocity curve thus obtained was similar to the double-hyperbolic force–velocity curve found in single muscle fibers. When the beads were subjected to increasing centrifugal force in the direction of their movement ("negative" loads), their velocity first decreased by 20–60% and then increased until the beads were detached from the actin cables.

When the ATP was removed from the medium, all the myosin-coated beads attached to the actin cables and did not move, even when positive or negative loads were applied by centrifugation, which suggests that firm, rigid links form between actin cables and myosin.

V. Concluding Remarks

The centrifuge microscope makes it possible to apply force to a specific region in a cell while the cell is being observed under appropriate magnification. Therefore, this instrument is useful for manipulating cells and cell components and measuring cytoplasmic consistency as well as the motive force generated at a specific region of the cell and the formation of cell fragment(s) with or without specific cell component(s).

However, the usefulness of the CM in cell biology has not been widely perceived, because these mostly handmade or custom-made microscopes have not generally been available for biologists. HL-type CMs placed on the market in the past are no longer available.

Developments in electronic techniques have made it possible to construct new types of CMs (KK, K, and H types). Image flicker is eliminated by a rotating video camera (KK type). In other types of CM, the flicker is reduced by electronic freezing of intermittent images. In CMs with a Nano Pulse Light (K and H type), the quality of the microscopic image is greatly improved by using simple microscopic optics in which various optical accessories of the microscope can be inserted. Video-microscopic techniques (Inoué, 1986), including contrast enhancement, image intensification, noise elimination by averaging multiple successive images (rolling average), and electric differentiation of the image, are useful in improving the quality of the microscopic images of the CM.

Although CMs are now custom made or handmade, we expect that they will become commercially available in the future, so that many researchers can use this instrument in various fields of cell biology.

References

Allen, R. D. (1960). The consistency of amoeba cytoplasm and its bearing on the mechanism of amoeboid movement. II. The effects of centrifugal acceleration observed in the centrifuge microscope. *J. Biophys. Biochem. Cytol.* **8,** 379–397.

Allen, R. D., David, G. B., and Nomarski, G. (1969). The Zeiss-Nomarski differential interference equipment for transmitted-light microscopy. *Z. Wiss. Mikrosk. Mikrosk. Tech.* **69,** 193–221.

Beams, H. W. (1937). The air turbine ultracentrifuge, together with some results upon ultracentrifuging the eggs of *Fucus serratus. J. Mar. Biol. Assoc. U.K.* **21,** 571–588.

Brown, R. H. (1940). The protoplasmic viscosity of *Paramecium. J. Exp. Biol.* **17,** 317–324.

Cole, K. S. (1932). Surface forces of the *Arbacia* egg. *J. Cell. Comp. Physiol.* **1,** 1–9.

Cole, K. S., and Michaelis, E. (1932). Surface forces of fertilized *Arbacia* eggs. *J. Cell. Comp. Physiol.* **2,** 121–126.

Danielli, J. F. (1952). Division of flattened egg. *Nature (London)* **170,** 496.

Goto, Y., and Ueda, K. (1988). Microfilament bundles of F-actin in *Spirogyra* observed by fluorescence microscopy. *Planta* **173,** 442–446.

Grolig, F. (1990). Actin based organelle movements in interphase *Spirogyra. Protoplasma* **155,** 29–42.

Harvey, E. B. (1933). Effects of centrifugal force on fertilized eggs of *Arbacia punctulata* as observed with the centrifuge microscope. *Biol. Bull. (Woods Hole, Mass.)* **65,** 389–396.

Harvey, E. B. (1934). Effects of centrifugal force on the ectoplasmic layer and nuclei of fertilized sea urchin eggs. *Biol. Bull. (Woods Hole, Mass.)* **66,** 228–245.

Harvey, E. B. (1935). The mitotic figure and cleavage plane in the egg of *Parechinus microtuberculatus,* as influenced by centrifugal force. *Biol. Bull. (Woods Hole, Mass.)* **69,** 287–297.

Harvey, E. B. (1938). Parthenogenetic merogony or development without nuclei of the eggs of sea urchins from Naples. *Biol. Bull. (Woods Hole, Mass.)* **75,** 170–188.

Harvey, E. B. (1940). A comparison of the development of nucleate and non-nucleate eggs of *Arbacia punctulata. Biol. Bull. (Woods Hole, Mass.)* **79,** 166–187.

Harvey, E. B. (1943). Rate of breaking and size of the "halves" of the *Arbacia punctulata* egg when centrifuged in hypo- and hypertonic sea water. *Biol. Bull.* (*Woods Hole, Mass.*) **85**, 141–150.

Harvey, E. B. (1946). Structure and development of the clear quarter of the *Arbacia punctulata* egg. *J. Exp. Zool.* **102**, 253–276.

Harvey, E. B. (1951). Cleavage in centrifuged eggs, and in parthonogenetic merogones. *Ann. N.Y. Acad. Sci.* **51**, 1336–1348.

Harvey, E. B. (1960). Cleavage with nucleus intact in sea urchin eggs. *Biol. Bull.* (*Woods Hole, Mass.*) **119**, 87–89.

Harvey, E. N. (1931a). Observation on living cells, made with the microscope-centrifuge. *J. Exp. Biol.* **8**, 267–274.

Harvey, E. N. (1931b). A determination of the tension at the surface of eggs of the anelid, *Chaetopterus. Biol. Bull.* (*Woods Hole, Mass.*) **60**, 67–71.

Harvey, E. N. (1931c). The tension at the surface of marine eggs, especially those of the sea urchin, *Arbacia. Biol. Bull.* (*Woods Hole, Mass.*) **61**, 273–279.

Harvey, E. N. (1932). The centrifuge-microscope for supercentrifugal forces. *Science* **75**, 267–268.

Harvey, E. N. (1933a). The flattening of marine eggs under the influence of gravity. *J. Cell. Comp. Physiol.* **4**, 35–47.

Harvey, E. N. (1933b). A new form of centrifuge-microscope for simultaneous observation of control and experimental materials. *Science* **77**, 430–431.

Harvey, E. N. (1934). The air turbine for high speed centrifuging of biological material, together with some observation of centrifuged eggs. *Biol. Bull.* (*Woods Hole, Mass.*) **66**, 48–54.

Harvey, E. N. (1954). Tension at the cell surface. *Protoplasmatologia* **2E**(5), 1–30.

Harvey, E. N., and Danielli, J. F. (1938). Properties of the cell surface. *Biol. Rev. Cambridge Philos. Soc.* **13**, 319–341.

Harvey, E. N., and Fankhauser, G. (1933). The tension at the surface of the eggs of the salamander, *Tritrus (Diemyctylus) viridescens. J. Cell. Comp. Physiol.* **3**, 463–475.

Harvey, E. N., and Loomis, A. L. (1930). A microscope-centrifuge. *Science* **72**, 42–44.

Harvey, E. N., and Marsland, D. A. (1932). The tension at the surface of *Amoeba dubia* with direct observations on the movement of cytoplasmic particles at high centrifugal speeds. *J. Cell. Comp. Physiol.* **2**, 75–97.

Harvey, E. N., and Schopfle, G. (1939). The interfacial tension of intracellular oil drops in the eggs of *Daphnia pulex* and *Amoeba proteus. J. Cell. Comp. Physiol.* **13**, 383–389.

Harvey, E. N., and Shapiro, H. (1934). The interfacial tension between oil and protoplasm within living cells. *J. Cell. Comp. Physiol.* **5**, 255–267.

Hayama, T., and Tazawa, M. (1980). Ca^{2+} reversibly inhibits active rotation of chloroplasts in isolated cytoplasmic drops of *Chara. Protoplasma* **102**, 1–9.

Hayashi, T. (1957). Some dynamic properties of the protoplasmic streaming in *Chara. Bot. Mag.* **70**, 168–174.

Hayashi, T. (1960). Experimental studies on protoplasmic streaming in Characeae. *Sci. Pap. Coll. Gen. Educ., Univ. Tokyo* **10**, 245–282.

Hayashi, T. (1964). Role of the cortical gel layer in cytoplasmic streaming. *In* "Primitive Motile Systems in Cell Biology" (R. D. Allen and N. Kamiya, eds.), pp. 19–29. Academic Press, New York.

Heilbrunn, L. V. (1926). The absolute viscosity of protoplasm. *J. Exp. Zool.* **44**, 255–278.

Heilbrunn, L. V. (1958). The viscosity of protoplasm. *Protoplasmatologia* **2C**(1), 1–109.

Hiramoto, Y. (1956). Cell division without mitotic apparatus in sea urchin eggs. *Exp. Cell Res.* **11**, 630–636.

Hiramoto, Y. (1963). Mechanical properties of sea urchin eggs. I. Surface force and elastic modulus of the cell membrane. *Exp. Cell Res.* **32**, 59–75.

Hiramoto, Y. (1965). Further studies on cell division without mitotic apparatus in sea urchin eggs. *J. Cell Biol.* **25**(1–2), 161–168.

Hiramoto, Y. (1967). Observations and measurements of sea urchin eggs with a centrifuge microscope. *J. Cell. Physiol.* **69**, 219–230.

Hiramoto, Y. (1970). Rheological properties of sea urchin eggs. *Biorheology* **6**, 201–234.

Hiramoto, Y. (1981). Mechanical properties of dividing cells. *In* "Mitosis/Cytokinesis" (A. M. Zimmerman and A. Forer, eds.), pp 397–418. Academic Press, London.

Hiramoto, Y. (1987). Evaluation of cytomechanical properties. *In* "Cytomechanics" (J. Bereiter-Hahn, O. R. Anderson, and W. E. Reif, eds.), pp. 31–46. Springer-Verlag, Berlin and New York.

Hiramoto, Y., and Yoneda, M. (1986). Determination of mechanical properties of the eggs by the sessile drop method. *Methods Cell Biol.* **27**, 443–456.

Inoué, S. (1986). "Videomicroscopy." Plenum, New York.

Jarosch, R. (1956). Plasmaströmung und Chloroplastenrotation. *Phyton (Buenos Aires)* **6**, 86–108.

Kamitsubo, E. (1966a). Motile protoplasmic fibrils in cells of *Characeae*. I. Movement of fibrillar loops. *Proc. Jpn. Acad.* **42**, 507–511.

Kamitsubo, E. (1966b). Motile protoplasmic fibrils in cells of *Characeae*. II. Linear fibrillar structure and its bearing of protoplasmic streaming. *Proc. Jpn. Acad.* **42**, 640–643.

Kamitsubo, E. (1972). Motile protoplasmic fibrils in cells of *Characeae*. *Protoplasma* **74**, 53–70.

Kamitsubo, E. (1979). Rotation of cytoplasmic protrusions in the centrifuged internodal cell of *Nitella*. *In* "Cell Motility: Molecules and Organization" (S. Hatano *et al.*, eds.), pp. 241–246. Univ of Tokyo Press, Tokyo.

Kamitsubo, E., and Kikuyama, M. (1992). Immobilization of endoplasm flowing contiguous to the actin cables upon electrical stimulus in *Nitella* internodes. *Protoplasma* **168**, 82–86.

Kamitsubo, E., Kikuyama, M., and Kaneda, I. (1988). Apparent viscosity of endoplasm of characean internodal cell measured by centrifuge method. *Protoplasma, Suppl.* **I**, 10–14.

Kamitsubo, E., Ohashi, Y., and Kikuyama, M. (1989). Cytoplasmic streaming in internodal cells of *Nitella* under centrifugal acceleration: A study done with a newly constructed centrifuge microscope. *Protoplasma* **152**, 148–155.

Kamiya, N. (1981). Physical and chemical basis of cytoplasmic streaming. *Annu. Rev. Plant Physiol.* **32**, 205–236.

Kamiya, N. (1986). Cytoplasmic streaming in giant algal cells: A historical survey of experimental approaches. *Bot. Mag.* **99**, 441–467.

Kamiya, N., and Kuroda, K. (1956). Velocity distribution of the protoplasmic streaming in *Nitella* cells. *Bot. Mag.* **69**, 544–554.

Kamiya, N., and Kuroda, K. (1958). Measurement of the motive force of the protoplasmic rotation in *Nitella*. *Protoplasma* **50**, 144–148.

Kamiya, N., and Kuroda, K. (1965). Rotational protoplasmic streaming in *Nitella* and some physical properties of the endoplasm. *Proc. Int. Congr. Rheol. 4th, 1963*, Part 4, pp. 157–171.

Kaneda, I., Kamitsubo, E., and Hiramoto, Y. (1990). The mechanical structure of the cytoplasm of the echinoderm egg determined by "gold particle method" using a centrifuge microscope. *Dev., Growth Differ.* **32**, 15–22.

Kikuyama, M., and Tazawa, M. (1982). Ca^{2+} ion reversibly inhibits the cytoplasmic streaming of *Nitella*. *Protoplasma* **113**, 241–243.

Kuroda, K. (1964). Behavior of naked cytoplasmic drops isolated from plant cells. *In* "Primitive Motile Systems in Cell Biology" (R. D. Allen and N. Kamiya, eds.), pp. 31–41. Academic Press, New York.

Kuroda, K. (1990). Cytoplasmic streaming in plant cells. *Int. Rev. Cytol.* **121**, 267–307.

Kuroda, K., and Kamiya, N. (1981). Behavior of cytoplasmic streaming in *Nitella* during centrifugation as revealed by the television centrifuge-microscope. *Biorheology* **18**, 633–641.

Kuroda, K., and Kamiya, N. (1989). Propulsive force of *Paramecium* as revealed by the video centrifuge microscope. *Exp. Cell Res.* **184**, 268–272.

Kuroda, K., and Kamiya, N. (1991). Cytoplasmic movement in *Spirogyra* during and after centrifugation. *Proc. Jpn. Acad., Ser. B* **67**, 78–82.

Kuroda, K., Kamiya, N., Yoshimoto, Y., and Hiramoto, Y. (1986). Paramecium behavior during centrifuge-microscopy. *Proc. Jpn. Acad., Ser. B* **62**, 117–121.

Nagai, R., and Rebhun, L. (1966). Cytoplasmic microfilaments in streaming *Nitella* cells. *J. Ultrastruct Res.* **14**, 571–589.

Oiwa, K., Chaen, S., Kamitsubo, E., Shimmen, T., and Sugi, H. (1990). Steady-state force-velocity relation in the ATP-dependent sliding movement of myosin-coated beads on actin cables *in vitro* studied with a centrifuge microscope. *Proc. Natl. Acad. Sci. U.S.A.* **87**, 7893–7897.

Palevitz, B. A., and Hepler, P. K. (1975). Identification of actin *in situ* at the ectoplasm-endoplasm interface of *Nitella*. *J. Cell Biol.* **65**, 29–38.

Palevitz, B. A., Ash, J. E., and Hepler, P. K. (1974). Actin in the green alga, *Nitella*. *Proc. Natl. Acad. Sci. U.S.A.* **71**, 363–366.

Rappaport, R. (1971). Cytokinesis in animal cells. *Int. Rev. Cytol.* **31**, 169–213.

Rappaport, R. (1986). Establishment of the mechanism of cytokinesis in animal cells. *Int. Rev. Cytol.* **105**, 245–281.

Shimmen, T. (1992a). Mechanism of cytoplasmic streaming and amoeboid movement. *Adv. Comp. Environ. Physiol.* **12**, 172–205.

Shimmen, T. (1992b). The characean cytoskeleton: Dissecting the streaming mechanism. *In* "The Cytoskeleton of the Algae" (D. Menzel, ed.), pp. 297–314. CRC Press, Boca Raton, FL.

Swann, M. M., and Mitchison, J. M. (1953). Cleavage of sea-urchin eggs in colchicine. *J. Exp. Biol.* **30**, 506–514.

Takagi, S., Kamitsubo, E., and Nagai, R. (1991). Light-induced change in the behavior of chloroplasts under centrifugation in *Vallisneria* epidermal cells. *J. Plant Physiol.* **138**, 257–262.

Takagi, S., Kamitsubo, E., and Nagai, R. (1992). Visualization of a rapid, red/far-red light-dependent reaction by centrifuge microscopy. *Protoplasma* **168**, 153–158.

Tazawa, M., and Kishimoto, U. (1968). Cessation of cytoplasmic streaming of *Chara* internode during action potential. *Plant Cell Physiol.* **9**, 361–368.

Williamson, R. E. (1974). Actin in the alga, *Chara corallina*. *Nature (London)* **248**, 801–802.

Williamson, R. E., and Ashley, C. C. (1982). Free Ca^{2+} and cytoplasmic streaming in the alga *Chara*. *Nature (London)* **296**, 647–651.

Yoneda, M. (1964). Tension at the surface of sea-urchin egg: A critical examination of Cole's experiment. *J. Exp. Biol.* **41**, 893–906.

Yoshimoto, Y., Pudles, J., and Hiramoto, Y. (1986). Cyclic changes of stiffness in artificially activated sea urchin eggs measured by centrifuge-microscope. *Proc. Jpn. Acad., Ser. B* **62**, 321–324.

Murine B Cell Development: Commitment and Progression from Multipotential Progenitors to Mature B Lymphocytes

Barbara L. Kee and Christopher J. Paige
The Wellesley Hospital Research Institute and the Department of Immunology, University of Toronto, Toronto, Ontario, Canada M4Y 1J3

B lymphocytes, the cellular source of antibody, are critical components of the immune response. They develop from multipotential stem cells, progressively acquiring the traits that allow them to function as mature B lymphocytes. This developmental program is dependent on appropriate interactions with the surrounding environment. These interactions, mediated by cell–cell and cell–matrix interactions, provide the growth and differentiation signals that promote progression along the developmental pathway. This chapter addresses the properties of developing B lineage cells and the nature of the environmental signals that support B lineage progression.

KEY WORDS: B cell, Immunoglobulin, Growth factors and receptors, B cell ontogeny, Stem cells.

I. Introduction

The hematopoietic system consists of distinct cell lineages that arise from a common multipotential stem cell. Stem cells have extensive proliferative and differentiative capacity as well as the ability to self-renew (Wu *et al.*, 1967; Abramson *et al.*, 1977). The balance of self-renewal and differentiation replenishes the hematopoietic system throughout life. Differentiation from a multipotential cell to its mature progeny is accompanied by a progressive loss in lineage options. The mechanisms of lineage restriction and differentiation have been a central question in cell biology for more than 30 years. This aspect of hematopoietic differentiation has been difficult to study for a number of reasons. First, stem cells and multilineage progenitors are present in very low numbers in hematopoietic tissues (<1 in 10^4–10^5 cells). Second, the phenotypic and morphological characteristics of these cells have not been fully identified. In the past, studies of stem cells have been limited to retrospective assays *in vivo,* in which the cell of interest is identified by its clonal progeny. These reconstitution assays relied on tagging individual stem cells with unique genetic markers, such as integrated retroviruses or chromosomal aberrations, and identification of the progeny of those cells after expansion in irradiated hosts (Wu *et al.,* 1967; Abramson *et al.,* 1977; Dick *et al.,* 1985; Keller *et al.,* 1985; Lemischka *et al.,* 1986). A number of advances have been made that have increased accessibility to the population of lineage-unrestricted hematopoietic progenitors. These include (1) the identification of cell surface markers expressed on lineage-restricted and -unrestricted progenitors, and (2) the establishment of *in vitro* systems that support lineage commitment and differentiation. *In vitro* assays provide a unique opportunity, lacking in *in vivo* assays, to identify intermediates in the commitment process. In addition, experimental manipulation of microenvironmental conditions and direct quantitation of progenitor populations are feasible.

The mature progeny of multilineage progenitors have been extensively studied, owing to their relative abundance and accessibility in adult bone marrow and peripheral tissues. Particularly well characterized are B lymphocytes, the producers of immunoglobulin (Ig), which is the mediator of humoral immune responses. Multiple stages in the developmental pathway of B lymphocyte differentiation have been identified and analyzed, both *in vitro* and *in vivo.* *In vitro* systems have been established that allow lineage-unrestricted progenitors to commit to the B lineage and progress through the entire B cell developmental pathway. This chapter examines our current understanding of murine B cell development.

II. Development of Hematopoietic System

A. Hematopoietic Ontogeny

B lymphocytes develop from hematopoietic stem cells found in distinct sites during ontogeny (Fig. 1) (Tyan and Herzenberg, 1968). The first site of definitive hematopoiesis is the blood islands of the embryonic yolk sac (Moore and Metcalf, 1970). In this site, erythropoiesis can be detected between days 7 and 11 of gestation, declining thereafter as hematopoiesis becomes established in the embryo. Hematopoietic progenitors are found in the fetal liver by day 10 of gestation and this is the dominant site of hematopoiesis until birth (day 19–20) (Owen *et al.,* 1975). By day 15,

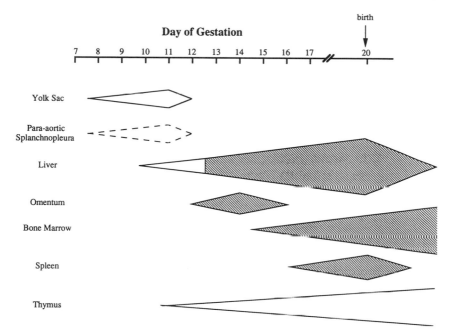

FIG. 1 Sites of hematopoiesis and B lymphopoiesis during ontogeny. The presence of hematopoietic progenitors in embryonic tissues, from day 7 until day 20, is represented. Hatched areas represent the presence of committed B lineage progenitors.

hematopoietic progenitors can be detected in fetal bone marrow, which becomes the primary site of hematopoiesis in the adult animal. The spleen has also been identified as a transient site for hematopoiesis from day 18 of gestation until 1 month after birth (Owen *et al.*, 1975). T cells develop in the specialized environment of the thymus, which is seeded by hematopoietic progenitors starting at day 11 of gestation (Owen and Ritter, 1969).

Despite extensive study, the origin of stem cells remains uncertain. It was initially suggested that hematopoietic stem cells first arise in the extraembryonic endoderm of the yolk sac and subsequently migrate, through the vascular system, to the developing hematopoietic tissues in the embryo (Moore and Metcalf, 1970). However, experiments using chick/quail chimeras revealed an intraembryonic source of stem cells that contributes to hematopoiesis after a first round of erythropoiesis in the yolk sac (Dieterlen-Lievre, 1975; LeDouarin *et al.*, 1975). In the mouse, cells with hematopoietic potential have been identified in the paraaortic splanchnic mesoderm of the embryo at the same time as they arise in the yolk sac, the 10- to 12-somite stage (day 8.5) (Godin *et al.*, 1993, 1994;

Medvinsky *et al.*, 1993). Because this is the time when circulation is first established it has been difficult to identify the source of these cells.

It is not known whether stem cells migrate, from one site to the next, during ontogeny or whether they arise *de novo* from mesoderm in each tissue. Stem cells from the fetal liver and yolk sac have the capacity to contribute to long-term hematopoiesis when injected into immunocompromised mice (Toles *et al.*, 1989; Jordan *et al.*, 1990; Huang and Auerbach, 1993). This suggests that stem cells in these sites have full developmental potential and can migrate into the hematopoietic tissues of adult mice. However, it is not clear that these are the stem cells that maintain the adult hematopoietic system under normal developmental conditions.

B. B Lymphocyte Ontogeny

The first progenitors committed to the B lineage are detected in fetal liver between days 12 and 13 of gestation (Raff *et al.*, 1976; Kamps and Cooper, 1982). At this stage 100–500 B lineage cells are present in this site and their numbers increase until birth (Cumano and Paige, 1992). Although the fetal liver is the dominant site of B lymphopoiesis during ontogeny, other sites such as the omentum and spleen may contribute to B cell development (Owen *et al.*, 1975; Solvason *et al.*, 1991; Rolink *et al.*, 1993b). B lineage-committed progenitors have not been detected prior to day 12 of gestation in either the yolk sac or the embryo body. However, the hematopoietic progenitors in these sites can develop into B lymphocytes *in vitro* and *in vivo* (Paige *et al.*, 1979; Ogawa *et al.*, 1988; Cumano *et al.*, 1993; Godin *et al.*, 1993, 1994). Their failure to do so *in situ* suggests that either the microenvironment is not able to support B lineage progression, or that the stem cells may require a period of time before they develop the capacity to commit to the B lineage.

There is evidence in favor of both of these hypotheses. The ability of the hematopoietic microenvironment to support B lymphopoiesis prior to day 12 of gestation has not been tested directly. However, transplantation of the paraaortic splanchnopleura, isolated from embryos at day 8.5 of gestation, under the kidney capsule results in long-term B lymphopoiesis in that site (Godin *et al.*, 1993). Excluding host influences at the site of transplantation, this suggests that the microenvironment in this tissue can support B cell differentiation but this capacity may have developed only after a period of time after transplantation. Additional studies have suggested that lymphoid commitment may occur prior to day 12 of gestation. Marcos *et al.* (1994) detected mRNA, encoded by a lymphoid-associated gene, in the paraaortic splanchnopleura by days 9–11 of gestation and in the yolk sac by day 11. However, several other B lineage-associated genes were not detected and, to date, definitive relationships between gene expression and lineage commitment have not been established. There is also evidence to suggest that differentiative events require a defined amount of time after the "birth" of a multipotential progenitor. B lineage differentiation is delayed by 3–4 days in cultures initiated with yolk sac progenitors from day 9 of gestation, compared with cultures initiated with fetal liver progenitors from day 12 of gestation (Cumano and Paige, 1992; Cumano *et al.*, 1993, 1994; B. L. Kee, unpublished). This raises the possibility that multipotential progenitors present at day 7 through 11 of gestation are not developmentally identical. They may be following a temporally constrained course of differentiation culminating in the ability to progress down multiple hematopoietic lineages.

In the adult mouse, B cell populations can be identified that have distinct phenotypic and functional characteristics (Hardy *et al.*, 1982). Greater than 90% of B cells in adult spleen are $B220^+IgM^+IgD^{hi}$ whereas many B cells resident in the peritoneal cavity are $B220^{lo}IgM^{++}IgD^{lo}$ and express CD5. It has been reported that transplantation of bone marrow-derived progenitors results in efficient reconstitution of the $CD5^-$ B cell population whereas $CD5^+$ cells are poorly reconstituted. In contrast, fetal liver-derived progenitors are better able to reconstitute cells with the $CD5^+$ phenotype (Hardy and Hayakawa, 1991). The reasons for these differences are not clear but a number of hypotheses have been discussed (Haughton *et al.*, 1993; Herzenberg and Kantor, 1993; Hardy *et al.*, 1994). Some hematopoietic tissues, such as the omentum and paraaortic splanchnopleura, give rise exclusively to $CD5^+$ B cells after transplantation (Solvason *et al.*, 1991; Godin *et al.*, 1993). However, these progenitors have the capacity to give rise to both $CD5^+$ and $CD5^-$ cells *in vitro* (Godin *et al.*, 1994). This suggests that factors external to the progenitors, such as the microenvironment, can determine the characteristics of the B cell. It has been shown that, *in vitro*, conventional $CD5^-$ B cells can assume the phenotypic characteristics of $CD5^+$ B cells after stimulation with anti-IgM antibody and interleukin 6 (Cong *et al.*, 1991). This raises the possibility that the CD5 phenotype may be the result of activation by some antigens. The developmental distinction between these populations could then be explained by differences in the antibody repertoire arising during fetal and adult life (Hayakawa *et al.*, 1988; Feeney, 1990). Alternatively, these populations may be the result of developmental changes in the progenitors present in these sites.

III. Markers of B Cell Differentiation

In the past 10 years a number of markers of B lineage development have been identified. These markers include the products of the immunoglobulin (Ig) receptor gene loci and proteins whose expression is regulated differentially during B lineage progression. Some of these proteins are B lineage specific whereas others are more widely distributed within the hematopoietic system. These markers help resolve distinct stages of B cell development and provide tools for the purification of progenitors. The function of many of these proteins in lineage determination and B cell differentiation is not known. However, in combination with other physical parameters, a model of B lineage progression emerges that provides a foundation on which to base future experimentation (Fig. 2).

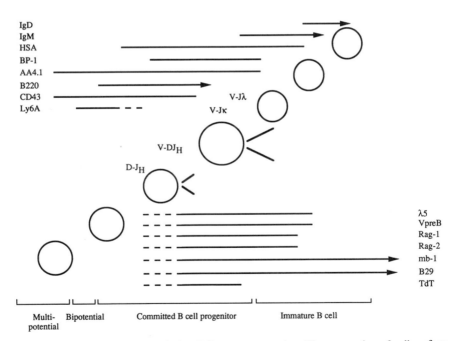

FIG. 2   Expression of markers during B lineage progression. The expression of cell surface and intracellular proteins during progression from multipotential progenitors to mature B lymphocytes is shown. The stage of expression of each marker listed at the top is based on studies detecting protein expression on the cell surface. The stage of expression of the markers listed at the bottom is based on mRNA and/or protein expression.

A. Immunoglobulin

The hallmark of cells committed to the B lineage is their ability to undergo somatic recombination of their Ig heavy and light chain genes (Hozumi and Tonegawa, 1976). Analysis of the Ig loci in Abelson virus-transformed pre-B cells lines suggested that these gene segments recombine in a predictable pattern (Alt *et al.*, 1984). The heavy chain D and J segments are joined before a V segment is joined to the DJ_H segment. Frequently DJ_H rearrangements are found on both heavy chain alleles before the V segment is rearranged. Rearrangement of the heavy chain loci normally precedes the light chain loci and, in most cells, kappa (κ) light chain rearrangement precedes lambda (λ). During B lymphopoiesis in the fetal liver a similar pattern of gene rearrangement is observed. The first $D-J_H$ rearrangements can be detected by day 12 of gestation, concomitant with the emergence of B lineage-restricted progenitors. At this time there are fewer than 100 $D-J_H$ rearrangements per fetal liver (Chang *et al.*, 1992). $V-DJ_H$ rearrangements can be detected 1 day later, on day 13, and the onset of light chain gene rearrangements is observed after an additional day (Cumano *et al.*, 1994; Ramsden *et al.*, 1994). In contrast to the observations in cell lines, κ and λ light chain gene rearrangements become detectable at the same stage of gestation, although the number of λ rearrangements is significantly lower than κ (Ramsden *et al.*, 1994). This order of rearrangement is also consistent with earlier experiments in which expression of intracellular heavy chain protein ($c\mu$) and surface Ig was detected by days 14 and 17 of gestation, respectively (Kamps and Cooper, 1982). A similar pattern of rearrangements is observed during B cell differentiation in the bone marrow (Hardy *et al.*, 1991; Ehlich *et al.*, 1993).

Rearrangement at the Ig loci has proved useful for tracking the developmental stage of a B cell progenitor. Pro-B cells have been defined as those that have undergone, or are in the process of undergoing, $D-J_H$ rearrangement. Pre-B cells have a $V-DJ_H$ rearrangement and may be undergoing light chain rearrangement (Blackwell and Alt, 1989). The mechanism of recombination allows imprecise joining of the V, D, and J segments, resulting in significant variability in the junctional sequences (Gellert, 1992). Although this mechanism has the potential to increase the antibody repertoire, it also leads to nonproductive rearrangements. Therefore, only a fraction of cells undergoing Ig rearrangement subsequently express protein products from the Ig gene segments. The first possible protein product observed is a $D\mu$ protein, as a consequence of some DJ_H joining events (Reth and Alt, 1984; Tsubata *et al.*, 1991). This product is thought to be deleterious for further development along the B lineage (Gu *et al.*, 1991a). Cytoplasmic μ proteins are observed after productive rearrangements of all three heavy chain segments. This protein is easily detected in the cytoplasm of developing pre-B cells and may be expressed in low amounts on the surface as well (Kamps and Cooper, 1982; Opstelten and Osmond, 1983). IgM is expressed after rearrangements of κ or λ light chain genes that result in productive Ig light chain proteins. Shortly thereafter IgD is coexpressed on the cell surface (Osmond, 1990).

B. Cell Surface Proteins Expressed on Lineage-Unrestricted Progenitors

AA4.1 is a cell surface protein that is expressed on long-term reconstituting hematopoietic stem cells from murine yolk sac and fetal liver, and on immature progenitors of many lineages including B cells (McKearn et al., 1984; Paige et al., 1984; Jordan et al., 1990; Huang and Auerbach, 1993). AA4.1 is expressed on multipotential cells by day 8.5 of gestation and bipotential B cell–macrophage progenitors present in fetal liver at day 12 of gestation (Jordan et al., 1990; Cumano et al., 1992, 1993; Huang and Auerbach, 1993; Godin et al., 1994). Stem cells in adult bone marrow do not express AA4.1 although it is present on multilineage and committed progenitors in this site. AA4.1 is expressed on all $c\mu^+$ cells and 50% of surface Ig-positive (sIg^+) cells in both fetal liver and adult bone marrow (McKearn et al., 1984). In contrast, peripheral B cells rarely, if ever, express this marker, suggesting that it is downregulated shortly after expression of sIg. The structure and function of AA4.1 are not known.

Ly6A/Sca-1 is a member of a complex family of cell surface molecules located on chromosome 15 in the mouse (van de Run et al., 1989; Rock et al., 1989; Shevach and Korty, 1989). This family consists of small proteins, ranging from 10 to 18 kDa, which are tethered to the cell membrane via a carboxy-terminal glycosyl phosphatidylinositol linkage (type III membrane protein). The Ly6 complex contains six known proteins denoted A, B, C, D, E, and ThB, which share a high degree of sequence similarity. Southern blot hybridization indicates that as many as 10 family members may exist. In mice, two Ly6 haplotypes have been identified and designated $Ly6^a$ and $Ly6^b$. Ly6A and Ly6E are allelic variants of the same protein expressed in the $Ly6^b$ and $Ly6^a$ haplotype, respectively. Ly6A is expressed on hematopoietic stem cells in the bone marrow of $Ly6^b$ mice (Spangrude et al., 1988). However, Ly6E is expressed by only 25% of stem cells in $Ly6^a$ mice (Spangrude and Brooks, 1993). The differential expression of allelic variants within the Ly6 locus has been well characterized but the basis for this variable regulation is not known. In fetal liver at day 12 of gestation, Ly6A identifies a population of AA4.1$^+$ cells that has the capacity to give rise to both B lymphocytes and macrophages in vitro (Cumano et al., 1992). It is not detected on multipotential progenitors from yolk sac or embryo body prior to day 11 of gestation (Cumano *et al.*, 1993). Ly6A is expressed on the most immature committed B lineage cells but is downregulated as progression through the lineage proceeds (other Ly6 family members become expressed at later stages of differentiation). The absence of Ly6A on pre-B cells *in vivo* is in contrast to its high expression on pre-B cells cultured *in vitro*. The expression of this protein is known to be inducible under some conditions, particularly in response to interferon γ (Bothwell *et al.*, 1988). The function of Ly6A on hematopoietic progenitors has not been established. It was originally identified by its ability to activate T cells when cross-linked by antibody in the presence of phorbol 12-myristate 13-acetate (PMA). Activation, in T cells, may be mediated through its association with the Src-related tyrosine kinases $p59^{fyn}$ and $p56^{lck}$ (Bohuslav *et al.*, 1993; Lee *et al.*, 1994). Ly6A may play a fundamental role in hematopoietic differentiation by activating intracellular kinases in response to extracellular stimuli.

CD43/leukosialin is a highly glycosylated, type I integral membrane protein of 90–120 kDa, expressed on mature lymphocytes, monocytes, neutrophils, platelets, and pre-B lymphocytes (Pallant *et al.*, 1992). CD43 is expressed on lineage-unrestricted progenitors and on large, cycling, pre-B lymphocytes but is downregulated prior to the expression of immunoglobulin light chain (Hardy *et al.*, 1991). However, it is expressed on activated and terminally differentiated B cells (Wiken *et al.*, 1988). CD43 has been postulated to mediate homotypic adhesion of T cells through its ability to bind to intracellular adhesion molecule type 1 (ICAM-1)/CD54 (Rosenstein *et al.*, 1991). However, in some cell lines it can function as an antiadhesion molecule and has been suggested to play additional roles in T cell activation (Manjunath *et al.*, 1993). It has been reported that CD43 is expressed at low levels in patients with Wiskott–Aldrich syndrome, an X-linked disorder resulting in eczema, thrombocytopenia, and a severe immunodeficiency. However, dysregulation of CD43 expression does not appear to be the primary cause of the disease but may be responsible for some of the clinical features (Shelley *et al.*, 1989; Remold-O'Donnell and Rosen, 1990). The function of this protein on pre-B lymphocytes is unknown but it may mediate adhesion to components of the extracellular microenvironment.

C. Cell Surface Proteins Defining Stages of B Cell Development

B220 is expressed on all B lineage cells except terminally differentiated plasma cells (Kincade *et al.*, 1981b). B220 is a 220-kDa isoform of CD45, a type I integral membrane protein with an intracellular protein tyrosine phosphatase activity. CD45 is expressed on all nucleated hematopoietic cells; however, the various isoforms of this enzyme show lineage-, stage-, and activation state-specific patterns of expression. CD45 isoforms arise by alternative splicing of exons 4, 5, and 6, which encode residues in the extracellular amino-terminal region of the protein. The function of B220 on B cells progenitors is not known although it appears to play a critical role in Ig-mediated signaling events on mature B cells (Gold *et al.*, 1990; Justement *et al.*, 1991). In T cells CD45 participates in T cell receptor (TCR)-mediated signal transduction via dephosphorylation and consequent activation of p56^{lck} and p59^{fyn} (Ostergaard *et al.*, 1989; Mustelin and Altman, 1990; Shiroo *et al.*, 1992). Similarly, CD45 may regulate signal transduction through the B cell antigen receptor (Ig). Mice deficient in expression of B220 have been created by targeted deletion of exon 6 (Kishihara *et al.*, 1993). In these mice mature B lymphocytes do not proliferate in response to cross-linking of Ig, suggesting that B220 plays a crucial role in regulating the response to foreign antigens. However, the ability to respond to B cell mitogens such as lipopolysaccharide is not affected. A ligand for CD45 has not been identified although some isoforms can bind to CD22, owing to the ability of CD22 to recognize sialic acid moieties with an α-2,6 linkage (Sgori *et al.*, 1993).

Heat-stable antigen (HSA/CD24) is a 31- to 50-kDa, type III membrane protein that is expressed on progenitors of the erythroid, myeloid, T, and B lineages as well as neuronal progenitors (Rougon *et al.*, 1991). On mature B lymphocytes low levels of HSA correlate with the ability of cells to initiate secondary immune responses (Linton *et al.*, 1989). The intensity of expression during differentiation is often indicative of the maturational stage of the cell. Heat-stable antigen is detected on pre-B cells after the onset of B220 expression and increases as pre-B cells mature, reaching a maximum at the time when CD43 expression is downregulated. Heat-stable antigen expression then declines prior to expression of sIg (Hardy *et al.*, 1991). The function of this protein is currently not known but preliminary data suggest that it may play a role in cell adhesion and activation (Nielson *et al.*, 1993; Hough *et al.*, 1994).

BP-1/6C3 is a 110- to 150-kDa, type II integral membrane protein with an extracellular aminopetidase A activity (Wu *et al.*, 1990, 1991). BP-1/6C3 is expressed at low levels on pre-B lymphocytes from bone marrow and fetal liver and at high levels on pre-B cells transformed by Abelson virus or grown *in vitro* (Ramakrishnan *et al.*, 1990; Welch *et al.*, 1990). It is first detected on B cell progenitors that express B220, CD43, and HSA and is downregulated on sIg$^+$ B cells (Hardy *et al.*, 1991). In addition, BP-1/6C3 is expressed on stromal cells and epithelial cells in many organs such as the renal proximal tubules and small intestine (Whitlock *et al.*, 1987; Wu *et al.*, 1991). This pattern of expression is similar to that of neutral endopeptidase, CD10/common acute lymphoblastic leukemia antigen (CALLA), in humans. In mouse bone marrow, expression of CD10 is restricted to stromal cells and a small subpopulation of pre-B lymphocytes (Kee *et al.*, 1992; Salles *et al.*, 1992). It has been proposed that these peptidases may play a role in lymphopoiesis by cleaving regulatory peptides in the hematopoietic microenvironment. However, the substrate specificity of these peptidases suggests that they are unlikely to function by directly hydrolyzing the cytokines known to regulate B cell development (Kerr and Kenny, 1974). They may act on these cytokines in concert with other proteases. It has been reported that inhibition of the enzymatic activity of CD10 *in vivo* results in an increase in the number of bone marrow B cell progenitors (Salles *et al.*, 1993). Therefore, these enzymes may function as negative regulators of lymphopoiesis by inactivating peptides that promote B lineage progression (Shipp and Look, 1993).

D. Immunoglobulin-Associated Proteins

Rearrangement of the immunoglobulin heavy and light chain genes is dependent on lymphoid-specific recombinase proteins that have not yet been definitively characterized. Two proteins have been identified that are essential for the recombination process. These proteins are the products of the recombinase activating genes 1 and 2 (*rag-1* and *rag-2*) and are expressed in B and T lineage cells (Schatz *et al.*, 1989; Oettinger *et al.*, 1990). Mice deficient in expression of either of these genes have a severe combined immunodeficiency due to an inability to rearrange their B cell and T cell receptor (Ig and TCR) genes (Mombaerts *et al.*, 1992; Shinkai *et al.*, 1992). The cDNA sequence of *rag-1* revealed a 119-kDa protein with a potential DNA-binding domain and a topoisomerase-like domain. Mutagenesis of these domains does not affect the ability of Rag-1 to mediate V(D)J recombination on extrachromosomal substrates, suggesting that they are not critical for this function (Kallenbach *et al.*, 1993; Silver *et al.*, 1993). The mRNA for Rag-1 has a very short half-life likely owing to the presence of 12 copies of a nucleotide sequence (AUUUA) associated with RNA instability. Rag-2 is a 58-kDa protein with no known sequence similarity to other proteins. Its gene is located 8 kb upstream of the coding sequence for Rag-1. The genomic organization of both genes is unique in that they are encoded within a single exon. Rag-2 contains two phosphorylation sites, which appear to regulate the stability and/or function of the protein (Lin and Desiderio, 1993). Phosphorylation of Ser-356 is required for optimal recombination whereas phosphorylation of Thr-490 results in rapid degradation of the protein. Because Thr-490 can be phosphorylated by a cyclin-dependent kinase, $p34^{cdc2}$, this suggests a mechanism by which

Rag-2 expression can be restricted to the G_0/G_1 phase of the cell cycle, consistent with the predicted timing of V(D)J recombination (Lin and Desiderio, 1993; Schlissel *et al.*, 1993; Desiderio, 1994). The role(s) of these proteins in the recombination process are not known; they may function in the regulation of, or as a component of, the recombinase machinery.

Terminal deoxynucleotidyltransferase (TdT) is a 58-kDa DNA polymerase that can add nucleotides to the 3'-OH end of DNA in the absence of a template. During the V(D)J recombination process TdT can add nucleotides at the coding junctions (N additions), thereby increasing the diversity in the antigen receptor (Alt and Baltimore, 1982; Desiderio *et al.*, 1984). Terminal deoxynucleotidyltransferase is expressed in B cell and T cell progenitors in adult mice but is not expressed during fetal development (Grégoire *et al.*, 1979). This corresponds to the absence of N additions in the antigen receptors rearranged during fetal ontogeny (Feeney, 1990). Mice deficient in expression of TdT lack N additions in their Ig and TCR gene rearrangements but otherwise appear to function normally (Gilfillan *et al.*, 1993; Komori *et al.*, 1993).

λ5 is a 22-kDa protein with sequence identity, in its carboxy-terminal domain, to the constant and J region of the λ1 light chain. The amino-terminal domain shows no similarity to known proteins (Sakaguchi and Melchers, 1986). V_{pre-B} is a 16-kDa protein with approximately 50% identity, in its amino-terminal region, to the V regions of both heavy and light chain genes. The 26 carboxy-terminal amino acids show no similarity to known proteins (Kudo and Melchers, 1987). At least two distinct V_{pre-B} genes have been cloned (Shirasawa *et al.*, 1993). In contrast to the immunoglobulin genes, neither λ5 or V_{pre-B} undergo somatic recombination. These proteins are expressed exclusively in pre-B lymphocytes and together form a surrogate light chain that is expressed on the cell surface in association with μ heavy chain (Misener *et al.*, 1990; Cherayil and Pillai, 1991; Nishimoto *et al.*, 1991; Lassoued *et al.*, 1993; Karasuyama *et al.*, 1994). λ5 and V_{pre-B} were identified in conjunction with a 130-kDa protein on B cell progenitors prior to the expression of μ, suggesting that it may have a function at the pro-B cell stage of development (Karasuyama *et al.*, 1993). mRNAs for λ5 and V_{pre-B} have been detected in the most immature B lineage cells and are expressed until sIg can be detected (Kudo *et al.*, 1992; Li *et al.*, 1993). Mice deficient in expression of λ5 show an arrest in B cell differentiation at the $B220^+CD43^+$ pre-B cell stage. However, mature B cells do develop at a reduced rate, reaching normal levels after 6 months of age (Kitamura *et al.*, 1992). Antibody-induced cross-linking of the V_{pre-B}–λ5–μ complexes on pre-B cell lines results in calcium mobilization, suggesting that this receptor is able to mediate signal transduction (Misener *et al.*, 1991). Thus, the surrogate light chain complex may send a critical signal for survival or lineage progression in B cell progenitors.

Deposition of Ig complexes on the cell surface is dependent on the expression of two Ig-associated proteins called mb-1 (Ig-α) and B29 (Ig-β) (Hombach *et al.*, 1990; Reth, 1993). B29 and mb-1 form a disulfide-linked heterodimer that mediates signal transduction through the B cell antigen receptor (Kim *et al.*, 1993). Both mb-1 and B29 are type I integral membrane proteins with intracellular domains containing motifs character-istic of proteins involved in signal transduction, such as the CD3 compo-nents γ, δ, and ε (Hermanson *et al.*, 1988; Sakaguchi *et al.*, 1988). These domains become phosphorylated after stimulation through the Ig receptor on pre-B lymphocytes and mature B cells. The phophorylation sites func-tion by promoting interaction with the SH2 domains present on other proteins in the signaling cascade. A third Ig-associated protein, called Ig-γ, has been identified as a carboxy-terminal truncated form of B29 (Freidrich *et al.*, 1993). The cytoplasmic domain of B29 consists of 48 amino acids, of which 30–36 are deleted in Ig-γ. Ig-γ appears to be ex-pressed on activated but not resting B cells. The mRNAs for both mb-1 and B29 are expressed at all stages of B cell differentiation, from the time of commitment until terminal differentiation (Li *et al.*, 1993).

E. A Model of B Cell Progression

A model of progression from multilineage progenitors to mature B lympho-cytes, made on the basis of Ig gene rearrangement and protein expression, is presented in Fig. 2. Additional parameters have been used to follow B cell development including changes in cell size and division rate (Osmond, 1990; Karasuyama *et al.*, 1994). Cell size increases, from 7 to 15 μm, as progenitors differentiate from B220$^+$TdT$^+$ cells, which are initiating heavy chain gene rearrangement, to B220$^+$TdT$^-$ cells, some of which express cμ. Large, B220$^+$ pre-B lymphocytes give rise to small B220$^+$cμ^+ cells that differentiate into sIg$^+$ B cells (Opstelten and Osmond, 1983, 1985; Osmond, 1990). Stathmokinetic and cytometric analyses have shown that large, B220$^+$ pre-B cells are dividing rapidly whereas their small B220$^+$cμ^+ progeny are not (Opstelten and Osmond, 1983; Park and Osmond, 1987; Karasuyama *et al.*, 1994). On the basis of these studies, a minimum of six cell divisions is predicted to occur between the onset of B220 expres-sion and the emergence of sIg$^+$ cells. As differentiation proceeds from multipotential progenitors to large cycling pre-B cells the absolute number of progenitors increases. In contrast, only a proportion (25–30%) of the large cycling pre-B cells gives rise to small resting pre-B cells (Osmond, 1990, 1991). Approximately 70% of large pre-B cells die by apoptosis (programmed cell death) and survival to the small pre-B cell stage is dependent on the production of a functional μ heavy chain (see discussion in Section VI, A.) (Merino et al., 1994).

A number of models of B cell progression have been proposed on the basis of the correlation of cell surface and intracellular proteins, rearrangement status of the Ig loci, and growth requirements as determined in functional assays. Such models provide a mechanism to integrate experimental findings and reveal fundamental traits that underlie the differentiation process. There is, of course, some danger in model building because of the natural tendency to generate orderly schemes of differentiation based on generalized experimental findings. It is important to realize, for example, that the correlation of one or more traits does not prove a causal relationship. In Fig. 2, B cell progenitors are depicted as progressing from $B220^-HSA^-BP-1^- \rightarrow B220^+HSA^-BP-1^- \rightarrow B220^+ HSA^+ BP-1^- \rightarrow B220^+HSA^+BSP-1^+$, all of which express AA4.1 and CD43. However, it is not certain that a mature B cell must pass through each of these stages. In fact, dysregulation of B220 (deletion by targeted disruption) or HSA (altered expression in transgenic mice) does not appear to inhibit differentiation, suggesting that passage through these phenotypic stages is not a requirement (Kishihara et al., 1993; Nielson et al., 1993; Hough et al., 1994).

An essential feature of model building is to apply in vitro or in vivo methods that allow the frequency of progenitors to be functionally determined. However, similar caution must be applied to the correlation of phenotypic and functional properties. For example, the majority of bone marrow cells (>80%) with the phenotype $B220^+CD43^+HSA^-BP-1^-$ have their Ig loci in a germline configuration (Hardy et al., 1991). Initial estimates suggest that 1% of this population is capable of growth and differentiation in a short-term in vitro assay (Loffert et al., 1994). It remains to be determined whether the B cells are emerging from rearranged or unrearranged progenitors found within the identified population. Similarly, we have reported that a population enriched in expression of AA4.1 and Ly6A, and that lacks expression of B220 or Mac-1, contains bipotential progenitors (Cumano et al., 1992). However, under optimal conditions only one in three cells in this population gives rise to B cells and macrophages. Although this observation proves that at least some of the $AA4.1^+Ly6A^+$ cells are bipotent, it does not prove that they all are. Nor does it exclude the existence of bipotential cells with other phenotypes which are not detected by the cell surface markers utilized in our studies.

IV. The Hematopoietic Microenvironment

A. *In Vivo* Microenvironmental Associations

The central role of the hematopoietic microenvironment in supporting blood cell production has been recognized for nearly 40 years. At this time two mutations in mice were identified, *W* and *Sl,* that resulted in defective germ cells, melanocytes, and hematopoietic stem cells leading to sterility, albinism, and severe anemia, respectively (Russell, 1949; Sarveila and Russell, 1956). Transplantation of normal bone marrow into *W* mice could cure the hematopoietic defect whereas normal mice became anemic when transplanted with *W* bone marrow (McCulloch *et al.,* 1964). In contrast, stem cells from *Sl* bone marrow could reconstitute recipient mice but normal bone marrow could not cure the anemia in *Sl* mice (McCulloch *et al.,* 1965). These data suggested that the *W* mutation was intrinsic to the hematopoietic stem cells whereas *Sl* affected the microenvironment. It was hypothesized that *W* and *Sl* might represent a receptor–ligand pair that functioned in the development of stem cells and erythrocytes. Cloning of a cDNA for *W* revealed a transmembrane receptor tyrosine kinase (c-*kit*) (Chabot *et al.,* 1988; Geissler *et al.,* 1988). The ligand for this receptor has been cloned, and called by a number of names including mast cell growth factor (MGF), stem cell factor (SCF), or *kit* ligand (KL), and is encoded by *Sl* (Anderson *et al.,* 1990; Copeland *et al.,* 1990; Huang *et al.,* 1990; Zsebo *et al.,* 1990). This growth factor–receptor combination functions in the maintenance of hematopoietic stem cells and erythrocytes but can also influence committed progenitors of multiple lineages, including B lymphocytes.

In vivo, developing B lymphocytes are found in close physical association with microenvironmental cells known as "stromal cells." Stromal cells are a heterogeneous population of cells that includes fibroblasts, adventitial reticular cells, preadipocytes, and macrophages (Kincade *et al.,* 1989; Dorshkind, 1990; Kincade, 1991). Stromal cells support B lymphopoiesis by providing the factors necessary for growth and differentiation. As well, they provide supporting matrices on which migration to blood vessels can occur. The most immature B cell progenitors are found in the subendosteal region of the marrow in close association with reticular cells (Jacobsen and Osmond, 1990). These early progenitors are found in clustered foci, suggesting that the progeny of an individual progenitor remain closely associated with the stroma in the immediate environment. As differentiation proceeds there is a progression from the endosteal region to the centrally located sinuses. Immature B cells, expressing surface IgM but not IgD, can be seen to traverse the endothelial lining of the sinusoids. These immature cells are held in an intravascular compartment, possibly for a final stage of differentiation, before being released into the circulation (Jacobsen and Osmond, 1990).

The integrity of the bone marrow microenvironment is critical for the maintenance of B cell development. Mature B cells and their progenitors, but not macrophages, are reduced by 90% in the bone marrow of mice carrying a targeted mutation in c-*fos* (Okada *et al.*, 1994). This is not due to an intrinsic B cell defect because stem cells from these mice show full differentiative potential *in vitro* and *in vivo* after transfer into normal recipients. In addition, B cell differentiation in the fetal liver is not affected. This B cell deficiency is the result of a perturbation of the hematopoietic microenvironment resulting from severe osteopetrosis. The primary lesion in these mice is the failure to produce mature osteoclasts, which function in the degradation of calcified bone. Surprisingly, an increase in the number of $B220^+sIg^-$, IL-7-responsive, pre-B cells is found in the spleens of c-*fos*-deficient mice. This suggests that, under some circumstances, stromal elements in the spleen can support B cell progression. This has been confirmed by the identification of spleen-derived stromal cell lines that support B cell differentiation *in vitro* (Gimble *et al.*, 1989).

B cell development in the fetal liver is also dependent on association with microenvironmental stromal cells. During the early stages of B cell ontogeny, pre-B lymphocytes are randomly distributed throughout the extravascular liver parenchyme. By day 19 of gestation sIg^+ cells are found in a "star burst" distribution in the midst of numerous erythropoietic and myelomonocytic cells (Owen *et al.*, 1975; Kamps and Cooper, 1982). This diffuse clustering of sIg^+ cells suggests that they develop from a common progenitor, similar to the clonal foci of hematopoietic cells present in the postnatal liver (Rossant *et al.*, 1986). However, the nature of the stromal elements in these foci, or in the hematopoietic microenvironment prior to day 12 of gestation, has not been determined. It has been proposed that the development of supporting capacity by the microenvironment is the limiting factor in the emergence of B lineage-committed cells during ontogeny.

B. *In Vitro* Microenvironmental Associations

The establishment of *in vitro* systems that support hematopoiesis provided a tool for the identification of the microenvironmental interactions that regulate B cell development (Dexter *et al.*, 1977; Melchers, 1977; Kincade *et al.*, 1981a; Whitlock and Witte, 1982; Dorshkind and Phillips, 1983; Paige, 1983). Stromal cell clones were established from these cultures that could either support, or inhibit, B lineage progression (Collins and Dorshkind, 1987; Hunt et al., 1987; Whitlock et al., 1987; Nishikawa et al., 1988; Pietrangeli et al., 1988). In addition, clones were identified that could support B cell progenitors at different stages of development (see Section IV,C) (Cumano et al., 1990; Hayashi et al., 1990; Henderson et al., 1990; Gunji et al., 1991; Rolink et al., 1991a). The stromal cells that support lymphopoiesis are generally fibroblasts or preadipocytes whereas macrophages are frequently inhibitory. More than 20 stromal cell-derived growth factors have been identified that can mediate many of the effects attributed to the microenvironment. However, the full spectrum of functions provided by these cells is not yet known.

In vitro, B cell progenitors adhere to, and can become completely engulfed within, the stromal cell membrane. Adhesion molecules have been identified that may mediate these interactions. These include neural cell adhesion molecule (NCAM-1), vascular cell adhesion molecule (VCAM-1), and components of the extracellular matrix such as fibronectin, laminin, hyaluronate, and collagen (Kincade, 1991). Inhibition of stromal cell–B cell interactions can have devastating effects on B cell growth *in vitro.* For example, long-term bone marrow cultures established in the presence of a monoclonal antibody (MAb) that inhibits the interaction of CD44 with hyaluronate fail to support B cell development (Miyake et al., 1990). Similarily, MAbs that inhibit the interaction of VCAM-1 with VLA-4 abrogate B lymphopoiesis *in vitro* (Kina et al., 1991; Miyake et al., 1991). Some adhesion molecules are able to mediate signal transduction, raising the possibility that adhesion may send a critical developmental signal to the cell (Hynes, 1987). Alternatively, adhesion may provide a mechanism for the juxtaposition of pre-B and stromal cell membranes to allow exchange of soluble or membrane-associated growth factors. The requirements for cell contact, at many stages of B cell development, can be overcome using stromal cell-derived growth factors, suggesting that adhesion-mediated signal transduction is not required (Kee et al., 1994). However, some stages of B cell differentiation still require physical interaction with stromal cells.

The spatial segregation of B cell progenitors and their more mature progeny *in vivo* has led to the suggestion that stromal cells might be specialized in their ability to support distinct stages of B cell differentiation. According to this model, the centrally located stromal cells provide factors that support late events in B lymphopoiesis, whereas those located in the endosteum provide factors that support early events (Rolink and Melchers, 1991). In support of this model, stromal cell clones have been characterized that support either early or late stages of B lymphocyte differentiation *in vitro* (Nishikawa et al., 1988). However, much of the heterogeneity in support capacity appears to be attributable to production of a single cytokine, interleukin 7 (IL-7; Cumano *et al.*, 1994; Era *et al.*, 1994; Kee *et al.*, 1994). In the presence of exogenously added IL-7, many stromal cell types are able to support the differentiation of lineage-unrestricted progenitors into mature B cells (Kee *et al.*, 1994). This suggests that the factors that support B lineage progression can be provided by most stromal cells. The question arises as to how closely cloned stromal lines reflect the functions of their *in vivo* counterparts. *In vivo* and *in vitro* these cells show a remarkable degree of plasticity, responding to changes in their local environment by altering the expression or localization of a number of molecules (Sudo *et al.*, 1989; Jacobsen *et al.*, 1992). In addition, the same stromal cell lines can support lymphopoiesis or myelopoiesis under different experimental conditions, suggesting that their functional properties are altered in response to changes in their environment (Dexter *et al.*, 1977; Whitlock and Witte, 1982; Dorshkind and Phillips, 1983). With this in mind, models that propose that the location of a supporting stromal cell dictates the location of B lymphopoiesis might propose, as well, that the location of the progenitor dictates the functions of the local stromal cells.

C. Stromal Cell-Dependent B Lymphopoiesis

The establishment of *in vitro* systems that support all stages of B lineage progression provided an essential tool for the characterization of early B lineage progenitors. Particularly useful in this respect have been clonal assays that allow the analysis of precursor progeny relationships. Assays of this type have led to the identification of novel B cell progenitor populations and their growth requirements (Muller-Sieburg *et al.*, 1986; Ohara *et al.*, 1991; Cumano *et al.*, 1992). As illustrated in Fig. 3, all stages of B cell development are dependent on factors or interactions provided by stromal cells. Our experimental findings revealed two stromal cell-dependent phases, designated stroma A and stroma B (Fig. 3). Stroma A represents the requirements of multilineage progenitors as they differentiate into IL-7 responsive pre-B lymphocytes. Two stromal cell-derived factors have been identified that can overcome some of the functions provided by stroma A (Hirayama *et al.*, 1992; Kee *et al.*, 1994). However, these factors are not sufficient to support differentiation from multilineage progenitors, indicating that stroma A provides additional molecules required by these cells. Stroma B represents the growth requirements for progression of Il-7-responsive pre-B lymphocytes into mature B cells (Cumano *et al.*, 1990). The mechanisms of support provided by stroma B are unknown.

The first *in vitro* B cell assays to gain widespread use were based on

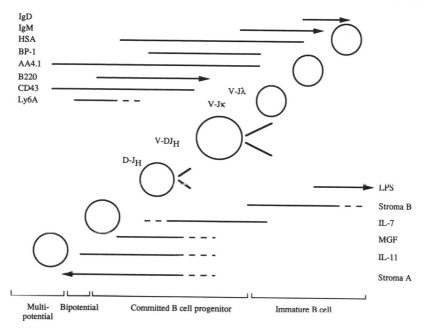

IgD
IgM
HSA
BP-1
AA4.1
B220
CD43
Ly6A

V-Jλ
V-Jκ
V-DJ~H~
D-J~H~

LPS
Stroma B
IL-7
MGF
IL-11
Stroma A

Multi-   Bipotential   Committed B cell progenitor   Immature B cell
potential

FIG. 3   Growth requirements during B lineage progression. The known growth requirements of multilineage progenitors during progression to mature B lymphocytes are shown. Stroma A and stroma B represent the requirements for stromal cell-derived products for the transition of multipotent progenitors to the stage of IL-7 responsiveness and for the transition of IL-7-responsive pre-B cells to the stage of mitogen responsiveness, respectively.

the ability of these cells to proliferate and secrete Ig in response to thymus-independent antigens such as bacterial lipopolysaccharide (LPS) (Cooper *et al.*, 1972; Melchers *et al.*, 1975). B lymphocytes capable of responding to LPS are detected in fetal liver by day 16 to 17 of gestation, concomitant with the development of sIg$^+$ cells. The ability of mature B cells to secrete Ig in response to LPS is still a fundamental component of many assays, providing conclusive evidence that a cell has progressed through the B lineage. Incorporation of adherent stromal cell layers into these assays allowed B lineage progenitors from earlier gestational and developmental stages to respond to LPS (Paige, 1983). Under these conditions, responsive progenitors could be detected in fetal liver at day 12 of gestation. The clonable population included sIg$^-$ cells, which were not yet committed to the expression of an Ig light chain isotype (Sauter and Paige, 1987) or heavy chain allotype (Sauter and Paige, 1988). Further, these assays were limiting only for the B cell progenitor, showing that all of the requirements for growth and differentiation could be provided by the adherent stromal layer (Paige *et al.*, 1985).

The cloning of IL-7 provided an invaluable reagent for the discussion of B cell progenitor growth requirements (see Section V,A). (Namen *et al.*, 1988). Interleukin 7 supports the proliferation of $B220^+sIg^-$ pre-B lymphocytes but does not promote their differentiation into mitogen-responsive B cells (Lee *et al.*, 1989; Cumano *et al.*, 1990). Pre-B cells, responding to IL-7, can differentiate into mitogen-responsive B cells in the presence of a stromal cell line called S17. Initially, the ability to support this transition was thought to be a unique property of the S17 cell line. However, in assays in which contact with stromal cells is permitted all stromal cell lines tested were able to support this transition (stroma B) (Cumano *et al.*, 1994). Cell contact does not, however, appear to be an absolute requirement for the B cell progenitor at this stage because S17 cells can support this transition even when separated from progenitor cells by a semisolid agar matrix. It is possible, therefore, that stromal cell lines utilize different mechanisms to support pre-B cell maturation. The relevant factors may occur in membrane, matrix, or soluble form on different cell lines. Alternatively, contact may be important for induction of the relevant factors in some of the stromal cell lines. To date no factor combination has been identified that can substitute for stroma B.

The utility of *in vitro* cultures for the study of B lineage progression from progenitors that have not yet undergone definitive commitment has been demonstrated. $AA4.1^+Ly6A^+B220^-Mac-1^-$ progenitors isolated from fetal liver at day 12 of gestation can give rise to both B lymphocytes and macrophages in the presence of stromal cells and Il-7 (Cumano *et al.*, 1992). Analysis of the progeny of single progenitors indicates that these cells most likely do not have the capacity to give rise to other hematopoietic lineages (Cumano *et al.*, 1992; Kee *et al.*, 1994). Their ability to give rise to T lymphocytes, predicted by some differentiative models, has not yet been tested under conditions that would promote T cell development. Similar experiments have led to the identification of cells with B cell potential in the yolk sac and the paraaortic splanchnopleura, at day 8.5 of gestation (Cumano *et al.*, 1993; Godin *et al.*, 1994). These progenitors are $AA4.1^+Ly6A^-$ and can give rise to multiple hematopoietic lineages under these conditions. This supports the conclusion that the progenitor isolated from fetal liver is bipotential. Comparison of the multilineage and bipotential progenitors *in vitro* suggests that the multilineage progenitors are delayed in their development by 4–5 days. The bipotential progenitor requires 8–9 days in the presence of stroma A and IL-7 before the first cells that can respond to stroma B are detected (Cumano *et al.*, 1992). Under the same conditions the multilineage progenitor requires 12–15 days before the first stroma B-responsive cells arise (Cumano *et al.*, 1993). Similarly, the multilineage progenitor undergoes Ig heavy chain allotype restriction 4 days after the bipotential progenitors (B. Kee, unpublished).

Thus, the developmental stage of a progenitor can be measured by the length of time in culture required to progress through the B lineage. Stromal cell-independent culture conditions have been established that support the growth and differentiation of the bipotential cells. These conditions do not alter the time in culture required for differentiation, suggesting that this parameter is an intrinsic property of the responding cell and is not directed by environmental factors (Kee *et al.*, 1994).

V. Cytokines and Growth Factors

A number of cytokines, including IL-2, IL-4, IL-5, and IL-6, are known to have effects on the later stages of B cell development. The role of cytokines in supporting the early events in B cell development has been actively investigated over the past 5 years. A complex story is emerging in which *in vitro* findings are not readily reconciled with *in vivo* data. The complexity of the *in vivo* environment, and the potential for secondary effects of cytokines on surrounding stromal cells, makes extrapolation from *in vitro* systems difficult. Furthermore, the ability of multiple cytokines to stimulate similar biological responses suggests that any single factor may not be crucial for differentiation. However, identification of the minimal requirements for growth and differentiation provides useful information about the mechanisms of lineage progression. In addition, a critical developmental stage in B lineage progression has been identified that can be supported by only a single factor.

A. Interleukin 7

The ability of stromal cells to support the growth and differentiation of pre-B lymphocytes *in vitro* was used as an assay to clone a pre-B cell growth factor called interleukin 7 (IL-7) (Namen *et al.*, 1988). Interleukin 7 is a 25-kDa soluble factor produced by stromal cells from many hematopoietic tissues such as bone marrow, spleen, thymus, and fetal liver. The receptor for IL-7 (IL-7R) is a 68-kDa protein belonging to the hematopoietic growth factor receptor family, which can exist in both a transmembrane and soluble form (Goodwin *et al.*, 1990). This family of receptors is characterized by the absence of an intracellular kinase domain and the presence of an extracellular 200-amino acid region containing 4 regularly spaced cysteine residues, two fibronectin III homology domains, and a Trp-Ser-X-Ser-Trp sequence (where X is any amino acid) (Bazan, 1990; Taga and Kishimoto, 1993). Both high-affinity ($K_a = 3.7 \times 10^{-9}\ M$) and low-affinity ($K_a = 6.7 \times 10^{-7} M$) receptors for IL-7 have been identified on pre-B lymphocytes but the molecular distinction between these has not been resolved (Park et al., 1990). The IL-7R shares a common component, the γ chain, with the receptors for IL-2 and IL-4, which functions in signal transduction (see Section V,E). (Kondo et al., 1993; Noguchi et al., 1993a; Russell et al., 1993). This receptor stimulates proliferation, in part, via recruitment and activation of p59^{fyn} and 1-phosphatidylinositol 3-kinase (PI 3-K) (Venkitaraman and Cowling, 1992; Dadi and Roifman, 1993). In humans, failure to express the common γ chain results in a severe combined immunodeficiency (SCID) characterized by a complete absence of mature B and T lymphocytes (Noguchi et al., 1993b).

Interleukin 7 promotes the proliferation of B220$^+$ pre-B lymphocytes found in bone marrow and fetal liver (Namen et al., 1988; Lee et al., 1989; Cumano et al., 1990). However, IL-7 is not a B lineage-specific factor, as it can influence the growth of myeloid progenitors, monocytes, T cell progenitors, and mature CD4$^+$ and CD8$^+$ T cells (Alderson et al., 1991; Samaridis et al., 1991; Jacobsen et al., 1993). Expansion of bone marrow or fetal liver cells in IL-7 results in a uniform population of B220$^+$sIg$^-$ pre-B lymphocytes that require stroma B for further differentiation (Cumano et al., 1990). The first IL-7-responsive progenitors, which are B220$^+$ and lineage restricted, arise in fetal liver by day 13 of gestation. During the course of IL-7-driven expansion pre-B lymphocytes undergo a number of discernible differentiative events. The clonable cells rearrange their Ig gene segments and, consequently, commit to the expression of an Ig heavy chain allotype and light chain isotype. Further, progenitors that initially were unable to secrete Ig in response to stromal cells and LPS acquire this capacity after expansion in IL-7 (Cumano and Paige, 1992). The precise role of this factor in B lineage progression is unclear. Interleukin 7 may support the survival and proliferation of progenitors that continue to differentiate through an internally programmed series of events. Alternatively, IL-7 may play a more active role in the differentiation process. Reports have suggested that IL-7 regulates the expression of rag-1 and rag-2 during T cell development (Appasamy et al., 1993; Muegge et al., 1993). However, dissociation of the survival and proliferative effects of this cytokine from induction of maturation will require further experimentation.

A central role for IL-7 in lymphopoiesis in vivo has been demonstrated. The number of mature B and T lymphocytes is elevated in IL-7 transgenic mice, and in mice administered IL-7 through subcutaneous injection (Morrissey et al., 1991; Samaridis et al., 1991; Fisher et al., 1992). The increase in B lineage cells is evident in the AA4.1$^+$B220$^+$ bone marrow population, suggesting that IL-7 promotes the growth of pre-B lymphocytes. In addition, IL-7 is essential for B lineage progression because administration of neutralizing anti-IL-7R MAb into adult mice completely abrogates B cell development at the pre-B cell stage (Grabstein *et al.*, 1993; Sudo *et al.*, 1993). This suggests that alternative pathways of differentiation, or compensatory growth factors, do not exist at this stage.

B. Mast Cell Growth Factor

Mast cell growth factor (MGF; *Sl* or c-*kit* ligand) is a pleiotropic factor that plays a crucial role in the maintenance of early hematopoietic progenitors. Mast cell growth factor can exist as a soluble, 25-kDa dimer or as a type I transmembrane protein containing a 189-amino acid extracellular domain, a 36-amino acid intracellular domain, and a 23-amino acid transmembrane domain (Anderson *et al.*, 1990; Copeland *et al.*, 1990; Huang *et al.*, 1990; Zsebo *et al.*, 1990). The soluble form can arise by alternative splicing of mRNA or by proteolytic cleavage of the transmembrane protein (Anderson *et al.*, 1990; Lu *et al.*, 1991). A third, alternatively spliced form of this factor exists that lacks the exon containing the extracellular cleavage site and can exist only as a membrane-associated protein (Flanagan *et al.*, 1991). The steel-dickie (*Sl^d*) mutation, which results in a complete absence of erythroid and mast cells, encodes a protein that can exist only in a soluble form, highlighting the importance of the transmembrane factor in hematopoiesis (Brannan *et al.*, 1991; Flanagan *et al.*, 1991). The MGF receptor (MGFR), c-*kit* or *W*, is a member of the protein tyrosine kinase family of receptors (Chabot *et al.*, 1988; Geissler *et al.*, 1988).

Mast cell growth factor, as the name implies, is a growth factor for murine mast cells. On its own, MGF does not stimulate the growth of B lineage cells. However, the MGF receptor can be detected on B220$^+$TdT$^+$ B lineage progenitors prior to the expression of cμ (Era *et al.*, 1994; Karasuyama *et al.*, 1994; Rico-Vargas *et al.*, 1994). Mast cell growth factor is synergistic with IL-7 in inducing the proliferation of pre-B lymphocytes (Billips *et al.*, 1992). This synergistic effect is due to an increase in the frequency of progenitors that can respond to IL-7 in the presence of MGF (Narendran *et al.*, 1993). Presumably, MGF recruits primitive B lineage progenitors to proliferate and ultimately differentiate into IL-7-responsive pre-B cells. The notion that MGF plus IL-7-responsive progenitors are less mature than IL-7 responsive progenitors is supported by two experimental findings. First, all IL-7-responsive cells are B220$^+$ whereas MGF plus IL-7 can support the growth of B220$^-$ cells (McNiece *et al.*, 1991; Kee *et al.*, 1994). Second, MGF plus IL-7-responsive progenitors develop 12–24 hr before IL-7-responsive progenitors during fetal ontogeny (Kee *et al.*, 1994).

In vitro, the requirement for stroma A (Fig. 3) can be attributed, in part, to MGF. Antagonistic anti-MGFR antibodies can inhibit the development of IL-7-responsive pre-B cells in some stromal cell-dependent culture systems (Rolink *et al.*, 1991b). However, the same MAb had a less dramatic effect on B lineage progression supported by a different stromal cell line (Era *et al.*, 1994). Thus MGF is not required for some stromal cells to support B cell differentiation. Consistent with this finding, *in vivo* administration of anti-MGFR antibodies into adult mice resulted in an increase in B cell progenitor numbers, while inhibiting development of progenitors in other lineages (Ogawa *et al.*, 1991; Era *et al.*, 1994; Rico-Vargas *et al.*, 1994). The *in vivo* data are not easily interpreted given the array of positive and negative regulatory molecules produced by a complex mixture of stromal cells. Removal of MGF may disrupt the cytokine balance, leading to secondary effects that could not be predicted on the basis of more simplistic, *in vitro* systems. However, these data show that MGF is not obligatory for the development of B lineage cells. This is consistent with the development of mature B cells in Sl/Sl^d and W/W^v mice (Mekori and Phillips, 1969). However, the absolute number of B220$^+$ cells is decreased in the fetal liver of W/W^x mice, indicating that MGF can influence B lymphopoiesis *in vivo* (Landreth *et al.*, 1984).

It is likely that additional stromal cell-derived factors exist that can substitute for MGF in the support of B lymphopoiesis. Flt-3/flk-2, a factor sharing many structural features with MGF, may fulfill this function (Lyman *et al.*, 1993). Flt-3/flk-2 is a 231-amino acid, transmembrane growth factor that can also exist as a soluble dimer. The flt-3/flk-2 receptor is a protein tyrosine kinase with a hematopoietic distribution similar to that of MGFR (Matthews *et al.*, 1991). Furthermore, flt-3/flk-2 can synergize with MGF and IL-7 to promote the proliferation of AA4.1$^+$Ly6A$^+$ progenitors from bone marrow. Stromal cell lines whose B lymphopoietic ability is not inhibited by anti-MGFR antibodies may be producing flt-3/flk-2. However, the function of this growth factor in B cell development remains to be investigated.

C. Interleukin 11

Interleukin 11 was cloned on the basis of its ability to support the proliferation of an IL-6-dependent plasmacytoma cell line in the presence of a neutralizing anti-IL-6 MAb (Paul *et al.*, 1990). It is a 20-kDa soluble protein secreted by fibroblastic stromal cells. In some stromal cell lines expression of this cytokine is inducible by IL-1α or PMA whereas in other lines it is expressed constitutively. The protein sequence of IL-11 lacks cysteine residues, unlike other known cytokines. The biological functions of IL-11 overlap significantly with those of IL-6. This is likely a reflection of the fact that IL-11 and IL-6 share a common receptor component, gp130 (see Section V,C) (Yin *et al.*, 1993; Zhang *et al.*, 1994). The ligand-binding component of the IL-11 receptor (IL-11R) is a single protein of 151 kDa and a single class of high-affinity IL-11R (K_d = 3.5 × 10^{-10} M) is expressed on stromal cell lines (Yin *et al.*, 1992a).

Interleukin 11 stimulates the growth of granulocyte–macrophage progenitors from murine bone marrow and synergizes with a number of factors including MGF, IL-3, and IL-4 (Musashi *et al.*, 1991a; Hangoc *et al.*, 1993). Interleukin 11, in concert with IL-3 and IL-4, augments colony formation by multilineage progenitors by shortening the G_0/G_1 phase of the cell cycle (Mushsashi *et al.*, 1991b). In combination with IL-3 or MGF, IL-11 supports the growth of platelet-producing megakaryocytes (Yonemura *et al.*, 1992). Antigen-specific antibody responses can be modulated by IL-11 but this appears to be mediated by CD4$^+$ T cells (Yin *et al.*, 1992b). Interleukin 11 has no known effects on pre-B lymphocytes but supports their formation from lineage-unrestricted progenitors in combination with MGF and IL-7. Interleukin 11, MGF, and IL-7 are sufficient to support the growth and differentiation of pre-B lymphocytes and macrophages from bipotential progenitors present in fetal liver at day 12 of gestation (Kee *et al.*, 1994). However, the requirement for differentiation of multilineage progenitors prior to day 10 of gestation is not provided by this factor combination. In contrast, IL-11, MGF, and IL-7 are sufficient to support B cell development from multipotential progenitors in bone marrow from mice treated with 5-fluorouracil (5-FU) (Hirayama *et al.*, 1992; Katayama *et al.*, 1993). This suggests that progenitors from bone marrow and early embryonic sites may have different growth factor requirements. Alternatively, treatment with 5-FU may alter the growth factor requirements of cell cycle-dominant progenitors.

D. Other Factors Regulating B Lymphopoiesis

In long-term cultures of pre-B lymphocytes factors of 10 and 40–60 kDa potentiate the differentiation of $c\mu^-$ pro-B cells to the $c\mu^+$ stage (Landreth and Dorshkind, 1988). This activity is attributed to insulin-like growth factor I (IGF-I) produced by stromal cells in a soluble form (10 kDa), or in conjunction with IGF-binding proteins (40–60kDa) (Landreth *et al.*, 1992). Insulin-like growth factor I synergizes with IL-7 in promoting pre-B cell proliferation. This is similar to the effects of MGF, although additive increases in proliferation are observed with both IGF-I and MGF in long-term cultures (Gibson *et al.*, 1993). Thus, IGF-I may function in a manner analogous to that proposed for flt-3/flk-2 in Section V,B. Insulin and IGF-

II also potentiate IL-7-dependent proliferation but none of these factors promote maturation to sIg expression.

A number of factors have been identified that inhibit the growth of pre-B lymphocytes. These include transforming growth factor β (TGF-β), interferon γ (IFN-γ), IL-1α, granulocyte–macrophage colony-stimulating factor (GM-CSF), and estrogen. Transforming growth factor β and IFN-γ inhibit proliferation of IL-7-responsive pre-B lymphocytes and result in death by apoptosis (Lee *et al.*, 1989; Gimble *et al.*, 1993). This effect may be specific to the IL-7-responsive stage because more mature pre-B cell lines can be induced to express sIg by treatment with IFN-γ (Paige *et al.*, 1978). Pre-B cells from *bcl-2* transgenic mice are rescued from the apoptotic effects of IFN-γ but proliferation in response to IL-7 is still inhibited (Grawunder *et al.*, 1993). Furthermore, treatment of Bcl-2-containing pre-B cells with IFN-γ did not result in sIg expression, suggesting that inhibition of proliferation is not sufficient for maturation.

The suppressive effects of some factors may be due to indirect effects on pre-B cells. For example, GM-CSF stimulates myelopoiesis *in vitro* and *in vivo* with a concomitant reduction in B cell production (Dorshkind, 1991; Lee *et al.*, 1993). This may be the result of interlineage competition for stromal cell-derived products or the production of a suppressive substance by myeloid cells. Furthermore, IFN-γ and IL-1α can inhibit B cell development by regulating the function of stromal cells (Dorshkind, 1988; Billips *et al.*, 1990; Gimble *et al.*, 1993). Estrogen also exerts a profound influence on B lymphopoiesis although its mechanism of action has not been resolved. Estrogen causes a 60% reduction in bone marrow pre-B cells at the stage of IL-7 responsiveness. However, more immature B220$^+$CD43$^+$HSA$^-$BP-1$^-$ cells are less affected by this hormone (Kincade *et al.*, 1994).

E. Cytokine Receptors and Signal Transduction

The precise function of cytokines in B lineage progression is largely unknown. Stimulation with a single cytokine can result in different biological consequences depending on the stage of differentiation or the responding cell type. As well, cytokines that operate through distinct receptors can result in overlapping, and yet distinct, cellular responses. Significant advances have been made in resolving the mechanisms by which growth factors mediate their biological effects. A comprehensive analysis of this area is beyond the scope of this chapter; however, we highlight here some of the receptors that function during B lineage progression (Schreurs *et al.*, 1992; Montminy, 1993; Stahl and Yancopoulos, 1993).

The receptors for a number of cytokines share components that can function in signal transduction. The receptors for IL-7 and IL-4 associate with a common γ chain, first identified in the high-affinity IL-2R (Kondo *et al.*, 1993; Noguchi *et al.*, 1993a; Russell *et al.*, 1993). This γ chain is involved directly in mediating signal transduction. Additional components of the trimeric IL-2R have been identified as components of other receptors. For example, IL-2R β chain is also a part of the receptor for IL-15, a cytokine that induces similar responses to IL-2 in mature T cells (Grabstein *et al.*, 1994). It is not known whether additional receptor components can associate with the IL-7R. Similarly, IL-11 shares a signal-transducing component, gp130, with the receptors for IL-6, leukemia inhibitory factor (LIF), ciliary neurotrophic factor (CNTFR), and oncostatin M (Ip *et al.*, 1992; Yin *et al.*, 1993; Zhang *et al.*, 1994). Additional complexity is added to this family of receptors by their ability to share ligand-binding chains (the LIFR is part of the CNTFR complex) (Davis *et al.*, 1993). It would appear that these cytokines can elicit similar biological responses through the use of common receptor components. Furthermore, the IL-11 and IL-6 receptors must share a common cellular distribution as their ligands support the growth of the same cell types. Signal transduction through many of these receptors results in survival or proliferation, suggesting that the growth requirements of a progenitor may be a simple reflection of receptor expression. Thus, knowledge of receptor expression on hematopoietic progenitors may provide insights for the establishment of *in vitro* systems that support their growth.

Signal transduction through many cytokine receptors leads to activation of *ras* and, consequently, mitogenesis (Ullrich and Schlessinger, 1990; Schreurs *et al.*, 1992). Activation of the same pathway by many growth factors has been difficult to reconcile with the distinct biological responses induced by different receptors. However, studies have identified a second, *ras*-independent, pathway that results in the activation of a family of latent cytoplasmic transcriptional regulators (Bonni *et al.*, 1993; Kotanides and Reich, 1993; Larner *et al.*, 1993; Ruff-Jamison *et al.*, 1993; Sadowski *et al.*, 1993; Shuai *et al.*, 1993; Silvennoinen *et al.*, 1993). Cytokine receptors that lack an intrinsic kinase activity, such as IL-7R or IL-11R, activate this pathway through recruitment of intracellular kinases such as Jak1, Jak2, and Tyk2 (Murakami *et al.*, 1993; Stahl and Yancopoulos, 1993; Lutticken *et al.*, 1994). These kinases, either directly or indirectly, phosphorylate different members of the p91STAT family of transcriptional activators. Phosphorylation of these factors results in nuclear translocation, activation, and their association in multiprotein transcription complexes. Distinct complexes are activated by different cytokine receptors, accounting for their divergent cellular responses. In the case of gp130, the signaling component of the IL-11R, a factor that has 52% homology to p91STAT is activated in response to ligand stimulation (Akira *et al.*,

1994). These transcriptional complexes regulate distinct but overlapping sets of genes. Identification of the genes that are regulated by these complexes will help to resolve the question of whether cytokines play an active role in directing hematopoietic differentiation.

VI. Requirements for B Lineage Progression

Progression through the B lineage is characterized by the appearance and disappearance of gene products, many of which have been identified and analyzed. However, it is not clear which of these gene products are critical for the differentiation process. Advances in this area have come from the creation of mice carrying mutated alleles of genes whose pattern of expression suggests a role in B cell development. For example, B220 is expressed on almost all B lineage cells and participates in signal transduction (Ostergaard et al., 1989; Mustelin and Altman, 1990; Shiroo et al., 1992). It has a clearly defined role in promoting the proliferative response to antigen receptor cross-linking (Gold et al., 1990; Justement et al., 1991). However, B220 is expressed much earlier than sIg and could function in lineage progression through regulation of additional signaling proteins. Mice that lack expression of B220 nonetheless develop normal numbers of pre-B and mature B lymphocytes (Kishihara et al., 1993). These pre-B cells appear to respond normally to stromal cell-derived signals and IL-7. As predicted, the mature B cells from these mice fail to proliferate after antigen receptor cross-linking induced by anti-μ antibody. B220 may influence B cell development; however, it is not an essential requirement for lineage progression.

A. The Central Role of Immunoglobulin

The principal function of the B cell in immune responses is the production of antibody. It is not surprising, therefore, that B cell development is dependent on the production of appropriately assembled Ig complexes. Mature B lymphocytes express functional Ig receptors composed of heavy chains in association with κ or λ light chains. The majority of peripheral B cells are antigen selected, indicating that the production of a functional Ig receptor may be required for survival (Gu et al., 1991b). However, pre-B lymphocytes are also selected on the basis of expression of a functional μ heavy chain protein. Selection at the pre-B cell stage cannot be dependent on the conventional antigenic specificity of the Ig receptor because κ or λ light chains are not yet expressed. At this stage B lineage cells express

μ in association with a surrogate light chain composed of $\lambda 5$ and V_{pre-B} (Cherayil and Pillai, 1991; Misener *et al.*, 1991; Lassoued *et al.*, 1993; Karasuyama *et al.*, 1994). The requirement for μ and $\lambda 5$ has been shown conclusively because mouse strains that fail to express these proteins have arrested development at the pre-B cell stage (Kitamura *et al.*, 1992; Ehlich *et al.*, 1993). In the absence of μ, $B220^+CD43^+HSA^+BP-1^+$ progenitors are present in normal numbers but $B220^+CD43^-$ pre-B cells and mature B lymphocytes do not develop.

The arrest in B lineage progression is also seen in mice lacking the membrane portion of μ (Kitamura *et al.*, 1991). The requirement for membrane expression of this pre-B cell receptor suggests that it may interact with a ligand in the surrounding microenvironment, or on the B cell itself. Ligand–receptor interaction would send a signal to the cell resulting in survival and progression to the next stage of differentiation. The postulated ligand may interact with V_{pre-B} or $\lambda 5$, as well, because B cell development in $\lambda 5$-deficient mice is also impaired. In contrast to μ, the requirement for $\lambda 5$ is not absolute because mature B cells do develop, albeit at a reduced rate, in these mice (Kitamura *et al.*, 1992). It has been suggested that a few cells may progress through this stage because they are able to express V_{pre-B} and μ on the surface in the absence of $\lambda 5$, or a minor developmental pathway exists that is not dependent on $\lambda 5$ (Desiderio, 1994; Rolink *et al.*, 1994). Alternatively, rearrangement and expression of conventional light chains at the pre-B cell stage may rescue some cells by performing the function of $V_{pre-B}-\lambda 5$. Occasional light chain gene rearrangement can be detected in normal pre-B cells at this stage of development, consistent with this latter hypothesis (Ehlich *et al.*, 1993; Loffert *et al.*, 1994).

The nature of the signal delivered by the $V_{pre-B}-\lambda 5-\mu$ complex has not been resolved. Similar to growth factor receptors, it could support survival and proliferation, allowing an intrinsic differentiation program to continue. Alternatively, this signal could promote lineage progression directly. Experiments addressing this question have been initiated in genetically altered mice (Strasser *et al.*, 1994). Severe combined immunodeficient (SCID) mice have a defect that results in the inability to repair double-strand breaks in DNA (Fulop and Phillips, 1990). SCID mice can make some normal rearrangements; however, functional μ heavy chains are generally not detected (Penneycook *et al.*, 1993). Predictably, these mice show an arrest in B cell development at the large $\rightarrow$ small pre-B cell transition (Osmond *et al.*, 1992). In contrast, pre-B lymphocytes in *bcl-2* transgenic mice do not die at this stage of development because apoptosis is inhibited (Korsmeyer, 1992). SCID/*bcl-2* mice have $B220^+$ cells in the peripheral lymphoid organs, which lack expression of sIg (Strasser *et al.*, 1994). These $B220^+$ cells appear to have progressed beyond the pre-B cell stage in that they lack BP-1 and CD43 and express markers of mature B cells such as CD40, CD21, and CD23. Furthermore, they have undergone extensive, nonproductive rearrangement at the heavy and light chain loci. These data raise the possibility that the signals derived from the $V_{pre-B}-\lambda5-\mu$ receptor promote survival rather than differentiation.

Additional roles for Ig in B cell differentiation have been proposed, on the basis of the observed pattern of Ig gene rearrangement. These models suggested that the products of a productive rearrangement at one locus would signal rearrangement at the next locus. In addition, it was envisioned that a similar signal could mediate allelic exclusion by inhibiting further rearrangement at the productive locus (Alt *et al.*, 1982; Reth *et al.*, 1987; Iglesias *et al.*, 1991). This model suggests that the observed pattern of rearrangement is a requirement for lineage progression. Although attractive in theory, exceptions to the ordered rearrangement process have been identified (Berg *et al.*, 1980; Schlissel *et al.*, 1991). Furthermore, this model is not supported by experimental findings in gene-targeted mice (Cumano *et al.*, 1994; Loffert *et al.*, 1994). Mice have been created that cannot produce an Ig heavy chain owing to deletion of $J_{H1}-J_{H4}$ (Ehlich *et al.*, 1993). Both κ and λ rearrangements occur in these mice, showing that a productive heavy chain rearrangement is not required for light chain rearrangement. Furthermore, mice that carry a functionally silenced κ gene (created by deletion of Cκ, J, and Cκ or the κ intron enhancer) have mature B cells, all of which express λ light chains (Chen *et al.*, 1993; Takeda *et al.*, 1993; Zou *et al.*, 1993). This result demonstrates that a nonproductive rearrangement of the κ locus is not a prerequisite for λ rearrangement. Therefore, the observed pattern of Ig rearrangement is not a requirement for B cell development and likely reflects a property of the recombination process.

B. The Role of Cytokines in Lineage Progression

All stages of B cell differentiation, from multipotent progenitors to mature B lymphocytes, are dependent on factors produced by stromal cells in the surrounding microenvironment. In many cases more than one factor can support the same stage of differentiation (see Section V). To date, the only factor identified that is uniquely required for B cell development is interleukin 7. Inhibition of IL-7 activity *in vivo* results in a block at the $B220^+CD43^+BP-1^-$, prior to that in stage mice which cannot express a functional $V_{pre-B}-\lambda5-\mu$ receptor (Grabstein *et al.*, 1993; Sudo *et al.*, 1993). It has been proposed that IL-7 is necessary for expression of *rag-1* and *rag-2* in pre-T cells (Appasamy *et al.*, 1993; Muegge *et al.*, 1993). If this is also true in B cell development, inhibition of IL-7 will inhibit rearrangement and expression of μ. However, the phenotype of IL-7-deficient mice suggests that IL-7 might regulate the expression of additional genes. Alternatively, IL-7 may support the growth of a population of cells that, independently, expresses the *rag* genes.

Comparison of B lymphopoiesis *in vivo* and *in vitro* may provide additional insights into the role of IL-7 in lineage progression. As noted, pre-B lymphocytes can be expanded *in vitro* for extended periods in the presence of IL-7. Although the vast majority of these cells eventually undergo senescence, IL-7-dependent cell lines can be derived routinely from rare progenitors that continue to divide (Ishihara *et al.*, 1991). It appears unlikely that a similar degree of proliferation occurs at this stage *in vivo*. Furthermore, *in vitro* expansion in IL-7 does not appear to be dependent on expression of a functional μ heavy chain. For example, the frequency of cells expressing $c\mu$, or that can secrete Ig in response to stromal cell and LPS, decreases with time in culture (Lee *et al.*, 1989; B. Kee, unpublished observation). Further, pre-B lymphocytes from SCID, $\lambda5$-, or Rag-deficient mice can proliferate in response to IL-7 *in vitro* (Witte *et al.*, 1987; Rolink *et al.*, 1993c; Cumano *et al.*, 1994). This suggests that the control mechanisms operative *in vivo*, at the pre-B cell stage, may not be operative *in vitro*, in the presence of excess IL-7. Presumably, if present in sufficient quantities, IL-7 could support cells that do not express μ *in vivo*. This is consistent with the observed increase in pre-B cell numbers in IL-7 transgenic mice (Samaridis *et al.*, 1991; Fisher *et al.*, 1992). This raises the possibility that, *in vivo*, IL-7 is limiting. Furthermore, acquisition of IL-7 may be dependent on expression of a functional pre-B receptor. One possibility is that expression of the $V_{pre-B}-\lambda5-\mu$ complex acts to bring pre-B lymphocytes in contact with IL-7-producing stromal cells. This may be mediated by adhesion to the appropriate cell, or by induction of IL-7 in a responsive stromal cell. This supposition would suggest that surface expression of the $V_{pre-B}-\lambda5-\mu$ complex is necessary but the capacity to mediate ligand-induced signal transduction in the pre-B cell is not. This prediction may soon be testable in gene-ablated mice.

Pre-B cells cultured in IL-7 undergo rearrangement of their Ig heavy and light chain genes but only a few sIg[+] cells are detected (Narendran *et al.*, 1993). It is possible that sIg[+] cells do develop regularly in these cultures and remain at low levels because they no longer respond to IL-7. Alternatively, under these conditions the cells may continue to rearrange their Ig loci, increasing the probability of acquiring a nonfunctional rearrangement (Rolink *et al.*, 1993a). Identification of the factors that support the transition of IL-7-responsive progenitors into mature B cells will allow the requirements for progression at this stage of development to be investigated more extensively.

VII. Lineage Commitment

Despite rapid advances in understanding the development of B cells, and other components of the hematopoietic system, the genetic basis for commitment to a particular lineage remains largely unknown. Over the past 30 years, a number of models for lineage determination have been proposed, ranging from inductive to stochastic (Till *et al.*, 1964; Curry and Trentin, 1967). The degree to which environmental factors direct the differentiation process, select amongst options intrinsic to developing stem cells, or regulate the choice between life and death is under intense investigation in numerous experimental systems. As noted previously in this chapter, studies of B cell development have revealed genes that, by virtue of their pattern of expression and potential function, might participate in a commitment process. Obvious candidates include genes that encode growth factor receptors, mediate cellular interactions, control intracellular signaling pathways, or regulate nuclear activities. However, placement of such genes into the context of commitment requires a better understanding of how commitment is achieved.

Figure 4 illustrate two models of lineage commitment based on gene expression. The first model, which we call the "directed" model, suggests that expression of a single gene is sufficient to initiate lineage determination. This postulated gene could encode a protein that regulates the expression of a set of genes required for commitment (Fig. 4a, top). Alternatively, this gene could initiate a cascade of events that eventually results in lineage restriction (Fig. 4a, bottom). The second model, which we call the "conjunction" model, suggests that commitment to a lineage is the result of the timely expression of a number of lineage-specifying genes (Fig. 4b). In this case, the precise temporal order of expression of each gene is not crucial and no one gene is more important than another. Possibly, more than one combination of genes could result in restriction to a given developmental pathway. Commitment would be dependent on the expression of a set of genes, within a given developmental window, in one cell. This model would allow (perhaps even require) the simultaneous expression of genes, within a given progenitor, that specify different lineages during the developmental process, an observation that is generally dismissed as anomalous when noted.

Studies with transformed cell lines have revealed an unexpected plasticity in the integrity of the differentiative state. There are numerous reports, for example, of B lineage cell lines that can lose some of their B cell properties and acquire macrophage properties (Bauer *et al.*, 1986; Davidson *et al.*, 1988; Klinken *et al.*, 1988; Borzillo and Sherr, 1989; Borzillo *et al.*, 1990; Hara *et al.*, 1990; Principato *et al.*, 1990). In the 70Z/3 cell

a   b

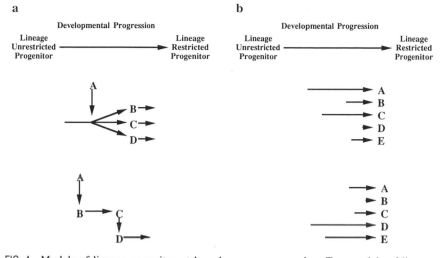

FIG. 4   Models of lineage commitment based on gene expression. Two models of lineage commitment, based on gene expression, are illustrated. Arrows indicate the expression of a given gene. (a) The directed model suggests that expression of a single gene (A) is sufficient to initiate commitment to the B lineage. This event may directly induce the expression of a number of genes (B, C, and D) that commit the progenitor to the B lineage (top). Alternatively, the product of this gene may initiate a cascade of events, the end result of which is expression of the genes required for commitment to the B lineage (bottom). (b) The conjunction model suggests that commitment is the result of expression of a set of lineage-specifying genes (A, B, C, D, and E) within a given developmental window, in one cell. The precise temporal order of expression of each gene is not critical and more than one combination of genes may result in commitment. The top and bottom panels represent two independent sequence of events that would result in lineage commitment.

line, we have found that this transition follows a pathway in which sets of genes, apparently functionally unrelated, are reproducibly lost or gained together (Tanaka *et al.*, 1994). This observation suggests that a significant degree of regulation is maintained during this highly abnormal transition. The existence of such cell lines raises the possibility that expression of lineage-associated gene products does not irrevocably commit a cell to progression through that developmental pathway. However, it remains to be determined whether lineage transitions are observed in cell lines because the large clone size affords an opportunity to detect rare events or whether these transitions are the result of genetic instability that may have led to the transformed state.

A number of B lineage-associated genes are expressed at an appropriate time to play a role in commitment. Their products include B220, Rag-

1, Rag-2, mb-1, B29, V_{pre-B}, λ5, IL-7R, and proteins that regulate their expression such as early B cell factor (EBF), B cell activator protein (BSAP), and lymphoid-specific enhancer factor (LEF-1) (Faust *et al.*, 1993; Li *et al.*, 1993; Hagman and Grosschedl, 1994; Marcos *et al.*, 1994). A directed model could predict that proteins that regulate gene expression may play a pivotal role in lineage determination. Transcription factors that regulate B lineage-specific genes have been identified; however, few of these factors are B lineage specific. For example, in addition to being expressed in B lineage cells LEF-1 is expressed in T cell progenitors and EBF is expressed in stromal cells (Travis *et al.*, 1991; Hagman *et al.*, 1993). Thus, additional parameters such as DNA accessibility (possibly owing to methylation or matrix attachment), or the combination of transcriptional activators present in one cell, may be crucial to establishing lineage-specific patterns of gene expression (Hagman and Grosschedl, 1994). For example, event A in the directed model (Fig. 4a) could be an intrinsic change in DNA conformation that alters the ability of transcriptional complexes to bind to the promoter of a lineage-specific gene. Alternatively, event A could be signal transduction through a growth factor receptor leading to posttranslational modifications in transcriptional complexes (see Section V,E). Because IL-7 responsiveness is one of the first detectable parameters of commitment to the B lineage it could be envisioned to function in this capacity. However, IL-7R is expressed in T cell progenitors, suggesting that it does not induce B lineage commitment directly. Expression of IL-7R may not be the commitment-initiating event; however, it could be a secondary event (for example, event C) such that subsequent signal transduction would result in further progression toward the committed state. The directed model suggests that each step in the pathway is essential, but not sufficient, for lineage commitment.

B220, λ5, Rag-1, and Rag-2 are not essential for large determination because mice deficient in the expression of these proteins develop committed B cell progenitors (Kitamura *et al.*, 1992; Mombaerts *et al.*, 1992; Shinkai *et al.*, 1992; Kishihara *et al.*, 1993). However, the conjunction model (Fig. 4b) suggests that they may play a role in commitment when they are expressed. The expression of Rag-1 and Rag-2 in T cell progenitors confirms the notion that they do not induce B lineage commitment. However, expression of numerous lineage-associated proteins in one cell may decrease the probability of progression through an alternate pathway. At the present time little information is available regarding gene expression in lineage-unrestricted progenitors. The ability to study lineage-unrestricted progenitors as they differentiate into mature B cells combined with techniques for the analysis of gene expression in small numbers of cells provide a powerful means by which to study gene expression during the commitment process.

VIII. Conclusions and Future Directions

Two decades of progress in cellular and molecular biology have provided a detailed view of B lymphocyte development. Differentiation from lineage-uncommitted progenitors to mature, immunoglobulin-secreting plasma cells has been observed *in vitro* under defined experimental conditions. Many of the central features of B lymphopoiesis such as the sites or origin, the physical properties of the developing cells, and the factors requirements for growth and differentiation are widely agreed on. These advances are incorporated into the outline of B cell development presented in this chapter. A model of B lineage development has emerged that provides a framework on which to build future models regarding the precise mechanisms of lineage commitment and differentiation. Experiments are now being directed toward delineating the mechanisms by which developmental signals are processed into nuclear actions plans, the end result being the development of a cell capable of mounting a functional immune response.

The identification of cell surface proteins, expressed on developing B cell progenitors, has been a fundamental advance in studies of lineage progression. These proteins have been used to define stages in the developmental program and have enabled investigators to localized and trace progenitor movement *in vitro* and *in vivo*. In many cases, precursor–progeny relationships have been defined. Although these tools have been invaluable in characterizing B cell development, the precise functions of many of these proteins remain to be defined. Knowledge of the functions fulfilled by these surface proteins is essential for a complete understanding of the B lineage, because they are likely to play a central role in the developmental process or in B cell effector functions.

Likewise, our current understanding of the role of microenvironmental interactions in the developmental process requires better definition in mechanistic terms. The nature of the cell surface interactions that lead to signal transduction, the intracellular signaling pathways, and the nuclear events that are a consequence of these signals is only beginning to be understood. Even the developmental role of the Ig molecule, independent of its antigen-binding properties, has resisted firm resolution. The central role of a few growth factors has been established; however, it will be important to ensure that the interactions and signals that operate in elegantly contrived *in vitro* assay systems perform similarly important functions *in vivo*.

Finally, the significance of the expression patterns, which reveal the onset and demise of particular gene products throughout B lineage differentiation, needs to be established. Expression of the products themselves must be determined as well as the precise role of these proteins in the regulation of developmental progression. It remains to be determined which products are critical for B lineage commitment versus those that are necessary for, or are a consequence of, lineage progression and function. This knowledge will lead to a better understanding of the normal process of development as well as of those dysregulated states that lead to immune system dysfunction.

Acknowledgments

We are grateful to A. Marshall for critical reading of this manuscript. We also thank the members of the laboratories of C. Paige and G. Wu for their continuous support and insightful discussions. This work was supported by the Medical Research Council of Canada and the National Cancer Institute of Canada.

References

Abramson, S., Miller, R. G., and Phillips, R. A. (1977). The identification in adult bone marrow of pluripotent and restricted stem cells of the myeloid and lymphoid systems. *J. Exp. Med.* **145,** 1567–1579.

Akira, S., Nishio, Y., Inoue, M., Wang, X.-J., Wei, S., Matsusaka, T., Yoshida, K., Sudo, T., Naruto, M., and Kishimoto, T. (1994). Molecular cloning of APRF, a novel IFN-stimulated gene factor 3 p91-related transcription factor involved in the gp130-mediated signaling pathway. *Cell (Cambridge, Mass.)* **77,** 63–71.

Alderson, M. R., Tough, R. W., Ziegler, S. F., and Grabstein, K. H. (1991). Interleukin 7 induces cytokine secretion and tumoricidal activity by human peripheral blood monocytes. *J. Exp. Med.* **173,** 923–930.

Alt, F. W., and Baltimore, D. (1982). Joining of immunoglobulin H chain gene segments, implications from a chromosome with evidence of three D-J_H fusions. *Proc. Natl. Acad. Sci. U.S.A.* **79,** 4118–4122.

Alt, F. W., Rosenberg, N., Casanova, R. J., Thomas, E., and Baltimore, D. (1982). Immunoglobulin heavy-chain expression and class switching in a murine leukaemia cell line. *Nature (London)* **296,** 325.

Alt, F. W., Yancopoulos, G. D., Blackwell, T. K., Wood, C., Thomas, E., Boss, M., Coffman, R., Rosenberg, M., Tonegawa, S., and Baltimore, D. (1984). Ordered rearrangement of immunoglobulin heavy chain variable region segments. *EMBO J.* **3,** 1209–1219.

Anderson, D. M., Lyman, S. D., Baird, A., Wignall, J., Eisenman, J., Rauch, C., March, C. J., Boswell, H. S., Gimpel, S. D., Cosman, D., and Williams, D. E. (1990). Molecular cloning of mast cell growth factor, a hematopoietin that is active in both membrane bound and soluble form. *Cell (Cambridge, Mass.)* **63,** 235–243.

Appasamy, P. M., Kenniston, T. W., Jr., Weng, Y., Holt, E. C., Kost, J., and Chambers, W. H. (1993). Interleukin 7 induced expression of specific T cell receptor γ variable region genes in murine fetal liver cultures. *J. Exp. Med.* **178,** 2201–2206.

Bauer, S. R., Holmes, K. L., Morse, H. C., III, and Potter, M. (1986). Clonal relationship of the lymphoblastic cell line P388 to the macrophage cell line P388D1 as evidenced by immunoglobulin gene rearrangements and expression cell surface antigens. *J. Immunol.* **136**, 4695–4699.

Bazan, J. F. (1990). Structural design and molecular evolution of a cytokine receptor super-family. *Proc. Natl. Acad. Sci. U.S.A.* **87**, 6934–6938.

Berg, J., McDowell, M., Jack, H.-M., and Wabl, M. (1980). Immunoglobulin λ gene re-arrangement can precede κ gene rearrangement. *Dev. Immunol.* **1**, 53.

Billips, L. G., Petitte, D., and Landreth, K. S. (1990). Bone marrow stromal cell regulation of B lymphopoiesis: Interleukin-1 (IL-1) and IL-4 regulate stromal cell support of pre-B cell production in vitro. *Blood* **75**, 611.

Billips, L. G., Petitte, D., Dorshkind, K., Narayanan, R., Chiu, C.-P., and Landreth, K. S. (1992). Differential roles of stromal cells, interleukin-7, and *kit*-ligand in the regulation of B lymphopoiesis. *Blood* **79**, 1185–1192.

Blackwell, K., and Alt, F. (1989). Mechanism and developmental program of immunoglobulin gene rearrangement in mammals. *Annu. Rev. Genet.* **23**, 605–636.

Bohuslav, J., Cinek, T., and Horejsi, V. (1993). Large, detergent-resistant complexes con-taining murine antiges Thy-1 and Ly-6 and protein tyrosine kinase p56^{lck}. *Eur. J. Immunol.* **23**, 825–831.

Bonni, A., Frank, D. A., Schindler, C., and Greenberg, M. E. (1993). Characterization of a pathway for ciliary neurotrophic factor signaling to the nucleus. *Science* **262**, 1575–1579.

Borzillo, G. V., and Sherr, C. J. (1989). Early pre-B cell transformation induced by the v-*fms* oncogene in long-term mouse bone marrow cultures. *Mol. Cell. Biol.* **9**, 3973–3981.

Borzillo, G. V., Ashmun, R. A., and Sherr, C. J. (1990). Macrophage lineage switching of murine early pre-B lymphoid cells expressing transduced *fms* genes. *Mol. Cell. Biol.* **10**, 2703–2714.

Bothwell, A., Pace, P. E., and LeClair, K. P. (1988). Isolation and expression of an IFN-responsive Ly-6C chromosomal gene. *J. Immunol.* **140**, 2815.

Brannan, C. I., Lyman, S. D., Williams, D. E., Eisenman, J., Anderson, D. M., Cosman, D., Bedell, M. A., Jenkins, N. A., and Copeland, N. G. (1991). Steel-Dickie mutation encodes a c-kit ligand lacking transmembrane and cytoplasmic domains. *Proc. Natl. Acad. Sci. U.S.A.* **88**, 4671–4674.

Chabot, S., Stephenson, D. A., Chapman, V. M., Besmer, P., and Bernstein, A. (1988). The protooncogene c-*kit* encoding a transmembrane tyrosine kinase receptor maps to the mouse W locus. *Nature (London)* **335**, 88–89.

Chang, Y., Paige, C. J., and Wu, G. E. (1992). Enumeration and characterization of DJ$_H$ structures in mouse fetal liver. *EMBO J.* **11**, 1891–1899.

Chen, J., Trounstine, M., Kuahara, C., Young, F., Kuo, C. C., Xu, Y., Loring, J. F., Alt, F. W., and Huszar, D. (1993). B cell development in mice that lack one or both immunoglobulin κ light chain genes. *EMBO J.* **12**, 821–830.

Cherayil, B. J., and Pillai, S. (1991). The ω/λ5 surrogate immunoglobulin light chain is expressed on the surface of transitional B lymphocytes in murine bone marrow. *J. Exp. Med.* **173**, 111–116.

Collins, L. S., and Dorshkind, K. (1987). A stromal cell line from myeloid long-term bone marrow cultures can support myelopoiesis and B lymphopoiesis. *J. Immunol.* **138**, 1082–1087.

Cong, Y. Z., Rabin, E., and Wortis, H. H. (1991). Treatment of murine CD5- B cells with anti-immunoglobulin, but not LPS, induces surface CD5: Two B cell activation pathways. *Int. Immunol.* **3**, 467–476.

Cooper, M. D., Lawton, A. L., and Kincade, P. W. (1972). A two stage models for develop-ment of antibody-producing cells. *Clin. Exp. Immunol.* **11**, 143–149.

Copeland, N. J., Gilbert, G. J., Cho, B. C., Donovan, P. J., Jenkins, B. A., Cosman, D., Anderson, D., Lyman, S. D., and Williams, D. E. (1990). Mast cell growth factor maps near the *Sl* locus and is structurally altered in a number of Steel alleles. *Cell (Cambridge, Mass.)* **63**, 175–183.

Cumano, A., and Paige, C. J. (1992). Enrichment and characterization of uncommitted B-cell precursors from fetal liver a day 12 of gestation. *EMBO J.* **11**, 593–601.

Cumano, A., Dorshkind, K., Gillis, S., and Paige, C. J. (1990). The influence of S17 stromal cells and interleukin 7 on B cell development. *Eur. J. Immunol.* **20**, 2183–2189.

Cumano, A., Paige, C. J., Iscove, N. N., and Brady, G. (1992). Bipotential precursors of B cells and macrophages in murine fetal liver. *Nature (London)* **356**, 612–615.

Cumano, A., Furlonger, C., and Paige, C. J. (1993). Differentiation and characterization of B-cell precursors detected in the yolk sac and embryo body of embryos beginning at the 10- to 12-somite stage. *Proc. Natl. Acad. Sci. U.S.A.* **90**, 6429–6433.

Cumano, A., Kee, B. L., Ramsden, D. A., Marshall, A., Paige, C. J., and Wu, G. E. (1994). Development of B lymphocytes from lymphoid committed and uncommitted progenitors. *Immunol. Rev.* **137**, 5–33.

Curry, J. L., and Trentin, J. J. (1967). Haemopoietic spleen colony-stimulating factors. *Science* **236**, 1229–1237.

Dadi, H. K., and Roifman, C. M. (1993). Interleukin 7 receptor mediates activation of phosphatidylinositol-3 kinase in human B-cell precursors. *Biochem. Biophys. Res. Commun.* **192**, 459–464.

Davidson, W. F., Pierce, J. H., Rudikoff, S., and Morse, H. C., III (1988). Relationships between B cell and myeloid differentiation. *J. Exp. Med.* **168**, 389–407.

Davis, S., Aldrich, R. H., Stahl, N., Pan, L., Yaga, T., Kishimoto, T., Ip, N. Y., and Yancopoulos, G. D. (1993). LIFRβ and gp130 as heterodimerizing signal transducers of the tripartite CNTF receptor. *Science* **260**, 1805–1808.

Desiderio, S. V. (1994). The B cell antigen receptor in B-cell development. *Curr. Opin. Immunol.* **6**, 248–256.

Desiderio, S. V., Yancopoulos, G. D., Paskind, M., Thomas, E., Boss, M. A., Landau, N., Alt, F. W., and Baltimore, D. (1984). Insertion of N regions into heavy chain genes is correlated with expression of terminal deoxytransferase in B cells. *Nature (London)* **311**, 752–755.

Dexter, T. M., Allen, T. D., and Lajtha, L. G. (1977). Conditions controlling the proliferation of haemopoietic stem cells in vitro. *J. Cell. Physiol.* **91**, 334–344.

Dick, J. E., Magli, M. C., Huszar, D., Phillips, R. A., and Bernstein, A. (1985). Introduction of a selectable gene into primitive stem cells capable of long term reconstitution of the hemopoietic system of W/W^v mice. *Cell (Cambridge, Mass.)* **42**, 71–79.

Dieterlen-Lievre, F. (1975). On the origin of haemopoietic stem cells in the avian embryo: An experimental approach. *J. Embryol. Exp. Morphol.* **3**, 607–619.

Dorshkind, K. (1988). IL-1 inhibits B cell differentiation in long term bone marrow cultures. *J. Immunol.* **141**, 531.

Dorshkind, K. (1990). Regulation of hemopoiesis by bone marrow stromal cells and their products. *Annu. Rev. Immunol.* **8**, 111–137.

Dorshkind, K. (1991). In vivo administration of recombinant granulocyte-macrophage colony stimulating factor results in a reversible inhibition of primary B lymphopoiesis. *J. Immunol.* **146**, 4202.

Dorshkind, K., and Phillips, R. A. (1983). Characterization of early B lymphocyte precursors present in long-term bone marrow cultures. *J. Immunol.* **129**, 2444–2450.

Ehlich, A., Schaal, S., Gu, H., Kitamura, D., Muller, W., and Rajewsky, K. (1993). Immunoglobulin heavy and light chain genes rearrange independently at early stages of B cell development. *Cell (Cambridge, Mass.)* **72**, 695–704.

Era, T., Nishikawa, S., Sudo, F., Fu-Ho, W., Ogawa, M., Kunisada, R., Hayashi, S.-I., and

Nishikawa, S.-I. (1994). How B-precursor cells are driven to cycle. *Immunol. Rev.* **137**, 35–51.

Faust, E. A., Saffran, D. C., Toksoz, D., Williams, D. A., and Witte, O. N. (1993). Distinctive growth requirements and gene expression patterns distinguish progenitor B cells from pre-B cells. *J. Exp. Med.* **177**, 915–923.

Feeney, A. J. (1990). Lack of N regions in fetal and neonatal mouse immunoglobulin V-D-J junctional sequences. *J. Exp. Med.* **172**, 1377–1390.

Fisher, A. G., Burdet, C., LeMeur, M., Haasner, D., Gerber, P., and Ceredig, R. (1992). Lymphoproliferative disorders in an IL-7 transgenic mouse line. *Leukemia* **2**, 566–568.

Flanagan, J. G., Chan, D. C., and Leder, P. (1991). Transmembrane form of the *kit* ligand growth factor is determined by alternative splicing and is missing in the *Sl^d* mutant. *Cell (Cambridge, Mass.)* **64**, 1025–1035.

Freidrich, R. J., Campbell, K. S., and Cambier, J. C. (1993). The gamma subunit of the B cell antigen-receptor complex is a C-terminally truncated product of the B29 gene. *J. Immunol.* **150**, 2814–2822.

Fulop, G. M., and Phillips, R. A. (1990). The scid mutation causes a general defect in DNA repair. *Nature (London)* **347**, 479–482.

Geissler, E. N., Ryan, M. A., and Houseman, D. E. (1988). The dominant white spotting locus of the mouse encodes the c-*kit* protooncogene. *Cell (Cambridge, Mass.)* **55**, 185–192.

Gellert, M. (1992). Molecular analysis of V(D)J recombination. *Annu. Rev. Genet.* **22**, 425–426.

Gibson, L. F., Piktel, D., and Landreth, K. S. (1993). Insulin-like growth factor-1 potentiates expansion of inteleukin-7 dependent pro-B cells. *Blood* **82**, 3005–3011.

Gilfillan, S., Dierich, A., Lemeur, M., Benoist, C., and Mathis, D. (1993). Mice lacking TdT: Mature animals with an immature lymphocyte repertoire. *Science* **261**, 1175–1178.

Gimble, J. M., Pietrangeli, C., Henley, A., Dorheim, M. A., Silver, J., Namen, A., Takeichi, M., Goridis, C., and Kincade, P. W. (1989). Characterization of murine bone marrow and spleen-derived stromal cells: Analysis of leukocyte marker and growth factor mRNA transcript levels. *Blood* **74**, 303–311.

Gimble, J. M., Medina, K., Hudson, J., Robinson, M., and Kincade, P. W. (1993). Modulation of lymphohematopoiesis in long-term cultures by gamma interferon: Direct and indirect action on lymphoid and stromal cells. *Exp. Hematol.* **21**, 224–230.

Godin, I., Garcia Porrero, J. A., Coutinho, A., Dieterlen-Lievre, F., and Marcos, M. A. R. (1993). Para-aortic splanchnopleura from early mouse embryos contains B1a cell progenitors. *Nature (London)* **364**, 67–70.

Godin, I., Dieterlen-Lievre, F., and Cumano, A. (1994). Emergence of multipotent hematopoietic cells in the yolk sac and para-aortic splanchnopleura of 8.5 dpc mouse embryos. In press.

Gold, M. R., Law, D. A., and DeFranco, A. L. (1990). Stimulation of protein tyrosine phosphorylation by the B-lymphocyte antigen receptor. *Nature (London)* **345**, 810–813.

Goodwin, R. G., Friend, D., Ziegler, S. F., Jerzy, R., Falk, B. A., Gimpel, S., Cosman, D., Dower, S. L., March, C. J., Namen, A. E., and Park, L. S. (1990). Cloning of the human and murine interluekin-7 receptors: Demonstration of a soluble form and homology to a new receptor superfamily. *Cell (Cambridge, Mass.)* **60**, 941–951.

Grabstein, K. H., Waldschmidt, T. J., Finkelman, F. D., Hess, B. W., Alpert, A. R., Boiani, N. E., Namen, A. E., and Morrissey, P. J. (1993). Inhibition of murine B and T lymphopoiesis *in vivo* by an anti-interleukin 7 monoclonal antibody. *J. Exp. Med.* **178**, 257–264.

Grabstein, K. H., Eisenman, J., Shanebeck, K., Rauch, C., Srinivasan, S., Fung, V., Beers, C., Richardson, J., Schoenborn, M. A., Ahdieh, M., Johnson, L., Alderson, M. R.,

Watson, J. D., Anderson, D. M., and Giri, J. G. (1994). Cloning of a T cell growth factor that interacts with the β chain of the interleukin-2 receptor. *Science* **264,** 965–968.

Grawunder, U., Melchers, F., and Rolink, A. (1993). Interferon-γ arrests proliferation and causes apoptosis in stromal cell/interleukin-7-dependent normal murine pre-B cell lines and clones *in vitro,* but does not induce differentiation of surface immunoglobulin-positive B cells. *Eur. J. Immunol.* **23,** 544–551.

Grégoire, K. E., Goldschneider, I., Barton, R. W., and Bollum, F. J. (1979). Ontogeny of terminal deoxynucleotidyl transferase-positive cells in lymphopoietic tissues of rat and mouse. *J. Immunol.* **123,** 1347.

Gu, H., Kitamura, D., and Rajewsky, K. (1991a). B cell development regulated by gene rearrangement: Arrest of maturation by membrane-bound Dμ protein and selection of D_H element reading frames. *Cell (Cambridge, Mass.)* **65,** 47–54.

Gu, H., Tarlington, D., Muller, W., Rajewsky, K., and Forster, I. (1991b). Most peripheral B cells are ligand-selected. *J. Exp. Med.* **173,** 1357–1371.

Gunji, Y., Sudo, T., Suda, J., Yamaguchi, Y., Nakauchi, H., Nishikawa, S.-I., Yanai, N., Obinata, M., Yanagisawa, M., Miura, Y., and Suda, T. (1991). Support of early B cell differentiation in mouse fetal liver by stromal cells and interleukin-7. *Blood* **77,** 2612–2617.

Hagman, J., and Grosschedl, R. (1994). Regulation of gene expression at early stages of B-cell differentiation. *Curr. Opin. Immunol.* **6,** 222–230.

Hagman, J., Bélanger, C., Travis, A., Turck, C. W., and Grosschedl, R. (1993). Cloning and functional characterization of early B-cell factor, a regulator of lymphocyte-specific gene expression. *Genes Dev.* **7,** 760–773.

Hangoc, G., Yin, T., Cooper, S., Schendel, P., Yang, Y.-C., and Broxmeyer, H. E. (1993). *In vivo* effects of recombinant interleukin-11 on myelopoiesis in mice. *Blood* **81,** 965–972.

Hara, H., Sam, M., Maki, R. A., Wu, G. E., and Paige, C. J. (1990). Characterization of a 70Z/3 pre-B cell derived macrophage clone. Differential expression of hox family genes. *Int. Immunol.* **2,** 691–696.

Hardy, R. R., and Hayakawa, K. (1991). A developmental switch in B lymphopoiesis. *Proc. Natl. Acad. Sci. U.S.A.* **88,** 11550–11554.

Hardy, R. R., Hayakawa, K., Haaijman, J., and Herzenberg, L. A. (1982). B-cell subpopulations identified by two-color fluorescence analysis. *Nature (London)* **297,** 589–591.

Hardy, R. R., Carmack, C. E., Shinton, S. A., Kemp, J. D., and Hayakawa, K. (1991). Resolution and characterization of pro-B and pre-pro-B cell stages in normal mouse bone marrow. *J. Exp. Med.* **173,** 1213–1225.

Hardy, R. R., Carmack, C. E., Li, Y. S., and Hayakawa, K. (1994). Distinctive developmental origins and specificities of murine CD5$^+$ B cells. *Immunol. Rev.* **137,** 91–118.

Haughton, G., Arnold, L. W., Whitmore, A. C., and Clarke, S. H. (1993). B-1 cells are made, not born. *Immunol. Today* **14,** 84–87.

Hayakawa, K., Carmack, C. E., and Hardy, R. R. (1988). Autoantibody specificities in a discrete B cell lineage. *Mol. Cell. Biol.* **85,** 69–75.

Hayashi, S.-I., Kunisada, T., Ogawa, M., Sudo, T., Kodama, H., Suda, T., Nishikawa, S., and Nishikawa, S.-I. (1990). Stepwise progression of B lineage differentiation supported by interleukin 7 and other stromal cell molecules. *J. Exp. Med.* **171,** 1683–1695.

Henderson, A. J., Johnson, A., and Dorshkind, K. (1990). Functional characterization of two stromal cell lines that support B lymphopoiesis. *J. Immunol.* **145,** 423–428.

Hermanson, G. G., Eisenberg, D., Kincade, P. W., and Wall, R. (1988). B29: A member of the immunoglobulin gene superfamily exclusively expressed on B-lineage cells. *Proc. Natl. Acad. Sci. U.S.A.* **85,** 6890–6894.

Herzenberg, L. A., and Kantor, A. B. (1993). B-cell lineage exist in the mouse. *Immunol. Today* **14,** 79–83.

Hirayama, F., Shih, J. P., Awgulewitsch, A., Warr, G. W., Clark, S. C., and Ogawa, M.

(1992). Clonal proliferation of murine lymphohemopoietic progenitors in culture. *Proc. Natl. Acad. Sci. U.S.A.* **89,** 5907–5911.

Hombach, J., Tsubata, T., Leclercq, L., Stappert, H., and Reth, M. (1990). Molecular components of the B-cell antigen receptor complex of the IgM class. *Nature (London)* **343,** 760–762.

Hough, M. R., Takei, F., Humphries, K. R., and Kay, R. (1994). Defective development of thymocytes overexpressing the co-stimulatory molecule, heat stable antigen. *J. Exp. Med.* **179,** 177–184.

Hozumi, N., and Tonegawa, S. (1976). Evidence for somatic rearrangement of immunoglobulin genes coding for variable and constant regions. *Proc. Natl. Acad. Sci. U.S.A.* **73,** 3628–3632.

Huang, E., Nocka, K., Bier, D. R., Chu, T., Buck, J., Lahm, H., Wellner, D., Leder, P., and Besmer, P. (1990). The hematopoietic growth factor KL is encoded by the steel locus and is the ligand for the c-*kit* receptor, the product of the *W* locus. *Cell (Cambridge, Mass.)* **63,** 223–233.

Huang, H., and Auerbach, R. (1993). Identification and characterization of hematopoietic stem cells from the yolk sac and early mouse embryo. *Proc. Natl. Acad. Sci. U.S.A.* **90,** 10110–10114.

Hunt, P., Robertson, D., Weiss, D., Rennick, D., Lee, F., and Witte, O. N. (1987). A single bone marrow-derived stromal cell type supports the *in vitro* growth of early lymphoid and myeloid cells. *Cell (Cambridge, Mass.)* **48,** 997–1007.

Hynes, R. O. (1987). Integrins: A family of cell surface receptors. *Cell (Cambridge, Mass.)* **48,** 549–554.

Iglesias, A., Kopf, M., Williams, G. S., Buhler, B., and Kohler, G. (1991). Molecular requirements for the μ-induced light chain gene rearrangement in pre-B cells. *EBMO J.* **10,** 2147–2156.

Ip, N. Y., Nye, S. H., Boulton, T. G., Davis, S., Taga, T., Li, Y., Birren, S. J., Yasukawa, K., Kishimoto, T., Anderson, D. J., Stahl, N., and Yancopoulos, G. D. (1992). CNTF and LIF act on neuronal cells via shared signaling pathways that involve the IL-6 signal transducing receptor component gp130. *Cell (Cambridge, Mass.)* **69,** 1121–1132.

Ishihara, K., Medina, K., Hayashi, S., Pietrangeli, C., Namen, A. E., Miyake, K., and Kincade, P. W. (1991). Stromal-cell and cytokine-dependent lymphocyte clones which span the pre-B to B-cell transition. *Dev. Immunol.* **1,** 149–161.

Jacobsen, F. W., Veiby, O. P., Skjonsberg, C., and Jacobsen, S. E. W. (1993). Novel role of interleukin 7 in myelopoiesis: Stimulation of primitive murine hematopoietic progenitor cells. *J. Exp. Med.* **178,** 1777–1782.

Jacobsen, K., and Osmond, D. G. (1990). Microenvironmental organization and stromal cell associations of B lymphocyte precursor cells in mouse bone marrow. *Eur. J. Immunol.* **20,** 2395–2404.

Jacobsen, K., Miyake, K., Kincade, P. W., and Osmond, D. G. (1992). Highly restricted expression of a stromal cell determinant in mouse bone marrow in vivo. *J. Exp. Med.* **176,** 927–935.

Jordan, C. T., McKearn, J. T., and Lemischka, I. R. (1990). Cellular and developmental properties of fetal hematopoietic stem cells. *Cell (Cambridge, Mass.)* **61,** 953–963.

Justement, L. B., Campbell, K. S., Chien, N. C., and Cambier, J. C. (1991). Regulation of B cell antigen receptor signal transduction and phosphorylation by CD45. *Science* **252,** 1839–1842.

Kallenbach, S., Brinkman, T., and Rougeon, F. (1993). Rag-1: A topoisomerase? *Int. Immunol.* **5,** 231–232.

Kamps, W. A., and Cooper, M. D. (1982). Microenvironmental studies of pre-B and B cell development in human and mouse fetuses. *J. Immunol.* **129,** 526–531.

Karasuyama, H., Rolink, A., and Melchers, F. (1993). A complex of glycoproteins is associated with VpreB/λ5 surrogate light chain on the surface of μ heavy chain-negative early precursor B cell lines. *J. Exp. Med.* **178**, 469–478.

Karasuyama, H., Rolink, A., Shinkai, Y., Young, F., Alt, F. W., and Melchers, F. (1994). The expression of VpreB/λ5 surrogate light chain in early bone marrow precursor B cells of normal and B cell-deficient mutant mice. *Cell (Cambridge, Mass.)* **77**, 133–143.

Katayama, N., Clark, S., and Ogawa, M. (1993). Growth factor requirements for survival in cell-cycle dormancy of primitive murine lymphohematopoietic progenitors. *Blood* **81**, 610–616.

Kee, B. L., Paige, C. J., and Letarte, M. (1992). Characterization of murine CD10, an endopeptidase expressed on bone marrow adherent cells. *Int. Immunol.* **4**, 1041–1047.

Kee, B. L., Cumano, A., Iscove, N. N., and Paige, C. J. (1994). Stromal cell independent growth of bipotent B cell-macrophage precursors from murine fetal liver. *Int. Immunol.* **6**, 401–407.

Keller, G., Paige, C. J., Gilboa, E., and Wagner, E. F. (1985). Expression of a foreign gene in myeloid and lymphoid cells derived from multipotent haematopoietic precursors. *Nature (London)* **318**, 149–154.

Kerr, M. A., and Kenny, A. J. (1974). The purification and specificity of a neutral endopeptidase from rabbit kidney brush border. *Biochem. J.* **137**, 477–488.

Kim, K.-M., Alber, G., Weiser, P., and Reth, M. (1993). Differential signaling through the Ig-α and Ig-β components of the B cell antigen receptor. *Eur. J. Immunol.* **23**, 911–916.

Kina, T., Majumdar, A. S., Heimfeld, S., Kaneshima, H., Holzmann, B., Katsura, Y., and Weissman, I. L. (1991). Identification of a 107-kD glycoprotein that mediates adhesion between stromal cells and hematolymphoid cells. *J. Exp. Med.* **173**, 373–381.

Kincade, P. W. (1991). Molecular interactions between stromal cells and B lymphocyte precursors. *Semin. Immunol.* **3**, 379–390.

Kincade, P. W., Lee, G., Paige, C. J., and Scheid, M. P. (1981a). Cellular interactions affecting the maturation of murine B lymphocyte precursors *in vitro. J. Immunol.* **127**, 255–260.

Kincade, P. W., Lee, G., Watanabe, S., and Scheid, M. P. (1981b). Antigens displayed on murine B lymphocyte precursors. *J. Immunol.* **127**, 2262–2268.

Kincade, P. W., Lee, G., Pietrangeli, C. E., Hayashi, S.-I., and Gimble, J. M. (1989). Cells and molecules that regulate B lymphopoiesis in bone marrow. *Annu. Rev. Immunol.* **7**, 111–143.

Kincade, P. W., Medina, K. L., and Smithson, G. (1994). Sex hormones as negative regulators of lymphopoiesis. *Immunol. Rev.* **137**, 119–134.

Kishihara, K., Penninger, J., Wallace, V. A., Kunkig, R. M., Kawai, K., Wakeham, A., Timms, E., Pfeffer, K., Ohashi, P. S., Thomas, M. L., Furlonger, C., Paige, C. J., and Mak, T. W. (1993). Normal B lymphocyte development but impaired T cell maturation in CD45-exon 6 protein tyrosine phosphatase-deficient mice. *Cell (Cambridge, Mass.)* **74**, 143–156.

Kitamura, D., Roes, J., Kuhm, R., and Rajewsky, K. (1991). A B cell deficient mouse generated through targeted disruption of the membrane exon of the immunoglobulin μ chain. *Nature (London)* **350**, 423–426.

Kitamura, D., Kudo, A., Schaal, S., Muller, W., Melchers, F., and Rajewsky, K. (1992). A critical role of λ5 protein in B cell development. *Cell (Cambridge, Mass.)* **69**, 823–831.

Klinken, S. P., Alexander, W. S., and Adams, J. M. (1988). Hemopoietic lineage switch: v-*raf* oncogene converts Eμ-*myc* transgenic B cells into macrophages. *Cell (Cambridge, Mass.)* **53**, 857–867.

Komori, T., Okada, A., Stewart, V., and Alt, F. W. (1993). Lack of N regions in antigen receptor variable region genes of TdT-deficient lymphocytes. *Science* **261**, 1171–1174.

Kondo, M., Takeshita, T., Ishii, N., Nakamura, M., Watanabe, S., Arai, K.-I., and Sugamura, K. (1993). Sharing of the interleukin-2 (IL-2) receptor γ chain between receptors for IL-2 and IL-4. *Science* **262**, 1874–1876.

Korsmeyer, S. J. (1992). Bcl-2: A repressor of lymphocyte death. *Immunol. Today* **13**, 285–288.

Kotanides, H., and Reich, N. C. (1993). Requirement of tyrosine phosphorylation for rapid activation of a DNA binding factor by IL-4. *Science* **262**, 1265–1267.

Kudo, A., and Melchers, F. (1987). A second gene, VpreB in the λ5 locus of the mouse, which appears to be selectively expressed in pre-B lymphocytes. *EMBO J.* **6**, 2267–2272.

Kudo, A., Thalmann, P., Sadaguchi, N., Davidson, W. F., Pierce, J. H., Kearney, J. F., Reth, M., Rolink, A., and Melchers, F. (1992). The expression of the mouse VpreB/λ5 locus in transformed cell lines and tumors of the B lineage differentiation pathway. *Int. Immunol.* **4**, 831–840.

Landreth, K. S., and Dorshkind, K. (1988). Pre-B cell generation potentiated by soluble factors from a bone marrow stromal cell line. *J. Immunol.* **140**, 845–852.

Landreth, K. S., Kincade, P. W., Lee, G., and Harrison, D. E. (1984). B lymphocyte precursors in embryonic and adult W anemic mice. *J. Immunol.* **132**, 2724–2729.

Landreth, K. S., Narayanan, R., and Dorshkind, K. (1992). Insulin-like growth factor-1 regulates pro-B cell differentiation. *Blood* **80**, 1207–1212.

Larner, A. C., David, M., Feldman, G. M., Igarashi, K.-I., Hackett, R. H., Webb, D. S. A., Sweitzer, S. M., Petricoin, E. F., III, and Finbloom, D. S. (1993). Tyrosine phosphorylation of DNA binding proteins by multiple cytokines. *Science* **261**, 1730–1733.

Lassoued, K., Nunez, C. A., Billips, L., Kubagawa, H., Monteiro, R. C., LeBien, T. W., and Cooper, M. D. (1993). Expression of surrogate light chain receptors is restricted to a late stage in pre-B cell differentiation. *Cell (Cambridge, Mass.)* **73**, 73–86.

LeDouarin, N. M., Houssaint, E., Joteriau, F. V., and Belo, M. (1975). Origin of hemopoietic stem cells in embryonic bursa of Fabricius and bone marrow studied through interspecific chimeras. *Proc. Natl. Acad. Sci. U.S.A.* **72**, 2701–2705.

Lee, G., Namen, A. E., Gillis, S., Ellingsworth, L. R., and Kincade, P. W. (1989). Normal B cell precursors responsive to recombinant murine IL-7 and inhibition of IL-7 activity by transforming growth factor-β. *J. Immunol.* **142**, 3875–3883.

Lee, M. Y., Fevold, K. L., Dorshkind, K., Fukunaga, R., Nagata, S., and Rosse, C. (1993). In vivo and in vitro suppression of primary B lymphocytopoiesis by tumor-derived and recombinant granulocyte colony-stimulating factor. *Blood* **82**, 2062.

Lee, S. K., Su, B., Maher, S. E., and Bothwell, A. L. M. (1994). Ly-6A is required for T cell receptor expression and protein tyrosin kinase fyn activity. *EMBO J.* **13**, 2167–2176.

Lemischka, I. R., Raulet, D. H., and Mulligan, R. C. (1986). Developmental potential and dynamic behavior of hemopoietic stem cells. *Cell (Cambridge, Mass.)* **45**, 917–927.

Li, Y.-S., Hayakawa, K., and Hardy, R. R. (1993). The regulated expression of B lineage associated genes during B cell differentiation in bone marrow and fetal liver. *J. Exp. Med.* **178**, 951–960.

Lin, W.-C., and Desiderio, S. (1993). Regulation of V(D)J recombination activator protein RAG-2 by phosphorylation. *Science* **260**, 953–959.

Linton, P. J., Decker, D. J., and Klinman, N. R. (1989). Primary antibody-forming cells and secondary B cells are generated from separate precursor subpopulations. *Cell (Cambridge, Mass.)* **59**, 1045–1059.

Loffert, D., Schaal, S., Ehlich, A., Hardy, R. R., Zou, Y.-R., Muller, W., and Rajewsky, K. (1994). Early B-cell development in the mouse: Insights from mutations introduced by gene targeting. *Immunol. Rev.* **137**, 135–153.

Lu, H. S., Clogston, C. L., Wypych, J., Faussett, P. R., Lauren, S., Mendiaz, E. A., Zsebo, K. M., and Langley, K. E. (1991). Amino acid and post-translational modification of stem cell factor isolated from Buffalo rat liver conditioned medium. *J. Biol. Chem.* **266**, 8102–8107.

Lutticken, C., Wegenka, U. M., Yuan, J., Buschmann, J., Schinkler, C., Ziemiecki, A., Harpur, A. G., Wilks, A. F., Yasukawa, K., Taga, T., Kishimoto, T., Barbieri, G., Pelligrini, S., Sendtner, M., Heinrich, P. C., and Horn, F. (1994). Association of transcription factor APRF and protein kinase Jak1 with interleukin-6 signal transducer gp130. *Science* **263**, 89–92.

Lyman, S. D., James, L., Banden Bos, T., de Vries, P., Brasel, K., Gliniak, B., Hooingsworth, L. T., Picha, K. S., McKenna, H. J., Splett, R. R., Fletcher, F. A., Maraskovsky, E., Farrah, T., Foxworthe, D., Williams, D. E., and Beckmann, M. P. (1993). Molecular cloning of a ligand for the flt3/flk-2 tyrosine kinase receptor: A proliferative factor for primitive hematopoietic cells. *Cell* (*Cambridge, Mass.*) **75**, 1157–1167.

Manjunath, N., Johnson, R. S., Staunton, D. E., Pasqualini, R., and Ardman, B. (1993). Targeted disruption of CD43 gene enhances T lymphocyte adhesion. *J. Immunol.* **151**, 1528–1534.

Marcos, M. A. R., Godin, I., Cumano, A., Morales, S., Garcia-Porrero, J. A., Dieterlen-Lievre, F., and Gaspar, M.-L. (1994). Developmental events from hemopoietic stem cells to B-cell populations and Ig repertoires. *Immunol. Rev.* **137**, 155–171.

Matthews, W., Jordan, C. T., Gavin, M., Jenkins, N. A., Copeland, N. G., and Lemischka, I. R. (1991). A receptor tyrosine kinase cDNA isolated from a population of enriched primitive hematopoietic cells and exhibiting close genetic linkage to c-*kit*. *Proc. Natl. Acad. Sci. U.S.A.* **88**, 9026–9030.

McCulloch, E. A., Siminovitch, L., and Till, J. E. (1964). Spleen-colony formation in anemic mice of genotype W/W^v. *Science* **144**, 844–846.

McCulloch, E. A., Siminovitch, L., Till, J. E., Russell, E. S., and Bernstein, S. E. (1965). The cellular basis of the genetically determined hematopoietic defect in anemic mice of genotype Sl/Sl^d. *Blood* **26**, 399–410.

McKearn, J. P., Baum, C., and Davie, J. M. (1984). Cell surface antigens expressed by subsets of pre-B cells and B cells. *J. Immunol.* **132**, 332–339.

McNiece, I. K., Langley, K. E., and Zsebo, K. M. (1991). The role of recombinant stem cell factor in early B cell development—synergistic interaction with IL-7. *J. Immunol.* **146**, 3785.

Medvinsky, A. L., Samoylina, N. L., Muller, A. M., and Dzierzak, E. A. (1993). An early pre-liver intraembryonic sourse of CFU-S in the developing mouse. *Nature* (*London*) **364**, 64–67.

Mekori, T., and Phillips, R. A. (1969). The immune response in mice of genotypes W/W^v and Sl/Sl^d. *Proc. Soc. Exp. Biol. Med.* **132**, 115.

Melchers, F. (1977). B lymphocyte development in the fetal liver. I. Development of precursor B cells during gestation. *Eur. J. Immunol.* **7**, 476–481.

Melchers, F., Braun, V., and Galanos, C. (1975). The lipoprotein of the outer membrane of *Escherichia coli*: A B-lymphocyte mitogen. *J. Exp. Med.* **142**, 473–482.

Merino, R., Ding, L., Veis, D. J., Korsmeyer, S. J., and Nunez, G. (1994). Developmental regulation of the bcl-2 protein and susceptibility to cell death in B lymphocytes. *EMBO J.* **13**, 683–691.

Misener, V., Jongstra-Bilen, J., Young, A. J., Atkinson, M. J., Wu, G. E., and Jongstra, J. (1990). Association of Ig L chain-like protein λ5 with a 16-kilodalton protein in mouse pre-B cell lines is not dependent on the presence of Ig H chain protein. *J. Immunol.* **145**, 905–909.

Misener, V., Downey, G. P., and Jongstra, J. (1991). The immunoglobulin light chain related protein λ5 is expressed on the surface of mouse pre-B cell lines and can function as a signal transducing molecule. *Int. Immunol.* **3**, 1129–1136.

Miyake, K., Medina, K. L., Hayashi, S.-I., Ono, S., Hamaoka, T., and Kincade, P. W. (1990). Monoclonal antibodies to Pgp-1/CD44 block lympho-hemopoiesis in long-term bone marrow cultures. *J. Exp. Med.* **171,** 477–488.

Miyake, K., Weissman, I. L., Greenberger, J. S., and Kincade, P. W. (1991). Evidence for a role of the integrin VLA-4 in lympho-hemopoiesis. *J. Exp. Med.* **173,** 599–607.

Mombaerts, P., Iacomini, J., Johnson, R. S., Herrup, K., Tonegawa, S., and Papaioannou, V. E. (1992). Rag-1 deficient mice have no mature B and T lymphocytes. *Cell (Cambridge, Mass.)* **68,** 869–877.

Montminy, M. (1993). Trying on a new pair of SH2s. *Science* **261,** 1694–1695.

Moore, M. A. S., and Metcalf, D. (1970). Ontogeny of the haemopoietic system: Yolk sac origin of *in vivo* and *in vitro* colony forming cells in the developing mouse embryo. *Br. J. Haematol.* **18,** 279–295.

Morrissey, P. J., Conlon, P., Charrier, K., Braddy, S., Alpert, A., Williams, D., Namen, A. E., and Mochizuki, D. (1991). Administration of IL-7 to normal mice stimulates B-lymphopoiesis and peripheral lymphadenopathy. *J. Immunol.* **147,** 561–568.

Muegge, K., Vila, M. P., and Durum, S. K. (1993). Interleukin-7: A cofactor for (V(D)J rearrangement of the T cell receptor β gene. *Science* **261,** 93–95.

Muller-Sieburg, C. E., Whitlock, C. A., and Weissman, I. L. (1986). Isolation of two early B lymphocyte progenitors from mouse marrow: A committed pre-pre-B cell and clonogenic Thy-1^{lo} hematopoietic stem cell. *Cell (Cambridge, Mass.)* **44,** 653–662.

Murakami, M., Hibi, M., Nakagawa, N., Nakagawa, T., Yasukawa, K., Yamanishi, K., Taga, T., and Kishimoto, T. (1993). IL-6-induced homodimerization of gp 130 and associated activation of a tyrosine kinase. *Science* **260,** 1808–1810.

Musashi, M., Yang, Y.-C., Paul, S. R., Clark, S. C., Sudo, T., and Ogawa, M. (1991a). Direct and synergistic effects of interleukin 11 on murine hematopoiesis in culture. *Proc. Natl. Acad. Sci. U.S.A.* **88,** 765–769.

Musashi, M., Clark, S. C., Sudo, T., Urdal, D., and Ogawa, M. (1991b). Synergistic interactions between interleukin-11 and interleukin-4 in support of proliferation of primitive hematopoietic progenitors of mice. *Blood* **78,** 1448–1451.

Mustelin, T., and Altman, A. (1990). Dephosphorylation and activation of the T cell tyrosine kinase p56^{lck} by the leukocyte common antigen (CD45). *Oncogene* **5,** 101–105.

Namen, A. E., Lupton, S., Hjerrild, K., Wignall, J., Mochizuki, D. Y., Schmierer, A., Mosley, B., March, C. J., Urdal, K., Gillis, S., Cosman, D., and Goodwin, R. G. (1988). Stimulation of B cell progenitors by cloned murine interleukin 7. *Nature (London)* **333,** 571–573.

Narendran, A., Ramsden, D., Cumano, A., Tanaka, T., Wu, G. E., and Paige, C. J. (1993). B cell developmental defects in X-linked immunodeficiency. *Int. Immunol.* **5,** 139–144.

Nielson, P. J., Eichmann, K., Kohler, G., and Iglesias, A. (1993). Constitutive expression of transgenic heat stable antigen (mCD24) in lymphocytes can augment secondary antibody responses. *Int. Immunol.* **5,** 1355–1364.

Nishikawa, S.-I., Ogawa, M., Nishikawa, S., Kunisada, T., and Kodama, H. (1988). B lymphopoiesis on stromal cell clone: Stromal cell clones acting on different stages of B cell differentiation. *Eur. J. Immunol.* **18,** 1767–1771.

Nishimoto, N., Kubagawa, H., Ohno, T., Gartland, G. L., Stankovic, A. K., and Cooper, M. D. (1991). Normal pre-B cells express a receptor complex of μ heavy chains and surrogate light-chain proteins. *Proc. Natl. Acad. Sci. U.S.A.* **88,** 6284–6288.

Noguchi, M., Nakamura, Y., Russell, S. M., Ziegler, S. F., Tsang, M., Cao, X., and Leonard, W. J. (1993a). Interleukin-2 receptor γ chain: A functional component of the interleukin-7 receptor. *Science* **262,** 1877–1880.

Noguchi, M., Yi, H., Rosenblatt, H. M., Fillpovich, A. H., Adelstein, S., Modi, W. S., McBride, O. W., and Leonard, W. J. (1993b). Interleukin-2 receptor γ chain mutation results in X-linked severe combined immunodeficiency in humans. *Cell (Cambridge, Mass.)* **73**, 147–157.

Oettinger, M. A., Schatz, D. G., Gorka, C., and Baltimore, D. (1990). RAG-1 and RAG-2, adjacent genes that synergistically activate V(D)J recombination. *Science* **248**, 1517–1523.

Ogawa, M., Nishikawa, S., Ikuta, K, Yamamura, F., Naito, M., Takahashi, K., and Nishikawa, S. I. (1988). B cell ontogeny in murine embryo studied by a culture system with the monolayer of a stromal cell clone, ST2: B cell progenitor develops first in the embryonal body rather than in the yolk sac. *EMBO J.* **7**, 1337–1343.

Ogawa, M., Matsuzaki, Y., Nishikawa, S., Hayashi, S. I., Kunisada, T., Sudo, T., Kina, T., Nakauchi, H., and Nishikawa, S. I. (1991). Expression and function of c-*kit* in hemopoietic progenitor cells. *J. Exp. Med.* **174**, 63–71.

Ohara, A., Suda, T., Tokuyama, N., Suda, J., Nakayama, K.-I., Miura, Y., Nishikawa, S.-I., and Nakauchi, H. (1991). Generation of B lymphocytes from a single hemopoietic progenitor cell in vitro. *Int. Immunol.* **3**, 703–709.

Okada, S., Wang, Z.-Q., Grigoriadis, A. W., Wagner, E. F., and von Ruden, T. (1994). Mice lacking c-*fos* have normal hematopoietic stem cells but exhibit altered B-cell differentiation due to an impaired bone marrow environment. *Mol. Cell. Biol.* **14**, 382–390.

Opstelten, D., and Osmond, D. G. (1993). Pre-B cells in mouse bone marrow: Immunofluorescence stathmokinetic studies of the proliferation of cytoplasmic μ-chain-bearing cells in normal mice. *J. Immunol.* **131**, 2635–2640.

Opstelten, D., and Osmond, D. G. (1985). Regulation of pre-B cell proliferation in bone marrow: Immunofluorescence stathmokinetic studies of cytoplasmic μ chain-bearing cells in anti-IgM-treated mice, hematologically deficient mutant mice and mice given sheep red blood cells. *Eur. J. Immunol.* **15**, 599–605.

Osmond, D. G. (1990). B cell development in the bone marrow. *Semin. Immunol.* **2**, 173–180.

Osmond, D. G. (1991). Proliferation kinetics and the lifespan of B cells in central and peripheral lymphoid organs. *Curr. Opin. Immunol.* **3**, 179–185.

Osmond, D. G., Kim, N., Manoukian, R., Phillips, R. A., Rico-Vargas, S. A., and Jacobsen, K. (1992). Dynamics and localization of early B-lymphocyte precursor cells (pro-B cells) in the bone marrow of *scid* mice. *Blood* **79**, 1695–1703.

Ostergaard, H. L., Shackelford, D. A., Hurley, T. R., Johnson, P., Hyman, R., Sefton, B. M., and Trowbridge, I. S. (1989). Expression of CD45 alters phosphorylation of the Lck tyrosine protein kinase in murine lymphoma T cell lines. *Proc. Natl. Acad. Sci. U.S.A.* **86**, 8959–8963.

Owen, J. J. T., and Ritter, M. A. (1969). tissue interactions in the development of thymus lymphocytes. *J. Exp. Med.* **129**, 431–442.

Owen, J. J. T., Raff, M. C., and Cooper, M. D. (1975). Studies on the generation of B lymphocytes in the mouse embryo. *Eur. J. Immunol.* **5**, 468–473.

Paige, C. J. (1983). Surface immunoglobulin-negative B-cell precursors detected by formation of antibody-secreting colonies in agar. *Nature (London)* **302**, 711–713.

Paige, C. J., Kincade, P. W., and Ralph, P. (1978). Murine B cell leukemia line with inducible surface immunoglobulin expression. *J. Immunol.* **121**, 641–647.

Paige, C. J., Kincade, P. W., Moore, M. A. S., and Lee, G. (1979). The fate of fetal and adult B-cell progenitors grafted into immunodeficient CBA/N mice. *J. Exp. Med.* **150**, 548–563.

Paige, C. J., Gisler, R. H., McKearn, J. P., and Iscove, N. N. (1984). Differentiation of murine B cell precursors in agar culture. I. Frequency, surface marker analysis and requirements for growth of clonable pre-B cells. *Eur. J. Immunol.* **14**, 979–987.

Paige, C. J., Skarvall, H., and Sauter, H. (1985). Differentiation of murine B cell precursors in agar culture. II. Response of precursor-enriched populations to growth stimuli and demonstration that the clonable pre-B cell assay is limiting for the B cell precursor. *J. Immunol.* **134**, 3699–3704.

Pallant, A., Eskenazi, A., Mattei, M. G., Fournier, R. E. K., Carlsson, R., Fukuda, M., and Frelinger, J. G. (1992). Characterization of cDNAs encoding human leukosialin and localization of the leukosialin gene to chromosome 16. *Proc. Natl. Acad. Sci. U.S.A.* **86**, 1328.

Park, L. S., Friend, D. J., Schmierer, A. E., Dower, S. K., and Namen, A. E. (1990). Murine interleukin 7 (IL-7) receptor: Characterization of and IL-7 dependent cell line. *J. Exp. Med.* **171**, 1073–1089.

Park, Y.-H., and Osmond, D. G. (1987). Phenotype and proliferation of early B lymphocyte precursor cells in mouse bone marrow. *J. Exp. Med.* **165**, 444–458.

Paul, S. R., Bennett, F., Calvetti, J. A., Kelleher, K., Wood, C. R., O'Hara, R. M., Leary, A. C., Sibley, B., Clark, S. C., Williams, D. A., and Yang, Y.-C. (1990). Molecular cloning of a cDNA encoding interleukin 11, a stromal cell-derived lymphopoietic and hematopoietic cytokine. *Proc. Natl. Acad. Sci. U.S.A.* **87**, 7512–7516.

Penneycook, J. L., Chang, Y., Celler, J., Phillips, R. A., and Wu, G. E. (1993). High frequency of normal DJ_H joints in B cell progenitors in severe combined immunodeficient mice. *J. Exp. Med.* **178**, 1007–1016.

Pietrangeli, C. E., Hayashi, S.-I., and Kincade, P. W. (1988). Stromal cell lines which support lymphocyte growth: Characterization, sensitivity to radiation and responsiveness to growth factors. *Eur. J. Immunol.* **18**, 863–872.

Principato, M., Cleveland, J. L., Rapp, U. R., Holmes, K. L., Pierce, J. C., Morse, H. C., III, and Klinken, S. P. (1990). Transformation of murine bone marrow cells with combined v-*raf*-v-*myc* oncogenes yields clonally related mature B cells and macrophages. *Mol. Cell. Biol.* **10**, 3562–3568.

Raff, M. C., Megson, M., Owen, J. J. T., and Cooper, M. D. (1976). Early production of intracellular IgM by B-lymphocyte precursors in mouse. *Nature (London)* **259**, 224–226.

Ramakrishnan, L., Wu, Q., Yue, A., Cooper, M. D., and Rosenberg, N. (1990). BP-1/6C3 expression defines a differentiation stage of transformed pre-B cells and is not related to malignant potential. *J. Immunol.* **145**, 1603–1608.

Ramsden, D. A., Paige, C. J., and Wu, G. E. (1994). Kappa light chain rearrangement in mouse fetal liver. *J. Immunol.* **153**, 1150–1160.

Remold-O'Donnell, E., and Rosen, F. S. (1990). Sialophorin (CD43) and the Wiscott-Aldrich syndrome. *Immunodefic. Rev.* **2**, 151–174.

Reth, M. (1993). Antigen receptors on B lymphocytes. *Annu. Rev. Immunol.* **10**, 98–121.

Reth, M., and Alt, F. W. (1984). Novel immunoglobulin heavy chains are produced from DJH gene segment rearrangements in lymphoid cells. *Nature (London)* **312**, 418–423.

Reth, M., Petrac, E., Wiese, P., Lobel, L., and Alt, F. W. (1987). Activation of Vκ gene rearrangement in pre-B cells follows the expansion of membrane-bound immunoglobulin heavy chains. *EMBO J.* **6**, 3299–3305.

Rico-Vargas, S. A., Weiskopf, B., Nishikawa, S.-I., and Osmond, D. G. (1994). c-*kit* expression by B cell precursors in mouse bone marrow. *J. Immunol.* **152**, 2845–2851.

Rock, K. L., Reiser, H., Bamezai, A., McGrew, J., and Benacerraf, B. (1989). The LY-6 locus: A multigene family encoding phosphatidylinositol-anchored membrane proteins concerned with T cell activation. *Immunol. Rev.* **111**, 195–224.

Rolink, A., and Melchers, F. (1991). Molecular and cellular origins of B-lymphocyte diversity. *Cell (Cambridge, Mass.)* **66**, 1081–1083.

Rolink, A., Kudo, A., Karasuyama, H., Kikuchi, Y., and Melchers, F. (1991a). Long-term proliferating early pre-B cell lines and clones with the potential to develop to surface Ig-positive, mitogen reactive B cells *in vitro* and *in vivo*. *EMBO J.* **10**, 327–336.

Rolink, A., Streb, M., Nishikawa, S.-I., and Melchers, F. (1991b). The c-*kit* encoded tyrosine kinase regulates the proliferation of early pre-B cells. *Eur. J. Immunol.* **21,** 2069.

Rolink, A., Grawunder, U., Haasner, D., Strasser, A., and Melchers, F. (1993a). Immature surface Ig⁺ B cells can continue to rearrange κ and λ L chain gene loci. *J. Exp. Med.* **178,** 1263–1270.

Rolink, A., Haasner, D., Nishikawa, S.-I., and Melchers, F. (1993b). Changes in frequencies of clonable pre B cells during life in different lymphoid organs of mice. *Blood* **81,** 2290–2300.

Rolink, A., Karasuvama, H., Grawunder, U., Haasner, D., Kudo, A., and Melchers, F. (1993c). B cell development in mice with a defective λ5 gene. *Eur. J. Immunol.* **23,** 1284–1288.

Rolink, A., Karasuyama, H., Haasner, D., Grawunder, U., Martensson, I.-L., Kudo, A., and Melchers, F. (1994). Two pathways of B-lymphocyte development in mouse bone marrow and the roles of surrogate L chain in this development. *Immunol. Rev.* **137,** 185–202.

Rosenstein, Y., Park, J. K., Hahn, W. C., Rosen, F. S., Bierer, B. E., and Burakoff, S. J. (1991). CD43, a molecule defective in Wiskott-Aldrich syndrome, binds ICAM-1. *Nature (London)* **354,** 233–235.

Rossant, J., Vijh, K. M., Grossi, C. E., and Cooper, M. D. (1986). Clonal origin of haematopoietic colonies in the postnatal mouse liver. *Nature (London)* **319,** 507–511.

Rougon, G., Alterman, L. A., Dennis, K., Guo, X.-J., and Kinnon, C. (1991). The murine heat-stable antigen: A differentiation antigen expressed in both the hematolymphoid and neural cell lineages. *Eur. J. Immunol.* **21,** 1397–1402.

Ruff-Jamison, S., Chen, K., and Cohen, S. (1993). Induction by EGF and interferon-γ of tyrosine phosphorylated DNA binding proteins in mouse liver nuclei. *Science* **261,** 1733–1736.

Russell, E. S. (1949). Analysis of pleiotropism at the *W* locus in the mouse. Relationship between the effects of *W* and *Wᵛ* substitution on hair pigmentation and on erythrocytes. *Genetics* **34,** 708–722.

Russell, S. M., Keegan, A. D., Harada, N., Nakamura, Y., Noguchi, M., Leland, P., Friedmann, M. C., Miyajima, A., Puri, R. K., Paul, W. E., and Leonard, W. J. (1993). Interleukin-2 receptor γ chain: A functional component of the interleukin-4 receptor. *Science* **262,** 1880–1883.

Sadowski, H. B., Shuai, K., Darnell, J. E., Jr., and Gilman, M. Z. (1993). A common nuclear signal transduction pathway activated by growth factor and cytokine receptors. *Science* **261,** 1739–1744.

Sakaguchi, N., and Melchers, F. (1986). λ5, a new light chain-related locus selectively expressed in pre-B lymphocytes. *Nature (London)* **324,** 579–582.

Sakaguchi, N., Kashiwamura, S.-I., Kimoto, M., Thalmann, P., and Melchers, F. (1988). By lymphocyte lineage-restricted expression of mb-1, a gene with CD3-like structural properties. *EMBO J.* **7,** 3457–3464.

Salles, G., Chen, C.-Y., Reinherz, E., and Shipp, M. A. (1992). CD10/NEP is expressed on Thy-1ᶫᵒʷ B220⁺ murine B-cell progenitors and functions to regulate stromal cell-dependent lymphopoiesis. *Blood* **80,** 2021–2029.

Salles, G., Rodewald, H.-R., Chin, B. s., Reinherz, E. L., and Shipp, M. A. (1993). Inhibition of CD10/neutral endopeptidase 24.11 promotes B-cell reconstitution and maturation in vivo. *Proc. Natl. Acad. Sci. U.S.A.* **90,** 7618–7622.

Samaridis, J., Casorati, G., Traunecker, A., Iglesias, A., Gutierrez, J., Muller, U., and Palacios, R. (1991). Development of lymphocytes in interleukin 7 transgenic mice. *Eur. J. Immunol.* **21,** 453–460.

Sarveila, P. A., and Russell, L. B. (1956). Steel, a new dominant gene in the mouse. *J. Hered.* **47,** 123–131.

Sauter, H., and Paige, C. J. (1987). Detection of normal B-cell precursors that give rise to colonies producing both κ and λ light immunoglobulin chains. *Proc. Natl. Acad. Sci. U.S.A.* **84**, 4989–4993.

Sauter, H., and Paige, C. J. (1988). B cell progenitors have different growth requirements before and after immunoglobulin heavy chain commitment. *J. Exp. Med.* **168**, 1511–1516.

Schatz, D. G., Oettinger, M. A., and Baltimore, D. (1989). The V(D)J recombination activating gene, RAG-1. *Cells* **59**, 1035–1048.

Schlissel, M. S., Constantinescu, A., Morrow, T., Baxter, M., and Peng, A. (1993). Double-strand signal sequence breaks in V(D)J recombination are blunt, 5'-phosphorylated, RAG-dependent, and cell cycle regulated. *Genes Dev.* **7**, 2520–2532.

Schlissel, M. S., Corcoran, L. M., and Baltimore, D. (1991). Virus-transformed pre-B cells show ordered activation but not inactivation of immunoglobulin gene rearrangement and transcription. *J. Exp. Med.* **173**, 711.

Schreurs, J., Gorman, D. M., and Miyajima, A. (1992). Cytokine receptors: A new superfamily of receptors. *Int. Rev. Cytol.* **137**, 121–155.

Sgori, D., Varki, A., Braesch-Anderson, S., and Stamenkovic, I. (1993). CD22, a B cell-specific immunoglobulin superfamily member, is a sialic acid-binding lectin. *J. Biol. Chem.* **268**, 7011–7018.

Shelley, C. S., Remold-O'Donnell, E., Davis, A. D., III, Bruns, G. A. P., F. S., R., Carroll, M. C., and Whitehead, A. S. (1989). Molecular characterization of sialophorin (CD43), the lymphocyte surface sialoglycoprotein defective in Wiskott-Aldrich syndrome. *Proc. Natl. Acad. Sci. U.S.A.* **86**, 2819.

Shevach, E. M., and Korty, P. E. (1989). Ly-6: A multigene family in search of a function. *Immunol. Today* **10**, 195–200.

Shinkai, Y., Rathburn, G., Lam, K.-P., Oltz, E. M., Stewart, V., Mendelsohn, M., Charron, J., Datta, M., Young, F., Stall, A. M., and Alt, F. W. (1992). Rag-2 deficient mice lack mature lymphocytes owing to inability to initiate V(D)J rearrangement. *Cell (Cambridge, Mass.)* **68**, 855–867.

Shipp, M. A., and Look, A. T. (1993). Hematopoietic differentiation antigens that are membrane-associated enzymes: Cutting is the key! *Blood* **82**, 1052–1070.

Shirasawa, T., Ohnishi, K., Hagiwara, S., Shigemoto, K., Takevi, Y., Rajewsky, K., and Takemori, T. (1993). A novel gene product associated with μ chains in immature B cells. *EMBO J.* **12**, 1827–1834.

Shiroo, M., Goff, L., Biffen, M., Shivnan, E., and Alexander, D. (1992). CD45 tyrosine phosphatase activated p59^{fyn} couples the T cell antigen receptor to pathways of diacylglycerol production, protein kinase C activation and calcium influx. *EMBO J.* **11**, 4887–4897.

Shuai, K., Stark, G. R., Kerr, I. M., and Darnell, J. E., Jr. (1993). A single phosphotyrosine residue of Stat91 required for gene activation by interferon-γ. *Science* **261**, 1744–1746.

Silvennoinen, O., Schindler, C., Schlessinger, J., and Levy, D. E. (1993). Ras-independent growth factor signaling by transcription factor tyrosine phosphorylation. *Science* **261**, 1736–1739.

Silver, D. P., Spanopoulou, E., Mulligan, R. C., and Baltimore, D. (1993). Dispensable sequence motifs in the RAG-1 and Rag-2 genes for plasmid V(D)J recombination. *Proc. Natl. Acad. Sci. U.S.A.* **90**, 6100–6104.

Solvason, N., Lehuen, A., and Kearney, J. F. (1991). An embryonic source of Ly1 but not conventional B cells. *Int. Immunol.* **3**, 543–550.

Spangrude, G. J., and Brooks, D. M. (1993). Mouse strain variability in the expression the hematopoietic stem cell antigen Ly-6A/E by bone marrow cells. *Blood* **82**, 3327–3332.

Spangrude, G. J., Heimfeld, S., and Weissman, I. L. (1988). Purification and characterization of mouse hematopoietic stem cells. *Science* **241**, 58–62.

Stahl, N., and Yancopoulos, G. D. (1993). The alphas, betas, and kinases of cytokine receptor complexes. *Cell (Cambridge, Mass.)* **74**, 587–590.

Strasser, A., Harris, A. W., Corcoran, L. M., and Cory, S. (1994). Bcl-2 expression promotes B- but not T-lymphoid development in scid mice. *Nature (London)* **368**, 457–460.

Sudo, R., Nishikawa, S., Ohno, M., Akiyama, N., Tamakoshi, M., Yoshida, H., and Nishikawa, S.-I. (1993). Expression and function of the interleukin 7 receptor in murine lymphocytes. *Proc. Natl. Acad. Sci. U.S.A.* **90**, 9125–9129.

Sudo, T., Ito, M., Ogawa, Y., Iizuka, M., Kodama, H., Kunisada, T., Hayashi, S., Ogawa, M., Sakai, K., Nishikawa, S., and Nishikada, S.-I. (1989). Interleukin 7 production and function in stromal cell-dependent B cell development. *J. Exp. Med.* **170**, 333–342.

Taga, R., and Kishimoto, T. (1993). Cytokine receptors and signal transduction. *FASEB J.* **7**, 3387–3396.

Takeda, S., Zou, Y.-R., Bluethmann, H., Kitamura, D., Muller, U., and Rajewsky, K. (1993). Deletion of the immunoglobulin κ chain intron enhancer abolishes κ chain gene rearrangements in trans. *EMBO J.* **12**, 2329–2336.

Tanaka, T., Wu, G. E., and Paige, C. J. (1994). Characterization of the B cell-macrophage lineage transition in 70Z/3 cells. *Eur. J. Immunol.* **24**, 1544–1548.

Till, J. E., McCulloch, E. A., and Siminovitch, L. (1964). A stochastic model of stem cell proliferation, based on the growth of spleen colony-forming cells. *Proc. Natl. Acad. Sci. U.S.A.* **51**, 29–36.

Toles, J. F., Chui, D. H. K., Belbeck, L. W., Starr, E., and Barker, J. E. (1989). Hemopoietic stem cells in murine embryonic yolk sac and peripheral blood. *Proc. Natl. Acad. Sci. U.S.A.* **86**, 7456–7459.

Travis, A., Amsterdam, A., Bélanger, C., and Grosschedl, R. (1991). LEF-1, a gene encoding a lymphoid-specific protein, with an HMG domain, regulates T-cell receptor α enhancer function. *Genes Dev.* **5**, 880–894.

Tsubata, T., Tsunata, R., and Reth, M. (1991). Cell surface expression of the short immunoglobulin μ chain (Dμ protein) in murine pre-B cells is differentially regulated from that of the intact μ chain. *Eur. J. Immunol.* **221**, 1359–1363.

Tyan, M. L., and Herzenberg, L. A. (1968). Studies on the ontogeny of the mouse immune system. *J. Immunol.* **101**, 446–450.

Ullrich, A., and Schlessinger, J. (1990). Signal transduction by receptors with tyrosine kinase activity. *Cell (Cambridge, Mass.)* **61**, 203–212.

van de Run, M., Heimfeld, S., Spangrude, G. J., and Weissman, I. L. (1989). Mouse hematopoietic stem-cell antigen Sca-1 is a member of the Ly-6 antigen family. *Proc. Natl. Acad. Sci. U.S.A.* **86**, 4634–4638.

Venkitaraman, A. R., and Cowling, R. J. (1992). Interleukin 7 receptor functions by recruiting the tyrosine kinase p59^{fyn} through a segment of its cytoplasmic tail. *Proc. Natl. Acad. Sci. U.S.A.* **89**, 12083–12087.

Welch, P. A., Burrows, P. D., Namen, A., Gillis, S., and Cooper, M. D. (1990). Bone marrow stomal cells and interleukin-7 induce coordinate expression of the BP-1/6C3 antigen and pre-B cell growth. *Int. Immunol.* **2**, 697–705.

Whitlock, C. A., and Witte, O. N. (1982). Long-term culture B lymphocytes and their precursors from murine bone marrow. *Proc. Natl. Acad. Sci. U.S.A.* **79**, 3608–3612.

Whitlock, C. A., Tidmarsh, G. F., Muller-Sieberg, C., and Weissman, I. L. (1987). Bone marrow stromal cell lines with lymphopoietic activity express high levels of a pre-B neoplasia-associated molecule. *Cell (Cambridge, Mass.)* **48**, 1009–1021.

Wiken, M., Bjorck, P., Axelsson, B., and Perlmann, P. (1988). Induction of CD43 expression during activation and terminal differentiation of human B cells. *Scand. J. Immunol.* **28**, 457–464.

Witte, P. L., Burrows, P. D., Kincade, P. W., and Cooper, M. D. (1987). Characterization of

B lymphocyte lineage progenitor cells from mice with severe combined immune deficiency disease (SCID) made possible by long term culture. *J. Immunol.* **138,** 2698–2705.

Wu, A. M., Till, J. E., Siminovitch, L., and McCulloch, E. A. (1967). Cytological evidence for a relationship between normal hemopoietic colony-forming cells and cells of the lymphoid system. *J. Exp. Med.* **127,** 455–463.

Wu, Q., Lahti, J. M., Air, G. M., Burrows, P. D., and Cooper, M. D. (1990). Molecular cloning of the murine BP-1/6C3 antigen: A member of the zinc-dependent metallopeptidase family. *Proc. Natl. Acad. Sci. U.S.A.* **87,** 993–997.

Wu, Q., Li, L., Cooper, M. D., Pierres, M., and Gorvel, J. P. (1991). Aminopeptidase A activity of the murine B-lymphocyte differentiation antigen BP-1/6C3. *Proc. Natl. Acad. Sci. U.S.A.* **88,** 676–680.

Yin, T., Miyazawa, K., and Yang, Y.-C. (1992a). Characterization of interleukin-11 receptor and protein tyrosine phosphorylation induced by interleukin-11 in mouse 3T3-L1 cells. *J. Biol. Chem.* **267,** 8347–8351.

Yin, T., Schendel, P., and Yang, Y.-C. (1992b). Enhancement of in vitro and in vivo antigen-specific antibody responses by interleukin 11. *J. Exp. Med.* **175,** 211–216.

Yin, T., Yaga, T., Tsang, M. L.-S., Yasukawa, D., Kishimoto, T., and Yang, Y.-C. (1993). Involvement of IL-6 signal transducer gp130 in IL-11 mediated signal transduction. *J. Immunol.* **151,** 2555–2561.

Yonemura, Y., Kawakita, M., Masuda, T., Fujimoto, K., Kato, K., and Takatsuki, K. (1992). Synergistic effects of interleukin 3 and interleukin 11 on murine megakaryopoiesis in serum-free culture. *Exp. Hematol.* **20,** 1011–1016.

Zhang, X.-G., Gu, J.-J., Liu, Z.-Y., Yasukawa, K., Yancopoulos, G. D., Turner, K., Shoyab, M., Taga, T., Kishimoto, T., Bataille, R., and Klein, B. (1994). Ciliary neurotropic factor, interleukin 11, leukemia inhibitory factor, and oncostatin M are growth factors for human myeloma cell lines using the interleukin 6 signal transducer gp130. *J. Exp. Med.* **177,** 1337–1342.

Zou, Y.-R., Takeda, S., and Rajewsky, K. (1993). Gene targeting in the Igκ locus: Efficient generation of λ chain-expressing B cells, independent of gene rearrangements in Igκ. *EMBO J.* **12,** 811–820.

Zsebo, K. M., Williams, D. A., Geissler, E. N., Broudy, V. C., Martin, F. H., Atkins, H. L., Hsu, R.-Y., Birkett, N. C., Okino, K. H., Murdock, D. C., Jacobsen, F. W., Langley, K. E., Smith, K. A., Takeishi, T., Cattanach, B. M., Galli, S. J., and Suggs, S. V. (1990). Stem cell factor is encoded at the *Sl* locus of the mouse and is the ligand for the c-*kit* tyrosine kinase receptor. *Cell (Cambridge, Mass.)* **63,** 213–224.

Signal Transduction by the Antigen Receptors of B and T Lymphocytes

Michael R. Gold* and Linda Matsuuchi†

* Department of Microbiology and Immunology and † Department of Zoology,
University of British Columbia, Vancouver, British Columbia, Canada V6T 1Z3

B and T lymphocytes of the immune system recognize and destroy invading microorganisms but are tolerant to the cells and tissues of one's own body. The basis for this self/non-self-discrimination is the clonal nature of the B and T cell antigen receptors. Each lymphocyte has antigen receptors with a single unique antigen specificity. Multiple mechanisms ensure that self-reactive lymphocytes are eliminated or silenced whereas lymphocytes directed against foreign antigens are activated only when the appropriate antigen is present. The key element in these processes is the ability of the antigen receptors to transmit signals to the interior of the lymphocyte when they bind the antigen for which they are specific. Whether these signals lead to activation, tolerance, or cell death is dependent on the maturation state of the lymphocytes as well as on signals from other receptors. We review the role of antigen receptor signaling in the development and activation of B and T lymphocytes and also describe the biochemical signaling mechanisms employed by these receptors. In addition, we discuss how signal transduction pathways activated by the antigen receptors may alter gene expression, regulate the cell cycle, and induce or prevent programmed cell death.

KEY WORDS: Lymphocytes, Antigen receptors, Signal transduction, Phosphorylation, Kinases.

I. Introduction

The primary function of the T and B lymphocytes of the immune system is to eliminate invading microorganisms such as pathogenic bacteria and viruses. Lymphocytes detect the presence of such invaders by means of cell surface receptors that recognize antigens, small portions of the proteins or carbohydrates of the microorganism. The antigen receptors on the lymphocyte inform the cell that it has bound an antigen by activating intracellular enzymes, in particular protein tyrosine kinases (PTKs). The PTKs activate signal transduction pathways that transmit the signal from the cell membrane to the nucleus of the cell. Signals emanating from the antigen receptors induce the expression of a number of genes and regulate the activity of proteins that control the cell cycle. The net result is that a normally quiescent lymphocyte enters the cell cycle and becomes activated. On receiving additional signals through other cell surface receptors, these activated B and T lymphocytes differentiate into effector cells that contribute to the elimination of invading microorganisms. B lymphocytes differentiate into antibody-secreting plasma cells. Antibodies facilitate the removal and destruction of pathogenic organisms by activating the complement cascade and by acting as tags that promote endocytosis, phagocytosis, and/or killing by macrophages, neutrophils, and other cells. There are several types of T lymphocytes that perform different functions. On activation, one type of helper T cell produces soluble mediators (cytokines), as well as cell surface molecules, that help B cells differentiate into antibody-secreting cells. A second class of helper T cell produces cytokines that increase the ability of macrophages to kill bacteria. Finally, cytolytic T lymphocytes (CTLs) directly kill cells infected with viruses as well as some types of tumor cells.

The remarkable ability of the immune system to recognize and respond to a large number of antigens, yet mount responses only to the invading microorganism, is due to the clonal nature of the antigen receptors on T and B cells. In general, each lymphocyte (or clone of lymphocytes derived from a single precursor cell) has a single antigen-binding specificity that is unique to that lymphocyte. In response to a given antigen, only those lymphocytes whose antigen receptors bind that antigen with high affinity will be activated. The antigen-binding sites of the T and B cell antigen receptors are each formed by two polypeptides. These polypeptides are encoded by multiple gene segments that are joined by DNA recombinations during lymphocyte development. There are multiple sequences for each gene segment and the random combining of these sequences is the basis for the vast diversity of antigen-binding specificities. Because the repertoire of antigen receptors is randomly generated in an antigen-independent manner during lymphocyte development, lymphocytes with antigen receptors that recognize components of one's own body will be generated as well. However, one of the hallmarks of the immune system is that it usually responds only to foreign antigens and is tolerant to "self" components. Self-reactive lymphocytes that are generated must be silenced or eliminated. Indeed, both T and B cells go through stages in which the binding of antigen to the antigen receptors results in programmed cell death (apoptosis). Thus, signals through the antigen receptor can promote either activation or cell death. As we shall describe in the first part of this chapter, antigen receptor signaling plays a critical role at multiple points in lymphocyte development and activation and can be either a positive or a negative signal, depending on the maturation state of the lymphocyte and on signals coming from other receptors.

The second aim of this chapter is to summarize our current understanding of how the antigen receptors of T and B lymphocytes transmit signals from the cell membrane to nucleus. The high degree of relatedness between T and B lymphocytes is reflected in the common design of their antigen receptors and the common signal transduction mechanisms these receptors use. Both the T cell antigen receptor (TCR) and the B cell antigen receptor (BCR) consist of an antigen-binding subunit and signaling subunits. The signaling subunits interact with members of at least two distinct families of PTKs, resulting in the rapid phosphorylation on tyrosine residues of a number of proteins. Both the TCR and BCR also interact with other cell surface molecules, termed coreceptors, that bring additional PTKs into the complex. In both T and B cells, the substrates of these PTKs include enzymes that generate intracellular second messengers (e.g., free Ca^{2+}) and proteins that control protein kinase cascades. Although a considerable amount is known about the initial signaling events that occur within minutes of engaging the antigen receptors, much less is known about the downstream events that contribute to the cellular responses, which often occur more than 24 hr after the initial contact with antigen. Thus, the focus of signal transduction research is turning toward investigations of how cytoplasmic signaling events are connected to transcription factors and proteins that control the cell cycle.

There are several reasons why elucidating the signaling pathways that link the antigen receptors to the nucleus are of great interest. First, it is likely that many receptors in different cell types will use common signaling pathways. Thus, identification of new signaling pathways could yield insights into how many receptors work. Second, mutations in signaling components that link the antigen receptors to the nucleus are likely to have profound effects on the function of the immune system. Activating mutations could result in inappropriate lymphocyte activation and autoimmune diseases. In contrast, there are several examples of loss-of-function mutations that result in severe immunodeficiency diseases. In patients with such diseases, identification of the defective signaling component may allow therapeutic strategies involving gene transfer.

II. Structure of B and T Cell Antigen Receptors: Modular Receptors

A number of receptors on cells of the immune system have a modular design in which ligand binding and signal transduction are mediated by separate subunits (Keegan and Paul, 1993). The receptors that fall into this class include the B and T cell antigen receptors and several receptors that bind the constant portions of immunoglobulin (Ig) molecules, including the mast cell FcεRI receptor, which binds IgE (Fig. 1). The ligand-binding subunits of these receptors all belong to the Ig superfamily of proteins, which are characterized by having globular domains that are formed by an intrachain disulfide bond. The signaling subunits of these receptors are also highly related and their cytoplasmic domains all contain a conserved motif [the antigen recognition activation motif (ARAM)] that mediates interactions with PTKs (see Section V,A). The common design of these receptors points to a common evolutionary origin. The structure of the B and T cell antigen receptors has been reviewed in detail elsewhere (Gold and DeFranco, 1994; Reth, 1992; Clevers *et al.*, 1988; Raulet, 1989) and only the salient points are summarized here.

A. The B Cell Antigen Receptor

The antigen-binding subunit of the B cell antigen receptor (BCR) is a membrane immunoglobulin (mIg). Like secreted antibodies, the mIgs consist of four polypeptides (two identical heavy chains and two identical light chains; see Fig. 1 and Table I) that are joined by disulfide bonds. The Ig heavy chain genes contain different exons that encode the carboxy termini of membrane versus secreted Igs. Membrane and secreted Igs are generated by alternative mRNA splicing and the exons used only in the membrane forms of the heavy chains encode an extracellular spacer region, a transmembrane domain of about 25 amino acids, and a short cytoplasmic domain (only 3 amino acids for mIgM and mIgD). Each of the heavy chain isotypes (IgM, IgD, IgG, IgA, and IgE) can be produced in both membrane and secreted forms.

The amino termini of both the Ig heavy and light chains are highly variable and together these variable regions form the antigen-binding pocket of the Ig molecule. During early B cell development, the DNA encoding the Ig heavy chain variable region is assembled by combining one each of the multiple V_H, D_H, and J_H gene segments. Similarly, the Ig light chain variable region gene is assembled by combining one each of the multiple V_L and J_L gene segments. Each B cell progenitor exhibits

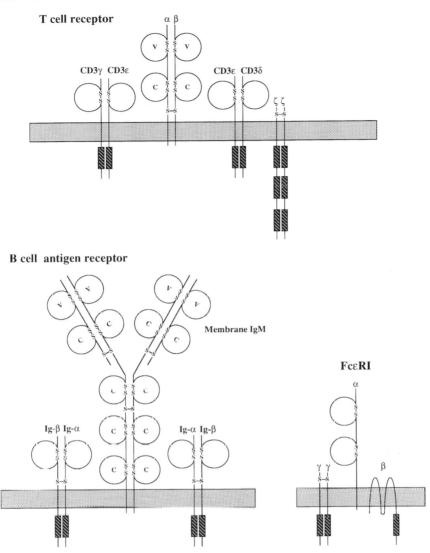

FIG. 1 Modular design of immune recognition receptors. Each receptor has a ligand-binding subunit [TCR $\alpha\beta$, membrane IgM (mIgM), and FCϵRIα] and associated signal-transducing subunits. Many of the chains are composed of Ig-like domains that are formed by intrachain disulfide (S–S) bonds. The domains that comprise the variable (V) and constant (C) regions of the TCR and BCR antigen-binding subunits are indicated. The antigen recognition activation motifs (ARAMs) that interact with tyrosine kinases are represented as shaded boxes. The BCR is presumed to have two copies of the Ig-α : Ig-β heterodimer, but this has not been proved. The organization of the various polypeptide chains for these receptors is also not known. For example, it is not clear whether Ig-α or Ig-β interacts with the membrane IgM component of the BCR. Adapted with permission from Weiss and Littman (1994).

TABLE I

Polypeptides of the T Cell Receptor and B Cell Receptor Complexes

TCR antigen-binding subunits	BCR antigen-binding subunits

$\alpha\beta$ TCR
  α, 40–45 kDa[a]
  β, 38–45 kDa
 	
$\gamma\delta$ TCR
  γ, 45–60 kDa
  δ, 40–45 kDa
	Ig heavy chains
	  μ, 72 kDa
	  δ, 63 kDa
	  γ, 55 kDa
	  α, 57 kDa
	  ε, 72 kDa
	
	Ig light chains and surrogate light chains
	  κ and λ, 25–28 kDa
	  V_{pre-B}, 15kDa
	  $\lambda5$, 21kDa

TCR signaling subunits	BCR signaling subunits
CD3γ, 21 kDa (Hu: 25–28 kDa) CD3δ, 26 kDa (Hu: 20kDa) CD3ε, 25 kDa (Hu: 20kDa) ζ, 16 kDa η, 22 kDa	Ig-α, 32–34 kDa (Hu: 47 kDa)

TCR coreceptors	BCR coreceptors

CD4, 55–59 kDa
 	
CD8α,[b] 32–34 kDa
CD8β, 32–34 kDa
	CD19 coreceptor complex
	  CD19, 69 kDa (Hu: 95 kDa)
	  CD21 (CR2), 140 kDa
	  TAPA-1, 20 kDa
	  Leu-13, 16 kDa
	CD22, 150 (Hu: CD22α, 130 kDa;
	  CD22β, 140 kDa[c])

[a] Molecular masses reported are for murine polypeptides. The molecular masses of corresponding human polypeptides are similar unless indicated otherwise. Molecular mass ranges reflect heterogeneous glycosylation. All chains listed are glycosylated except for ζ.

[b] CD8 can exist as an $\alpha\alpha$ homodimer or as an $\alpha\beta$ heterodimer.

[c] Human CD22 exists as a noncovalently associated $\alpha\beta$ heterodimer. The α and β chains are derived by alternative splicing of the same gene. It is not clear if mouse CD22 is also a heterodimer.

allelic exclusion, assembling a complete and in-frame heavy chain variable region gene on only one of the two chromosomes bearing the heavy chain locus and a complete and in-frame light chain variable region gene on only one of the chromosomes that bear the Ig κ and λ light chain loci. These DNA rearrangements commit that cell and all its daughter cells to use only those rearranged variable region genes to produce Ig heavy and light chain polypeptides. Together, the amino acids of the heavy and light chains variable regions form a unique antigen-binding site. Because the mIg has two identical heavy chains and two identical light chains, each mIg molecule has two identical antigen-binding sites. The combinatorial joining of the heavy and light chain variable region gene segments, together with the random pairing of different heavy and light chain variable regions, generates a large repertoire of monospecific B cells with different antigen-binding sites.

The heavy chain variable region gene segments are initially assembled just upstream of the μ constant region gene and immature B cells express exclusively mIgM on their surface. Mature B cells express on their surface both mIgM and mIgD, which on a single cell have the same antigen-binding site. The μ and δ constant regions genes are closely spaced on the chromosome and can be joined to the same rearranged variable region gene by alternative splicing of a long RNA transcript containing the variable region, μ, and δ genes. B cell activation can give rise to memory B cells that have undergone isotype switching and now express on their surface mIgG, mIgA, or mIgE instead of mIgM and mIgD. Isotype switching reflects additional DNA rearrangements in which the μ and δ constant region genes are excised and the same complete variable region is placed upstream of the gene encoding the constant region of one of these isotypes.

The mIg of the BCR is noncovalently associated with two invariant chains, Ig-α and Ig-β, which form a disulfide-linked heterodimer (Fig. 1) (Reth, 1992). Ig-α and Ig-β are both transmembrane glycoproteins which, unlike mIgM and mIgD, have substantial cytoplasmic domains (61 and 48 amino acids, respectively). Because the mIg has two identical heavy chains, it is likely that each mIg is complexed to two Ig-α : Ig-β heterodimers, although this has not been demonstrated experimentally. The membrane forms of all five classes of mIg can associate with the same Ig-α : Ig-β molecules (Venkitaraman et al., 1991). The Ig-α : Ig-β heterodimer is essential for cell surface expression of the BCR and for signal transduction by the BCR. In the absence of either Ig-α or Ig-β, mIgM is retained in the endoplasmic reticulum and is not expressed on the cell surface (Reth, 1992; Matsuuchi et al., 1992). Experiments in which mIgM heavy and light chain genes were transfected into fibroblasts or into a pituitary cell line showed that coexpression of Ig-α and Ig-β was sufficient to induce cell surface expression of mIgM in these non-B cells (Venkitaraman et al., 1991; Matsuuchi et al., 1992). Thus, Ig-α and Ig-β are the only lymphoid-specific proteins required for cell surface expression of mIgM. Because mIgM and mIgD have cytoplasmic tails of only three amino acids, it was proposed that Ig-α and Ig-β are the portions of the BCR that interact with the intracellular signal transduction machinery. This hypothesis has now been confirmed. Chimeric proteins containing a short conserved motif (the ARAM) from the cytoplasmic domain of either Ig-α or Ig-β can mediate many of the signaling reactions characteristic of the intact BCR (Law et al.,1993; Kim et al., 1993b; Sanchez et al., 1993).

B. The T Cell Antigen Receptor

The structure of the T cell antigen receptor complex (TCR) is analogous to that of the BCR in that it consists of a clonally variable antigen-binding subunit and noncovalently associated invariant polypeptides that are involved in signal transduction (Clevers et al., 1988). The antigen-binding portion of the TCR on most T cells is formed by two polypeptides, the TCR α and β chains, which are joined by a disulfide bond (Fig. 1). Both of these chains have two extracellular Ig-like domains with an intrachain disulfide bond as well as a transmembrane domain and a short (5–12 amino acids) cytoplasmic domain. The amino-terminal Ig-like domain of each chain is highly variable, similar to the Ig heavy and light chain variable regions. Each TCR molecule has one α chain and one β chain, so the TCR has only one antigen-binding site, unlike the BCR. Like the Ig variable region genes, the variable region genes of the TCR α and β chains are assembled by DNA rearrangements that join multiple gene segments. The α chain variable region gene is assembled from V_α and J_α gene segments whereas the β chain variable region is composed of V_β, D_β, and J_β gene segments. The combinatorial joining of gene segments and the random combining of different α and β chains create a large repertoire of antigen-binding sites. Although each progenitor T cell correctly assembles only one β variable region gene, it is common for T cells to rearrange both α chain loci correctly. When both α chain loci are correctly rearranged, the T cell will express on its surface two TCRs that have the same β chain but different α chains (Padovan et al., 1993). This T cell and its daughter cells may be able to recognize two different antigens.

About 5% of the T cells in humans have TCRs that lack the α and β chains and instead have alternative polypeptides, γ and δ, that form the antigen-binding subunit of the receptor (Raulet, 1989). The γ chain is analogous to the α chain in that it is composed of only V and J gene segments whereas the δ chain is analogous to the β chain, being comprised of V, D, and J gene segments. There are relatively few of each of these gene segments that make up the γ and δ chains and the so-called $\gamma\delta$ T cells exhibit a limited repertoire of antigen-binding sites. $\gamma\delta$ T cells appear to be a separate lineage of T cells that split off from the $\alpha\beta$ TCR lineage early in T cell development (Haas and Tonegawa, 1992). The $\gamma\delta$ cells are highly enriched in the gut-associated lymphoid tissues and in the skin.

Their function is not known, but it has been proposed that they are stimulated by antigens present on common pathogenic microorganisms or by tissue damage caused by pathogens (Kronenberg, 1994).

Both the $\alpha\beta$ and $\gamma\delta$ antigen-binding subunits of the TCRs are associated with the same set of invariant polypeptides that include the CD3γ, CD3δ, and CD3ε chains and two ζ chains (Clevers et al., 1988). The two ζ chains are linked by a disulfide bond. Some TCRs contain a ζ–η heterodimer instead of a ζ–ζ homodimer. The η chain is a truncated version of the ζ chain generated by alternative splicing of ζ chain mRNA (Jim et al., 1990). Although the precise stoichiometry is not known, it is thought that there may be two CD3ε chains per TCR, one of which forms a noncovalently associated pair with CD3δ and one that pairs with CD3γ (Fig. 1) (Manolios et al., 1991). As for the BCR, all of the TCR chains are required for efficient assembly and cell surface expression of the TCR. Incompletely assembled complexes are retained in the endoplasmic reticulum or targeted for degradation (Klausner et al., 1990). These invariant chains, in particular the ζ chain, mediate signaling by the TCR. Chimeric proteins containing the cytoplasmic domain of the ζ chain can induce all of the signal transduction events that are stimulated by the intact TCR (Irving and Weiss, 1991). Chimeric proteins containing the cytoplasmic tail of CD3ε can also induce some of the responses characteristic of the TCR (Letourneur and Klausner, 1992).

C. Antigen Recognition by B and T Cell Antigen Receptors

Although the TCR and BCR share many structural features, they recognize fundamentally distinct forms of antigen that reflect the different roles of T and B cells in protecting the host. B cells protect the host from extracellular pathogens and toxins and the BCR recognizes soluble and cell-bound proteins or carbohydrates. Antigen binding by the BCR initiates intracellular signals that promote the differentiation of that cell into a plasma cell that secretes antibodies with the identical antigen specificity as the BCR. Antibodies contribute to the elimination of antigens in several ways. Gramnegative bacteria coated with antibodies can be lysed by complement proteins. Coating bacteria or viruses with antibodies promotes their uptake and digestion by macrophages, which possess receptors for the constant portions of antibodies (Fc receptors). Toxins that are bound by antibodies are also targeted for elimination. While B cells protect the host against extracellular antigens, T cells protect the host against intracellular pathogens. Cytolytic T cells (CTLs) kill cells infected with viruses. Another type of T cell, the Th1 cell, secretes cytokines that increase the ability of macrophages to kill parasites (e.g., *Leishmania major*) and bacteria (e.g., *Mycobacterium tuberculosis*) that normally replicate inside the macrophage. The TCR recognizes a short peptide (8–15 amino acids) derived from a foreign antigen that is bound in the groove of a major histocompatibility complex (MHC) molecule (see below) on the surface of acell. Antigens are proteolytically processed into peptides inside antigen-presenting cells. The peptides are bound by MHC molecules and these MHC–peptide complexes are transported to the cell surface and presented to T cells. The antigen-binding site of the TCR binds the peptide as well as a portion of the self-MHC molecule (Jorgensen *et al.*, 1992).

There are two distinct pathways of antigen processing and presentation, depending on the source of the antigen (Germain, 1994). Proteins produced inside of a cell, such as viral proteins, are processed into peptides by large cytoplasmic complexes called proteasomes that possess multiple proteolytic activities (Goldberg and Rock, 1992). The resulting peptides are imported into the endoplasmic reticulum via peptide transporters, where they bind to MHC class I molecules. The MHC class I–peptide complex is then transported to the cell surface. All cells of the body express MHC class I and can act as antigen-presenting cells for intracellular antigens. In contrast, extracellular antigens (e.g., bacteria that are taken up by phagocytosis) are presented to T cells only by specialized antigen-presenting cells such as dendritic cells, macrophages, and B cells that express MHC class II proteins. Antigens taken up by endocytosis or by phagocytosis are degraded into peptides in an endocytic compartment. The MHC class II protein is initially assembled in the endoplasmic reticulum with a polypeptide called the invariant chain blocking its peptide-binding cleft. This same polypeptide is thought to direct the MHC class II protein to the intracellular compartment that contains the antigen-derived peptides (Lotteau *et al.*, 1990). In this specialized compartment, the invariant chain dissociates from the MHC class II protein, allowing peptides to bind (Schmid and Jackson, 1994). The MHC class II–peptide complex is then transported to the cell surface.

The two types of MHC–peptide complexes are recognized by different subsets of T cells, one of which expresses the cell surface protein CD4, the other of which expresses the CD8 protein. $CD4^+$ T cells recognize peptides presented on MHC class II proteins whereas $CD8^+$ cells recognize peptides presented on MHC class I proteins. CD4 and CD8 bind the MHC class II and class I proteins, respectively, and efficient signaling by the TCR requires that the CD4 or CD8 protein bind to the same MHC–peptide complex to which the TCR binds (Janeway, 1992). Thus, the TCR on T cells expressing CD4 is restricted to recognizing peptides presented by MHC class II proteins whereas the TCR on T cells expressing CD8 is restricted to recognizing peptides presented by MHC class I proteins. The role of the CD4 and CD8 "coreceptors" in TCR signaling is discussed in Section IV,C.

Biochemical studies of TCR and BCR signal transduction often require the ability to stimulate large numbers of cells through their antigen receptors. Given the great diversity of antigen-binding specifities of cells from normal animals, antigen-specific stimulation of significant numbers of normal cells could not be achieved until recently. Many T and B cell lines exist, but in general their antigen specificities are not known. Fortunately, the signaling functions of the TCR and BCR can be triggered by antibodies against the constant portions of the TCR or BCR, which cross-link or aggregate the receptors into clusters on the cell surface. This clustering activates the antigen receptors on all cells, regardless of their antigen specificity. At least for B cells, antigen receptor aggregation is in fact a requirement for BCR signaling, as monovalent antigens do not induce signaling reactions or activate B cells. Many studies have documented that anti-TCR α or β antibodies, anti-CD3ε antibodies, and anti-Ig antibodies can mimic the effects of antigens on T and B cells. One criticism of the use of polyclonal anti-receptor antibodies is that they can bind the antigen receptors at multiple sites and cross-link them much more extensively than real antigens, which can bind only to the antigen-binding sites of the TCR or BCR.

Two types of systems for antigen-specific stimulation of large numbers of lymphocytes have been developed. Immunoglobulin heavy and light chain genes from B cell hybridomas of known antigen specificity or TCR α and β chain genes from T cell clones of known antigen specificity can be cloned and then expressed in cell lines to generate antigen-specific B and T cell lines. Alternatively, transgenic mice expressing cloned Ig or TCR genes can be produced. The same allelic exclusion mechanisms that prevent single lymphocytes from expressing more than one antigen specificity results in all the T or B cells of the transgenic animal expressing only the antigen receptors encoded by the introduced transgenes. The transgenic system not only provides a way of producing normal cells of identical specificity, but provides a powerful system for studying the effects of antigen receptor signaling on lymphocyte development and activation *in vivo*.

III. Cellular Responses Mediated by Antigen Receptor Signaling

The goals of a properly functioning immune system are (1) to promote the development of lymphocytes that will recognize foreign antigens,

(2) to silence or eliminate lymphocytes that recognize self-antigens, and (3) to ensure that only lymphocytes with the proper antigen specificity are activated on challenge with an antigen. To accomplish these goals, the immune system has developed a number of checkpoints during lymphocyte development and activation that require signals through the antigen-specific receptors on lymphocytes. Lymphocyte development and activation can be thought of as a series of positive and negative selection events that are mediated by antigen receptor signaling. These steps are shown in Fig. 2.

A. Positive Selection of Pre-B and Pre-T Cells

During B cell development in the bone marrow, DNA rearrangement of the heavy chain variable region gene segments and expression of μ heavy chain polypeptides occur before rearrangement and expression of the light chain genes. In these pre-B cells, the μ heavy chain pairs with two invariant polypeptides called V_{pre-B} and $\lambda 5$ (see Table I) that act together as a surrogate light chain (Tsubata and Reth, 1990). V_{pre-B} resembles a variable region domain of an Ig light chain whereas $\lambda 5$ resembles the constant region domain of a light chain. The μ chain bound to the surrogate light chains associates with Ig-α and Ig-β and is expressed on the cell surface (Iglesias et al., 1991). Signaling through this pre-B cell receptor turns off further DNA rearrangements of the heavy chain loci, thereby assuring that only one heavy chain variable region is expressed by the cell (allelic exclusion). In transgenic mice, introduction of a rearranged μ chain gene containing the membrane exons suppresses all endogenous heavy chain rearrangement, giving rise to animals in which all B cells express only the introduced heavy chain. In contrast, a rearranged μ chain gene containing only the exons for the secreted form of IgM does not suppress endogenous Ig heavy chain rearrangement (Manz et al., 1988). This suggests that the pre-B cell detects a signal emanating from the pre-B cell receptor on the cell surface and does not have a mechanism for detecting correct in-frame rearrangement of the variable region gene segments at the DNA level. Signals transmitted by the pre-B cell receptor also promote the survival and maturation of pre-B cells and accelerate the rearrangement of the light chain genes (Kitamura et al., 1992; Iglesias et al., 1991; Reth et al., 1987). In transgenic mice in which the $\lambda 5$ gene has been disrupted ("$\lambda 5$ knockout"), the number of pre-B cells is greatly reduced and there are few mature B cells (Kiramura et al., 1992). Evidence suggesting that this failure of pre-B cells to develop is a signal transduction defect comes from the study of patients with X-linked agammaglobulinemia, who have a phenotype similar to that of the $\lambda 5$ knockout mice.

The genetic basis for this disease has been shown to be mutations in the gene encoding the Btk PTK (Tsukada *et al.*, 1993; Vetrie *et al.*, 1993). Activation of Btk by the pre-B cell receptor may initiate positive signals that are essential for B cell development. As described in Section IV, PTK activation is an essential component of BCR signaling.

Analogous events occur during early T cell development in the thymus (see Fig. 2). The TCR β chain rearranges before the α chain. Successful rearrangement of the TCR β chain turns off further rearrangement of the β chain loci, accelerates rearrangement of the α chain loci, and promotes the survival and further development of pre-T cells (Groettrup and von Boehmer, 1993; von Boehmer, 1994). In transgenic mice deficient in TCR β expression, T cell development is arrested at a very early stage (prior to expression of the CD4 and CD8 cell surface proteins) and the number of T cells and T cell precursors is greatly decreased. In contrast, the introduction of a TCR β transgene into severe combined immunodeficiency (SCID) mice, which are unable to rearrange their own TCR or Ig genes, allows pre-T cells to differentiate into immature thymocytes that express both CD4 and CD8. This positive selection of pre-T cells expressing the TCR β chain is mediated by a pre-T cell receptor, analogous to the pre-B cell receptor, in which the TCR β chain is linked to an invariant surrogate α chain called gp33 (Groettrup *et al.*, 1993). Evidence that the pre-T cell receptor initiates signals that regulate T cell development come from experiments showing that the Lck PTK is essential for TCR β chain-driven T cell maturation. Disruption of the *lck* gene in transgenic mice causes an early block in T cell development that cannot be overcome by expression of a rearranged TCR β chain gene (Perlmutter *et al.*, 1993). Expression of catalytically inactive Lck in transgenic mice also blocks T cell development. The essential role of Lck in pre-T cell receptor signaling is also supported by experiments showing that activated Lck can bypass the requirement for the pre-T cell receptor. In trangenic mice in which the TCR β gene has been disrupted, expression of a constitutively active version of Lck can promote T cell development up to the stage of $CD4^+CD8^+$ thymocytes, even in the absence of a pre-T cell receptor (Perlmutter, 1994).

If the pre-T cell and pre-B cell receptors deliver signals required for lymphocyte development, what are the ligands for these receptors? Do they bind molecules on the surface of cells in the thymus (for pre-T cells) or bone marrow (for pre-B cells) or are these receptors constitutively active when they reach the cell surface? If there are ligands for these receptors, they are likely to be the same for all pre-T or pre-B cells. In that case, they would probably bind to invariant portions of these receptors such as the gp33 or the CD3 chains for the pre-T cell receptor or the surrogate light chains or Ig-α : β of the pre-B cell receptor. Alternatively,

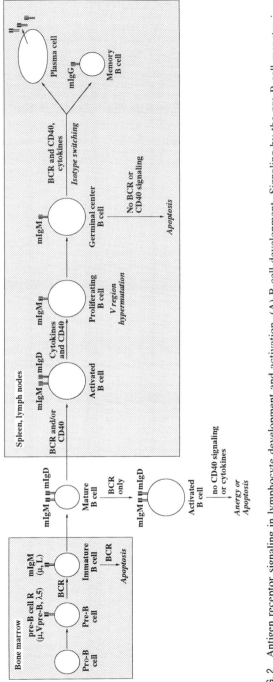

FIG. 2 Antigen receptor signaling in lymphocyte development and activation. (A) B cell development. Signaling by the pre-B cell receptor is required for pre-B cells to differentiate into immature B cells expressing mIgM. Immature B cells that bind antigen via their BCR are negatively selected and undergo apoptosis. Immature B cells that are not negatively selected become mature B cells that express both IgM and mIgD. B cell antigen receptor signaling activates mature B cells, but in the absence of CD40 engagement or cytokines this leads to anergy or apoptosis. B cells activated through the BCR and/or CD40 can be induced to proliferate by cytokines such as interleukin 4 (IL-4). In T cell-dependent responses in which CD40 is involved, the proliferating B cells undergo somatic hypermutation of their Ig variable region genes and migrate to germinal centers of lymphoid organs. Cells that still bind antigen with high affinity are positively selected by binding antigen via their BCR. Germinal center B cells require signaling through the BCR and CD40 in order to survive and become either antibody-secreting plasma cells or memory B cells. In the absence of signaling through the BCR or CD40, germinal B cells undergo apoptosis. In T-independent responses, cytokines are probably required for B cells activated through their BCRs to differentiate into IgM-secreting cells. In this case, hypermutation, isotype switching, and memory cell formation do not occur. (B) T cell development. Signaling by the pre-T cell receptor is required for differentiation of pre-T cells into double-positive (CD4⁺CD8⁺) thymocytes expressing the αβ TCR. The stochastic model in which thymocytes randomly turn off expression of either CD4 or CD8 is shown and the development and activation of CD4⁺ thymocytes is depicted. Single-positive thymocytes undergo both positive and negative selection. Only those undergoing positive selection are exported from the thymus. In mature T cells, TCR

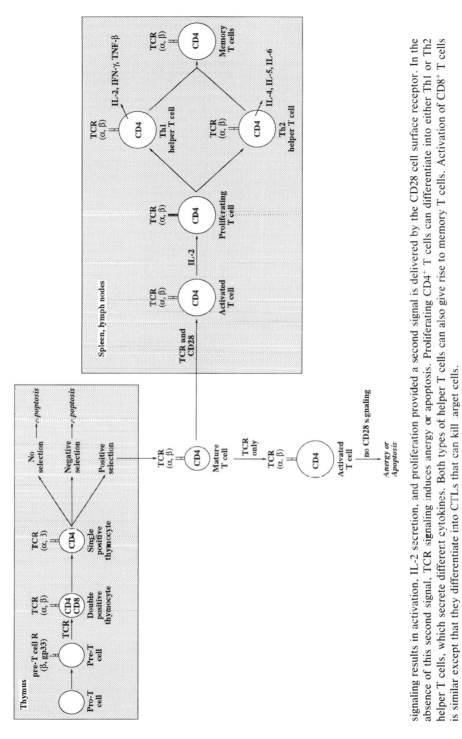

B

signaling results in activation, IL-2 secretion, and proliferation provided a second signal is delivered by the CD28 cell surface receptor. In the absence of this second signal, TCR signaling induces anergy or apoptosis. Proliferating CD4⁺ T cells can differentiate into either Th1 or Th2 helper T cells, which secrete different cytokines. Both types of helper T cells can also give rise to memory T cells. Activation of CD8⁺ T cells is similar except that they differentiate into CTLs that can kill target cells.

the ligands could bind to framework regions of the TCR β or Ig heavy chain, in much the same way that superantigens do. Identification of such ligands would be an important step in understanding the early steps in lymphocyte development.

B. Negative Selection of Immature B Cells

Positive signals from the pre-B cell receptor induce the rearrangement of the Ig light chain genes and promote the differentiation of pre-B cells into immature B cells that express a conventional mIgM molecule (one that contains either the κ or λ light chain instead of V_{pre-B} and $\lambda 5$) on their cell surface. Owing to the random nature of the Ig gene rearrangements, some of these immature B cells will have mIgs with high affinity for self-antigens. The majority of these self-reactive B cells, however, are deleted at this stage of development. Antigen binding by mIgM in immature B cells generally results in cell death (Nossal, 1994). Thus, BCR signaling leads to negative selection of immature B cells. The molecular basis for BCR-induced clonal deletion can be studied using B cell lines such as WEHI-231 and CH31. These cell lines have surface markers characteristic of immature B cells and cross-linking their BCRs with anti-IgM antibodies results in growth arrest in the G_1 phase of the cell cycle followed by programmed cell death (see Section VI,C) (Page and DeFranco, 1990; hasbold and Klaus, 1990; Yao and Scott, 1993). At a first approximation, the initial BCR-induced signaling events in these cell lines are identical to those in more mature B cells (see Section IV). Immature B cells that are not negatively selected by self-antigen develop into mature B cells that can be distinguished by surface markers such as mIgD, CD21, and high levels of the B220 form of CD45.

The development of transgenic mouse strains in which all B cells have the same antigen specificity has allowed dissection of the clonal deletion process *in vivo*. Nemazee and Burki (1989) expressed rearranged μ and κ chain genes derived from a monoclonal antibody against an MHC class I protein (H-2K^b) in transgenic mice. These rearranged μ and κ genes suppress rearrangement of the endogenous Ig genes, and in mice that do not express H-2K^b all B cells are specific for this antigen. When these Ig transgenes are expressed in mice that do express H-2K^b, the immature B cells encounter the antigen for which they are specific while developing in the bone marrow, because all cells of the body express MHC class I proteins. This results in clonal deletion of these self-reactive immature B cells in the bone marrow. Only a small number of mature B cells develop. Analysis of the B cells that escaped negative selection revealed that they expressed the λ light chain instead of the transgenic κ chain (Tiegs *et al.*, 1993). With a new light chain variable region, the immature B cell now has a BCR with a different antigen specificity; it no longer recognizes the self-antigen and escapes negative selection to become a mature B cell. Antigen binding to the BCR apparently reactivates the machinery involved in Ig gene rearrangements [e.g., the recombinase-activating genes RAG1 and RAG2 (reviewed by Schatz *et al.*, 1992) are induced], allowing a newly rearranged endogenous λ gene to replace the κ transgene. This process has been termed receptor editing.

Goodnow and colleagues have also used transgenic mice to study negative selection of immature B cells. They have employed doubly transgenic mice in which all of the B cells express mIg specific for hen egg lysozyme and the mice also produce a membrane-bound form of this antigen that is expressed on all cells. In these mice, large numbers of self-reactive immature B cells are found in the bone marrow, but no mature B cells are found in the secondary lymphoid organs such as the spleen or lymph node. This indicates that the immature B cells are deleted in the bone marrow before they can be exported. The BCR-mediated clonal deletion appears to involve two discrete steps: arrested development and then programmed cell death (Hartley *et al.*, 1993). Antigen-induced developmental arrest is characterized by a failure to acquire surface markers found on mature B cells (e.g., mIgD and CD21) and is reversible if the cells are cultured *in vitro* without the self-antigen. In the presence of self-antigen, the cells survive in this arrested state for about 15 hr and presumably attempt receptor editing. More prolonged exposure to the self-antigen leads to programmed cell death. In transgenic mice, the antigen-induced apoptosis, but not the developmental arrest, can be delayed by expression of the *bcl-2* gene, which prevents apoptosis in a number of situations (see Section VI,D). Thus, the developmental arrest is independent of the commitment to cell death.

Whereas membrane-bound antigens that cause extensive BCR cross-linking induce rapid clonal deletion of self-reactive B cells in the bone marrow, soluble forms of self-antigens result in clonal anergy. Clonal anergy is characterized by the development of mature self-reactive B cells that are nonfunctional and short-lived. In the double-transgenic system developed by Goodnow *et al.* (1988), expressing a soluble form of self-antigen (hen egg lysozyme) that can occur as small aggregates does not prevent the appearance of mature self-reactive B cells in the spleen and lymph nodes. However, these mature B cells are anergic (or tolerant) in that they cannot be activated *in vitro* by hen egg lysozyme or by anti-IgD antibodies coupled to dextran, a potent polyclonal B cell activator that causes extensive BCR cross-linking (Cooke *et al.*, 1994). Consistent with their inability to be activated by BCR cross-linking, BCR signaling is impaired in these anergic B cells. Moreover, anergic B cells survive only 3–4 days *in vivo,* as opposed to 4–5 weeks for B cells that do not encounter self-antigen during their development (Fulcher and Basten, 1994). Thus, clonal anergy may be a delayed form of clonal deletion that is induced by self-antigens that cause only a limited degree of BCR cross-linking.

C. Positive and Negative Selection of Immature T Cells: Learning to Distinguish Self from Foreign

The editing of the T cell repertoire is more complicated than that for B cells because T cells recognize peptides derived from foreign antigens that are bound to "self" MHC molecules. Immature T cells in the thymus that express both the CD4 and CD8 markers (double-positive thymocytes) undergo positive and negative selection (see Fig. 2). The negative selection process eliminates double-positive thymocytes with TCRs that recognize self-MHC with high enough affinity to be activated regardless of the peptide bound. Thymocytes that recognize self-MHC molecules complexed with peptides derived from self-proteins are also eliminated. Because all proteins made inside a cell are degraded to some extent, peptides derived from self-proteins will occupy a significant portion of the MHC class I molecules on the surface of cells. Self-reactive thymocytes are deleted via antigen-induced programmed cell death. The signals involved in negative selection can be studied *in vitro* using thymic organ cultures. Cross-linking the TCR with anti-TCR antibodies does not by itself cause clonal deletion. Instead, it downregulates expression of CD4 and CD8 and renders the immature T cells sensitive to deletion induced by a protein expressed on the surface of antigen-presenting cells (Page *et al.,* 1993). The nature of this second signal required for negative selection is not clear.

T cells must also undergo positive selection. Double-positive thymocytes with no affinity for self-MHC are not useful and are discarded. In contrast, thymocytes that bind self-MHC with modest affinity are positively selected by signals through their TCR. These signals turn off expression of the recombinase-activating genes RAG1 and RAG2 so that receptor editing due to rearrangement of the α chain genes can no longer occur (Turka *et al.,* 1991). The antigen specificity of the T cell is now fixed and it is exported from the thymus to become a long-lived mature T cell.

There is now experimental evidence supporting the concept that the avidity of the TCR–MHC–peptide interaction determines whether TCR engagement on CD4$^+$CD8$^+$ thymocytes leads to positive or negative selection. The avidity is the product of the intrinsic affinity of the TCR for

MHC–peptide, the density of the TCR on the thymocyte, and the density of the relevant MHC–peptide complex on the antigen-presenting cell. Using an *in vitro* fetal thymic organ culture system, Ashton-Rickardt *et al.* (1994) showed that low concentrations of a viral peptide induced positive selection whereas high concentrations caused negative selection. This suggested that no selection (cell death due to "neglect") occurs below a certain threshold of avidity, positive selection occurs at low avidity (in which there is modest TCR signaling), and negative selection occurs at high avidity (in which there is prolonged or excessive TCR signaling) (Allen, 1994). It is not clear how quantitative differences in TCR signaling lead to opposite outcomes. Pathways that override the positive signals and induce apoptosis may require high levels of signaling in order to be activated. Consistent with this interpretation, activated mature T cells are also killed by immobilized anti-TCR antibodies that induce extensive TCR cross-linking (J. H. Russell *et al.*, 1991). It is possible that other signals initiated by cell–cell interactions between the double-positive thymocyte and the antigen-presenting cells may also influence whether the TCR signal is perceived as positive or negative.

The CD4 and CD8 T cell surface proteins play an important role in the positive selection of CD4$^+$CD8$^+$ thymocytes. CD4 and CD8 act as "coreceptors" for the TCR. A pair of coreceptors (e.g., the TCR and CD4) are two receptors that bind simultaneously to the same ligand, resulting in synergistic signaling. Efficient activation of T cells by antigen-presenting cells requires that CD4 or CD8 binds to invariant portions of the same MHC molecule to which the TCR has bound (Fig. 3) (Janeway, 1992). As described in Section IV,C, the cytoplasmic tails of both CD4 and CD8 are tightly associated with the Lck PTK. Juxtaposition of this PTK with the cytoplasmic tails of the TCR chains amplifies TCR signaling. T cell precursors in the thymus undergoing selection bear both CD4 and CD8. In contrast, mature T cells express either CD4 or CD8. Which surface marker is expressed determines the function of the mature T cell. CD4 binds to MHC class II and CD4$^+$ T cells are generally helper T cells that make products that activate B cells or macrophages. This is fitting because MHC class II proteins present peptides derived from antigens taken up from outside of the cell (e.g., bacteria, parasites, and toxins), the type of antigens best attacked by antibodies or by phagocytic macrophages. In contrast, CD8 binds to MHC class I and CD8$^+$ T cells are generally killer T cells. Again this make sense, because MHC class I presents peptides derived from intracellular proteins such as viral proteins. Thus, T cells must express the proper coreceptor for the type of antigen (intracellular or extracellular) for which they are specific.

The choice between CD4 expression and CD8 expression is intimately

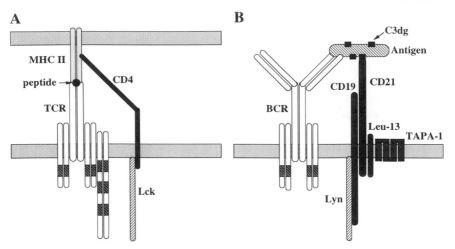

FIG. 3   Coreceptors for the T and B cell antigen receptors. Coreceptors are two receptors that simultaneously bind the same ligand, resulting in synergistic signaling. (A) In T cells, the antigen-binding subunit of the TCR binds an MHC class II–peptide complex on the surface of an antigen-presenting cell. The peptide is bound in a groove formed by the two chains of the MHC class II protein. Each chain has multiple alleles and the differences between the various alleles lie mainly in the regions that form the peptide-binding groove. The CD4 coreceptor binds to an invariant portion of the same MHC molecule to which the TCR is bound, and in doing so brings the Lck PTK in close proximity to the ARAMs of the TCR (shaded regions in TCR chains). Phosphorylation of tyrosine residues in the ARAMs is required for TCR signaling (see Fig. 7). The CD8 coreceptor acts in an analogous manner except that CD8 and the TCR on a CD8$^+$ cell bind to an MHC class I–peptide complex. (B) In B cells, the mIg portion of the BCR binds to antigenic determinants on the surface of an antigen (e.g., a bacterium). One component of the CD19 coreceptor complex is CD21, also known as complement receptor 2 (CR2). CD21 binds several complement components, including C3dg. *In vivo*, antigens such as bacteria are likely to be coated with complement proteins and this would bring the CD19–CD21 complex in close proximity to the BCR. CD19 is associated with the Lyn PTK, which could phosporylate tyrosine residues in the ARAMs of the BCR (shaded areas). Adapted with permission from Weiss and Littman (1994).

related to the process of positive selection. There are two theories as to how T cells differentiate from CD4$^+$CD8$^+$ cells into single positive cells (von Boehmer, 1994). In the instructive theory, the nature of the MHC molecule to which the TCR binds during positive selection (MHC class I vs MHC class II) determines whether CD4 or CD8 is brought into the complex. Coengagement of CD4, for example, with the TCR would transmit a signal that instructs that cell to turn off expression of CD8. Similarly, the binding of CD8 and the TCR to an MHC class I protein would deliver a qualitatively different signal that would commit the cell thereafter to expressing only CD8. In contrast, the stochastic theory proposes that thymocytes randomly turn off either CD4 or CD8 as they mature and that a thymocyte can be positively selected only if it has the correct coreceptor for the type of MHC molecule to which its TCR binds.

Both positive and negative selection of thymocytes can be demonstrated in transgenic mouse models, one of the most elegant of which has been developed by von Boehmer and colleagues (reviewed in von Boehmer, 1994). In this system, the mice are made transgenic for rearranged TCR α and β chain genes that encode a TCR that recognizes a peptide derived from the male-specific antigen H-Y only when it is bound to the product of a specific MHC class I gene allele, H-2D^b (the MHC class I genes have many alleles). In mice that do not express H-2D^b (which would be the self-MHC in this case), none of the T cells bearing the transgenic TCR survive because there is no positive selection. In female mice that express H-2D^b, but not the male antigen, there is positive selection and all of the T cells mature into CD8$^+$ CTLs. The fact that none develop into CD4$^+$ T cells shows that the specificity of the TCR for either MHC class I or class II determines whether the cells become committed to the CD4 lineage or the CD8 lineage. Finally, in H-2D^b male mice, the thymocytes expressing the transgenic TCR undergo programmed cell death because they encounter self-antigen.

D. Consequences of T Cell Receptor Signaling in Mature T Cells

1. Differentiation of CD4$^+$ T Cells into Functional Helper T Cells: Activation versus Anergy

Once exported from the thymus, mature naive CD4$^+$ T cells are capable of differentiating into one of two types of helper T cells (Th1 versus Th2) when their TCR binds peptide–MHC class II complexes on antigen-presenting cells (Swain et al., 1991). Th1 cells produce IL-2 and tumor necrosis factor β (TNF-β), as well as interferon-γ (IFN-γ), which increases the ability of macrophages to kill bacteria and other intracellular parasites such as Leishmania major. In contrast, Th2 cells produce a number of cytokines involved in activating B cells including interleukin (IL-4), IL-5, IL-6, IL-10, and IL-13 (Paul and Seder, 1994). T cell receptor signaling also induces both Th1 cells and Th2 cells to express on their surface the ligand for the CD40 receptor protein, which is present on B cells and several other cell types. Although the function of the CD40 ligand (CD40L) on Th1 cells is not clear, CD40L plays a critical role in Th2-dependent B cell activation (Noelle *et al.*, 1992a) (see Section III,E).

T cell receptor signaling alone cannot induce differentiation of resting T cells into Th1 or Th2 effector cells. The TCR must act in concert with other receptors on the T cell. The CD4 coreceptor amplifies the TCR signal whereas the CD28 T cell surface protein delivers a second or "costimulatory" signal. Because a single antigen-presenting cell is likely to have only a few hundred of the relevant MHC class II–peptide complexes on its surface (Harding and Unanue, 1990), only a small number of TCRs will normally be stimulated to deliver signals to the interior of the cell. The TCR signal must therefore be amplified in order to achieve levels of signaling sufficient to induce T cell responses. This amplification is achieved by having the CD4 coreceptor bind to invariant portions of the same MHC class II molecule as the TCR (Fig. 3). The cytoplasmic tail of CD4 binds Lck, a PTK that is required for TCR signaling (see Section V,C). Coclustering of CD4 with the TCR brings Lck in contact with the TCR and this has been shown to potentiate TCR signaling (Turka *et al.*, 1992).

In addition to signals from the TCR–CD4 complex, T cell activation requires a second costimulatory signal delivered by CD28 (Schwartz, 1992; Jenkins and Johnson, 1993). CD28 binds to at least two ligands on antigen-presenting cells, B7-1 and B7-2 (Azuma *et al.*, 1993; Freeman *et al.*, 1993). Activated T cells also express another CD28-like molecule, CTLA-4, that binds B7-1 and B7-2 with higher affinity than CD28 (Linsley *et al.*, 1991). When both the TCR–CD4 signal and the CD28 costimulatory signal are received by the T cell, transcription of a number of genes, including those encoding IL-2 and the α chain of the IL-2 receptor, is turned on (Ullman *et al.*, 1990). The induced IL-2 receptor α chain, together with the constitutively expressed β and γ chains, form a high-affinity IL-2 receptor and allows the secreted IL-2 to act as an autocrine growth factor that causes the cells to proliferate (Meuer *et al.*, 1984). Although TCR–CD4 signals alone induce the expression of many cytokine genes, little or no IL-2 is produced. Engagement of CD28 synergizes with TCR signaling to induce cytokine gene expression and stabilizes cytokine mRNAs, which are normally highly unstable (Fraser *et al.*, 1993; June *et al.*, 1990b). Thus, high-level IL-2 production and subsequent T cell proliferation requires a second or costimulatory signal from the CD28 T cell surface protein.

Proliferating CD4$^+$ T cells differentiate into either Th1 or Th2 cells. Whether TCR signaling leads to expression of cytokine genes characteristic of Th1 or Th2 cells depends primarily on signals coming from cytokines present locally at the time of TCR stimulation (Paul and Seder, 1994). Interleukin 12 and interferon γ promote differentiation into Th1 cells whereas IL-4 promotes differentiation into Th2 cells. The type of antigen-presenting cell and the duration and magnitude of TCR signaling may also affect Th1 versus Th2 differentiation (Rocken *et al.*, 1992). The molecular mechanisms that commit CD4$^+$ T cells to either the Th1 or Th2 pattern of gene expression are not known.

In the absence of CD28 engagement, TCR ligation results in long-term unresponsiveness. The cell cannot be activated even if it subsequently receives both signals. This anergic state is characterized by an inability to produce IL-2 (Schwartz, 1992), which is required for the T cells to proliferate before differentiating into effector cells. T cell receptor signaling may induce the initial steps of both the activation and anergy pathways, whereas CD28 signals override the anergy signals and commit the cell to the activation/proliferation pathway (Jenkins and Johnson, 1993). T cell receptor-induced anergy may be important for establishing tolerance to self-antigens that are not present in the thymus and that therefore cannot induce negative selection of immature T cells. Self-antigens expressed on cells that do not express costimulators such as B7-1 and B7-2 (e.g., kidney cells) would engage only the TCR of the naive T cell and induce anergy. The ability to silence mature self-reactive T cells that have been exported from the thymus is referred to as peripheral tolerance. Thus, TCR signaling in naive T cells can lead to either activation or anergy, depending on the context in which it is received.

2. Differentiation of CD8$^+$ T Cells into Functional Cytolytic T Cells

Cytolytic T lymphocytes play an essential role in tumor surveillance and in eliminating cells infected with viruses. When the TCR on a naive CD8$^+$ T cell binds to MHC class I–peptide complexes, signals are initiated that activate the T cell and promote its differentiation into a mature CTL effector capable of killing target cells. The TCR-mediated activation requires that the CD8 coreceptor bind the same MHC class I molecule as the TCR (Janeway, 1992). As for CD4$^+$ cells, coclustering of the CD8 coreceptor with the TCR potentiates TCR signaling, presumably due to the Lck PTK that is tightly associated with CD8. Signals delivered by CD28 are also required for activation of CD8$^+$ T cells (Azuma *et al.*, 1992). Together, these signals induce expression of the high-affinity IL-2 receptor, allowing the cells to proliferate in response to IL-2 made by helper T cells. After several days, proliferation ceases and the cells become mature CTLs. During this time, CTL-specific genes involved in cell killing are induced. The products of these CTL-specific genes include pore-forming proteins called perforins and a variety of serine proteases called granzymes (Doherty, 1993).

When a mature CTL effector binds the appropriate MHC class I–peptide complex on a target cell, it rapidly kills that cell (Doherty, 1993). Granules containing perforin and the granzymes are released preferentially at the point of contact between the CTL and the target. The TCRs that are bound to the target cell cause localized increases in second messengers, such as free Ca^{2+}. These signals cause cytoskeletal elements to orient themselves toward the target and this facilitates the movement of granules toward the target cell. Once released, perforin creates channels in the target cell membrane that allow the granzymes to enter the cell. The granzymes destroy cell proteins and may contribute to the activation of endonucleases that trigger the DNA fragmentation associated with apoptosis. Cytolytic T lymphocytes can also kill target cells in a perforin-independent manner. This is carried out by the Fas ligand, a cell surface protein that is induced on CTLs by TCR cross-linking. When the Fas ligand binds to its receptor (the Fas antigen) on the target cell, it induces apoptotic cell death (Suda *et al.*, 1993).

E. Consequences of Antigen Receptor Signaling in Mature B Cells

1. Activation of Mature Quiescent B Cells

Immature B cells that escape negative selection develop into quiescent mature B cells that circulate in the blood. As these cells pass through the secondary lymphoid organs such as the spleen, they scan for the presence of trapped antigens that their mIgM and mIgD recognize. The binding of antigen to the BCR initiates signals that activate the B cell and prepare it to proliferate and secrete antibodies. The process of going from a resting B cell to an antibody-secreting cell can be divided into three stages: early activation, proliferation, and differentiation (reviewed in Gold and DeFranco, 1994). Early activation is characterized by (1) entry into the G_1 phase of the cell cycle; (2) increased synthesis of biosynthetic/secretory machinery (such as ribosomes and endoplasmic reticulum) that is needed for high-rate antibody secretion; (3) increased expression of surface proteins involved in the activation of T cells such as MHC class II molecules and B7-1, the ligand for CD28; and (4) increased expression of receptors for T cell-derived cytokines that promote B cell proliferation and antibody secretion. The proliferative phase that follows early activation greatly increases the number of cells capable of producing antibodies against the stimulating antigen. This clonal expansion is followed by differentiation into plasma cells that secrete IgM. For some types of antigens (T dependent; see below), additional differentiative events occur, including affinity maturation, the induction of memory B cells, and isotype switching. As for T cells, antigen receptor signals alone are not sufficient to induce this entire program of events. B cell antigen receptor signaling induces the early activation events and makes B cells receptive to T cell-derived signals required for proliferation and antibody production.

The early activation events can be induced by two different mechanisms: one that requires contact with activated helper T cells and one that can be stimulated by antigen in the absence of helper T cells. Whether or not T cells are required for early activation depends on the structure of the antigen, in particular whether it can bind to and cross-link multiple BCR molecules. T-independent and T-dependent antigens represent a continuous spectrum from multivalent antigens that cause extensive BCR clustering (T independent) to monovalent antigens that cause little or no BCR clustering (T dependent). Because BCR signaling is induced by cross-linking or clustering of the BCR, the T-independent antigens are those that stimulate high levels of BCR signaling whereas the T-dependent antigens cause little or no BCR signaling.

T-independent antigens are highly repetitious antigens that cause extensive and persistent cross-linking of the BCR. Examples of T-independent antigens include repeating polysaccharides found on bacterial or viral surfaces, protein polymers such as bacterial flagellin, and anti-Ig antibodies immobilized on high molecular mass dextran polymers. These highly multivalent antigens can induce the early activation events described above, as well as B cell proliferation, in the absence of T cells. Antigens that cross-link the BCR less extensively than these highly repetitive antigens (e.g., soluble anti-Ig antibodies or low molecular weight haptens coupled to erythrocytes) induce only the early activation events by themselves. B cells activated by soluble anti-Ig antibodies require T cell-derived cytokines such as IL-4 in order to proliferate to a significant degree. The ability of highly multivalent antigens to induce proliferation in the absence of T cells or T cell-derived cytokines suggests that persistent, high levels of BCR signaling can reduce or eliminate the cytokine requirement for B cell proliferation. Although multivalent antigens can induce early activation, and in some cases proliferation, differentiation into antibody-secreting cells *in vitro* requires cytokines produced by helper T cells such as IL-2, IL-4, and IL-5 (Pecanha *et al.*, 1991). *In vivo*, T-independent antigens such as bacterial flagellin can stimulate IgM production in the absence of T cells, but this response is likely to require cytokines that are produced by dendritic cells or macrophages. Cytokines can stimulate activated B cells, but not resting B cells, to proliferate and secrete antibodies. B cell antigen receptor cross-linking by T-independent antigens makes resting B cells more responsive to IL-2 and IL-4 by upregulating the expression of receptors for these cytokines (Paul and Ohara, 1987; Prakash *et al.*, 1985). Thus, BCR signaling stimulates early activation events that are required for subsequent cytokine-driven differentiation of B cells.

In contrast to multivalent antigens, soluble protein antigens are usually monovalent, having only one copy of a particular antigenic determinant per molecule. They do not cause extensive BCR cross-linking and may induce little or no BCR signaling. Activation of resting B cells by these antigens is T cell dependent and requires contact with an activated T cell. Supernatants of activated T cells or mixtures of recombinant cytokines cannot substitute for activated T cells (Noelle and Snow, 1991). Activated T cells express a cell surface protein called the CD40 ligand (CD40L), which binds to the CD40 protein on the surface of B cells (Noelle *et al.*, 1992a). Engagement of CD40 by its ligand delivers a signal that can induce all of the responses characteristic of early B cell activation, even in B cells that have not received any signal through the BCR (Banchereau *et al.*, 1994). The CD40–CD40L interaction is essential for T cell-dependent B cell activation. Blocking this interaction with antibodies against the CD40L or with soluble CD40 prevents the activation of resting B cells by plasma membrane fractions derived from activated T cells (Noelle *et al.*, 1992b). Moreover, antibody responses to T-dependent antigens are greatly decreased in hyper-IgM syndrome patients who have loss-of-function mutations in the CD40L (Notarangelo *et al.*, 1992). B cells activated through CD40 can be driven to proliferate and secrete antibodies by the combination of IL-4 and IL-5 (Noelle *et al.*, 1991,1992b).

T cell-dependent B cell activation involves the formation of a T : B cell pair in which the T and B cell activate each other (reviewed in Clark and Ledbetter, 1994). B cells express MHC class II proteins and can present antigen to T cells. The formation of the T : B cell pair is initiated by the binding of the T cell TCR to MHC class II–peptide complexes on the surface of the B cell. This interaction is then strengthened by the binding of CD4 to the MHC class II protein on the B cell and by the interaction of adhesion molecules such as LFA-1 and ICAM-1 on the surfaces of the T and B cells. If the B cell has been activated by BCR cross-linking, it will also express B7-1 and B7-2 (see below), ligands that stimulate CD28 signaling and thereby deliver a costimulatory signal for T cell activation. If no BCR cross-linking has occurred, the costimulatory signal must be delivered by cells that constitutively express B7-1 and B7-2, such as a macrophage or a dendritic cell. In either case, by acting as an antigen-presenting cell, the B cell contributes to activation of the T cell, resulting in expression of the CD40L and secretion of IL-4 and IL-5. The CD40L activates the B cell and subsequent signals from the cytokines cause the B cell to proliferate and differentiate into an antibody-secreting plasma cell.

The critical event in T cell-dependent B cell activation is the interaction of an activated T cell bearing the CD40L with a resting B cell. *In vitro,* T cells expressing the CD40L can activate all B cells, regardless of their antigen specificity and in the absence of BCR signaling. Antibody responses *in vivo,* however, are highly specific. What ensures that only B cells specific for the relevant antigen are activated and go on to produce antibodies *in vivo?* Current thinking is that the B cell that forms a pair with a T cell is most likely to be one whose BCR is specific for the same antigen that is being presented to the T cell. In their capacity as antigen-presenting cells, B cells can take up any antigen by nonspecific fluid phase endocytosis. However, B cells can also use their BCR to take up the antigen for which they are specific and this process is 10^3–10^4 times more efficient than nonspecific uptake by fluid phase endocytosis (Lanzavecchia, 1990). B cell antigen receptor-mediated antigen uptake reflects continual internalization of the BCR that is independent of BCR signaling and does not require BCR cross-linking. Antigens taken up by the BCR are thought to be delivered to a specialized intracellular compartment, where they are processed into peptides that are then bound by MHC class II proteins (Schmid and Jackson, 1994). Because BCR-mediated antigen uptake is so much more efficient than nonspecific uptake, B cells with BCRs specific for a given antigen will be much more likely to have MHC class II proteins bearing peptides derived from that antigen on their surface than are B cells with other antigen specificities. These antigen-specific B cells will therefore be more likely to form T · B cell pairs with T cells specific for that antigen and will be activated by these T cells much more frequently than B cells with other antigen specificities. Thus, BCR-mediated antigen uptake is likely to maintain the antigen specificity of the immune response, even when B cells are activated in a BCR-independent manner via CD40 and other surface receptors.

T cell-dependent and BCR-dependent early activations of B cells are not mutually exclusive and may occur simultaneously for protein antigens that are oligomeric or that form small aggregates. Even monovalent antigens may be able to cross-link BCRs under certain conditions. Dendritic cells, which trap antigens that pass through the lymphoid organs, may display multiple copies of a monovalent antigen on its surface, effectively creating a multivalent array that can cross-link BCRs on an antigen-specific B cell. The BCR signaling makes T cell-dependent B cell activation much more efficient in several ways. First, BCR cross-linking, even at low levels, greatly improves the ability of B cells to present antigens to T cells (Casten *et al.,* 1985). B cell antigen receptor cross-linking upregulates the expression of MHC class II proteins (Mond *et al.,* 1981) and increases the affinity and/or expression of adhesion molecules such as LFA-1, ICAM-1, and CD44 (Dang and Rock, 1991; Murphy *et al.*, 1990) that maintain the contact between the B cell and the T cell. Second, BCR cross-linking induces expression of the CD28 ligands B7-1 and B7-2 on B cells (Freeman *et al.*, 1989;1993). T cell activation requires a costimulatory signal through CD28 in addition to signaling by the TCR–CD4 complex. Finally, CD40 signaling and BCR signaling act synergistically to induce vigorous and prolonged B cell proliferation (Lane *et al.*, 1993). Thus, *in vivo*, efficient B cell activation is likely to involve both BCR signaling and T cell-dependent signaling through CD40.

2. Positive Selection of Germinal Center B Cells

B cells that are activated in a T cell-dependent manner by the CD40L and T cell-derived cytokines migrate to the lymphoid follicles in the spleen or lymph nodes and proliferate extensively. This clonal expansion of antigen-specific B cells results in the formation of a germinal center, a structure consisting of activated B cells, activated T helper cells, and follicular dendritic cells that trap antigens. In the germinal center, a number of differentiative events that are unique to T-dependent antibody responses occur: hypermutation of the Ig variable region genes, Ig isotype switching, and the formation of long-lived memory B cells.

The proliferation of germinal center B cells is accompanied by somatic hypermutation of the Ig heavy and light chain variable region genes (Jacob *et al.*, 1991; Berek *et al.*, 1991). This is an important genetic mechanism for generating high-affinity antibodies that effectively target antigens for elimination. Random mutations in the Ig variable region genes alter the three-dimensional shape of the antigen-binding site of the BCR. This results in some B cells that bind the original antigen with much greater affinity than the parent B cell as well as cells whose BCR no longer binds the original antigen at all (Weiss *et al.*, 1992). Because only those cells that still bind the antigen will secrete useful antibodies after they differentiate into plasma cells, the germinal center B cells must be retested for their ability to bind the antigen. *In vivo*, germinal center B cells are positively selected for their ability to bind to antigen immobilized on the surface of the follicular dendritic cells (MacLennan, 1994). Cells that cannot bind antigen undergo apoptosis. Thus, signaling through the BCR selects cells that still bind antigen with high affinity and rescues them from programmed cell death. This can be demonstrated *in vitro*. Germinal center B cells die rapidly when cultured, but the apoptosis can be delayed for up to 24 hr by BCR signaling induced by anti-IgM antibodies (Y.-J Liu *et al.*,1989). Longer survival requires additional signals through CD40. *In vivo*, if the amount of antigen is limiting, the germinal center B cells with the highest affinity antigen-binding sites will preferentially bind the antigen and be positively selected. Only these cells will go on to make antibodies. This results in an increase in the average affinity of antibodies produced over time, a process referred to as affinity maturation of the antibody response.

Two other differentiative events that are important for a property functioning immune system occur only in T-dependent antibody responses: class switching and memory cell formation. Germinal center B cells undergo extensive class switching to Ig isotypes other than IgM. Whether cells switch to IgG, IgA, or IgE is determined by cytokines such as IL-4, interferon γ, and transforming growth factor β (TGF-β). Although some of the germinal center B cells differentiate into plasma cells, others differentiate into long-lived memory B cells. Both class switching and memory B cell formation require activated helper T cells expressing CD40L. *In vitro,* signals through CD40 are required for long-term survival of germinal center B cells. In the absence of agents that induce CD40 signaling, such as soluble CD40L or anti-CD40 antibodies, germinal center B cells undergo apoptosis (Y.-J. Liu *et al.,* 1989). The essential role of CD40L in class switching is illustrated by hyper-IgM syndrome patients who lack functional CD40L. These individuals have normal or elevated levels of IgM, but make very little IgG or IgA (Allen *et al.,* 1993). Because expression of the CD40L on activated T cells is transient, germinal center B cells must reinduce the expression of the CD40L on T cells in germinal center. As described in the previous section, this involves the antigen-presenting cell function of the B cell, which is facilitated by efficient antigen uptake by the BCR and is potentiated by BCR signaling.

3. Peripheral B Cell Tolerance

As discussed in Section III,B, immature B cells that encounter self-antigen in the bone marrow are deleted. Because many self-antigens are not present in the bone marrow, there must be a mechanism for removing self-reactive mature B cells that escape deletion in the bone marrow. Nemazee and colleagues showed that peripheral deletion of mature self-reactive B cells does occur outside of the bone marrow (D. M. Russell *et al.,* 1991). In transgenic mice in which all the B cells are specific for the H-2K^b MHC class I allele, tissue-specific expression of H-2K^b in the liver has no effect on B cell formation in the bone marrow but results in clonal deletion of peripheral B cells. These mice have almost no B cells specific for H-2K^b in their spleens or lymph nodes. Thus, BCR signaling in mature B cells can promote either deletion or activation leading to antibody secretion.

Whether BCR signaling leads to activation or cell death depends on whether or not the B cell receives a second, costimulatory signal, analogous to the CD28-mediated costimulatory signal in T cells. In the presence of activated T cells, B cells receive second signals that stabilize the activated state and allow the B cells to proliferate and secrete antibodies. In the absence of T cells specific for the same antigen, the B cells does not receive these second signals and BCR signaling leads to abortive activation and cell death. Most mature self-reactive B cells will be deleted because they encounter their specific antigen in the absence of activated T cells specific for the same antigen. The T cells specific for those self-antigens will have been rendered anergic by binding to antigens expressed on cells that do not express costimulatory ligand such as B7-1. *In vitro* experiments by Klaus and colleagues support this two-signal model for B cell activation (Parry *et al.*, 1994). They showed that extensive cross-linking of the BCR on mature B cells results in cell death. This BCR-induced cell death could by overcome by using the combination of IL-4 and anti-CD40 antibodies (which induce CD40 signaling) to mimic a second signal delivered by a helper T cell.

IV. Activation of Tyrosine Kinases and Tyrosine Kinase-Regulated Signaling Pathways by Antigen Receptors

A. Protein Tyrosine Phosphorylation

Both the TCR and BCR belong to the family of receptors that activate PTKs. Unlike growth factor receptors [e.g., the receptors for platelet-derived growth factor (PDGF) or epidermal growth factor (EGF)], which have PTK activities in their cytoplasmic domains, the polypeptides that make up the antigen receptors do not possess intrinsic PTK activity. Instead, the antigen receptors associate with discrete, cytoplasmic PTKs. Activation of PTKs by the TCR and BCR results in a rapid increase in the tyrosine phosphorylation of a number of proteins (reviewed in Gold and DeFranco, 1994; Perlmutter *et al.*, 1993; Chan *et al.*, 1994a). Increased protein tyrosine phosphorylation can be detected within seconds of antigen receptor ligation and usually peaks after 2 to 5 min. The levels of tyrosine phosphorylation then decline gradually over a period of hours.

Tyrosine phosphorylation is a relatively rare event in cells, but one that is strongly tied to receptor signaling and control of cell growth. As we discuss in Section IV,D, many of the PTK substrates that have been identified in lymphocytes and in other cells are involved in signal transduction pathways (Cantley *et al.*, 1991). Tyrosine phosphorylation of these proteins can regulate their function in two ways. The activity of several key signaling enzymes, for example, phospholipase C-γ (see Section IV,D,1), is increased by phosphorylation of critical tyrosine residues. In addition, tyrosine phosphorylation initiates protein–protein interactions that are important for signal transduction. Many signaling proteins contain Src homology 2 (SH2) domains, structural domains of approximately 100 amino acids that bind phosphotyrosine-containing regions of proteins with high affinity (Pawson and Schlessinger, 1993). The ligands for SH2 domains consist of a phosphotyrosine residue and one or two specific amino acids in positions +1 to +3 (i.e., C terminal) relative to the phosphotyrosine. Although all SH2 domains recognize phosphotyrosine, different SH2 domains prefer different amino acids in the +1 to +3 positions, conferring specificity to SH2-mediated binding (Songyang *et al.*, 1994). Because regulatory proteins involved in signaling cascades are likely to be of low abundance, the high-affinity SH2-dependent interactions may be essential for colocalizing components of signaling pathways and facilitating signal transmission from one component to the next in the pathway.

B. Tyrosine Kinases Activated by B and T Cell Antigen Receptors

The B and T cell antigen receptors each activate members of the Src family of PTKs as well as a member of the Syk family of PTKs. The Src family PTKs are characterized by a number of shared structural features (reviewed in Chan *et al.*, 1994a) that are illustrated in Fig. 4. These include (1) myristylation of an N-terminal glycine residue that directs the localization of these PTKs to the inner face of the plasma membrane (and perhaps other cell membranes), (2) a unique N-terminal region of approximately 80 amino acids that may direct specific interactions of the PTK with other proteins, (3) a Src homology 3 (SH3) domain of approximately 60 amino acids that can interact with specific proline-rich sequences in other proteins, (4) an SH2 domain of approximately 100 amino acids that can direct binding to other tyrosine-phosphorylated proteins, (5) a C-terminal catalytic domain, and (6) a conserved tyrosine residue close to the C-terminus that acts as a negative regulatory site when phosphorylated. The members of the Syk family, p72^{syk} and ZAP-70, are structurally distinct from the Src family PTKs (Chan *et al.*, 1994a). Syk and ZAP-70 have two SH2 domains but no SH3 domain (Fig. 4). In addition, they are not myristylated and therefore do not associate directly with the plasma membrane. They also lack the C-terminal negative regulatory site characteristic of Src PTKs.

The BCR activates Syk, as well as multiple members of the Src family,

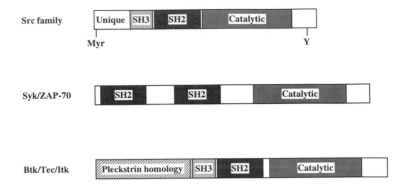

FIG. 4 Schematic representation of different PTK families. Indicated are the SH2 and SH3 domains, the catalytic domains, and the pleckstrin homology domain. Myr, Site of N-terminal myristylation of the Src family PTKs; Y, the negative regulatory site found in the Src family PTKs.

including $p55^{blk}$ (Blk), $p53/p56^{lyn}$ (Lyn), $p59^{fyn}$ (Fyn), and $p56^{lck}$ (Lck) (Burkhardt *et al.*, 1991; Hutchcroft *et al.*, 1992; Li *et al.*, 1992; Yamada *et al.*, 1993). Of the Src family PTKs, Blk and Lyn are expressed throughout B cell development whereas the expression of Fyn and Lck may be developmentally regulated, at least in mice. Immature murine B cell lines such as CH31 and WEHI-231 express Lck, but not Fyn, whereas mature B cell lines express Fyn but not Lck (Law *et al.*, 1992). This putative developmental exchange of Fyn for Lck remains to be confirmed using normal B cells at different stages of development. The significance of this switch depends on whether or not the different Src kinases activated by the BCR have distinct functions.

The BCR also activates Btk (de Weers *et al.*, 1994), a PTK that belongs to a third distinct class of PTKs that includes the Itk and Tec PTKs (Fig. 4). These PTKs have one SH2 domain and one SH3 domain, like the Src PTKs, but they are not myristylated and lack the C-terminal negative regulatory site (Desiderio, 1994). Btk, Itk, and Tec are also unique among PTKs in that they contain a pleckstrin homology (PH) region, a protein sequence of unknown function that is also found in signal transduction proteins such as phospholipase C-γ, the Ras GTPase-activating protein (GAP), and $p95^{vav}$ (see Sections IV,D,1 and IV,D,2).

There is considerable evidence that activation of PTKs is required for many of the cellular responses induced by the BCR. Loss-of-function mutations in Btk block B cell development at the pre-B cell stage and yield the same phenotype as mutations that prevent cell surface expression of the pre-B cell receptor (Kitamura *et al.*, 1992; Tsukada *et al.*, 1993; Vetrie *et al.*, 1993). In mature B cells, PTK inhibitors block the ability of high concentrations of anti-Ig antibodies (acting as a surrogate for T-independent antigens) to induce proliferation (Padeh *et al.*, 1991). Finally, ablating the expression of specific PTKs with antisense oligonucleotides prevents anti-Ig-induced apoptosis in two different experimental systems. Decreasing the expression of Blk in the CH31 B cell line prevents anti-Ig-induced apoptosis in this model for BCR-induced tolerance of immature B cells (Yao and Scott, 1993). Similarly, decreasing the expression of the Lyn PTK blocks anti-Ig-induced cell cycle arrest in the mature B cell lines Daudi and BCL_1 (Scheuerman *et al.*, 1994). Daudi and BCL_1 may be models for peripheral deletion of mature B cells that receive a signal through the BCR in the absence of a second signal from T cells.

Three PTKs have been implicated in TCR signaling: the Src family PTKs Fyn and Lck and ZAP-70, the T cell homolog of Syk (Chan *et al.*, 1994a). Protein tyrosine kinase inhibitors block TCR-induced secretion of IL-2 as well as expression of the high-affinity IL-2 receptor (June *et al.*, 1990a). Induction of this IL-2/IL-2 receptor autocrine loop is essential for TCR-induced clonal expansion and differentiation of naive T cells into effector CTLs and helper T cells. ZAP-70 appears to have a critical role in this process. In human patients with loss-of-function mutations in ZAP-70, $CD8^+$ cells fail to develop and TCR signaling is severely impaired in the $CD4^+$ cells (Arpaia *et al.*, 1994; Chan *et al.*, 1994b). In $CD4^+$ cells from the patients lacking ZAP-70, TCR-induced tyrosine phosphorylation is markedly reduced, the cells do not produce IL-2, and TCR cross-linking does not induce proliferation. Genetic evidence suggests that Fyn is involved in TCR signaling, at least in thymocytes. Disruption of the *fyn* gene in transgenic mice interferes with TCR signaling in thymocytes and, in some cases, prevents negative selection (Stein *et al.*, 1992; Appleby *et al.*, 1992). The absence of Fyn expression has less pronounced effects on TCR signaling in mature T cells. In contrast, overexpression of Fyn in transgenic mice or in mature T cell lines enhances responses to TCR ligation (Cooke *et al.*, 1991; Davidson *et al.*, 1992). Fyn associates with the ζ chain of the TCR (Samelson *et al.*, 1990) and there is some biochemical evidence that TCR ligation increases Fyn activity (Tsygankov *et al.*, 1992).

Lck plays a critical role in T cell development. Disruption of the *lck* gene in transgenic mice blocks T cell development at a very early stage (Molina *et al.*, 1992). Similarly, expressing a dominant negative version of Lck (one that cannot be activated and that also prevents endogenous Lck from being activated by virtue of its ability to compete for upstream components required for activation) in transgenic mice prevents a re-arranged TCR β chain transgene from turning off further TCR β gene rearrangement (Perlmutter, 1994). In contrast, expression of Lck with activating mutations in transgenic mice turns off TCR β gene re-arrangement in thymocytes and allows these cells to differentiate to the

CD4$^+$CD8$^+$ stage, even in the absence of the TCR β chain (Perlmutter, 1994). This suggests that activated Lck can bypass the requirement for signaling by the pre-T cell receptor. Lck is also essential for TCR signaling in mature T cells. In two different T cell lines lacking functional Lck, TCR ligation fails to induce tyrosine phosphorylation, IL-2 secretion, and other responses, even though the cells have normal levels of Fyn (Straus and Weiss, 1992; Karnitz et al., 1992). Expression of wild-type Lck in these cell lines restores all TCR-mediated responses. In mature T cells, Lck is associated primarily with the CD4 and CD8 coreceptors (see Section IV,C), which greatly potentiate TCR signaling when coclustered with the TCR. Lck may also be able to associate with the TCR. Early thymocytes expressing the pre-T cell receptor do not express CD4 or CD8. Similarly, the Jurkat cell line in which loss of Lck blocks all TCR-induced responses expresses little or no CD4 or CD8 (Straus and Weiss, 1992).

C. Coreceptors

Coreceptors are two receptors that simultaneously bind to the same ligand and in doing so elicit a synergistic response that is much greater than that caused by engaging either receptor individually (Fig. 3). CD4 or CD8 can act as coreceptors for the TCR by binding to the same MHC molecule as the TCR. Although high concentrations of immobilized anti-TCR antibodies can induce maximal responses, coclustering of CD4 or CD8 can reduce the ligand density required for T cell activation by 30- to 300-fold (Janeway, 1992). This potentiation by the coreceptor is especially important *in vivo*, where an antigen-presenting cell may have as few as 10–100 of the relevant MHC–peptide complexes on its surface (Harding and Unanue, 1990). *In vitro*, the coreceptor effect has been demonstrated by showing that coligation of CD4 to the TCR with biotinylated anti-TCR and anti-CD4 antibodies, followed by avidin as a secondary cross-linking reagent, greatly enhances TCR signaling (Turka et al., 1992). In contrast, creating separate TCR and CD4 clusters with antibodies inhibits TCR signaling (Haughn et al., 1992). The mechanism by which CD4 and CD8 regulate TCR signaling involves their association with the Lck PTK. Lck is essential for TCR signaling (see Section IV,B) and the cytoplasmic tails of both CD4 and CD8 bind Lck with high affinity (Rudd et al., 1988; Veillette et al., 1988). Coclustering of CD4 or CD8 with the TCR therefore brings Lck close to the TCR complex and potentiates TCR signaling whereas sequestering CD4 or CD8 away from the TCR deprives the TCR of Lck and inhibits TCR signaling.

The CD19 complex of B cells may be an analogous coreceptor for the

BCR (reviewed in Fearon, 1993). This complex consists of four non-covalently associated transmembrane polypeptides: (1) CD19 (95 kDa), (2) CD21 (140 kDa), which is also known as complement receptor 2 (CR2), (3) Leu 13 (16 kDa), and (4) TAPA-1, a 20-kDa polypeptide with four membrane-spanning domains (Fig. 3). The effects of CD19 ligation on BCR signaling are similar to those of CD4 or CD8 ligation on TCR signaling. Coligating the CD19 complex to the BCR potentiates BCR signaling (Carter et al., 1991b). Moreover, the concentration of anti-Ig antibodies required to induce proliferation of tonsil B cells is reduced 100-fold when the CD19–CR2 complex is coligated to the BCR (Carter and Fearon, 1992). The cytoplasmic tail of CD19 is tightly associated with the Lyn PTK and 10–15% of the Lyn in the Daudi B cell line is associated with CD19 (van Noesel et al., 1993b). Like CD4 and CD8 in T cells, the CD19 coreceptor complex in B cells may potentiate BCR signaling by bringing an essential PTK into the BCR–coreceptor complex (see Section V,D). Consistent with this hypothesis, creating separate CD19 complex clusters and BCR clusters with anti-CD19 and anti-BCR antibodies inhibits BCR-induced B cell activation (Pezzutto et al., 1987), presumably by sequestering CD19 and the associated Lyn away from the BCR. Thus, the CD19 complex is likely to play an important role in activation of B cells via the BCR.

While CD19 appears to be the primary signaling component of the CD19 complex because of its association with Lyn, CD21 (CR2) is likely to be the primary ligand-binding subunit of the CD19 coreceptor complex. Four types of ligands for CD21 have been identified: (1) the iC3b, C3dg, and C3d fragments of the C3 component of complement, (2) Epstein–Barr virus, (3) CD23, (Aubry et al., 1992), and (4) interferon α (Delcayre et al., 1991). The ability of CD21 to bind complement components may be its most important function. Bacteria can activate complement by the alternative pathway and by binding low-affinity cross-reactive IgM antibodies. Bacterial antigens are therefore likely to be coated with complement components and would coligate the CD19–CD21 complex to the BCR. The ability of the CD19–CD21 complex to potentiate BCR signaling would reduce the concentration of antigen required for B cell activation. The binding of CD21 to CD23 may also potentiate B cell activation in vivo. CD23 is a cell surface protein expressed on follicular dendritic cells and its ability to engage the CD19–CR2 complex may allow small amounts of antigen immobilized on these cells to activate B cells (Burton et al., 1993). The essential role of the CD19–CD21 complex in B cell activation via the BCR in vivo is supported by observations that complement-deficient animals have impaired antibody responses (reviewed in van Noesel et al., 1993a). Moreover, soluble forms of CD21 strongly inhibit antibody responses *in vivo* (Hebell *et al.*, 1991), presumably by competing with cell-bound CD21 for ligands.

The CD22 protein on B cells may also function as a coreceptor for the BCR. There is some evidence that CD22 associates with the BCR and cross-linking CD22 with antibodies enhances anti-Ig-induced signaling and proliferation (Pezzutto *et al.*, 1988). CD22 may be required for some BCR-induced signaling responses. CD22$^+$ B cells exhibit anti-Ig-induced increases in cytosolic Ca^{2+} concentrations whereas CD22$^-$ B cells do not. This suggests that CD22 may be involved in activation of phospholipase C (see Section IV,D,1). Protein tyrosine kinase activity has been detected in anti-CD22 immunoprecipitates (Leprince *et al.*, 1993), but it is not clear whether the PTK binds directly to CD22 or indirectly via the BCR. CD22 is a lectin that binds to α2,6-linked sialic acid residues on glycoproteins (Sgroi *et al.*, 1993). Because α2,6-linked sialic acids are common on cell surface proteins, it is likely that CD22 acts as an adhesion molecule that strengthens the interactions between B cells and T cells or dendritic cells.

D. Signaling Pathways Regulated by B and T Cell Antigen Receptors

In subsequent sections we describe our current understanding of the mechanisms by which the antigen receptor–coreceptor complexes activate PTKs, the potential roles of the different PTKs, and the regulation of these PTKs by other enzymes. Before doing so, we describe the PTK-regulated signal transduction pathways that are activated by the T and B cell antigen receptors. These pathways presumably transmit the signal from the antigen receptor to the nucleus and induce nuclear events that lead to cellular responses. The key regulatory proteins in these pathway are (1) phospholipase C, (2) the p21^{ras} oncoprotein (Ras), and (3) phosphatidylinositol 3-kinase (PtdIns 3-kinase).

1. Phospholipase C Signaling Pathway

Both the TCR and BCR activate phospholipase C (PLC) (Gold and DeFranco, 1994; Weiss and Littman, 1994), an enzyme that cleaves the plasma membrane lipid phosphatidylinositol 4,5-biphosphate [PtdIns(4,5)P$_2$] into two second messengers, inositol 1,4,5-trisphosphate (InsP$_3$) and diacylglycerol (Berridge, 1993) (see Fig. 5). Inositol 1,4,5-trisphosphate increases cytosolic free Ca^{2+} concentrations by stimulating the release of Ca^{2+} from intracellular storage vesicles that possess an InsP$_3$-regulated Ca^{2+} channel (Berridge, 1993). Inositol 1,4,5-trisphosphate is rapidly converted into a large number of other inositol phosphate species

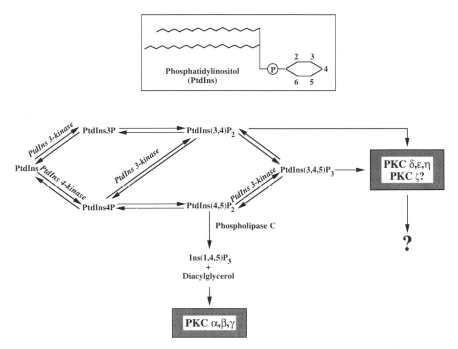

FIG. 5   Inositol phospholipid-dependent signaling pathways. The structure of PtdIns and the numbering of the inositol head group are depicted in the box. Phospholipase C preferentially cleaves PtdIns(4,5)P$_2$, yielding the second messengers Ins(1,4,5)P$_3$ and diacylglycerol. Phosphatidylinositol 3-kinase produces PtdIns3P, PtdIns(3,4)P$_2$, and PtdIns(3,4,5)P$_3$, which are not substrates for phospholipase C. There is some evidence that PtdIns(3,4)P$_2$ and PtdIns(3,4,5)P$_3$ regulate the activity of several different PKC isoforms (δ, ε, η, and perhaps ζ) that are distinct from the α, β, and γ isoforms of PKC, which are activated by diacylglycerol.

by the action of kinases and phosphatases. It is not known if these other inositol phosphate isomers also have signaling functions. In both T and B cells, increased levels of inositol phosphates and elevated levels of intracellular Ca^{2+} can be detected within seconds of antigen receptor ligation (Bijsterbosch *et al.*, 1985; Imboden and Stobo, 1985). Diacylglycerol, in contrast, activates the α, β, and γ subtypes of the serine/threonine kinase protein kinase C (PKC) (Nishizuka, 1988). Protein kinase C activation is usually accompanied by translocation of these enzymes from the cytosol to the membrane fraction of cells. B cell antigen receptor cross-linking has been shown to increase diacylglycerol levels in B cells (Bijsterbosch *et al.*, 1985) and cause a transient increase in the fraction of PKC associated with the membrane fraction (Chen *et al.*, 1986; Nel *et al.*, 1986). Ligation of the TCR by soluble anti-TCR antibody also induces a transient relocalization of PKC to the plasma membrane (Manger *et al.*,

1987). In contrast, immobilized anti-TCR antibodies, which may better mimic an array of peptide–MHC complexes displayed on the surface of an antigen-presenting cell, induce sustained translocation of PKC.

Activation of PLC by the T and B cell antigen receptors is dependent on tyrosine phosphorylation. T cell receptor cross-linking induces tyrosine phosphorylation of the PLC-γ1 isoform of PLC (Weiss et al., 1991) whereas BCR cross-linking induces tyrosine phosphorylation of both PLC-γ1 and PLC-γ2 (Carter et al., 1991a; Hempel et al., 1992). Moreover, PTK inhibitors block antigen receptor-induced inositol phosphate production and calcium increases in both T and B cells (Carter et al., 1991a; June et al., 1990a). Tyrosine phosphorylation activates PLC-γ1 by rendering it resistant to the inhibitory effects of profilin, an abundant cytoskeletal protein that binds to PtdIns(4,5)P$_2$ (Goldschmidt-Clermont et al., 1991). In the presence of profilin, PLC-γ1 must be tyrosine phosphorylated in order to cleave PtdIns(4,5)P$_2$. It is likely that PLC-γ2 is regulated in a similar manner because it has extensive homology to PLC-γ1. Phospholipase C-catalyzed PtdIns(4,5)P$_2$ hydrolysis may also involve translocation of PLC from the cytosol to the plasma membrane by means of its SH2 domains binding to phosphotyrosine residues on activated EGF and PDGF receptors (Cantley et al., 1991). With the exception of one report of PLC-γ1 interacting with the TCR (Dasgupta et al., 1992), association of PLC with the antigen receptors or with other membrane proteins in either T or B cells has not been demonstrated.

The PtdIns(4,5)P$_2$-derived second messengers play an essential role in TCR and BCR signaling. Many of the responses to BCR and TCR cross-linking can be induced, at least partially, by using Ca^{2+} ionophores to increase intracellular free Ca^{2+} concentrations and phorbol esters to activate PKC. These include entry of resting T and B cells into the cell cycle, induction of immediate early genes such as c-myc, c-fos, and egr-1 (see Section VI) in both T and B cells, and induction of IL-2 and interferon γ gene transcription in T cells (reviewed in Gold and DeFranco, 1994; Weiss and Imboden, 1987). Moreover, induction of these responses by antigen receptor cross-linking can be blocked by chelation of extracellular Ca^{2+} (required for refilling intracellular Ca^{2+} pools), by inhibitors of PKC, or by PKC downregulation induced by long-term exposure to phorbol esters.

Increases in intracellular free Ca^{2+} activate Ca^{2+}/calmodulin-dependent protein kinases as well as the Ca^{2+}/calmodulin-dependent serine/threonine phosphatase calcineurin (phosphatase 2B). Studies using the immunosuppressive drugs cyclosporin A and FK506 have shown that calcineurin is an essential component of TCR signaling. These drugs bind to cytosolic receptor proteins called immunophilins (Schreiber, 1991) and such complexes inhibit the activity of calcineurin (J. Liu et al., 1991b). In T cells, cyclosporin A and FK506 block TCR-induced activation of resting T cells and prevent the transcription of a number of genes associated with T cell activation including those encoding IL-2, IL-3, IL-4, and interferon γ (Sigal and Dumont, 1992). T cell receptor-mediated apoptosis in thymocytes is also blocked by cyclosporin A and FK506 (Fruman et al., 1992). Although the effects of these drugs on B cells have not been studied as extensively, there are reports that cyclosporin inhibits anti-Ig-induced proliferation of resting B cells from mouse spleen (O'Garra et al., 1986) and anti-Ig-induced apoptosis in the human B cell line B104 (Bonnefoy-Berard et al., 1994). Thus, calcineurin may play an important role in both TCR and BCR signaling.

The role of calcineurin in T cell activation is understood in some detail. Calcineurin controls the subcellular localization of NF-AT (see Section VI,A,3), a factor required for transcriptional activation of the IL-2 gene and other genes associated with T cell activation (Schreiber and Crabtree, 1992). Increases in intracellular free Ca^{2+} induce NF-AT to translocate from the cytosol to the nucleus in a calcineurin-dependent manner. Evidence that calcineurin is the most important Ca^{2+}-regulated signaling component in T cells comes from experiments showing that expression of constitutively active calcineurin in T cell lines eliminates the requirement for Ca^{2+} increases in T cell activation (Clipstone and Crabtree, 1992; O'Keefe et al., 1992). While the functional role of calcineurin in B cell activation is not as clear, recent work has shown that the BCR activates NF-AT (Choi et al., 1994; Venkitaraman et al., 1994).

Like the Ca^{2+} arm of the PLC pathway, PKC also regulates enzymes that control the activity of transcription factors. In both T and B cells, PKC activation contributes to activation of the mitogen-activated protein (MAP) kinases (also known as extracellular signal-regulated kinases or ERKs) (Fig. 6). The MAP kinases are a family of related serine/threonine kinases that are activated by virtually all receptors that signal via PLC and/or PTKs (Pelech and Sanghera, 1992). In both T and B cells, antigen receptor ligation or activation of PKC with phorbol esters leads to a rapid increase in MAP kinase activity that is maximal after 5 min and returns to near-basal levels by 30 min (Nel et al., 1990; Gold et al., 1992; Casillas et al., 1991). There are at least four members of the MAP kinase family, the best studied of which are p42^{mapk} (ERK2) and p44^{mapk} (ERK1) (Pelech and Sanghera, 1992). It is not known whether the various MAP kinases are regulated differently or have different substrates in vivo. B cell antigen receptor ligation selectively activates p42^{mapk} (ERK2) (Gold et al., 1992). Activated MAP kinases can translocate to the nucleus and, at least in vitro, they can phosphorylate a variety of transcription factors including

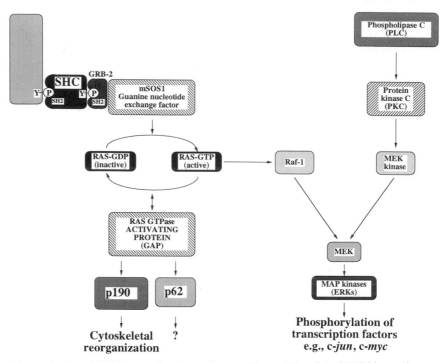

FIG. 6   The Ras–MAP kinase signaling pathways. The activity of the MAP kinases is controlled by the dual-specificity kinase MEK. MEK can be activated by two different pathways: the phospholipase C pathway and the Ras pathway. The Ras signaling pathway is regulated by controlling the amount of active GTP-bound Ras. This reflects a balance between guanine nucleotide exchange factors (GNEFs) that favor active GTP-bound Ras and GTPase-activating proteins (GAPs) that favor inactive GDP-bound Ras. Two major guanine nucleotide exchange factors for Ras have been found in lymphocytes: Vav and mSOS1. The role of Vav in regulating Ras is controversial. mSOS1 associates with an SH2-containing protein called GRB-2. In T cells, the GRB-2–mSOS1 complex binds to an unidentified 36-kDa membrane protein. This presumably brings mSOS1 to the plasma membrane where Ras is located. SHC may be an additional adapter protein that couples the GRB-2–mSOS1 complex to proteins that lack a GRB-2-binding site. In B cells, SHC is the major target for the SH2 domain of GRB-2. SHC binds to a cytosolic 145-kDa tyrosine-phosphorylated protein in B cells. It is not known how the p145–SHC–GRB-2–mSOS1 complex interacts with the plasma membrane. The role of SHC in T cells is unclear. In addition to MAP kinase, the Ras GTPase-activating protein (GAP) may be a downstream effector of Ras. GAP binds to two other tyrosine-phosphorylated proteins, p190 and p62, which may be involved in signal transduction.

c-*jun*, c-*myc*, c-*fos*, p62^{TCF} (Elk-1), and the p53 tumor suppressor protein (Davis, 1993; Milne *et al.*, 1994). The MAP kinases are therefore a direct link between cytoplasmic signaling events and the regulation of gene expression in the nucleus. MAP kinases also regulate protein synthesis by phosphorylating the p90^{rsk} protein kinase. p90^{rsk} phosphorylates the ribosomal S6 protein (Davis, 1993), which results in the increase in protein synthesis that accompanies cell activation.

Because the activation of MAP kinases is likely to be a critical event in signaling by many receptors, a great deal of attention has been focused on how MAP kinases are activated. Phosphorylation on threonine and tyrosine residues is required for maximal activation of ERK1 and ERK2. Both of these modifications are carried out by a dual-specificity (serine/threonine/tyrosine) kinase called MAP kinase kinase or MEK (for MAP or ERK kinase) (Rossomando *et al.*, 1992). There are multiple isoforms of MEK and again it is not clear if each is regulated differently or preferentially acts on different MAP kinase isoforms. What is clear is that there are multiple pathways that lead to the activation of MEK (Fig. 6). Protein kinase C phosphorylates and activates MEK kinase (S. L. Pelech, personal communication), a mammalian homolog of the yeast STE11 gene product, which in turn phosphorylates and activates MEK (Leevers and Marshall, 1992; Lange-Carter *et al.*, 1993). Studies using PKC inhibitors, however, have shown that activation of MAP kinases by the antigen receptors is only partly due to PKC (Nel *et al.*, 1990; Gold *et al.*, 1992). Thus, there must also be a PKC-independent pathway of MAP kinase activation. As described in the next section, activation of the p21^{ras} oncoprotein (Ras) can also lead to MAP kinase activation (Leevers and Marshall, 1992). Both the T and B cell antigen receptors have been shown to activate Ras in a PKC-independent manner (Harwood and Cambier, 1993; Downward *et al.*, 1992).

2. Ras Signaling Pathway

The P21^{ras} oncoprotein (Ras) acts as a molecular switch that is active when it has GTP bound to it and inactive when GDP is bound. Both the TCR (Downward *et al.*, 1990) and the BCR (Harwood and Cambier, 1993; Lazarus *et al.*, 1993; Torda *et al.*, 1994) cause a rapid (maximal after 2–3 min) increase in the fraction of cellular Ras that is in the active GTP-bound state. This response is likely to be important for T and B cell activation because Ras is a potent regulator of cell growth and differentiation in other systems. Mutant versions of Ras that accumulate in the active GTP-bound form can transform fibroblasts (Gibbs *et al.*, 1984), induce the differentiation of the neuronal PC12 cell line (Hagag *et al.*, 1986), and stimulate the maturation of *Xenopus* oocytes (Korn *et al.*, 1987). Ras activation is a common feature of PTK-mediated receptor signaling and a number of cytokines and growth factors activate Ras (Downward, 1992). In several cases, activation of Ras has been shown to be an essential component of signaling by these receptors. Blocking Ras activation by microinjecting neutralizing anti-Ras antibodies or by expressing mutant versions of Ras that cannot be activated abrogates growth factor-stimulated proliferation in fibroblasts (Mulcahy *et al.*, 1985; Feig and Cooper, 1988) as well as nerve growth factor-induced differentiation of PC12 cells (Hagag *et al.*, 1986). Genetic studies in *Drosophila* (Rubin, 1991) and in *Caernorhabditis elegans* (Sternberg and Horvitz, 1991) have also shown that Ras is essential for PTK receptors to exert their effects.

Ras controls a protein kinase cascade that leads to activation of the MAP kinases (Fig. 6), important signal-transducing enzymes that are activated by both the TCR and BCR. In fibroblasts, activated Ras associates with the Raf-1 serine/threonine kinase (Koide *et al.*, 1993) and this interaction promotes activation of Raf-1 (Cook and McCormick, 1993). Raf-1 phosphorylates and activates MEK (Kyriakis *et al.*, 1992), the kinase that in turn phosphorylates and activates the MAP kinases (Davis, 1993). Thus, MEK lies at the intersection of two signaling pathways because it is phosphorylated by Raf-1, which is a component of the Ras pathway, as well as by the STE11-like MEK kinase that is regulated by PKC, part of the PLC signaling pathway (Fig. 6). In lymphocytes, both TCR cross-linking and BCR cross-linking activate Raf-1 (Siegel *et al.*,1990; Torda *et al.*, 1994). B cell receptor cross-linking has also been shown to activate MEK (Torda *et al.*, 1994). The MAP kinases are likely to mediate many of the effects of Ras on cell growth and differentiation. Activated MAP kinases can translocate to the nucleus and *in vitro* they can phosphorylate a number of transcription factors such as c-*jun*, c-*myc*, c-*fos* (Davis, 1993), and the p53 tumor suppressor protein (Milne *et al.*, 1994).

Ras may also control the activity of other signaling proteins besides MAP kinases. The Ras GTPase-activating protein (rasGAP) participates in the regulation of Ras activity (see below) but may also transmit signals from Ras (Fig. 6). rasGAP contains a Src homology 2 (SH2) domain and a Src homology 3 (SH3) domain, both of which mediate interactions with other proteins. SH2 domains, as discussed in Section IV,A, bind phosphotyrosine residues in the context of specific neighboring amino acids. In contrast, SH3 domains are 60-amino-acid structural motifs that bind to proline-rich regions in other proteins (Pawson and Schlessinger, 1993). SH3 domains from different proteins have distinct sequences, suggesting that there is some specificity as to which proline-rich sequences a given SH3 domain will bind. Microinjection of a monoclonal antibody against the SH3 domain of GAP into *Xenopus* oocytes blocks the ability of activated Ras to stimulate germinal vesicle breakdown (Duchesne *et al.*, 1993). This is the strongest evidence that rasGAP is a downstream effector of Ras and implicates protein–protein interactions mediated by the SH3 domain of rasGAP.

If rasGAP is a downstream effector of Ras, then what does rasGAP regulate? Proteins that bind to the SH3 domain of rasGAP have not been identified. However, in a number of cell types, receptor signaling induces rasGAP to bind via its SH2 domains to two tyrosine-phosphorylated proteins, p62 and p190 (Moran et al., 1991). Although these interactions have not been observed in T cells (Downward et al., 1992), BCR signaling induces tyrosine phosphorylation of p62 (Kawauchi et al., 1994) and the association of GAP with tyrosine-phosphorylated p62 and p190 (Gold et al., 1993). Both p62 and p190 have been cloned from fibroblasts (Settleman et al., 1992b; Wong et al., 1992) and their functions are now being studied. p190 regulates the activity of Rac and Rho (Settleman et al., 1992a), two other members of the Ras family of guanine nucleotide-binding proteins. In fibroblasts, Rac is involved in membrane ruffling whereas Rho regulates the formation of actin-based cytoskeleton elements such as stress fibers (Ridley and Hall, 1992; Ridley et al., 1992). In addition to regulating the cytoskeleton, p190 may also be involved in gene regulation because it has homology to a transcriptional repressor of the glucocorticoid receptor gene and has been detected in the nuclei of rat fibroblasts (Settleman et al., 1992b). p62 has sequence homology to heteronuclear ribonucleoproteins (hnRNPs) proteins that are involved in mRNA splicing (Wong et al., 1992). Although p62 binds RNA in vitro, there is not evidence to support a role for p62 in regulating mRNA splicing, stability, or nuclear export. Work has shown that p62 contains multiple proline-rich regions that can bind in vitro to SH3 domains from signaling proteins such as the Src PTK, PtdIns 3-kinase, PLC-γ1, and GRB-2 (see below) (M. F. Moran, personal communication). Moreover, p62 associates with vimentin, a component of intermediate filaments. If these interactions also occur in vivo, it may represent (1) a mechanism by which receptors regulate intermediate filaments, (2) a mechanism for immobilizing signaling proteins on cytoskeletal elements in order to facilitate the formation of protein complexes, or (3) a mechanism for downregulating signaling pathways by sequestering proteins such PLC-γ1 away from the receptors and PTKs that regulate them.

Because Ras activation has strong effects on cell growth and differentiation, considerable attention has been focused on how Ras is activated. Two types of protein are involved in this process (Boguski and McCormick, 1993): proteins that stimulate the binding of GTP to Ras (guanine nucleotide exchange factors or GNEFs) and proteins that stimulate the intrinsic GTPase activity of Ras to convert bound GTP to GDP (GTPase-activating proteins or GAPs). Cellular Ras activity reflects a balance between GNEF activity and GAP activity. Receptor signaling can increase the amount of GTP-bound Ras either by stimulating GNEF activity and/or inhibiting GAP activity.

There are two distinct GAPs that favor the inactive GDP-bound form of Ras: p120 rasGAP and the NF-1 gene product, neurofibromin. Both are expressed in T and B cells. In permeabilized cells, both TCR and BCR ligation decrease total cellular GAP activity (Downward *et al.,* 1990; Lazarus *et al.,* 1993). Decreased GAP activity would increase the half-life of the GTP-bound form of Ras and lead to an increase in the amount of GTP-bound Ras in the cell. It is not clear how receptors regulate GAP activity. B cell receptor ligation, but not TCR ligation, causes tyrosine phosphorylation of rasGAP (Gold *et al.,* 1993; Downward *et al.,* 1992). There is no evidence, however, that this modification alters the ability of rasGAP to stimulate the GTPase activity of Ras. GAP activity may be controlled instead by regulating the association of rasGAP and neurofibromin with other proteins. The binding of rasGAP to p190 has been reported to decrease its ability to stimulate the GTPase activity of Ras *in vitro* (Moran *et al.,* 1991). rasGAP also binds tightly to the GTP-bound form of Rap 1A, another member of the Ras family (Frech *et al.,* 1990). Although rasGAP does not stimulate the GTPase activity of Rap 1A, binding to Rap 1A may prevent rasGAP from acting on Ras. While this model would explain how receptor signaling decreases rasGAP activity, it is not consistent with rasGAP being a downstream mediator of Ras action.

In contrast to GAPs, GNEFs activate Ras by promoting the release of GDP, allowing GTP to bind in its place (Boguski and McCormick, 1993). Several GNEFs have been described in mammalian cells including two homologs of the *Drosophila son of sevenless* gene product (mSOS1 and mSOS2 in mouse) (Bowtell *et al.,* 1992) and the $p95^{vav}$ (Vav) protooncogene (Bustelo and Barbacid, 1992). T cell receptor cross-linking and BCR cross-linking both increase total cellular GNEF activity as determined by the ability of cell lysates to induce the release of [^{3}H]GDP from recombinant Ras (Gulbins *et al.,* 1993,1994). Both the TCR and BCR also induce tyrosine phosphorylation of Vav, a modification that increases the *in vitro* GNEF activity of Vav toward Ras (Gulbins *et al.,* 1993,1994). Immunodepletion of Vav from B cell lysates removes approximately 80% of the anti-Ig-induced GNEF activity, arguing that Vav is the major GNEF regulated by the BCR (Gulbins *et al.,* 1994). Although these data appear convincing, other groups, have not been able to demonstrate that Vav has GNEF activity toward Ras (Buday *et al.,* 1994). Thus, further experiments are required to establish a role for Vav in activation of Ras by the BCR and TCR.

The 170-kDa mSOS1 GNEF is required for activation of Ras by EGF in fibroblasts and may also be involved in Ras activation in T and B cells (Fig. 6). mSOS1 exists in a complex with a 23-kDa adapter protein called

GRB-2. GRB-2 binds proline-rich regions in mSOS1 via its two SH3 domains (Rozakis-Adcock *et al.,* 1993) and also contains an SH2 domain that can direct the interaction of mSOS1 with other proteins (Pawson and Schlessinger, 1993). The *Drosophila* homolog of GRB-2, the *drk* gene product, binds to the *Drosophila* SOS GNEF and is required for activation of Ras by the *sevenless* PTK (Olivier *et al.,* 1993). In EGF-stimulated fibroblasts, GRB-2 binds via its SH2 domain to phosphotyrosine residues on activated EGF receptors (Buday and Downward, 1993; Rozakis-Adcock *et al.,* 1993). This results in translocation of mSOS1 from the cytosol to the plasma membrane where Ras is localized (Buday and Downward, 1993). Because there is no evidence that the intrinsic GNEF activity of mSOS1 is increased by EGF, mSOS1 activity toward Ras may be regulated solely by controlling its localization and access to Ras. Evidence supports the concept that the SH2 domain of GRB-2 directs interactions that are essential for activation of Ras by mSOS1. Epidermal growth factor-induced activation of Ras in permeabilized fibroblasts can be blocked by tyrosine-phosphorylated peptides that bind to the SH2 domain of GRB-2 (Buday and Downward, 1993). These peptides presumably prevent GRB-2 from binding to the EGF receptor and bringing mSOS1 to the plasma membrane. Similarly, GRB3-3, a naturally occurring variant form of GRB-2 that binds mSOS1 but lacks the SH2 domain (presumably due to alternate splicing of the GRB-2 gene primary transcript), is a dominant negative inhibitor of Ras activation in fibroblasts (Fath *et al.,* 1994). Controlling the relative levels of GRB-2 and GRB3-3 in the cell may be a mechanism for regulating the Ras pathway.

Although there is no direct evidence that mSOS1 has a role in Ras activation in T or B cells, both TCR and BCR signaling induces the formation of protein complexes containing mSOS1. The TCR cross-linking induces the GRB-2–mSOS1 complex to bind to an unidentified 36-kDa tyrosine-phosphorylated membrane-associated protein (Buday *et al.,* 1994). Although not directly demonstrated, it was assumed that the formation of these complexes resulted in translocation of mSOS1 to the plasma membrane where it could act on Ras. In B cells, BCR cross-linking induces the formation of complexes containing GRB-2, mSOS1, and SHC (46- and 52-kDa forms), a cytosolic protein that is tyrosine phosphorylated in response to BCR ligation (Saxton *et al.,* 1994). The SH2 domain of GRB-2 binds to tyrosine-phosphorylated SHC (Rozakis-Adcock *et al.,* 1992) and studies by Saxton *et al.* (1994) suggested that SHC is the primary target of the GRB-2 SH2 domain in B cells. This implies that SHC is essential for the function of the GRB-2–mSOS1 complex in B cells. SHC contains an SH2 domain (Pelicci *et al.,* 1992) and may act as an additional adapter protein that couples GRB-2–mSOS1 to regulatory proteins that do not contain a binding site for the SH2 domain of GRB-2. Consistent with this hypothesis, SHC couples GRB-2–mSOS1 to an unidentified 145-kDa tyrosine-phosphorylated (p145) in anti-Ig-stimulated B cells (Saxton *et al.,* 1994). p145 is a cytosolic protein, but p145–SHC–GRB-2–mSOS1 complexes are found in the membrane fraction of activated B cells. The membrane protein involved in attracting or anchoring this complex to the membrane in B cells is not known. Whereas SHC appears to be important for directing the interactions of the GRB-2–mSOS1 complex in B cells, the role of SHC in T cells is not clear. One group has shown that SHC associates with a 145-kDa tyrosine phosphoprotein in activated T cells but does not associate with GRB-2 (Buday *et al.,* 1994). However, another group has reported that complexes containing mSOS1 and SHC can be recovered from T cells stimulated with anti-TCR antibodies and that SHC associates with the ζ chain of the TCR in these activated T cells (Ravichandran *et al.,* 1993). The role of SHC in directing the interactions of the GRB-2–mSOS1 complex and in activating Ras in lymphocytes requires further analysis.

In both T and B cells, only a small fraction of the mSOS1 in the cell associates with SHC after antigen receptor cross-linking. This may be sufficient to perturb the balance between GNEFs and GAPs acting on Ras. Alternatively, SHC may play only a minor role in the regulation of Ras. Its major function may be to couple other signaling proteins to each other. For example, GRB-7, an SH2-containing protein of unknown function, binds to tyrosine-phosphorylated SHC in breast cancer cells (Stein *et al.,* 1994). Similarly, GRB-2 may couple receptors to other signaling proteins besides mSOS1. One report showed that the C-terminal SH3 domain of GRB-2 binds an unidentified 75-kDa protein that is tyrosine phosphorylated in response to TCR cross-linking (Reif *et al.,* 1994).

In summary, both the TCR and BCR activate Ras, a potent regulator of many cellular processes. In T cells, Ras has been implicated in activation of the NF-AT transcription factor (see Section VI,A,3). The role of Ras in B cell activation remains to be determined. Activation of Ras by the antigen receptors appears to involve both inhibition of GAP activity and stimulation of GNEF action on Ras. The mechanism by which the antigen receptors regulate GAP activity is not yet understood. Both antigen receptors activate the Vav GNEF and induce the formation of complexes containing the mSOS1 GNEF. However, the ability of Vav to act as a Ras GNEF *in vivo* needs to be confirmed and the relative contributions of Vav and mSOS1 to antigen receptor-induced Ras activation also need to be assessed.

3. Phosphatidylinositol 3-Kinase Signaling Pathway

A third signaling pathway associated with PTK activation involves phosphatidylinositol 3-kinase (PtdIns 3-kinase), an enzyme that produces inositol phospholipids that may act as second messengers (Auger and Cantley, 1991). Phosphatidylinositol 3-kinase consists of an 85-kDa regulatory subunit that contains two SH2 domains and a 110-kDa catalytic subunit. There are at least two isoforms of each subunit but there is no clear indication that they differ in function. Phosphatidylinositol 3-kinase phosphorylates the membrane lipids phosphatidylinositol (PtdIns), PtdIns(4)P, and PtdIns(4,5)P$_2$ on the 3-position of the inositol ring, yielding PtdIns(3)P, PtdIns(3,4)P$_2$, and PtdIns(3,4,5)P$_3$ (Fig. 5). Phosphatidylinositol 3,4-bisphosphate and PtdIns(3,4,5)P$_3$ are present at low levels in cells and are more likely to be second messengers or precursors of second messengers than the relatively more abundant PtdIns(3)P.

Both TCR and BCR ligation result in increased levels of PtdIns 3-kinase products. In the murine B cell lines BAL17 and WEHI-231, BCR ligation causes a five- to sixfold increase in the levels of PtdIns(3,4)P$_2$ after 1–2 min and PtdIns(3,4)P$_2$ levels remain elevated for at least 15 min (Gold and Aebersold, 1994). In contrast, PtdIns(3,4,5)P$_3$ levels increase by only 50% and this increase is evident only in the first 30 sec after BCR ligation. Thus, PtdIns(3,4)P$_2$ is likely to be the main PtdIns 3-kinase-derived second messenger in B cells. In the Jurkat T cell line, TCR cross-linking increases the levels of both PtdIns(3,4)P$_2$ and PtdIns(3,4,5)P$_3$ (Ward et al., 1992). The increases in the levels of PtdIns 3-kinase products in activated T and B cells are most likely due to activation of PtdIns 3-kinase, as opposed to inhibition of PtdIns 3-phosphatases. Both TCR (von Willebrand et al., 1994) and BCR (Pleiman et al., 1994) cross-linking increase the specific activity of PtdIns 3-kinase (see below). It should be noted that other groups have been unable to detect TCR-mediated activation of PtdIns 3-kinase (A. Weiss, personal communication). However, CD28 and the IL-2 receptor, two other surface receptors involved in T cell activation, have been shown to activate PtdIns 3-kinase (Ward et al., 1993; Remillard et al.,1991).

Several lines of evidence suggest an important role for PtdIns 3-kinase in receptor signaling. First, mutations in the PDGF receptor that destroy the YXXM motif to which the SH2 domains of PtdIns 3-kinase bind (see below) prevent PDGF-induced DNA synthesis in fibroblasts (Fantl et al., 1992). Similarly, mutated versions of the v-src and v-abl PTKs that have lost the ability to bind to PtdIns 3-kinase are unable to transform cells (Auger and Cantley, 1991). Second, wortmannin, an inhibitor of PtdIns 3-kinase activity, blocks neutrophil responses to chemotactic peptides

(Arcaro and Wymann, 1993; Okada *et al.*, 1994a) and insulin-stimulated glucose uptake in adipocytes (Okada *et al.*, 1994b). There are, however, alternative explanations for these results. SH2 domains of other critical signaling proteins may share the PtdIns 3-kinase-binding site on the PDGF receptor and wortmannin may also inhibit other enzymes. Overexpression of defective subunits of PtdIns 3-kinase that act as dominant suppressors of PtdIns 3-kinase activation should provide better insight into PtdIns 3-kinase function. Wennström *et al.* (1994) showed that PDGF-induced membrane ruffling in fibroblasts could be blocked by transient expression of a defective PtdIns 3-kinase p85 subunit that is unable associate with the catalytic subunit of the enzyme.

The signaling functions of PtdIns 3-kinase and its products are not well understood. The lipids produced by PtdIns 3-kinase are not substrates for PLC [which cleaves $PtdIns(4,5)P_2$] (Serunian *et al.*, 1989) (Fig. 5). There is some evidence, however, that $PtdIns(3,4)P_2$ and $PtdIns(3,4,5)P_3$ act directly as second messengers. Both of these lipids have been reported to activate the ζ isoform of PKC *in vitro* (Nakanishi *et al.*, 1993). Protein kinase C ζ differs from the classic α, β, and γ isoforms of PKC in that it is not activated by diacylglycerol [a product of PLC-catalyzed $PtdIns(4,5)P_2$ hydrolysis] or by phorbol esters (Nakanishi and Exton, 1992). Although little is known about the function of PKC ζ, microinjection of a mutated version of PKC ζ lacking kinase activity into fibroblasts does block serum-induced proliferation (Berra *et al.*, 1993). Protein kinase C ζ also regulates an NF-κB-like transcription factor in *Xenopus* oocytes (Diaz-Meco *et al.*, 1993). The role of PKC ζ as a downstream effector of PtdIns 3-kinase, however, is still controversial. Cantley and colleagues find that $PtdIns(3,4)P_2$ and $PtdIns(3,4,5)P_3$ are potent activators of the Ca^{2+}-independent PKC isoforms (PKC δ, PKC ε, and PKC η) but have little effect on PKC ζ activity (Liscovitch and Cantley, 1994). One of the downstream targets of PtdIns 3-kinase-regulated PKC isoforms may be the p70 S6 kinase. The PtdIns 3-kinase inhibitor wortmannin blocks the ability of PDGF, insulin, and IL-2 to induce the phosphorylation and activation of this enzyme (Chung *et al.*, 1994). p70 S6 kinase phosphorylates the S6 protein of the 40S ribosomal subunit and is important for growth factor-induced proliferation in a variety of cell types.

PtdIns 3-kinase may be involved in receptor internalization. In fibroblasts stimulated with PDGF, complexes containing the PDGF receptor and PtdIns 3-kinase are found in clathrin-coated endocytic vesicles. These complexes associate with microtubules, and become concentrated at the centrosome (Kapeller *et al.*, 1993). Moreover, mutant forms of the PDGF receptor that lack PtdIns 3-kinase-binding sites fail to undergo internalization by endocytosis (Joly *et al.*, 1994). Thus, PtdIns 3-kinase may affect the duration of receptor signaling by controlling the rate at which receptors are internalized. If PtdIns 3-kinase is involved in the intracellular trafficking of receptor proteins, it may also control whether receptors are degraded after they are internalized as opposed to being recycled to the cell surface.

There is some evidence that PtdIns 3-kinase regulates cytoskeletal elements and cytoskeleton-dependent movement of intracellular vesicles. Platelet-derived growth factor-induced membrane ruffling in fibroblasts, a cytoskeleton-dependent event, appears to involve PtdIns 3-kinase (Wennström et al., 1994). A role for PtdIns 3-kinase in intracellular protein trafficking is suggested by observations that the SH3 domain of PtdIns 3-kinase binds to and activates dynamin (Gout et al., 1993), a protein thought to be involved in vesicle movement along microtubules (Obar et al., 1990). GRB-2, a component of the Ras signaling pathway, also binds to dynamin via its SH3 domain (Miki et al., 1994), suggesting that other signaling proteins may have a similar dual role in signal transduction and protein trafficking.

Activation of the PtdIns 3-kinase signaling pathway in cells involves (1) recruiting PtdIns 3-kinase to the plasma membrane, where its substrates are located, and (2) increasing the specific activity of the enzyme. These processes can be mediated by three distinct types of protein–protein interactions. The first type of protein–protein interaction that contributes to activation of PtdIns 3-kinase involves the two SH2 domains of the p85 regulatory subunit of PtdIns 3-kinase. Both SH2 domains bind with high affinity to the motif phospho-YXXM (single-letter code; X is any amino acid) in other proteins. SH2-mediated binding of PtdIns 3-kinase to membrane proteins containing this motif can bring PtdIns 3-kinase to the plasma membrane. In anti-Ig-stimulated B cells, PtdIns 3-kinase binds to the CD19 coreceptor (Tuveson et al., 1993). CD19 contains two YXXM motifs in its cytoplasmic domain and is tyrosine phosphorylated after BCR ligation (Chalupny et al., 1993). Changing the tyrosine residues in these two motifs of CD19 to phenylalanines prevents the association of PtdIns 3-kinase with CD19 (Tuveson et al., 1993). In vitro, the binding of peptides containing the phospho-YXXM motif to the SH2 domains of PtdIns 3-kinase increases PtdIns 3-kinase activity severalfold (Backer et al., 1992; Carpenter et al., 1993). Peptides containing two phospho-YXXM motifs are much more potent activators of PtdIns 3-kinase than those containing a single motif, suggesting that both SH2 domains of PtdIns 3-kinase must be engaged for maximal activation (Carpenter et al., 1993). Thus, the SH2 domains of PtdIns 3-kinase can mediate both translocation of PtdIns 3-kinase to the plasma membrane and activation of the enzyme.

The second type of protein-protein interaction that regulates PtdIns 3-kinase activity and subcellular localization is the binding of a proline-rich region in the p85 subunit of PtdIns 3-kinase to the SH3 domains of the

Lyn, Fyn, and Lck PTKs (Pleiman *et al.*, 1994; Prasad *et al.*, 1993). *In vivo*, PtdIns 3-kinase has been shown to associate with Lyn in anti-Ig-stimulated B cells (Yamanishi *et al.*, 1992) and with Lck in anti-CD3-stimulated T cells (Thompson *et al.*, 1992). In B cells, 10–15% of the Lyn in the cells is tighly associated with CD19 (van Noesel *et al.*, 1993b), suggesting that PtdIns 3-kinase associates both directly (see above) and indirectly with CD19. Because Lyn, Fyn, and Lck are localized to the inner face of the plasma membrane, this interaction would bring PtdIns 3-kinase to the plasma membrane where its substrates are located. Binding of PtdIns 3-kinase to the SH3 domain of Src family PTKs may also activate PtdIns 3-kinase. *In vitro*, peptides containing only the SH3 domain of Lyn or Fyn cause a substantial increase in the specific activity of PtdIns 3-kinase (Pleiman *et al.*, 1994). Moreover, introducing proline-rich peptides corresponding to the SH3 target sequence in the p85 subunit into permeabilized B cells prevents anti-Ig-induced increases in the specific activity of PtdIns 3-kinase (Pleiman *et al.*, 1994). This presumably reflects the ability of the proline-rich peptides to compete with PtdIns 3-kinase for the SH3 domains of the Src family PTKs and suggests that this interaction is critical for activation of PtdIns 3-kinase *in vivo*. Because PtdIns 3-kinase associates with Src family PTKs only in activated T and B cells, it implies that antigen receptor cross-linking induces conformational changes in the PTKs that expose the SH3 domains and make them available for binding to PtdIns 3-kinase.

Recently, a third type of protein-protein interaction involved in regulation of PtdIns 3-kinase has been described. *In vitro*, the active GTP-bound form of Ras can bind specifically to the p110 catalytic subunit of PtdIns 3-kinase (Rodriguez-Viciana *et al.*, 1994). The inactive GDP-bound form of Ras does not bind to PtdIns 3-kinase, suggesting that receptor-induced activation of Ras initiates the formation of Ras/PtdIns 3-kinase complexes *in vivo*. Consistent with a role for Ras in activation of PtdIns 3-kinase, expression of a dominant negative Ras mutant partially blocks EGF- and nerve growth factor-induced increases in PtdIns(3,4)P$_2$ and PtdIns(3,4,5)P$_3$ levels in PC-12 cells. The partial inhibition of PtdIns 3-kinase activation by the inactive mutant Ras protein suggests that these receptors invoke multiple mechanisms of PtdIns 3-kinase activation that include SH2 binding and interaction with Src kinases in addition to Ras-mediated activation of PtdIns 3-kinase.

Both subunits of PtdIns 3-kinase are tyrosine phosphorylated after BCR ligation (Law *et al.*, 1993) and after TCR ligation in Jurkat cells expressing transfected p85 and p110 (von Willebrand *et al.*, 1994). There is, however, no evidence that this modification regulates PtdIns 3-kinase activity. In a mast cell line, a variety of cytokines activate PtdIns 3-kinase without causing detectable tyrosine phosphorylation of the enzyme (Gold *et al.*,

1994b). It is possible that tyrosine phosphorylation of PtdIns 3-kinase creates binding sites for other SH2-containing proteins. It is tempting to speculate that such proteins could be targets for regulation by the products of PtdIns 3-kinase. T cell receptor ligation also causes serine phosphorylation of p110 and threonine phosphorylation of the $p85\beta$ isoform, but not $p85\alpha$ (Reif *et al.*, 1993). The significance of these modifications is not known.

In summary, PtdIns 3-kinase is activated by the BCR and perhaps also by the TCR. This enzyme may be an important component of PTK-mediated signaling. Alternatively, PtdIns 3-kinase may be involved in receptor internalization and intracellular protein trafficking. In this case PtdIns 3-kinase could regulate signaling indirectly by determining whether receptors are rapidly internalized and degraded or recycles to the cell surface to allow for prolonged signaling. If PtdIns 3-kinase is indeed part of a signal transduction pathway, the downstream components of that pathway remain to be elucidated.

V. How Antigen Receptors Activate Tyrosine Kinases

A. Antigen Recognition Activation Motifs

In 1989, Reth pointed out that the amino acid motif D/EXXYXXL-$(X)_{6-8}$YXXL/I (where X is any amino acid) was present in the cytoplasmic domains of the invariant chains of the TCR and BCR complexes and in other multichain immune recognition receptors (Table II). Ig-α and Ig-β each have one such motif (Fig. 1), as do CD3γ, CD3δ, and CD3ε. The ζ chain of the TCR has three of these motifs, which are also found in the β and γ subunits of the mast cell receptor for IgE (FcεRI), the LMP2A membrane protein of Epstein–Barr virus, and the gp30 envelope protein of bovine leukemia virus. Because all of these receptors activate PTKs, it was suggested that these motifs were "signaling domains" involved in PTK activation and signal transduction. This has been confirmed experimentally and this motif is now referred to as the antigen recognition activation motif (ARAM), the antigen receptor homology 1 (ARH1) motif, or the tyrosine-based activation motif (TAM). Chimeric proteins with the extracellular and transmembrane domains of another receptor (such as CD8, CD16, or the IL-2 receptor α chain) and a cytoplasmic domain consisting solely of one of these ARAMs are able to activate PTKs as well as subsequent signaling events. For example, chimeric proteins containing only a single ARAM from either CD3ε or the TCR ζ chain can induce tyrosine phosphorylation, Ca^{2+} mobilization, IL-2 production, and induc-

TABLE II

Conserved Antigen Recognition Activation Motifs[a]

Receptor	Sequence
Consensus	**D** X X **Y** X X **L** X X X X X X X X **Y** X X **L**
	E **I**
hIg-α and mIg-α	E N L Y E G L N L D D C S M – Y E D I
hIg-β	D H T Y E G L D I D Q T A T – Y E D I
mIg-β	D H T Y E G L N I D Q T A T – Y E D I
hζ/η1 and mζ/η1	N Q L Y N E L N L G R R E E – Y D V L
hζ/η2	E G L Y N E L Q K D K M A E A Y S E I
mζ/η2	E G V Y N A L Q K D K M A E A Y S E I
hζ/3	D G L Y D G L S T A T K D T – Y D A L
mζ3	D G L Y Q G L S T A T K D T – Y D A L
mFcεR1-γ and rFcεR1-γ	D A V Y T G L N T R N Q E T – Y E T L
mFcεR1-β	D R L Y E E L N H V Y S P I – Y S E L
rFcεR1-β	D R L Y E E L – H V W S P I – Y S A L
BLV gp30	D S D Y Q A L L P S A P E I – Y S H L
EBV LMP2A	P E I Y S H L S P V K P D – – Y I N L

[a] Single-letter amino acid code. The ARAM consensus sequence is indicated by bold letters. X, Any amino acid. Human, mouse, and rat sequences are designated as h, m, and r, respectively. EBV, Epstein–Barr virus; BLV, bovine leukemia virus.

tion of CTL activity when expressed in T cells (Letourneur and Klausner, 1992; Romeo *et al.*, 1992; Irving *et al.*, 1993). Similarly, when expressed in B cells, chimeric proteins containing the ARAM from either Ig-α or Ig-β can induce all of the signaling reactions characteristic of the BCR, including tyrosine phosphorylation of specific substrates and increases in intracellular Ca^{2+} concentrations (Law *et al.*, 1993; Kim *et al.*, 1993b; Sanchez *et al.*, 1993). Chimeric proteins containing either the bovine leukemia virus gp30 ARAM or the Epstein–Barr virus LMP2A ARAM can also mimic the effects of BCR ligation in B cells (Alber *et al.*, 1993). Both of these viruses transform B cells, perhaps by mimicking continual BCR signaling. The conserved nature of the ARAMs is further illustrated by experiments showing that chimeric proteins containing either the Ig-α or Ig-β ARAM can induce tyrosine phosphorylation and IL-2 production when expressed in a T cell line (Taddie *et al.*, 1994).

The ARAMs contain two conserved YXXL/I sequences. Mutational analysis has indicated that the two tyrosine (Y) residues are essential for signal transduction. Changing either of the conserved tyrosines in the CD3ε, TCR ζ chain, Ig-α, or Ig-β ARAMs to phenylalanine ablates the signaling capacity of chimeric proteins containing these domains (Romeo *et al.*, 1992; Irving *et al.*, 1993; Flaswinkel and Reth, 1994; D. A. Law and A. L. DeFranco, personal communication). Because phenylalanine is similar in structure to tyrosine, but cannot be phosphorylated, it suggests that tyrosine phosphorylation of antigen receptor components is essential for signal transduction. Consistent with this hypothesis, both Ig-α and Ig-β are tyrosine phosphorylated after BCR ligation (Gold *et al.*, 1991) and TCR ligation induces tyrosine phosphorylation of CD3γ, CD3δ, CD3ε, and the TCR ζ chain (Wange *et al.*, 1992; Straus and Weiss, 1993; Qian *et al.*, 1993). At least one of the two conserved leucine or isoleucine (L or I) residues in the ARAM is also important for signaling. In chimeric proteins containing the Ig-β ARAM, altering the leucine in the N-terminal YXXL sequence of the ARAM abolishes signaling whereas mutation of the isoleucine in the C-terminal YXXI sequence has little effect (Taddie *et al.*, 1994). The essential role of at least one of the conserved leucine residues in the ARAM suggests that a phospho-YXXL/I sequence may be important for SH2-mediated binding of proteins to the ARAM.

Phosphopeptide mapping of antigen receptor components isolated from ^{32}P-labeled cells can be used to determine which of the ARAM tyrosine residues are phosphorylated after antigen receptor ligation and may therefore be SH2-binding sites. In B cell lines, both tyrosine residues in the cytoplasmic domain of Ig-β are phosphorylated after BCR ligation (Gold *et al.*, 1994c). The cytoplasmic tail of Ig-α has four tyrosine residues, of which Tyr-2 and Tyr-3 form the ARAM. Although Tyr-2 is phosphorylated after BCR ligation, no labeling of Tyr-3, the most C terminal of the two tyrosines in the ARAM, was detected (Gold *et al.*, 1994c). The same conclusion was reached by Flaswinkel and Reth (1994), who expressed mutated versions of Ig-α in a B cell line that lacks endogenous Ig α expression. Altering Tyr-2 drastically reduces anti-Ig-induced tyrosine phosphorylation of Ig-α whereas changing Tyr-3 to phenylalanine has little effect. Thus, in the context of an intact BCR, the second tyrosine of the Ig-α ARAM is not phosphorylated and probably does not participate in SH2-mediated interactions. However, changing Tyr-3 to phenylalanine abolishes the signaling capacity of fusion proteins containing the Ig-α ARAM fused to the extracellular and transmembrane domains of CD8 (Flaswinkel and Reth, 1994; D. A. Law and A. L. DeFranco, personal communication). It is possible that optimal BCR signaling requires two tandem SH2-binding sites and that Tyr-3 of the Ig-α cytoplasmic tail is phosphorylated when the Ig-α ARAM is expressed as part of a fusion protein in the absence of Ig-β, but not when it is part of an intact BCR. In the intact BCR, the Ig-β ARAM, which is phosphorylated at both tyrosines, may fulfill the requirement for tandemly arranged SH2-binding sites.

B. Tyrosine Kinase Recruitment

Because the SH2 domains of the Src family PTKs, Syk and ZAP-70, bind the sequence phospho-YXXL/I (Songyang *et al.*, 1994), tyrosine phosphorylation of the YXXL/I sequences in the ARAMs creates binding sites for these PTKs. There is considerable evidence that phosphorylation of the ARAMs recruits PTKs to the antigen receptor complexes. ZAP-70 association with the TCR is undetectable in resting T cells, but after TCR crosslinking the ζ chain and CD3ε ARAMs become tyrosine phosphorylated and bind ZAP-70 (Chan *et al.*, 1992; Wange *et al.*, 1992; Straus and Weiss, 1993). *In vitro* experiments have shown that ZAP-70 binds only to ζ ARAM-containing peptides in which both tyrosine residues are phosphorylated (Wange *et al.*, 1993; Iwashima *et al.*, 1994). Moreover, both SH2 domains of ZAP-70 are required for stable binding to the ζ ARAM. Mutating critical arginine residues in either of the ZAP-70 SH2 domains significantly decreases its binding to the doubly-phosphorylated ζ ARAM. Thus, the stable interaction of ZAP-70 with the ζ chain (and probably CD3ε) requires that both SH2 domains bind to phospho-YXXL sequences in the ARAMs. It is not known if engaging the SH2 domains of ZAP-70 increases its enzymatic activity.

Tyrosine phosphorylation of the BCR ARAMs also recruits PTKs to the BCR. In resting B cells, only 1–3% of the BCRs have Src family PTKs associated with them and there is no detectable association of Syk with the BCR (Lin and Justement, 1992; Law *et al.*, 1993). However, crosslinking chimeric proteins with cytoplasmic domains containing the ARAMs of either Ig-α or Ig-β results in tyrosine phosphorylation of the these ARAMs and significantly increases the amount of Lyn, Fyn, and Syk bound to these chimeric proteins (Law *et al.*, 1993). *In vitro*, the binding of Fyn to peptides containing the Ig-α and Ig-β ARAMs is enhanced by phosphorylation of the ARAM tyrosine residues (Clark *et al.*, 1994). Tyrosine-phosphorylated Ig-α and Ig-β ARAMs bound Fyn equally well. Moreover, the enzymatic activity of Fyn is increased threefold by the binding of phosphorylated Ig-α ARAMs to its SH2 domain *in vitro* (Clark *et al.*, 1994). Thus, tyrosine phosphorylation of the BCR ARAMs not only recruits Src family PTKs to the BCR but also, at least in the case of Fyn, increases their activity.

Although Syk associates only with cross-linked Ig-α or Ig-β chimeric proteins that are tyrosine phosphorylated (Law *et al.*, 1993), it has not been determined whether binding of Syk to the ARAMs requires that both tyrosines in the ARAM be phosphorylated as is the case for ZAP-70. In that respect, it is interesting to note that BCR ligation results in phosphorylation of both tyrosines in the Ig-β ARAM, but only one of the tyrosines in the Ig-α ARAM (see above). If engagement of both SH2 domains of Syk is required for stable binding, then Syk may bind only to Ig-β. However, chimeric proteins containing the Ig-α ARAM signal as well or better than those containing the Ig-β ARAM (Law *et al.,* 1993; Kim *et al.,* 1993b; Sanchez *et al.,* 1993), suggesting that Ig-α can activate Syk. The interaction of Syk with Ig-α may involve only one SH2 domain of Syk or, as discussed above, it is possible that both tyrosines in the Ig-α ARAM can be phosphorylated when it is expressed as part of a fusion protein that forms homodimers. It is also possible that the two SH2 domains of Syk could bind to ARAMs on two adjacent polypeptide chains that are brought together by ligating their external domains. This last possibility is less likely because the spacing between the two tyrosine residues in the ARAM appears to be critical for signaling.

C. Interplay of Src Family and Syk/ZAP-70 Family Tyrosine Kinases

Studies by Weiss and colleagues have indicated that both Lck and ZAP-70 are required for TCR-induced phosphorylation and that these PTKs interact sequentially with the TCR. ZAP-70 is clearly essential for TCR signaling because T cells from patients lacking ZAP-70 exhibit markedly reduced tyrosine phosphorylation, Ca^{2+} mobilization, and IL-2 production in response to TCR ligation (Arpaia *et al.,* 1994; Chan *et al.,* 1994b). Activation of ZAP-70 by the TCR, however, requires the presence of Lck. In a variant of the Jurkat T cell line that lacks Lck, no TCR-induced tyrosine phosphorylation occurs, the ζ chain is not phosphorylated, and ZAP-70 does not associate with the ζ chain (Iwashima *et al.,* 1994). Expression of wild-type Lck in these cells restores all of these responses whereas an Lck mutant lacking kinase activity cannot (Straus and Weiss, 1992). Fyn is expressed in Jurkat cells, and even though it associates with the TCR, it cannot effectively substitute for Lck. A model in which Lck phosphorylates the ζ chain (and perhaps CD3ε), allowing ZAP-70 to bind to the ARAMs, is supported by observations that Lck, but not ZAP-70, can phosphorylate the TCR ζ chain *in vitro* (Chan *et al.,* 1994a). In addition to the sequential interaction of Lck and ZAP-70 with the TCR, Lck and ZAP-70 may interact directly. Lck can phosphorylate ZAP-70 *in vitro* (I. Watts and R. Aebersold, personal communication). Although it is not known if this regulates the activity of ZAP-70, ZAP-70 may regulate the activity of Lck. Lck can bind via its SH2 domain to tyrosine-phosphorylated ZAP-70 (Duplay *et al.,* 1994). Engagement of the Lck SH2 domain could activate Lck, in much the same way that Fyn is activated by the binding of phosphorylated Ig-α ARAMs to its SH2 domain (Clark *et al.,*

1994). Thus, assembly of complexes containing the TCR ζ or CD3ε ARAM, Lck, and ZAP-70 could lead to activation of both PTKs.

B cell receptor signaling also requires the collaboration of Src family PTKs and Syk. When the complete BCR is expressed in the AtT20 endocrine cell line that expresses Fyn, but not Lyn, Blk, Lck, or Syk, BCR ligation induces tyrosine phosphorylation of Ig-α and Ig-β, but none of the other signaling events characteristic of the BCR (Matsuuchi *et al.*, 1992). Expression of Syk reconstitutes the ability of the BCR to induce tyrosine phosphorylation of other cellular proteins (Richards *et al.*, 1994). These results demonstrate the importance of Syk in BCR signaling and suggest a model in which a Src family PTK phosphorylates the BCR, Syk then binds to the BCR, and phosphorylation of other substrates occurs (Fig. 7). A critical role for the Src family PTKs in BCR-induced tyrosine phosphorylation is supported by studies using a chicken B cell line, DT40, which expresses Syk and Lyn, but not other Src family PTKs. When the *lyn* gene is disrupted by homologous recombination, BCR ligation does not activate Syk (Kurosaki *et al.*, 1994) and anti-Ig-induced tyrosine phosphorylation of cellular proteins is greatly reduced (Takata *et al.*, 1994). Expression of Lck or Fyn, but not Src, restores anti-Ig-induced tyrosine phosphorylation of cellular proteins in the Lyn-negative cells, indicating that Lyn, Fyn, and Lck are to some extent functionally redundant in B cells (Takata *et al.*, 1994). Consistent with this interpretation, both Lck and Fyn can phosphorylate the Ig-α and Ig-β ARAMs *in vitro* (Flaswinkel and Reth, 1994; Gold *et al.*,1994a).

FIG. 7   Model of B cell antigen receptor signaling. (A) In resting B cells, a small fraction of the BCRs have Src family PTKs associated with the ARAMs (shaded regions of the BCR) of the Ig-α and Ig-β chains. This association is not dependent on the SH2 domain of the Src family PTK. In contrast, as much as 10–15% of the Lyn PTK in the cell is associated with CD19. (B) A multivalent antigen coated with complement components will cross-link BCRs to each other and to the CD19 coreceptor complex. Bringing additional ARAMs in close proximity to the PTKs associated with the BCR and CD19 results in phosphorylation of tyrosine residues in the BCR ARAMs as well as in the cytoplasmic tail of CD19. The phosphorylated tyrosine residues are indicated by asterisks. (C) Additional Src family PTKs, as well as the Syk PTK, then bind via their SH2 domains to the phosphorylated ARAMs of the BCR. The formation of these complexes activates the PTKs and allows them to phosphorylate substrates. Phosphorylated tyrosine residues in the BCR or CD19 may attract substrates to the BCR–CD19 complexes and facilitate the phosphorylation of these proteins. In particular, PtdIns 3-kinase binds to phosphotyrosine residues on CD19 via its SH2 domains. An analogous series of events is likely to occur during T cell activation. Binding of the TCR and CD4 to the same MHC class II–peptide complex brings the CD4-associated PTK Lck into close contact with the ARAMs of the TCR and allows Lck to phosphorylate tyrosine residues in the ARAMs. The ZAP-70 PTK can then bind via its SH2 domains to the phosphotyrosine residues of the TCR ARAMs. The TCR–CD4 complex containing Lck and ZAP-70 can then phosphorylate other substrates.

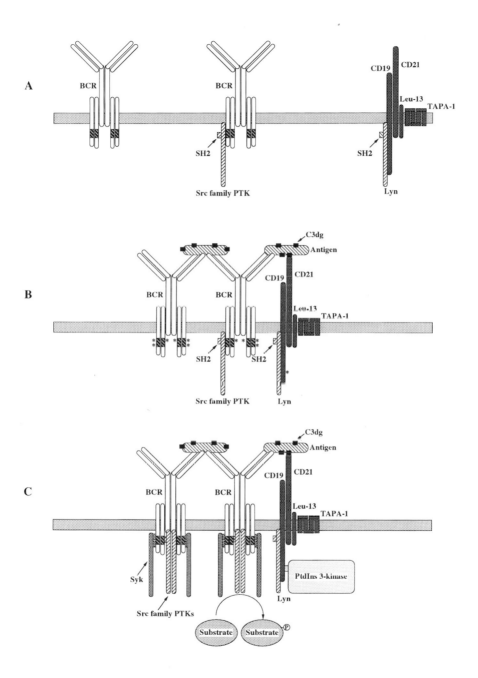

D. The First Steps

Tyrosine phosphorylation of ARAMs is essential for antigen receptor signaling because it is required for recruiting Syk or ZAP-70. An important question then is how antigen receptor ligation induces tyrosine phosphorylation of the ARAMs. The CD4 and CD8 coreceptors in T cells bind Lck with high stoichiometry whereas the CD19 coreceptor in B cells binds significant amounts of Lyn. Clustering of these coreceptors with the antigen receptors would allow the coreceptor-associated PTK to phosphorylate the ARAMs in the antigen receptors (Fig. 7). However, both TCR and BCR ligation appear to be able to induce signaling without engaging the coreceptors. For example, the Jurkat T cell line expresses little or no CD4 or CD8, suggesting that Lck interacts directly with the TCR. Depending on the T cell line, as much as 80% of the Lck is not associated with CD4 or CD8 (Chan *et al.*, 1994a). It is possible that antigen receptor cross-linking nonspecifically traps Src family PTKs and that this can lead to tyrosine phosphorylation of the ARAMs. However, there is now some evidence that Src family PTKs can bind with modest affinity to unphosphorylated ARAMs. The N-terminal 10 amino acids of Fyn and Lyn bind to the unphosphorylated Ig-α ARAM (Clark *et al.*, 1994). Similarly, the N-terminal 10 amino acids of Fyn mediate binding to chimeric proteins containing the cytoplasmic tails of the TCR ζ chain, CD3γ, or CD3ε (Timson-Gauen *et al.*, 1992). The weak association of Src family PTKs with unphosphorylated ARAMs is consistent with observations that only a small fraction of antigen receptors in resting lymphocytes have associated PTKs (Samelson *et al.*, 1990; Lin and Justement, 1992; Law *et al.*, 1993). Antigen receptor ligation would bring more ARAMs into contact with these PTKs, resulting in phosphorylation of the ARAMs. Once the ARAMs are phosphorylated, additional PTKs are recruited to the antigen receptors by means of their SH2 domains binding strongly to the phosphorylated ARAMs.

Clark *et al.* (1994) have mapped the residues in the Ig-α ARAM required for Fyn binding. They had previously observed that beads coupled with the unphosphorylated Ig-α ARAM could bind Fyn and Lyn present in lysates of the K46 B lymphoma cell line whereas beads coupled with the Ig-β ARAM could not (Clark *et al.*, 1992). The Ig-α and Ig-β ARAMs differ in 8 of the 18 amino acids. In particular, the Ig-α ARAM has the sequence DCSM (single-letter code) preceding the second (C-terminal) tyrosine of the ARAM whereas the Ig-β ARAM has the sequence QTAT in the corresponding positions (see Table II). When these four amino acids were exchanged in the Ig-α and Ig-β ARAMs, Fyn now bound to Ig-β and not to Ig-α (Clark *et al.*, 1994). Changing the tyrosine that follows the DCSM motif to phenylalanine had no effect. Thus, the DCSM portion of the ARAM appears to specify the interaction of Src PTKs with the unphosphorylated Ig-α ARAM.

The interaction between the unique N-terminal region of the Src PTKs and the Ig-α ARAM may not be direct. *In vitro,* peptides containing the Ig-α ARAM do not bind recombinant Fyn that has been produced in bacteria. The interaction may instead involve a linker protein that couples the PTK to the Ig-α ARAM. A 38-kDa protein found in K46 B lymphoma cell lysates is a possible candidate. This protein binds the cytoplasmic tail of the Ig-α chain, but not the Ig-β chain, *in vitro* (Clark *et al.,* 1992). N-Terminal sequencing indicates that p38 is a previously unidentified protein (Cambier *et al.,* 1994). Further characterization of this protein and its potential role in linking PTKs to the unphosphorylated BCRs of resting B cells is needed.

The selective binding of Fyn to the unphosphorylated Ig-α ARAM, as opposed to the Ig-β ARAM, may explain why in some B cell lines (e.g., A20) chimeric proteins containing the Ig-α ARAM deliver a stronger signal than similar chimeric proteins containing the Ig-β ARAM (Kim *et al.,* 1993b; Sanchez *et al.,* 1993). However, in other B cell lines (e.g., 2PK3), chimeric proteins containing the Ig-α and Ig-β ARAMs are equally capable of eliciting all of the responses characteristic of the BCR (Law *et al.,* 1993). The 2PK3 cell line may express Src family PTKs at higher levels than A20, giving the Ig-β fusion a better chance to trap these PTKs nonspecifically when it is cross-linked. The situation may be more limiting in A20 cells, giving the Ig-α ARAM a significant advantage owing to its ability to bind PTKs when not phosphorylated. Alternatively, the ability of Ig-β ARAM-containing chimeric proteins to induce stronger signals when expressed in 2PK3 cells than in A20 cells may reflect differences in the expression of a linker protein that specifically couples PTKs to the unphosphorylated Ig-β ARAM. In addition to the 38-kDa protein that binds the unphosphorylated Ig-α ARAM, but not to the Ig-β ARAM (see above), Clark *et al.* (1992) found novel proteins of 40 and 42 kDa in K46 B lymphoma cell lysates that bound to Ig-β, but not to Ig-α. The p38, p40, and p42 proteins may represent a family of linker proteins that couple Src family PTKs to different unphosphorylated ARAMs. If such linker proteins couple unphosphorylated BCRs to Src family PTKs, either they are not absolutely required for BCR signaling or they are not lymphocyte specific, because the BCR can efficiently stimulate protein tyrosine phosphorylation when expressed in the AtT20 endocrine cell line (Matsuuchi *et al.,* 1992).

The TCR contains 10 ARAMs whereas the BCR contains 4 ARAMs (provided that a single BCR has two Ig-α : Ig-β heterodimers). Having multiple ARAMs per receptor would allow a single antigen receptor to bind and activate multiple PTKs, thereby amplifying the signal (Fig. 7).

The different ARAMs within a single antigen receptor may also have different functions. Except for the conserved tyrosine, leucine/isoleucine, and aspartate residues, the various ARAM sequences are divergent (see Table II). For example, the first and second TCR ζ chain ARAMs (counting from the membrane-proximal ARAM) differ at 13 of 19 amino acids. The unique sequences of the ARAMs may bind different PTKs or other signaling proteins and thereby specify distinct functions for each of the ARAMs within an antigen receptor. In support of this hypothesis, chimeric proteins containing the CD3ε and ζ ARAMs induce overlapping, but different, patterns of tyrosine phosphorylation (Letourneur and Klausner, 1992). Thus, the presence of multiple ARAMs within a single antigen receptor may facilitate both signal amplification and signal diversification.

E. Synthesis

The data described above suggest the following model for PTK activation by the antigen receptors (Fig. 7). The first crucial event is the phosphorylation of the tyrosine residues in the ARAMs by Src family PTKs. Because the coreceptors bind Src family PTKs with high stoichiometry whereas the antigen receptors in resting lymphocytes have few prebound PTKs, the interaction of the coreceptors with the antigen receptors may play a key role in initiating antigen receptor signaling. The antigen receptors are, however, able to signal in the absence of coreceptor engagement, indicating that they can induce Src family PTKs to phosphorylate their ARAMs. The N-terminal unique regions of Fyn and Lyn interact with the unphosphorylated Ig-α ARAMs, perhaps through a linker protein. B cell antigen receptor ligation would bring these PTKs in contact with many more BCR ARAMs and allow the PTKs to phosphorylate them. Similarly, Fyn associates with the nonphosphorylated ARAMs of the TCR and TCR ligation may allow Fyn to phosphorylate other TCR ARAMs. Antigen receptor cross-linking may also nonspecifically trap Src family PTKs among the clustered cytoplasmic tails of the antigen receptor polypeptides. There is no evidence that Lck interacts with the unphosphorylated TCR, yet Lck is essential for TCR signaling and TCR signaling can occur in cell lines that lack CD4 and CD8. All of these mechanisms may contribute to phosphorylation of the antigen receptor ARAMs.

Phosphorylation of the antigen receptor ARAMs creates binding sites for the SH2 domains of the Src family and Syk family PTKs and attracts these PTKs to the antigen receptors. In B cells, the CD22 coreceptor may also contribute PTK-binding sites to the BCR complex because it has two YXXL sequences and is tyrosine phosphorylated on BCR cross-linking (Leprince et al., 1993). Syk and ZAP-70 appear to bind only to phosphorylated ARAMs. Binding of Src PTKs to phosphorylated ARAMs increases their enzymatic activity. It is not known if the activity of Syk or ZAP-70 is affected by binding to phosphorylated ARAMs via their SH2 domains. Substantial evidence indicates that both the Src family PTKs and Syk family PTKs are required for antigen receptor-induced tyrosine phosphorylation. The juxtaposition of these two types of PTKs may allow them to regulate each other either through phosphorylation or by SH2-mediated interactions. The relative roles of the two families of PTKs are not clear. The sole function of the Src PTKs may be to phosphorylate the antigen receptors so that Syk or ZAP-70 can bind, become activated, and phosphorylate other substrates. Altneratively, full activity of both the Src and Syk family PTKs, and/or their ability to access substrates, may require the formation of a complex containing both PTKs.

Cross-linking of antigen receptor/coreceptor complexes results in tyrosine phosphorylation of multiple sites in the TCR and BCR chains and in the associated PTKs. Moreover, the cytoplasmic tails of CD19 and CD22 contain many tyrosine residues and these coreceptors are tyrosine phosphorylated to a substantial degree after BCR ligation (Chalupny *et al.*, 1993; Leprince *et al.*, 1993). Tyrosine phosphorylation of the antigen receptor–receptor–PTK complexes may create binding sites for SH2-containing proteins that are substrates for the PTKs in the complex. For example, PtdIns 3-kinase binds via its SH2 domains to tyrosine-phosphorylated CD19 (Tuveson *et al.*, 1993), the SHC protein associates with the TCR ζ chain (Ravichandran *et al.*, 1993), and one group has reported that PLC-γ1 associates with the TCR (Dasgupta *et al.*, 1992). In addition, the SH2-mediated binding of Src family PTKs to phosphorylated ARAMs may induce conformational changes in the PTKs that allow them to bind potential substrates. *In vitro,* binding of Lyn or Fyn to phosphorylated ARAMs allows PtdIns 3-kinase to bind to their SH3 regions (Pleiman *et al.*, 1994) and phospholipase C-γ2 and rasGAP to bind to their unique regions (J. Cambier, personal communication). Such interactions may determine which signaling pathways are activated by the antigen receptors and may specify different functions for each of the Src family PTKs. In general, however, stable binding of PTK substrates to the antigen receptor complexes *in vivo* has not been observed. This may indicate that the PTK substrates are rapidly released once they are phosphorylated so that they can interact with other signaling proteins.

F. Regulation of Src Family Tyrosine Kinases by CD45 and Csk

The activity of Src Family PTKs is controlled, at least in part, by phosphorylation of a tyrosine residue near the carboxy terminus that is conserved in all members of the Src family. Phosphorylation at this site inhibits kinase activity, whereas changing this tyrosine residue to a phenylalanine (which cannot be phosphorylated) results in constitutive activation (Amrein and Sefton, 1988; Okada *et al.*, 1991; Bergman *et al.*,1992). When phosphorylated, this tyrosine residue may bind to the SH2 domain of the PTK and maintain the kinase in an inactive conformation (Amrein *et al.*, 1993; X. Liu *et al.*, 1993). Two opposing enzymes control the phosphorylation of this negative regulatory site, the p50^{csk} PTK (Csk) and the CD45 tyrosine phosphatase.

Csk is expressed in both T and B cells. *In vitro,* Csk can phosphorylate Src, Lck, Lyn, and Fyn at the negative regulatory site and inhibit their activity (Okada *et al.*, 1991; Bergman *et al.*, 1992). It is likely that Csk has similar effects on other Src family PTKs such as Blk. Thus, Csk could simultaneously downregulate the activity of all of the Src family PTKs. Overexpression of Csk in T cells decreases TCR-induced tyrosine phosphorylation and IL-2 production (Chow *et al.*, 1993). In both T and B cells, antigen receptor-induced tyrosine phosphorylation of cellular proteins peaks after 2 to 5 min and then declines to basal levels over a period of several hours. Csk may contribute to the slow decline in tyrosine phosphorylation by inactivating the Src family PTKs. However, there is currently no evidence that receptor signaling regulates Csk activity.

While Csk is likely to be involved in turning off antigen receptor signaling, CD45 plays a critical role in initiating antigen receptor signaling, at least in T cells. CD45 is required for activation of PTKs by the TCR. CD45 can remove the phosphate from the negative regulatory tyrosine residues of Lck and Fyn and activate these PTKs *in vitro* (Mustelin *et al.*, 1989, 1992). In CD45-negative T cell lines, there is increased phosphorylation of Lck at the negative regulatory site (Ostergaard *et al.*, 1989), Fyn activity is greatly reduced (Shiroo *et al.*, 1992), and TCR-induced tyrosine phosphorylation, Ca^{2+} increases, and IL-2 production are severely impaired (Pingel and Thomas, 1989; Koretzky *et al.*, 1991; Shiroo *et al.*, 1992). These results are compatible with a model in which dephosphorylation of Lck by CD45 must occur before Lck can phosphorylate the TCR ARAMs and allow ZAP-70 to bind. There is some evidence that CD45 associates with the TCR (Volarevic *et al.*, 1990). T cell receptor ligation could therefore bring CD45 into contact with TCR-associated PTKs and with coreceptor-associated Lck. In this case the TCR would regulate the access of CD45 to the Src family PTKs without necessarily altering the enzymatic activity of CD45.

T cell receptor ligation has also been reported to induce transient tyrosine phosphorylation of CD45 (Stover *et al.*, 1991). Autero *et al.* (1994) showed that Csk can phosphorylate CD45 *in vitro* and that this increases CD45 activity severalfold. Moreover, they found that tyrosine phosphorylation of CD45 creates a binding site for the SH2 domain of Lck. The phosphotyrosine residue on CD45 could displace the C-terminal phospho-tyrosine of Lck from the SH2 domain, allowing Lck to assume an active conformation. Subsequent removal of the phosphate group from the negative regulatory site by CD45 could stabilize the activated state of Lck. Coclustering of this CD45–Lck complex with the TCR would facilitate phosphorylation of the TCR ARAMs by the activated Lck. This model is highly speculative and it would be necessary to explain how the TCR activates Csk prior to activation of the Src family PTKs or ZAP-70. Moreover, there is no evidence that the TCR or any other receptor regulates the activity of Csk. In this model, Csk would be involved in the initial activation of PTKs by the TCR as well as in downregulation of Src family PTK activity at later times.

In contrast to T cells, CD45 is not required for activation of PTKs by the BCR (Kim *et al.*, 1993a). However, CD45 does regulate signaling pathways activated by the BCR. In a CD45-negative cell line, BCR cross-linking fails to induce Ca^{2+} increases (Justement *et al.*, 1991). Expression of CD45 in these cells restores this response. The Ras pathway may also be regulated by CD45. Ligation of CD45 on B cells with anti-CD45 antibodies inhibits BCR-induced activation of Ras (Kawauchi *et al.*, 1994). Thus, in B cells CD45 may dephosphorylate downstream signaling proteins such as phospholipase C-γ or components of the Ras signaling pathway.

There are likely to be additional mechanisms for controlling the activity of antigen receptor-regulated PTKs. For example, a distinct regulatory pathway that controls Lck activity has been described. Both TCR and BCR ligation induce phosphorylation of Lck at Ser-59 (Watts *et al.*, 1993; Gold *et al.*, 1994a). Phosphorylation of Lck at this site *in vitro* by MAP kinase decreases Lck activity (Watts *et al.*, 1993). This suggests that activation of MAP kinase by the antigen receptors leads to downregulation of Lck activity *in vivo*. The only other Src family PTK with a potential phosphorylation site analogous to Ser-59 of Lck is Blk. It is not known if Blk is regulated in a similar manner. Another potential mechanism for downregulating PTK activity is the internalization and degradation of receptor–PTK complexes.

VI. Signaling to the Nucleus

A. Regulation of Gene Expression

A large number of genes whose expression is induced during T cell activation have been identified, including those encoding IL-2, the IL-2 receptor, and a variety of cytokines that act on T cells, B cells, and macrophages

(Ullman *et al.*, 1990). Similarly, BCR ligation increases the expression of a number of genes important for B cell function such as MHC class II genes, B7-1, and the IL-2 receptor. This section focuses on how antigen receptor signaling regulates the activity of transcription factors which, in turn, control the expression of genes that are important for lymphocyte activation.

Transcription of one class of genes can be induced within 30–120 min of receptor engagement. Both TCR and BCR signaling induce the expression of a number of these "immediate early" genes including c-*myc*, c-*fos*, *egr-1*, *jun*B, and *nur77* (McCormack *et al.*, 1984; Ullman *et al.*, 1990; Kelly *et al.*, 1983; Monroe, 1988; Seyfert *et al.*, 1989; Tilzey *et al.*, 1991; Mittelstadt and DeFranco, 1993). Most of the known immediate early gene products are themselves transcription factors, each of which is likely to regulate a battery of genes. Thus, expression of the immediate early genes leads to a second, more extensive wave of new gene expression. For example, c-Fos dimerizes with c-Jun, resulting in the formation of AP-1 transcription factor complexes. AP-1 is a potent transcriptional activator that binds to sites called TPA-responsive elements (TREs) (Chiu *et al.*, 1988) that are present in the 5' regulatory regions of a number of genes. Anti-Ig treatment of B cells has been shown to induce the appearance of protein complexes that bind to TREs *in vitro* (Chiles *et al.*, 1991; Chiles and Rothstein, 1992).

Receptor-induced expression of the immediate early genes is not blocked by the protein synthesis inhibitor cycloheximide. Thus, induction of these genes must involve the regulation of preexisting transcription factors by antigen receptor signaling pathways. Two types of mechanisms for regulating preexisting transcription factors have been described. In one case, a component of a protein kinase cascade migrates to the nucleus and either directly phosphorylates transcription factors or activates a nuclear kinase that phosphorylates transcription factors. Alternatively, covalent modification of cytoplasmic complexes containing latent transcription factors releases an active component that translocates to the nucleus. We now describe examples of these mechanisms for regulating transcription factor activity.

1. Regulation of Transcription Factors by Phosphorylation

Phosphorylation is likely to be an important mechanism by which the antigen receptors control the activity of transcription factors. Many transcription factors must dimerize with other proteins in order to form an active complex that can bind DNA and induce transcription. Phosphorylation of transcription factor components may regulate complex formation, the ability of these complexes to bind DNA, or their ability to induce transcription (reviewed by Meek and Street, 1992; Hunter and Karin, 1992). Mitogen-activated protein kinase, which is activated by both the TCR and BCR, may be a key regulator of transcription factors. On activation, MAP kinase can translocate to the nucleus and, at least *in vitro*, it can phosphorylate a number of transcription factors. One potential target of MAP kinase is a constitutively expressed transcription factor, $p62^{TCF}$ (Davis, 1993). $p62^{TCF}$ binds to a protein called serum response factor (SRF) (Berk, 1989) and this complex regulates the expression of genes that contain serum response elements (SREs) in their 5' regulatory regions. Both c-Fos and Egr-1, the products of two immediate early genes whose expression is induced by BCR cross-linking (Monroe, 1988; Seyfert *et al.*, 1990), contain an SRE that the $p62^{TCF}$/SRF complex could bind to. In addition to phosphorylating pre-existing transcription factors such as $p62^{TCF}$ and c-Jun, MAP kinase can also phosphorylate the products of some of the immediate early genes including c-Myc and c-Fos (Davis, 1993). Thus, MAP kinase may have profound effects on gene expression by regulating the activity of multiple transcription factors. Recently, a distant relative of the MAP kinases called JNK (for c-Jun NH_2-terminal kinase) has been shown to enhance c-Jun transcriptional activity by phosphorylating it two critical on serine residues (Derijard *et al.*, 1994). TCR signaling contributes to the activation of JNK and phosphorylation of c-Jun by JNK is essential for induction of IL-2 gene expression (Su *et al.*, 1994).

In addition to MAP kinase, the TCR and BCR also activate other kinases that phosphorylate transcription factors. Both antigen receptors stimulate the activity of a c-Fos kinase that is distinct from MAP kinase, but which also appears to be regulated by PKC (Nel *et al.*, 1994). T cell receptor and BCR signaling also activate a serine kinase that phosphorylates the Ets-1 transcription factor (Pognonec *et al.*, 1988; Fisher *et al.*, 1991). In fibroblasts, Ets-1 acts as a transcriptional activator that cooperates with c-Fos and c-Jun (Wasylyk *et al.*, 1990). Antigen receptor-induced phosphorylation of Ets-1 is dependent on increases in intracellular Ca^{2+} concentrations (Pognonec *et al.*, 1988; Fisher *et al.*, 1991) but the identity of the kinase that phosphorylates Ets-1 is not known, nor is it known whether this kinase is regulated directly by Ca^{2+} or via a cascade of protein kinases. Phosphorylation of Ets-1 can be detected within 2 to 5 min of antigen receptor ligation, suggesting that Ets-1 could be involved in the induction of immediate early gene transcription. The c-*myc* gene does in fact have an Ets-1-binding site in its 5' regulatory region (Roussel *et al.*, 1994). A number of other genes whose transcription is induced by TCR cross-linking also contain potential Ets-1 binding sites. These include the genes encoding IL-2, IL-3, and IL-4 (Ho *et al.*, 1990).

2. Regulation of Transcription Factor Nuclear Localization: NF-κB and STATs

The nuclear factor kappa B (NF-κB) transcription factor is activated by both TCR and BCR ligation (J. Liu et al., 1991a; Rooney et al., 1991; Kang et al., 1992). NF-κB is a heterodimer, consisting of p50 and p65 (a member of the c-Rel family), that binds DNA and has trans-activating activity (Bauerle and Henkel, 1994). Many genes that are transcriptionally activated after antigen receptor ligation are regulated by NF-κB. These include the TCRα and Ig κ chain genes, which are induced in response to signaling by the pre-T cell receptor and pre-B cell receptor, respectively. A number of cytokine genes that are induced by TCR ligation (e.g., those encoding IL-2 and granulocyte/macrophage colony-stimulating factor) also contain NF-κB-binding sites. Perhaps most significant is the role of NF-κB in induction of the immediate early gene c-myc. Anti-Ig-induced transcription of c-myc in the WEHI-231 B cell line depends on an upstream NF-κB-binding site (Duyao et al., 1990). c-Myc is a transcription factor that plays an important role in both cell proliferation and apoptosis (see Sections VI,B and D).

NF-κB is regulated primarily by controlling its subcellular localization. In resting cells, NF-κB is found mostly in the cytosol complexed with an inhibitor, I-κB, that prevents translocation of NF-κB to the nucleus. Receptor signaling results in dissociation and rapid degradation of I-κB, allowing NF-κB to migrate to the nucleus and activate transcription (Lenardo and Baltimore, 1989). In B cells, some NF-κB is constitutively active and plays a role in maintaining expression of the Ig κ chain gene. Nevertheless, BCR cross-linking can activate the latent cytosolic NF-κB, which may contain slightly different subunits than the constitutively active NF-κB. The mechanism by which NF-κB is activated is not well understood, but there is considerable evidence that activation of PKC can lead to NF-κB translocation to the nucleus (Bauerle and Henkel, 1994). Although some in vitro experiments have suggested that PKC can directly phosphorylate I-κB and cause it to release active NF-κB, other work suggests that a distinct PKC-regulated kinase may phosphorylate I-κB in vivo. In either case, it is clear that NF-κB is regulated primarily through its association with I-κB.

Another class of transcription factors, called STAT (signal transducer and activator of transcription) proteins, are also regulated by controlling their subcellular localization (reviewed in Darnell et al., 1994). A number of cytokine receptors stimulate tyrosine phosphorylation of a member of the STAT family (Larner et al., 1993; Sadowski et al., 1993). Once phosphorylated, STAT proteins form complexes with other cytosolic proteins and then migrate to the nucleus where they induce transcription of cytokine-responsive genes. The STAT proteins contain SH2 domains that may facilitate their interaction with receptors and associated PTKs. It is not known if the lymphocyte antigen receptors regulate similar proteins that act as direct activators of transcription. An intriguing observation is that HS-1, a 75-kDa protein that is tyrosine phosphorylated after BCR cross-linking (Yamanashi et al., 1993), has an SH2 domain as well as a helix–turn–helix motif characteristic of a DNA-binding protein. Whether HS-1 acts as a transcription factor, analogous to the STAT proteins, remains to be determined.

3. Regulation of a Single Transcription Factor by Multiple Mechanisms: NF-AT

The nuclear factor of activated T cells (NF-AT) complex consists of two components that illustrate different ways in which transcription factors can be regulated (Northrop et al., 1993; Jain et al., 1993). One component, NF-AT$_C$, is found in the cytosol of resting cells but translocates to the nucleus when the TCR is cross-linked. The other component consists of two immediate early gene products that are found only in the nuclei of activated T cells. NF-AT plays a critical role in T cell activation (reviewed in Schreiber and Crabtree, 1992; Crabtree, 1989). The regulatory regions of genes encoding a number of cytokines, including IL-2, contain NF-AT-binding sites. Moreover, NF-AT is required for TCR-induced expression of these genes. Deletion of the NF-AT-binding site from the IL-2 promoter significantly reduces the ability of this promoter to drive transcription of a reporter gene. In contrast, reporter gene constructs expressed in T cell lines can be made responsive to TCR signaling by fusing an NF-AT-binding site to their promoters (Durand et al., 1988). Thus, NF-AT controls the expression of a number of genes that are important for T cell activation. In B cells, recent work has shown that BCR signaling contributes to the induction of NF-AT activity (Choi et al., 1994; Venkitaraman et al., 1994).

Like NF-κB, the cytosolic component of NF-AT, NF-AT$_C$, is regulated by controlling its ability to migrate to the nucleus. The TCR-induced translocation of NF-AT$_C$ to the nucleus is dependent on calcineurin, a calcium-dependent serine/threonine phosphatase that is inhibited by cyclosporin A (Jain et al., 1993; Flanagan et al., 1991). Consistent with this model, artifically increasing intracellular calcium concentrations in T cells with calcium ionophores induce nuclear translocation of NF-AT$_C$ and this can be blocked by cyclosporin A. NF-AT$_C$ is a phosphoprotein and calcineurin has been shown to dephosphorylate NF-AT$_C$ in vitro (Jain et al., 1993). It is not clear how dephosphorylation of NF-AT$_C$ induces it to translocate to the nucleus. NF-AT$_C$ has a potential nuclear localization sequence that may be revealed by conformational changes induced by dephosphorylation. Alternatively, dephosphorylation of NF-AT$_C$ may cause it to dissociate from an inhibitory component analogous to I-κB that retains NF-AT$_C$ in the cytosol.

The nuclear component of the active NF-AT complex is an AP-1-type dimer consisting of c-Fos- and c-Jun-like proteins (Northrop *et al.*, 1993; Jain *et al.*, 1993; Boise *et al.*, 1993). These proteins are absent in resting T cells, but can be detected within 30 min of adding anti-TCR antibodies to T cells (Ullman *et al.*, 1990). Thus, these proteins are the products of immediate early genes. T cell receptor induction of this nuclear component of NF-AT can be inhibited by dominant negative versions of Ras (Baldari *et al.*, 1993; Woodrow *et al.*, 1993). It is likely that the Ras-dependent induction of the nuclear component of NF-AT involves the activation of MAP kinase and the subsequent phosphorylation of preexisting transcription factors by MAP kinase.

B. Cell Cycle Entry

Antigen receptor signaling induces resting T and B cells to enter the cell cycle. This transition from the G_0 to the G_1 phase of the cell cycle is marked by the induction of the immediate early genes. Although antigen stimulation causes lymphocytes to enter G_1, progression into S phase and subsequent proliferation usually requires another signal. For example, in T cells, the antigen receptor and costimulatory (CD28) signals are "competence factors" that induce entry into the G_1 phase of the cell cycle, production of IL-2, and expression of the IL-2 receptor. Interleukin 2 then acts as a "progression factor," driving the cells into S phase (Stern and Smith, 1986; Kumagai *et al.*, 1988). In B cells, anti-Ig and CD40L can cause B cells to enter G_1 and, at high concentrations, can also induce progression into S phase (DeFranco *et al.*, 1985; Hollenbaugh *et al.*, 1992). B cells induced to enter G_1 via BCR or CD40 signaling can be driven into S phase by IL-4 or in some cases IL-2. IL-4, however, does not act as a pure progression factor in that it is most effective when added to B cell cultures at the same time as anti-Ig. Although IL-4 does not by itself induce resting B cells to enter G_1, it increases the number of cells that can be driven into G_1 by anti-Ig and it lowers the concentration of anti-Ig required for cells to enter S phase (Kishimoto, 1985; Oliver *et al.*, 1985). Thus, whereas TCR ligands and IL-2 form a classic pair of competence and progression factors, the relation between anti-Ig and IL-4 is more complex. Nevertheless, it is clear that lymphocytes must receive multiple signals before they begin to proliferate.

The decision as to whether or not a cell will divide is made in late G_1

phase. The transition from G_1 to S phase involves an ordered series of events, and once a cell passes a certain point (termed the restriction point) it is committed to cell division and will divide even if the growth factor is removed. The molecular basis of the commitment to cell division has been studies extensively in a number of mammalian cell types and in yeast. We present a brief and simplified model of these events. Stimulation of cells with growth factors induces the expression of genes encoding the G_1 cyclins (D- and E-type cyclins) (Matsushime *et al.*, 1991). These proteins regulate progression through the G_1 phase of the cell cycle and entry into S phase. In fibroblasts, overexpression of cyclin E decreases the length of G_1 and accelerates entry into S phase (Ohtsubo and Roberts, 1993) whereas microinjection of anti-cyclin D1 antibodies prevents entry into S phase (Baldin *et al.*, 1993). The G_1 cyclins exert their effect on the cell cycle by controlling the activity of cyclin-dependent kinases (CDKs). Cyclin D proteins associate with and activate CDK2, CDK4, and CDK5 whereas cyclin E associates with CDK2 (Sherr, 1993). Cyclin-dependent kinase activity is also regulated by phosphorylation and dephosphorylation of the CDKs (Pines, 1993), as well as by inhibitory subunits that associate with these complexes (Pines, 1994). One of the key targets of the cyclin–CDK complexes is the RB tumor supressor protein (Hinds *et al.*, 1992). RB is phosphorylated at a low level in G_0 and early G_1 but becomes increasingly phosphorylated in mid-G_1 after active cyclin–CDK complexes are formed (Chen *et al.*, 1989). RB controls entry into S phase by regulating the activity of E2F-1, a transcription factor that promotes the expression of a number of genes required for DNA synthesis such as DNA polymerase α, dihydrofolate reductase, and thymidylate kinase (Nevins, 1992). The hypophosphorylated RB present in G_0 cells binds E2F-1 and suppresses its activity. Phosphorylation of RB by cyclin–CDK complexes causes the release of active E2F-1 and allows progression into S phase. In support of a role for E2F-1 in S phase entry, overexpression of E2F-1 is sufficient to induce quiescent fibroblasts to enter S phase (Johnson *et al.*, 1993). It is likely that the series of events described here also occurs when resting lymphocytes are activated by antigen receptor engagement in conjunction with cytokine stimulation. For example, T cell activation results in increased phosphorylation of RB (Terada *et al.*, 1991), which presumably releases active E2F-1 and promotes entry into S phase.

Although the G_1 cyclins are critical regulators of cell cycle progression, little is known about how their expression is regulated. mRNA for the G_1 cyclins usually appears 4 to 8 hr after growth factor stimulation. Immediate early genes that function as transcription factors are likely to be involved in inducing the expression of the G_1 cyclin genes. Although this remains to be demonstrated, it is clear that c-*myc* is essential for cells to progress from G_1 to S phase. For example, decreasing the expression of c-*myc* with antisense oligonucleotides prevents activated T cells from entering S phase (Heikkila *et al.*, 1987). The essential function of c-*myc* in promoting the G_1 to S transition has also been demonstrated in experiments by Roussel *et al.* (1991). They showed that a mutated colony-stimulating factor 1 receptor that can induce the expression of c-*fos* and *junB*, but not c-*myc*, is unable to stimulate proliferation when expressed in fibroblasts. The ability of this mutant receptor to induce entry into S phase can be restored by expression of a transfected c-*myc* gene. The role of c-*myc* in cell cycle progression is also supported by experiments showing that hyperphosphorylation of the RB protein, which is required for entry into S phase, depends on expression of c-*myc* (Fischer *et al.*, 1994). It is not known how c-*myc* controls RB phosphorylation. c-*myc* may promote the expression of active cyclin–CDK complexes capable of phosphorylating RB. Alternatively, c-*myc* may influence the capacity of RB to act as a substrate for cyclin–CDK complexes. Consistent with this possibility, c-Myc can bind to RB *in vitro* (Rustgi *et al.*, 1991). Much still needs to be learned about how c-*myc* is connected to cell cycle regulation.

C. Apoptosis

At certain stages of lymphocyte development, signals through the antigen receptor result in programmed cell death or *apoptosis* (see Section III). Apoptosis is distinguished from necrotic cell death by characteristic morphological features such as condensation of the cytoplasm, plasma membrane blebbing, and segmentation of the nucleus, as well as by endonuclease-mediated cleavage of genomic DNA into nucleosome-sized multimers (Cohen *et al.*, 1992). The fragmentation of the genomic DNA can easily be visualized on agarose gels as a ladder of bands spaced at 180-bp intervals. *In vivo,* apoptotic cells are rapidly ingested and destroyed by macrophage-like cells. Thymocytes undergoing negative selection die by apoptosis. Immature B cells undergoing BCR-mediated negative selection are also likely to die by apoptosis. Although this has not been demonstrated directly in immature B cells from bone marrow, BCR ligation causes apoptosis in several immature B cell lines such as WEHI-231 (Hasbold and Klaus, 1990). Apoptosis may also occur in mature T and B cells that receive signals through the antigen receptor in the absence of costimulatory signals or other signals that drive progression into S phase. For example, T cell clones that have been activated by antigen undergo apoptosis if they are deprived of IL-2. Similarly, mature B cells also undergo apoptosis if they are treated with high concentrations of anti-Ig antibodies in the absence of T cell-derived signals such as IL-4 and the CD40L (Parry *et al.*, 1994). Antigen–receptor programmed cell death thus has a critical role in the elimination of self-reactive cells and in regulation of the immune response.

The mechanism underlying antigen receptor-induced apoptosis can be studied *in vitro* using T and B cell lines that mimic antigen-induced negative selection of immature T and B cells. In a number of T cell hybridomas, TCR signaling induces growth arrest in the G_1 phase of the cell cycle followed by apoptosis (Mercep *et al.*, 1989). Similarly, BCR ligation causes growth arrest in the G_1 phase of the cell cycle followed by apoptosis in the immature B cell lines WEHI-231 and CH31 (Page and DeFranco, 1990; Hasbold and Klaus, 1990; Fischer *et al.*, 1994). Studies using these cell lines have revealed that antigen receptor-induced cell cycle arrest can be separated from the induction of programmed cell death in both T and B cells (Scheuerman *et al.*, 1994; Mercep *et al.*, 1989). For example, cyclosporin A blocks anti-TCR-induced apoptosis in the 2B4.11 T hybridoma cell line but does not block cell cycle arrest (Mercep *et al.*, 1989). Thus, different signaling pathways may contribute to cell cycle arrest versus induction of apoptosis.

Receptor-induced apoptosis is an active process that requires signal transduction and new gene transcription. In terms of signal transduction, TCR-induced apoptosis of thymocytes and T cell hybridomas involves the calcium-activated phosphatase calcineurin because it can be blocked by cyclosporin A both *in vitro* and *in vivo* (Mercep *et al.*, 1989; Shi *et al.*, 1989). The need for new gene transcription is demonstrated by experiments showing that TCR-induced expression of the immediate early gene *nur77* is required for apoptosis in thymocytes and in T cell hybridomas (Woronicz *et al.*, 1994; Z.-G Liu *et al.*, 1994). *nur77* encodes a zinc finger-containing transcription factor that has a DNA-binding domain and a domain that can act as a *trans*-activator of transcription. Antisense oligonucleotides that decrease *nur77* expression, as well as expression of a dominant negative version of *nur77* that contains only the DNA-binding domain, prevent TCR-induced apoptosis in T cell hybridomas. Identification of genes that are regulated by *nur77* may provide insights into the mechanisms underlying receptor-driven apoptosis. Although *nur77* is essential for TCR-induced apoptosis in T cells, it may not be involved in BCR-induced apoptosis. B cell antigen receptor ligation causes apoptosis in the WEHI-231 B cell line even though these cells do not express *nur77* (Mittelstadt and DeFranco, 1993). It is possible that WEHI-231 cells express a homolog of *nur77* that performs a similar function in committing the cells to programmed cell death.

Once antigen receptor signaling commits a cell to apoptosis, a cell death program must be activated. Recent work has implicated the interleukin-1β-converting enzyme (ICE) in the morphological changes associated with apoptosis (Vaux *et al.*, 1994). Interleukin-1β-converting enzyme is homologous to the *ced-3* gene product of *C. elegans,* which is required for programmed cell death during development. Overexpression of either ICE or the *ced-3* gene product in murine fibroblasts causes apoptosis. Interleukin-1β-converting enzyme is a protease that presumably acts on a structural elements, such as cytoskeletal proteins, whose degradation results in collapse of the cytosol. It is also likely to initiate DNA cleavage, which is thought to be carried out by DNase I. DNase I normally exists in an inactive state complexes with actin. Degradation of structural proteins by ICE may release active DNase I, which can migrate to the nucleus and degrade the DNA.

D. Activation versus Apoptosis

A question that has confronted immunologists for many years is how antigen receptor signaling can lead to either activation or apoptosis, depending on the maturation state of the lymphocyte, the presence or absence of signals coming from other receptors, or the magnitude of the antigen receptor signal. At a first approximation, the initial antigen receptor-induced signaling events described in Section IV appear to be similar throughout T and B cell development. Thus, it is likely that the context in which the signals are received alters how they are interpreted. In a model that accounts for contextual differences, for example whether or not a mature T cell has received a costimulatory signal in addition to the TCR signal, one could propose that antigen receptor signals in mature lymphocytes turn on the initial steps of both activation and death pathways. In the absence of the costimulatory signal the death pathway prevails, whereas signals coming from other receptors (e.g., CD28) can overcome the death pathway and rescue the cells. Cells may also be able to discriminate quantitative differences in signaling by having different thresholds for different response elements. For example, in double-positive thymocytes low-avidity TCR binding induces positive selection whereas high-avidity binding induces negative selection. One can postulate that low levels of TCR signaling turn on activation pathways whereas high levels of TCR signaling also initiate a pathway that overrides the activation pathways and leads to cell death. At different stages of development, the outcome of antigen receptor signaling may be skewed by differences in transcription factors or other downstream effectors that change the quality of the antigen receptor signal. For example, antigen receptor signaling pathways in immature T and B cells may be strongly skewed toward inducing cell death.

Although it is easy to propose models, there are few situations in which the molecular basis for the net outcome of antigen receptor signaling is truly understood. Perhaps the best understood example is how engagement of both the TCR and CD28 leads to activation and proliferation whereas TCR signaling alone leads to anergy or cell death. The production of interleukin 2 (IL-2), which acts as an autocrine growth factor, is required for activated T cells to survive and proliferate. Engaging either the TCR alone or CD28 alone results in negligible production of IL-2. The combination of these signals, however, results in substantial production of IL-2. Recent work has identified the JNK kinase as a signaling component that can integrate the signals coming from the TCR and CD28 (Su et al., 1994). The target of JNK is c-Jun, which together with c-Fos, comprises the AP-1 transcription factor. The IL-2 gene promoter contains both an AP-1 site and a site for the NF-AT complex which consists of $NF-AT_C$ and AP-1. Phosphorylation of critical serine residues in the amino terminus of c-Jun by JNK is required for c-Jun (and therefore AP-1) to act as a transcriptional activator (Su et al., 1994). The TCR and CD28 act synergistically to cause activation of JNK (Su et al., 1994). Engaging either the TCR alone or CD28 alone results in minimal activation of JNK. The requirement for both a TCR signal and a CD28 signal is specific for JNK. Maximal activation of MAP kinase, in contrast, requires only TCR cross-linking. Moreover, there is a complete correlation between the combination of stimuli that activate JNK and those that induce IL-2 production. While TCR signaling alone fails to activate JNK, it also renders AP-1, but not other transcription factors, refractory to subsequent signals from the TCR (Kang et al., 1992). Thus, AP-1 activity may be critical determinant of activation versus anergy in T cells. The molecular basis for AP-1 non-responsiveness is not understood.

It is not known if JNK is also activated in B cells. However, it is tempting to propose an analogous two signal model in which only the combination of BCR signaling and CD40 signaling (for example) would activate JNK and lead to B cell activation. One would still have to account for how BCR signaling alone would lead to cell death.

A model in which one receptor induces both an activation and a death pathway, while costimulatory signals from another receptor determine which prevails, could involve c-myc (Evan and Littlewood, 1993). In quiescent fibroblasts, overexpression of c-myc in the absence of growth factors leads to apoptosis. In contrast, growth factor-induced proliferation in fibroblasts is accompanied by and requires the expression of c-myc. This suggests that c-myc can activate both a death pathway and a proliferative pathway. Growth factors presumably allow the proliferative pathway to prevail by also inducing the expression of survival genes that rescue the cells from c-myc-induced death. In these experiments, serum could be considered a costimulatory signal for fibroblasts. In lymphocytes, antigen receptor ligation in both T and B cells induces c-myc expression and c-

myc is required for progression from G_1 to S, at least in T cells. In the absence of costimulatory signals, c-*myc* expression may lead to apoptosis.

Identifying survival genes that determine whether antigen receptor signaling leads to activation as opposed to cell death is important for understanding immune regulation. The *bcl-2* gene is a survival gene that prevents c-*myc*-induced apoptosis in fibroblasts (Bissonnette *et al.*, 1992). T and B cell mitogens that stimulate proliferation induce expression of *bcl-2* mRNA in lymphocytes (Reed *et al.*, 1987). However, it is not clear whether the Bcl-2 protein can protect lymphocytes from antigen receptor-induced cell death. Although antigen-induced apoptosis of immature B cells in transgenic mice can be delayed by expression of *bcl-2* in these cells (Hartley *et al.*, 1993), expression of a transfected *bcl-2* gene in the WEHI-231 B cell line does not prevent anti-Ig-induced apoptosis (Cuende *et al.*, 1993). Similarly, TCR-induced negative selection in immature thymocytes still occurs in transgenic mice expressing *bcl-2* (Sentman *et al.*, 1991; Strasser *et al.*, 1991).

Although *bcl-2* may not prevent antigen receptor-induced apoptosis, it is involved in situations in which antigen receptor signaling prevents apoptosis. The clearest example of this is in germinal center B cells that have undergone somatic hypermutation. These B cells must be retested for their ability to bind antigen before they can go on to become antibody-secreting cells or memory B cells. In the absence of signals through the BCR, germinal center B cells undergo programmed cell death. Consistent with this observation, long-term survival of memory B cells requires continual or repeated antigen stimulation. B cell receptor signaling in germinal center B cells induces the expression of *bcl-2* (Y.-J Liu *et al.*, 1991). Moreover, *bcl-2* expression has been shown to promote the survival of memory B cells. Memory B cells from mice expressing a *bcl-2* transgene survive much longer than those from normal mice when transferred to an unimmunized animal in which the antigen is not present (Nunez *et al.*, 1991). Developing B cells also undergo apoptosis if they do not receive a positive signal through the pre-B cell receptor and *bcl-2* may be involved in this process as well. Mice with severe combined immunodeficiency disease (SCID), which are unable to rearrange their Ig genes and do not produce functional BCRs, are virtually devoid of cells that express other markers characteristic of mature B cells, such as the B220 form of CD45. However, expression of a *bcl-2* transgene permits the development of B220[+] cells, even though they do not express the BCR (Strasser *et al.*, 1994). Induction of *bcl-2* expression may therefore be an essential component of BCR-induced survival or positive selection.

B cell antigen receptor signaling induces expression of *bcl-2* at all stages of B cell development. Why then does *bcl-2* expression promote the survival of pre-B cells and germinal center B cells, but not immature B cells, which are normally killed by BCR cross-linking? The Bcl-2 protein can form heterodimers with a related protein called Bax (Oltvai *et al.*, 1993). Unlike Bcl-2, overexpression of Bax accelerates apoptosis. This has led to a model in which the ratio of Bcl-2 to Bax determines whether or not cells become committed to apoptosis. When Bcl-2 is present in excess, Bcl-2 forms heterodimers with Bax and neutralizes the ability of Bax to promote apoptosis. In contrast, when the concentration of Bax is greater than that of Bcl-2, Bax–Bax homodimers form and induce cell death. In pre-B cells and in germinal center B cells, BCR signaling may increase BCL-2 levels so that Bax is neutralized and the cells survive. Perhaps in immature B cells BCR cross-linking also induces a large increase in the expression of Bax that exceeds BCR-induced increases in Bcl-2 levels and causes apoptosis. The inability of a transfected *bcl-2* gene to prevent BCR-induced apoptosis in immature B cells may simply be a consequence of extremely high levels of Bax induced by BCR cross-linking. Developmental differences in transcription factors or in BCR signal transduction between immature B cells and germinal center B cells may link Bax expression to BCR signaling only in the immature B cells. The role of Bax in antigen receptor-induced apoptosis has not been investigated and this model is purely hypothetical. The regulation of programmed cell death may in fact be much more complex, because there appear to be additional members of the Bcl-2/Bax family.

Developmental differences in the expression of other regulatory molecules may also determine whether antigen receptor signaling leads to activation or cell death. Two such differences that may favor TCR-induced cell death in immature T cells have been described. First, mRNA for the GRB3-3 protein is high in thymocytes (Fath *et al.*, 1994). This protein is a dominant negative inhibitor of the Ras pathway and also induces apoptosis when microinjected into fibroblasts. T cell receptor-induced activation of other signaling pathways (e.g., the PLC pathway) in the absence of Ras activation may deliver a negative signal instead of an activation signal. Alternatively, GRB3-3 may trigger an apoptosis signal that overrides the other TCR-derived signals. Differences in the regulation of the *nur77* gene in mature and immature T cells may also specify whether TCR signaling leads to activation or to cell death. In mature T cells, TCR signaling causes a small transient increase in *nur77* expression that peaks after 1 hr and disappears by 3 hr (Woronicz *et al.*, 1994). In contrast, in thymocytes undergoing apoptosis, much higher levels of *nur77* gene product are attained and *nur77* expression persists for at least 12 hr, at which point most of the cells are dead. Z.-G. Liu *et al.* (1994) have shown that *nur77* expression in thymocytes and mature T cells involves different 5' regulatory elements and that the *nur77* mRNA induced in thymocytes is polyadenylated whereas the mRNA found in mature T cells is not. These differences may influence the levels of *nur77* expression by controlling transcription rates, mRNA stability, and the ability of the mRNA to be translated. Although *nur77* expression is required for TCR-induced apoptosis in thymocytes, its role in mature T cells is not known. Nevertheless, these differences in *nur77* regulation may determine whether TCR signaling induces activation or apoptosis.

VII. Perspective

The T and B lymphocytes of the immune system are powerful weapons designed to eliminate pathogenic microorganisms that threaten the host. Like most powerful weapons, they can also be dangerous. In terms of the immune system, this danger lies in autoimmune reactions. Activation of the destructive mechanisms of the immune system is therefore subject to many checkpoints. The development, activation, and survival of lymphocytes directed against foreign antigens require positive signals through the antigen receptors while at the same time the elimination of self-reactive lymphocytes requires negative signals initiated by the antigen receptors. The antigen specificity of these divergent responses requires signaling through the clonally restricted antigen receptors. The ability to have different outcomes of antigen receptor signaling at different stages of lymphocyte development, or under different circumstances, has required the immune system to develop molecular strategies for distinguishing the context of antigen receptor signaling. These strategies may include developmental differences in signal transduction components or transcription factors or a requirement for signals from other receptors that determine how the antigen receptor signal should be interpreted. One can consider each of the second messengers or signaling reactions elicited by a receptor as words in a molecular language. The sense of the sentence is determined by which words are used and how they are arranged. We are only beginning to understand how signals from different receptors are integrated and how they are converted into the cellular responses we can observe.

The design of the antigen receptors is well suited for their function. The modular design allows vast flexibility in antigen-binding specificity to be connected to common signal transduction mechanisms. This design has also allowed diversification of immune receptors by coupling the same or related signaling subunits to different ligand-binding subunits. The development of coreceptors that amplify the antigen receptor signal is an evolutionary development that increases the sensitivity of antigen detection by lymphocytes. The ability to convert the binding of a few hundred antigen molecules into second messenger concentrations sufficient to alter gene expression also requires amplification of the signal. The activation of protein kinase cascades allows amplification at each step. Moreover, the use of phosphorylation, especially tyrosine phosphorylation, as the currency of the realm allows the phosphorylated and nonphosphorylated form of a protein to have different protein–protein interactions and different levels of activity. Protein phosphorylation is also a rapid and readily reversible response. This is a useful property for neurons that must make decisions on the microsecond level, but what about lymphocytes whose responses such as proliferation occur 24 or more hours later? The initial signaling events that we understand in reasonable detail, protein phosphorylation and activation of PTK-dependent signaling pathways, are transient and occur within the first 30 min. Our understanding of the events that occur in the subsequent hours is still fragmentary. The task for the future is to better understand the nuclear events that control the cell cycle and lead to changes in gene expression and to determine how these nuclear events are regulated by the cytoplasmic signaling reactions initiated by the antigen receptors.

References

Alber, G., Kim, K.-M., Weiser, P., Riesterer, C., Carsetti, R., and Reth, M. (1993). Molecular mimicry of the anitgen receptor signaling motif by transmembrane proteins of the Epstein-Barr virus and the bovine leukemia virus. *Curr. Biol.* **3**, 333–339.

Allen, P. M. (1994). Peptides in positive and negative selection: A delicate balance. *Cell (Cambridge, Mass.)* **76**, 593–596.

Allen, R. C., Armitrage, R. J., Conley, M. E., Rosenblatt, H., Jenkins, N. A., Copeland, N. G., Bedell, M. A., Edelhoff, S., Disteche, C. M., Simoneaux, D. K., Fanslow, W. C., Belmont, J., and Spriggs, M. K. (1993). CD40 ligand gene defects responsible for X-linked hyper-IgM syndrome. *Science* **259**, 990–993.

Amrein, K. E., and Sefton, B. M. (1988). Mutation of a site of tyrosine phosphorylation in the lymphocyte-specific tyrosine protein kinase, p56*lck*, reveals its oncogenic potential in fibroblasts. *Proc. Natl. Acad. Sci. U. S. A.* **85**, 4247–4251.

Amrein, K. E., Panholzer, B., Flint, N. A., Bannwarth, W., and Burn, P. (1993). The SH2 domain of the protein tyrosine kinase p56*lck* mediates both intermolecular and intramolecular interactions. *Proc. Natl. Acad. Sci. U. S. A.* **90**, 10285–10289.

Appleby, M. W., Gross, J. A., Cooke, M. P., Levine, S. D., Qian, X., and Perlmutter, R. M. (1992). Defective T cell receptor signaling in mice lacking the thymic isoform of p59*fyn*. *Cell (Cambridge, Mass.)* **70**, 751–763.

Arcaro, A., and Wymann, M. P. (1993). Wortmannin is a potent phosphatidylinositol 3-kinase inhibitor: The role of phosphatidylinositol 3,4,5-trisphosphate in neutrophil responses. *Biochem. J.* **296**, 297–301.

Arpaia, E., Shahar, M., Dadi, H., Cohen, A., and Roifman, C. M. (1994). Defective T cell receptor signaling and CD8+ thymic selection in humans lacking ZAP-70 kinase. *Cell (Cambridge, Mass.)* **76**, 947–958.

Ashton-Rickardt, P. G., Bandeira, A., Delaney, J. R., van Kaer, L., Pircher, H.-P., Zinkernagel, R. M., and Tonegawa, S. (1994). Evidence for a differential avidity model of T cell selection in the thymus. *Cell (Cambridge, Mass.)* **76**, 651–663.

Aubry, J.-P., Pochon, S., Graber, P., Jansen, K. U., and Bonnefoy, J.-Y. (1992). CD21 is a ligand for CD23 and regulates IgE expression. *Nature (London)* **358,** 505–507.

Auger, K. R., and Cantley, L. C. (1991). Novel polyphosphoinositides in cell growth and activation. *Cancer Cells* **3,** 263–270.

Autero, M., Saharinen, J., Pessa-Morikawa, T., Soula-Rothhut, M., Oetken, C., Gassmann, M., Bergmen, M., Alitalo, K., Burn, P., Gahmberg, C. G., and Mustelin, T. (1994). Tyrosine phosphorylation of CD45 phosphotyrosine phosphatase by $p50^{csk}$ kinase creates a binding site for $p56^{lck}$ tyrosine kinase and activates the phosphatase. *Mol. Cell. Biol.* **14,** 1308–1321.

Azuma, M., Cayabyab, M., Buck, D., Phillips, J. H., and Lanier, L. L. (1992). CD28 interaction with B7 costimulates primary allogeneic proliferative responses and cytotoxicity mediated by small, resting T lymphocytes. *J. Exp. Med.* **175,** 353–360.

Azuma, M., Ito, D., Yagita, H., Okumura, K., Phillips, J., Lanier, L., and Somoza, C. (1993). B70 antigen is a second ligand for CTLA-4 and CD28. *Nature (London)* **366,** 76–79.

Backer, J. M., Myers, M. G., Jr., Shoelson, S. E., Chin, D. J., Sun, X.-J., Miralpeix, M., Hu, P., Margolis, B., Skolnik, E. Y., Schlessinger, J., and White, M. F. (1992). Phosphatidylinositol 3'-kinase is activated by association with IRS-1 during insulin stimulation. *EMBO J.* **11,** 3469–3479.

Baldari, C. T., Heguy, A., and Telford, J. L. (1993). Ras protein activity is essential for T-cell antigen receptor signal transduction. *J. Biol. Chem.* **268,** 2693–2698.

Baldin, V., Likas, J., Marcote, M. J., Pagano, M., Bartek, J., and Draetta, G. (1993). Cyclin D1 is a nuclear protein required for cell cycle progression in G_1. *Genes Dev.* **7,** 812–821.

Banchereau, J., Bazan, F., Blanchard, D., Briere, F., Galizzi, J. P., van Kooten, C., Liu, Y. J., Rousset, F., and Saeland, S. (1994). The CD40 antigen and its ligand. *Annu. Rev. Immunol.* **12,** 881–922.

Bauerle, P. A., and Henkel, T. (1994). Function and activation of NF-κB in the immune system. *Annu. Rev. Immunol.* **12,** 141–179.

Berek, C., Berger, A., and Apel, M. (1991). Maturation of the immune response in germinal centers. *Cell (Cambridge, Mass.)* **67,** 1121–1129.

Bergman, M., Mustelin, T., Oetken, C., Partanen, J., Flint, N. A., Amrein, K. E., Autero, M., Burn, P., and Alitalo, K. (1992). The human $p50^{csk}$ tyrosine kinase phosphorylates $p56^{lck}$ at Tyr-505 and down regulates its catalytic activity. *EMBO J.* **11,** 2919–2924.

Berk, A. J. (1989). Regulation of eukaryotic transcription factors by post-translational modification. *Biochim. Biophys. Acta* **1009,** 103–109.

Berra, E., Diaz-Meco, M. T., Dominguez, I., Municio, M. M., Sanz, L., Lozano, J., Chapkin, R. S., and Moscat, J. (1993). Protein kinase C ζ isoform is critical for mitogenic signal transduction. *Cell (Cambridge, Mass.)* **74,** 555–563.

Berridge, M. J. (1993). Inositol triphosphate and calcium signaling. *Nature (London)* **361,** 315–325.

Bijsterbosch, M. K., Meade, C. J., Turner, G. A., and Klaus, G. G. B. (1985). B lymphocyte receptors and polyphosphoinositide degradation. *Cell (Cambridge, Mass.)* **41,** 999–1006.

Bissonnette, R. P., Echeverri, F., Mahboubi, A., and Green, D. R. (1992). Apoptotic cell death induced by c-*myc* is inhibited by *bcl*-2. *Nature (London)* **359,** 552–554.

Boguski, M. S., and McCormick, F. (1993). Proteins regulating Ras and its relatives. *Nature (London)* **366,** 643–654.

Boise, L. H., Petryniak, B., Mao, X., June, C. H., Wang, C.-Y., Lindsten, T., Bravo, R., Kovary, K., Leiden, J. M., and Thompson, C. B. (1993). The NFAT-1 DNA binding complex in activated T cells contains Fra-1 and JunB. *Mol. Cell. Biol.* **13,** 1911–1919.

Bolen, J. (1994). In preparation.

Bonnefoy-Berard, N., Genestier, L., Flacher, M., and Revillard, J. P. (1994). The phosphoprotein phosphatase calcineurin controls calcium-dependent apoptosis in B cell lines. *Eur. J. Immunol.* **24,** 325–329.

Bowtell, D., Fu, P., Simon, M., and Senior, D. (1992). Identification of murine homologues of the *Drosophila son of sevenless* gene: Potential activators of Ras. *Proc. Natl. Acad. Sci. U. S. A.* **89,** 6511–6515.

Buday, L., and Downward, J. (1993). Epidermal growth factor regulates p21^{ras} through the formation of a complex of receptor, Grb2 adapter protein, and Sos nucleotide exchange factor. *Cell (Cambridge, Mass.)* **73,** 611–620.

Buday, L., Egan, S. E., Viciana, R., Cantrell, D. A., and Downward, J. (1994). A complex of Grb2 adaptor protein, Sos exchange factor, and a 36-kDa membrane-bound tyrosine phosphoprotein is implicated in Ras activation in T cells. *J. Biol. Chem.* **269,** 9019–9023.

Burkhardt, A. L., Brunswick, M., Bolen, J. B., and Mond, J. J. (1991). Anti-immunoglobulin stimulation of B lymphocytes activates src-related protein-tyrosine kinases. *Proc. Natl. Acad. Sci. U. S. A.* **88,** 7410–7414.

Burton, G. F., Conrad, D. H., Szakal, A. K., and Tew, J. G. (1993). Follicular dendritic cells and B cell costimulation. *J. Immunol.* **150,** 31–38.

Bustelo, X., and Barbacid, M. (1992). Tyrosine phosphorylation of the *vav* proto-oncogene product in activated B cells. *Science* **256,** 1196–1199.

Cambier, J. C. (1994). In preparation.

Cambier, J. C., Pleiman, C. M., and Clark, M. R. (1994). Signal transduction by the B cell antigen receptor and its co-receptors. *Annu. Rev. Immunol.* **12,** 457–486.

Cantley, L. C., Auger, K. R., Carpenter, C., Duckworth, B., Graziani, A., Kapeller, R., and Soltoff, S. (1991). Oncogenes and signal transduction. *Cell (Cambridge, Mass.)* **64,** 281–302.

Carpenter, C. L., Auger, K. R., Chanudhuri, M., Yoakim, M., Schaffhausen, B., Shoelson, S., and Cantley, L. C. (1993). Phosphoinositide 3-kinase is activated by phosphopeptides that bind to the SH2 domains of the 85-kDa subunit. *J. Biol. Chem.* **268,** 9478–9483.

Carter, R. H., and Fearon, D. T. (1992). Lowering the threshold for antigen receptor stimulation of B lymphocytes. *Science* **256,** 105–107

Carter, R. H., Park, D. J., Rhee, S. G., and Fearon, D. T. (1991a). Tyrosine phosphorylation of phospholipase C induced by membrane immunoglobulin crosslinking in B lymphocytes. *Proc. Natl. Acad. Sci. U. S. A.* **88,** 2745–2749.

Carter, R. H., Tuveson, D. A., Park, D. J., Rhee, S. G., and Fearon, D. T. (1991b). The CD19 complex of B lymphocytes. Activation of phospholipase C by a protein tyrosine kinase-dependent pathway that can be enhanced by the membrane IgM complex. *J. Immunol.* **147,** 3663–3671.

Casillas, A., Hanekom, C., Williams, K., Katz, R., and Nel, A. E. (1991). Stimulation of B-cells via the membrane immunoglobulin receptor of with phorbol myristate 13-acetate induces tyrosine phosphorylation and activation of a 42-kDa microtubule-associated protein-2 kinase. *J. Biol. Chem.* **266,** 19088–19094.

Casten, L. A., Lakey, E. A., Jelachich, M. L., Margoliash, E., and Pierce, S. K. (1985). Anti-immunoglobulin augments the B-cell antigen-presentation function independently of receptor-antigen complex. *Proc. Natl. Acad. Sci. U. S. A.* **82,** 5890–5894.

Chalupny, N. J., Kanner, S. B., Schieven, G. L., Wee, S. F., Gilliland, L. K., Aruffo, A., and Ledbetter, J. A. (1993). Tyrosine phosphorylation of CD19 in pre-B and mature B cells. *EMBO J.* **12,** 2691–2696.

Chan, A. C., Iwashima, M., Turck, C. W., and Weiss, A. (1992). ZAP-70: A 70 kd protein-tyrosine kinase that is associated with the TCR zeta chain. *Cell (Cambridge, Mass.)* **71,** 649–662.

Chan, A. C., Desai, D. M., and Weiss, A. (1994a). The role of protein tyrosine kinases and protein tyrosine phosphatases in T cell antigen receptor signal transduction. *Annu. Rev. Immunol.* **12,** 555–592.

Chan, A. C., Kadlacek, T. A., Elder, M. E., Filipovich, A. H., Kuo, W.-L., Iwashima, M., Parslow, T. G., and Weiss, A. (1994b). ZAP-70 deficiency in an autosomal recessive form of severe combined immunodeficiency. *Science* **264,** 1599–1601.

Chen, P.-L., Scully, P., Shew, J.-Y., Wang, J. Y.-J., and Lee, W.-H. (1989). Phosphorylation of the retinoblastoma gene product is modulated during the cell cycle and cellular differentiation. *Cell (Cambridge, Mass.)* **58,** 1193–1198.

Chen, Z. Z., Coggeshall, K. M., and Cambier, J. C. (1986). Translocation of protein kinase C during membrane immunoglobulin-mediated transmembrane signaling in B lymphocytes. *J. Immunol.* **136,** 2300–2304.

Chiles, T. C., and Rothstein, T. L. (1992). Surface Ig receptor-induced nuclear AP-1-dependent gene expression in B lymphocytes. *J. Immunol.* **149,** 825–831.

Chiles, T. C., Liu, J., and Rothstein, T. L. (1991). Cross-linking of surface Ig receptors on murine B lymphocytes stimulates the expression of nuclear tetradecanoyl phorbol acetate-response element-binding proteins. *J. Immunol.* **146,** 1730–1735.

Chiu, R., Boyle, W. J., Meek, J., Smeal, T., Hunter, T., and Karin, M. (1988). The c-Fos protein interacts with c-Jun/AP-1 to stimulate transcription of AP-1 response genes. *Cell (Cambridge, Mass.)* **54,** 541–552.

Choi, M. S. K., Brines, R. D., Holman, M. J., and Klaus, G. G. B. (1994). Induction of NF-AT in normal B lymphocytes by anti-immunoglobulin or CD40 ligand in conjunction with IL-4. *Immunity* **1,** 179–187.

Chow, L. M. L., Fournel, M., Davidson, D., and Veillette, A. (1993). Negative regulation of T-cell receptor signalling by tyrosine protein kinase $p50^{csk}$. *Nature (London)* **365,** 156–160.

Chung, J., Grammer, T. C., Lemon, K. P., Kazlauskas, and Blenis, J. (1994). PDGF- and insulin-dependent $pp70^{S6k}$ activation mediated by phosphatidylinositol-3-OH kinase. *Nature (London)* **370,** 71–75.

Clark, E. A., and Ledbetter, J. A. (1994). How B and T cells talk to each other. *Nature (London)* **367,** 425–428.

Clark, M. R., Campbell, K. S., Kazlauskas, A., Johnson, S. A., Hertz, M., Potter, T. A., Pleiman, C., and Cambier, J. C. (1992). The B cell antigen receptor complex-association of Ig-α and Ig-β with distinct cytoplasmic effectors. *Science* **258,** 123–126.

Clark, M. R., Johnson, S. A., and Cambier, J. C. (1994). Analysis of Ig-α-tyrosine kinase interaction reveals two levels of binding specificity and tyrosine phosphorylated Ig-α stimulation of Fyn activity. *EMBO J.* **13,** 1911–1919.

Clevers, H., Alarcon, B., Wileman, T., and Terhorst, C. (1988). The T cell receptor/CD3 complex: A dynamic protein ensemble. *Annu. Rev. Immunol.* **6,** 629–662.

Clipstone, N. A., and Crabtree, G. R. (1992). Identification of calcineurin as a key signaling enzyme in T-lymphocyte activation. *Nature (London)* **357,** 695–697.

Cohen, J. J., Duke, R. C., Fadok, V. A., and Sellins, K. S. (1992). Apoptosis and programmed cell death in immunity. *Annu. Rev. Immunol.* **10,** 267–293.

Cook, S. J., and McCormick, F. (1993). Inhibition by cAMP of Ras-dependent activation of Raf. *Science* **262,** 1069.

Cooke, M. P., Abraham, K. M., Forbush, K. A., and Perlmutter, R. M. (1991). Regulation of T cell receptor signaling by the *src* family protein-tyrosine kinase ($p59^{fyn}$). *Cell (Cambridge, Mass.)* **65,** 281–291.

Cooke, M. P., Heath, A. W., Shokat, K. M., Zeng, Y., Finkelman, F. D., Linsley, P. S., Howard, M., and Goodnow, C. C. (1994). Immunoglobulin signal transduction guides the specificity of B cell-T cell interactions and is blocked in tolerant self-reactive B cells. *J. Exp. Med.* **179,** 425–438.

Crabtree, G. R. (1989). Contingent genetic regulatory events in T lymphocyte activation. *Science* **243**, 355–361.

Cuende, E., Alcs-Martinez, J. E., Ding, L., Gonzales-Garcia, M., Martinez-A. C., and Nunez, G. (1993). Programmed cell death by *bcl-2*-dependent and independent mechanisms in B lymphoma cells. *EMBO J.* **12**, 1555–1560.

Dang, L. H. and Rock, K. L. (1991). Stimulation of B lymphocytes through surface Ig receptors induces LFA-1 and ICAM-1-dependent adhesion. *J. Immunol.* **146**, 3273–3279.

Darnell, J. E., Kerr, I. M., and Stark, G. R. (1994). Jak-STAT pathways and transcriptional activation in response to IFNs and other extracellular signaling proteins. *Science* **264**, 1415–1421.

Dasgupta, J. D., Granja, C., Druker, B., Lin, L-L., Yunis, E. J., and Relias, V. (1992). Phospholipase C-γ1 association with CD3 structure in T cells. *J. Exp. Med.* **175**, 285–288.

Davidson, D., Chow, L. M., Fournel, M., and Veillette, A. (1992). Differential regulation of T cell antigen responsiveness by isoforms of the *src*-related tyrosine protein kinase p59*fyn*. *J. Exp. Med.* **175**, 1483–1492.

Davis, R. J. (1993). The mitogen-activated protein kinase signal transduction pathway. *J. Biol. Chem.* **268**, 14553–14556.

DeFranco, A. L., Raveche, E. S., and Paul, W. E. (1985). Separate control of B lymphocyte early activation and proliferation in response to anti-IgM antibodies. *J. Immunol.* **135**, 87–94.

Delcayre, A. X., Salas, F., Mathur, S., Kovats, K., Lotz, M., and Lernhardt, W. (1991). Epstein-Barr virus/complement C3d receptor is an inferferon α receptor. *EMBO J.* **10**, 919–926.

Derijard, B., Hibi, M., Wu, I.-H., Barrett, T., Su, B., Deng, T., Karin, M., and Davis, R. J. (1994). JNK1: a protein kinase stimulated by UV light and Ha-Ras that binds and phosphorylates the c-Jun activation domain. *Cell (Cambridge, Mass.)* **76**, 1025–1037.

Desiderio, S. (1994). The B cell antigen receptor in B-cell development. *Curr. Opin. Immunol.* **6**, 248–256.

deWeers, M., Brouns, G. S., Hinshelwood, S., Kinnon, C., Schleurman, R. K. B., Hendriks, R. W., and Borst, J. (1994). B-cell antigen receptor stimulation activates the human Bruton's tyrosine kinase, which is deficient in x-linked agamma-globulinemia. *J. Biol. Chem.* **269**, 23857–23860.

Diaz Meco, M. T., Berra, E., Municio, M. M., Sanz, L., Lozano, J., Dominguez, I., Diaz-Golpe, V., Lain de Lera, M. T., Alcami, J., Paya, C. V., Arenzana-Seisdedos, F., Virelizier, J.-L., and Moscat, J. (1993). A dominant negative protein kinase C ζ subspecies blocks NF-κB activation. *Mol. Cell. Biol.* **13**, 4770–4775.

Doherty, P. C. (1993). Cell-mediated cytotoxicity. *Cell (Cambridge, Mass.)* **75**, 607–612.

Downward, J. (1992). Regulatory mechanisms for *ras* proteins. *BioEssays* **14**, 177–182.

Downward, J., Graves, J. D., Warne, P. H. Rayter, S., and Cantrell, D. A. (1990). Stimulation of p21*ras* upon T-cell activation. *Nature (London)* **346**, 719–723.

Downward, J., Graves, J., and Cantrell, D. A. (1992). The regulation and function of p21*ras* in T cells. *Immunol. Today* **13**, 89–92.

Duchesne, M., Schweighoffer, F., Parker, F., Clerc, F., Frobert, Y., Thang, M. N., and Tocque, B. (1993). Identification of the SH3 domain of GAP as an essential sequence for Ras-GAP-mediated signaling. *Science* **259**, 525–528.

Duplay, P., Thome, M., Hervé, F., and Acuto, O. (1994). p56*lck* interacts via its src homology 2 domain with the ZAP-70 kinase. *J. Exp. Med.* **179**, 1163–1172.

Durand, D. B., Shaw, J.-P., Bush, M. R., Replogle, R. E., Belagaje, R., and Crabtree, G. R. (1988). Characterization of antigen receptor elements within the interleukin-2 enhancer. *Mol. Cell. Biol.* **8**, 1715–1724.

Duyao, M. P., Buckler, A. J., and Sonenshein, G. E. (1990). Interaction of an NF-κB-like factor with a site upstream of the c-*myc* promoter. *Proc. Natl. Acad. Sci. U.S.A.* **87**, 4727–4731.

Evan, G. I., and Littlewood, T. D. (1993). The role of c-*myc* in cell growth. *Curr. Opin. Genet. Dev.* **3**, 44–49.

Fantl, W. J., Escobedo, J. A., Martin G. A., Turck, C. W., del Rosario, M., McCormick, F., and Williams, L. T. (1992). Distinct phosphotyrosines on a growth factor receptor bind to specific molecules that mediate different signaling pathways. *Cell (Cambridge, Mass.)* **69**, 413–423.

Fath, I., Schweighoffer, F., Rey, I., Multon, M.-C., Boiziau, J., Duchesne, M., and Tocque, B. (1994). Cloning of a Grb2 isoform with apoptotic properties. *Science* **264**, 971–974.

Fearon, D. T. (1993). The CD19-CR2-TAPA-1 complex, CD45 and signaling by the antigen receptor of B lymphocytes. *Curr. Opin. Immunol.* **5**, 341–348.

Feig, L. A., and Cooper, G. M. (1988). Inhibition of NIH 3T3 cell proliferation by a mutant *ras* protein with preferential affinity for GDP. *Mol. Cell. Biol.* **8**, 3235–3243.

Fischer, G., Kent, S. C., Joseph, L., Green, D. R., and Scott, D. W. (1994). Lymphoma models for B cell activation and tolerance. X. Anti-μ-mediated growth arrest and apoptosis of murine B lymphomas is prevented by the stabilization of myc. *J. Exp. Med.* **179**, 221–228.

Fisher, C. L., Ghysdael, J., and Cambier, J. C. (1991). Ligation of membrane Ig leads to calcium-mediated phosphorylation of the proto-oncogene product, Ets-1. *J. Immunol.* **146**, 1743–1749.

Flanagan, W. M., Corthesy, B., Bram, R. J., and Crabtree, G. R. (1991). Nuclear association of a T-cell transcription factor blocked by FK-506 and cyclosporin A. *Nature (London)* **352**, 803–807.

Flaswinkel, H., and Reth, M. (1994). Dual role of the tyrosine activation motif of the Ig-α protein during signal transduction via the B cell antigen receptor. *EMBO J.* **13**, 83–89.

Fraser, J. D., Straus, D., and Weiss, A. (1993). Signal transduction events leading to T-cell lymphokine gene expression. *Immunol. Today* **14**, 357–362.

Frech, M., John, J., Pizon, V., Chardin, P., Tavitian, A., Clark, R., McCormick, F., and Wittinghofer, A. (1990). Inhibition of GTPase activating protein stimulation of Ras-p21 GTPase by the Krev-1 gene product. *Science* **249**, 169–171.

Freeman, G. J., Freedman, A. S., Segil, J. M., Lee, G., Whitman, J. F., and Nadler, L. M. (1989). B7, a new member of the Ig superfamily with unique expression on activated and neoplastic B cells. *J. Immunol.* **143**, 2714–2722.

Freeman, G. J., Gribben, J., Boussiotis, V., Ng, J., Restivo, V., Lombard, L., Gray, G., and Nadler, L. M. (1993). Cloning of B7-2: A CTLA-4 counter receptor that costimulates human T cell proliferation. *Science* **262**, 909–911.

Fruman, D. A., Mather, P. E., Burakoff, S. J., and Bierer, B. E. (1992). Correlation of calcineurin phosphatase activity and programmed cell death in murine T cell hybridomas. *Eur. J. Immunol.* **22**, 2513–2517.

Fulcher, D. A., and Basten, A. (1994). Reduced life span of anergic self-reactive B cells in a double-transgenic model. *J. Exp. Med.* **179**, 125–134.

Germain, R. N. (1994). MHC-dependent antigen processing and peptide presentation: Providing ligands for T lymphocyte activation. *Cell (Cambridge, Mass.)* **76**, 287–299.

Gibbs, J. B., Sigal, I. S., Poe, M., and Scolnick, E. M. (1984). Intrinsic GTPase activity distinguishes normal and oncogenic *ras* p21 molecules. *Proc. Natl. Acad. Sci. U. S. A.* **81**, 5704–5708.

Gold, M. R., and Aebersold, R. (1994). Both phosphatidylinositol 3-kinase and phosphatidyl-inositol 4-kinase products are increased by antigen receptor signaling in B lymphocytes. *J. Immunol.* **152**, 42–50.

Gold, M. R., and DeFranco, A. L. (1994). Biochemistry of B lymphocyte activation. *Adv. Immunol.* **55,** 221–295.

Gold, M. R., Matsuuchi, L., Kelly, R. B., and DeFranco, A. L. (1991). Tyrosine phosphorylation of components of the B cell antigen receptor following receptor crosslinking. *Proc. Natl. Acad. Sci. U. S. A.* **88,** 3436–3440.

Gold, M. R., Sanghera, J. S., Stewart, J., and Pelech, S. L. (1992). Selective activation of p42 MAP kinase in murine B lymphoma cell lines by membrane immunoglobulin crosslinking. Evidence for protein kinase C-independent and -dependent mechanisms of activation. *Biochem. J.* **287,** 269–276.

Gold, M. R., Crowley, M. T., Martin, G. A., McCormick, F., and DeFranco, A. L. (1993). Targets of B lymphocyte antigen receptor signal transduction include the p21^{ras} GTPase activating protein (GAP) and two GAP-associated proteins. *J. Immunol.* **150,** 377–386.

Gold, M. R., Chiu, R., Ingham, R. J., Saxton, T. M., van Oostveen, I., Watts, J. D., Affolter, M., and Aebersold, R. (1994a). Activation and serine phosphorylation of the p56^{lck} protein-tyrosine kinase in response to antigen receptor cross-linking in B lymphocytes. *J. Immunol.* **153,** 2369–2380.

Gold, M. R., Duronio, V., Saxena, S. P., Schrader, J. W., and Aebersold, R. (1994b). Multiple cytokines activate phosphatidylinositol 3-kinase in hemopoietic cell lines. Association of the enzyme with various tyrosine-phosphorylated proteins. *J. Biol. Chem.* **269,** 5403–5412.

Gold, M. R., Affolter, M., and Aebersold, R. (1994c). In preparation.

Goldberg, A. L., and Rock, K. L. (1992). Proteolysis, proteosomes, and antigen presentation. *Nature (London)* **357,** 375–379.

Goldschmidt-Clermont, P. J., Kim, J. W., Machesky, L. M., Rhee, S. G., and Pollard, T. D. (1991). Regulation of phospholipase C-γ1 by profilin and tyrosine phosphorylation. *Science* **251,** 1231–1233.

Goodnow, C. C., Crosbie, J., Adelstein, S., Lavoie, T. B., Smith-Gill, S. J., Brink, R. A., Pritchard-Briscoe, H., Wotherspoon, J. S., Loblay, R. H., Raphael, K., Trent, R. J., and Basten, A. (1988). Altered immunoglobulin expression and functional silencing of self-reactive B lymphocytes in transgenic mice. *Nature (London)* **334,** 676–682.

Gout, I., Dhand, R., Hiles, I. D., Fry, M. J., Panayotou, G., Das, P., Truong, O., Totty, N. F., Hsuan, J., Booker, G. W., Campbell, I. D., and Waterfield, M. D. (1993). The GTPase dynamin binds to and is activated by a subset of SH3 domains. *Cell (Cambridge, Mass.)* **75,** 25–36.

Groettrup, M., and von Boehmer, H. (1993). A role for a pre-T-cell receptor in T-cell development. *Immunol. Today* **14,** 610–614.

Groettrup, M., Ungewiss, K., Azogui, O., Palacios, R., Owen, M. J., Hayday, A. C., and von Boehmer, H. (1993). A novel disulfide-linked heterodimer on pre-T cells consists of the T cell receptor β chain and a 33 kd glycoprotein. *Cell (Cambridge, Mass.)* **75,** 283–294.

Gulbins, E., Coggeshall, K. M., Baier, G., Katzav, S., Burn, P., and Altman, A. (1993). Tyrosine kinase-stimulated guanine nucleotide exchange activity of Vav in T cell activation. *Nature (London)* **260,** 822–825.

Gulbins, E., Langlet, C., Baier, G., Bonnefoy-Berard, N., Herbert, E., Altman, A., and Coggeshall, K. M. (1994). Tyrosine phosphorylation and activation of Vav GTP/GDP exchange activity in antigen receptor-triggered B cells. *J. Immunol.* **152,** 2123–2129.

Haas, W., and Tonegawa, S. (1992). Development and selection of γδ T cells. *Curr. Opin. Immunol.* **4,** 147–155.

Hagag, N., Halegoua, S., and Viola, M. (1986). Inhibition of growth factor-induced differentiation of PC12 cells by microinjection of antibody to *ras* p21. *Nature (London)* **319,** 680–682.

Harding, C. V., and Unanue, E. (1990). Quantitation of antigen presenting cell MHC class II/peptide complexes necessary for T cell stimulation. *Nature (London)* **346,** 574–576.

Hartley, S. B., Cooke, M. P., Fulcher, D. A., Harris, A. W., Cory, S., Basten, A., and

Goodnow, C. C. (1993). Elimination of self-reactive B lymphocytes proceeds in two stages: Arrested development and cell death. *Cell (Cambridge, Mass.)* **72,** 325–335.

Harwood, A. E., and Cambier, J. C. (1993). B cell antigen receptor cross-linking triggers rapid PKC independent activation of p21^{ras}. *J. Immunol.* **151,** 4513–4522.

Hasbold, J., and Klaus, G. G. B. (1990). Anti-immunoglobulin antibodies induce apoptosis in immature B cell lymphomas. *Eur. J. Immunol.* **20,** 1685–1690.

Haughn, L., Gratton, S., Caron, L., Sekaly, R.-P., Veillette, A., and Julius, M. (1992). Association of tyrosine kinase p56^{lck} with CD4 inhibits the induction of growth through the $\alpha\beta$ T-cell receptor. *Nature (London)* **358,** 328–331.

Hebell, T., Ahearn, J. M., and Fearon, D. T. (1991). Suppression of the immune response by a soluble complement receptor of B lymphocytes. *Science* **254,** 102–105.

Heikkila, R., Schwab, G., Wickström, E., Loke, S. L., Pluznik, D. H., Watt, R., and Neckers, L. M. (1987). A c-*myc* antisense oligonucleotide inhibits entry into S phase but not progress from G_0 to G_1. *Nature (London)* **328,** 445–449.

Hempel, W. M., Schatzman, R. C., and DeFranco, A. L. (1992). Tyrosine phosphorylation of phospholipase C-γ2 upon crosslinking of membrane Ig on murine B lymphocytes. *J. Immunol.* **148,** 3021–3027.

Hinds, P. W., Mittnacht, S., Dulic, V., Arnold, A., Reed, S. I., and Weinberg, R. A. (1992). Regulation of retinoblastoma protein functions by ectopic expression of human cyclins. *Cell (Cambridge, Mass.)* **70,** 993–1006.

Ho, I.-C., Bhat, N. K., Gottschalk, L. R., Lindsten, T., Thompson, C. B., Papas, T. S., and Leiden, J. M. (1990). Sequence-specific binding of human Ets-1 to the T cell receptor α gene enhancer. *Science* **250,** 814–818.

Hollenbaugh, D., Grosmaire, L. S., Kullas, C. D., Chalupny, N. J., Braesch-Andersen, S., Noelle, R. J., Stamenkovic, I., Ledbetter, J. A., and Aruffo, A. (1992). The human T cell antigen gp39, a member of the TNF gene family, is a ligand for the CD40 receptor: Expression of a soluble form of gp39 with B cell co-stimulatory activity. *EMBO J.* **11,** 4313–4321.

Hunter, T., and Karin, M. (1992). The regulation of transcription by phosphorylation. *Cell (Cambridge, Mass.)* **70,** 375–387.

Hutchcroft, J. E., Harrison, M. L., and Geahlen, R. L. (1992). Association of the 72 kDa protein-tyrosine kinase PTK72 with the B cell antigen receptor. *J. Biol. Chem.* **267,** 8613–8619.

Iglesias, A., Kopf, M., Williams, G. S., Buhler, B., and Kohler, G. (1991). Molecular requirements for the μ-induced light chain rearrangement in pre-B cells. *EMBO J.***10,** 2147–2156.

Imboden, J. B., and Stobo, J. D. (1985). Transmembrane signaling by the T cell antigen receptor. Perturbation of the T3-antigen receptor complex generates inositol phosphates and releases calcium ions from intracellular stores. *J. Exp. Med.* **161,** 446.

Irving, B. A., and Weiss, A. (1991). The cytoplasmic domain of the T cell receptor ζ chain is sufficient to couple to receptor-associated signal transduction pathways. *Cell (Cambridge, Mass.)* **64,** 891–901.

Irving, B. A., Chan, A. C., and Weiss, A. (1993). Functional characterization of a signal transducing motif present in the T cell antigen receptor ζ chain. *J. Exp. Med.* **77,** 1093–1103.

Iwashima, M., Irving, B. A., van Oers, N. S. C., Chan, A. C., and Weiss, A. (1994). Sequential interactions of the TCR with two distinct cytoplasmic tyrosine kinases. *Science* **263,** 1136–1139.

Jacob, J., Kelsoe, G., Rajewsky, K., and Weiss, U. (1991). Intraclonal generation of antibody mutants in germinal centers. *Nature (London)* **354,** 389–392.

Jain, J., McCaffrey, P. G., Miner, Z., Kerppola, T. K., Lambert, J. N., Verdine, G. L., Curran, T., and Rao, A. (1993). The T-cell transcription factor NFATp is a substrate for calcineurin and interacts with Fos and Jun. *Nature (London)* **365,** 352–355.

Janeway, C. A. (1992). The T cell receptor as a multicomponent signaling machine: CD4/CD8 coreceptors and CD45 in T cell activation. *Annu. Rev. Immunol.* **10**, 645–674.

Jenkins, M. K., and Johnson, J. G. (1993). Molecules involved in T-cell costimulation. *Curr. Opin. Immunol.* **5**, 361–367.

Jin, Y.-J., Clayton, L. K., Howard, F. D., Koyasu, S., Sieh, M., Steinbrich, R., Tarr, G. E., and Reinherz, E. L. (1990). Molecular cloning of the CD3η subunit identifies a CD3ζ-related product in thymus-derived cells. *Proc. Natl. Acad. Sci. U. S. A.* **87**, 3319–3323.

Johnson, D. G., Schwarz, J. K., Cress, W. D., and Nevins, J. R. (1993). Expression of transcription factor E2F1 induces quiescent cells to enter S phase. *Nature (London)* **365**, 349–352.

Joly, M., Kazlauskas, A., Fay, F. S., and Corvera, S. (1994). Disruption of PDGF receptor trafficking by mutation of its PI-3 kinase binding sites. *Science* **263**, 684–687.

Jorgensen, J. L., Reay, P. A., Ehrich, E. W., and Davis, M. M. (1992). Molecular components of T-cell recognition. *Annu. Rev. Immunol.* **10**, 835–873.

June, C. H., Fletcher, M. C., Ledbetter, J. A., Schieven, G. L., Siegel, J. N., Phillips, A. F., and Samelson, L. E. (1990a). Inhibition of tyrosine phosphorylation prevents T cell receptor-mediated signal transduction. *Proc. Natl. Acad. Sci. U. S. A.* **87**, 7722–7726.

June, C. H., Ledbetter, J. A., Linsley, P. A., and Thompson, C. B. (1990b). Role of the CD28 receptor in T-cell activation. *Immunol. Today* **11**, 211–216.

Justement, L. B., Campbell, K. S., Chien, N. C., and Cambier, J. C. (1991). Regulation of B cell antigen receptor signal transduction and phosphorylation by CD45. *Science* **252**, 1839–1842.

Kang, S. M., Beverly, B., Tran, A. C., Brorson, K., Schwartz, R. H., and Lenardo, M. J. (1992). Transactivation by AP-1 is a molecular target of T cell clonal anergy. *Science* **257**, 1134–1138.

Kang, S. M., Tran, A. C., Grilli, M., and Lenardo, M. J. (1992). NF-κB subunit regulation in nontransformed CD4+ T lymphocytes. *Science* **256**, 1452–1456.

Kapeller, R., Chakrabarti, R., Cantley, L., Fay, F., and Corvera, S. (1993). Internalization of activated platelet-derived growth factor receptor-phosphatidylinositol-3' kinase complexes: Potential interactions with the microtubule cytoskeleton. *Mol. Cell. Biol.* **13**, 6052–6063.

Karnitz, L., Sutor, S. L., Torigoe, T., Reed, J. C., Bell, M. P., McKean, D. J., Leibson, P. J., and Abraham, R. T. (1992). Effects of p56^{lck} deficiency on the growth and cytolytic function of an interleukin-2-dependent cytotoxic T-cell line. *Mol. Cell. Biol.* **12**, 4521–4530.

Kawauchi, K., Lazarus, A. H., Rapoport, M. J., Harwood, A., Cambier, J. C., and Delovitch, T. L. (1994). Tyrosine kinase and CD45 tyrosine phosphatase activity mediate p21^{ras} activation in B cells stimulated through the antigen receptor. *J. Immunol.* **152**, 3306–3316.

Keegan, A., and Paul, W. E. (1993). Multichain immune recognition receptors: Similarities in structure and signaling pathways. *Immunol. Today* **14**, 111.

Kelly, K., Cochran, B. H., Stiles, C. D., and Leder, P. (1983). Cell-specific regulation of the c-*myc* gene by lymphocyte mitogens and platelet-derived growth factor. *Cell (Cambridge, Mass.)* **35**, 603–610.

Kim, K.-M., Alber, G., Weiser, P., and Reth, M. (1993a). Signaling function of the B cell antigen receptor. *Immunol. Rev.* **132**, 125–146.

Kim, K.-M., Albert, G., Weiser, P., and Reth, M. (1993b). Differential signaling through the Ig-α and Ig-β components of the B cell antigen receptor. *Eur. J. Immunol.* **23**, 911–916.

Kishimoto, T. (1985). Factors affecting B-cell growth and differentiation. *Annu. Rev. Immunol.* **3**, 133–157.

Kitamura, D., Kudo, A., Schall, S., Muller, W., Melchers, F., and Rajewsky, K. (1992). A critical role of λ5 protein in B cell development. *Cell (Cambridge, Mass.)* **69**, 823–831.

Klausner, R. D., Lippincott-Schwartz, J., and Bonifacino, J. S. (1990). The T cell antigen receptor: Insights into organelle biology. *Annu. Rev. Cell Biol.* **6**, 403–431.

Koide, H., Satoh, T., Nakafuku, M., and Kaziro, Y. (1993). GTP-dependent association of Raf-1 with Ha-Ras: Identification of Raf as a target downstream of Ras in mammalian cells. *Proc. Natl. Acad. Sci. U. S. A.* **90**, 8683–8686.

Koretzky, G., Picus, J., Schultz, T., and Weiss, A. (1991). Tyrosine phosphatase CD45 is required for both T cell antigen receptor and CD2 mediated activation of a protein tyrosine kinase and interleukin 2 production. *Proc. Natl. Acad. Sci. U. S. A.* **88**, 2037–2041.

Korn, L. J., Siebel, C. W., McCormick, F., and Roth, R. A. (1987). Ras p21 as a potential mediator of insulin action in *Xenopus* oocytes. *Science* **236**, 840.

Kronenberg, M. (1994). Antigens recognized by γδ T cells. *Curr. Opin. Immunol.* **6**, 64–71.

Kumagai, N., Benedict, S. H., Mills, G. B., and Gelfand, E. W. (1988). Induction of competence and progression signals in human T lymphocytes by phorbol esters and calcium ionophores. *J. Cell. Physiol.* **137**, 329.

Kurosaki, T., Takata, M., Yamanashi, Y., Inazu, T., Taniguchi, T., Yamamoto, T., and Yamamura, H. (1994). Syk activation by the Src-family tyrosine kinase in the B cell receptor signaling. *J. Exp. Med.* **179**, 1725–1729.

Kyriakis, J. M., App, H., Zhang, X.-F., Banerjee, P., Brautigan, D. L., Rapp, U. R., and Avruch, J. (1992). Raf-1 activates MAP kinase-kinase. *Nature (London)* **358**, 417–421.

Lane, P., Brocker, T., Hubele, S., Padovan, E., Lanzavecchia, A., and McConnell, F. (1993). Soluble CD40 ligand can replace normal T cell-derived CD40 ligand signal to B cells in T cell-dependent activation. *J. Exp. Med.* **177**, 1209–1213.

Lange-Carter, C. A., Pleiman, C. M., Gardner, A. M., Blumer, K. J., and Johnson, G. L. (1993). A divergence in the MAP kinase regulatory network defined by MEK kinase and Raf. *Science* **260**, 315–319.

Lanzavecchia, A. (1990). Receptor-mediated antigen uptake and its effect on antigen presentation to class II restricted T lymphocytes. *Annu. Rev. Immunol.* **8**, 773–793.

Larner, A. C., David, M., Feldman, G. M., Igarashi, K.-I., Hackett, R. H., Webb, D. S. A., Sweitzer, S. M., Petricoin, E. F. I., and Finbloom, D. S. (1993). Tyrosine phosphorylation of DNA binding proteins by multiple cytokines. *Science* **261**, 1730–1733.

Law, D. A., Gold, M. R., and DeFranco, A. L. (1992). Examination of B lymphoid cell lines for membrane immunoglobulin-stimulated tyrosine phosphorylation and *src*-family tyrosine kinase mRNA expression. *Mol. Immunol.* **29**, 917–926.

Law, D. A., Chan, V. W.-F., Datta, S. K., and DeFranco, A. L. (1993). B-cell antigen receptor motifs have redundant signalling capabilities and bind the tyrosine kinases PTK72, Lyn, and Fyn. *Curr. Biol.* **3**, 645–657.

Lazarus, A. H., Kawauchi, K., Rapoport, M. J., and Delovitch, T. J. (1993). Antigen-induced B lymphocyte activation involves the p21^{ras} and ras.GAP signaling pathway. *J. Exp. Med.* **178**, 1765–1769.

Leevers, S. J., and Marshall, C. J. (1992). Activation of extracellular signal-regulated kinase, ERK2, by p21*ras* oncoprotein. *EMBO J.* **11**, 569–574.

Lenardo, M., and Baltimore, D. (1989). NF-κB: A pleiotropic mediator of inducible and tissue-specific gene control. *Cell (Cambridge, Mass.)* **58**, 227–229.

Leprince, C., Draves, K. E., Geahlen, R. L., Ledbetter, J. A., and Clark, E. A. (1993). CD22 associates with the human surface IgM-B-cell antigen receptor complex. *Proc. Natl. Acad. Sci. U. S. A.* **90**, 3236–3240.

Letourneur, F., and Klausner, R. D. (1992). Activation of T cells by tyrosine kinase activation domain in cytoplasmic tail of CD3ε. *Science* **255**, 79–82.

Li, Z.-H., Mahajan, S., Prendergast, M. M., Faargnoli, J., Zhu, X., Klages, S., Adam, D.,

Schieven, G. L., Blake, J., Bolen, J. B., and Burkhardt, A. L. (1992). Cross-linking of surface immunoglobulin activates *src*-related tyrosine kinases in WEHI 231 cells. *Biochem. Biophys. Res. Commun.* **187**, 1536–1544.

Lin, J., and Justement, L. B. (1992). The MB-1/B29 heterodimer couples the B cell antigen receptor to multiple *src* family protein tyrosine kinases. *J. Immunol.* **149**, 1548–1555.

Linsley, P. S., Brady, W., Urnes, M., Grosmaire, L., Damle, N., and Ledbetter, J. (1991). CTLA-4 is a second receptor for the B cell activation antigen B7. *J. Exp. Med.* **174**, 561–569.

Liscovitch, M., and Cantley, L. C. (1994). Lipid second messengers. *Cell (Cambridge, Mass.)* **77**, 329–334.

Liu, J., Chiles, T. C., Sen, R., and Rothstein, T. L. (1991a). Inducible nuclear expression of NF-κB in primary B cells stimulated through the surface Ig receptor. *J. Immunol.* **146**, 1685–1691.

Liu, J., Farmer, J. D., Lane, W. S., Friedman, J., Weissman, I., and Schreiber, S. L. (1991b). Calcineurin is a common target of cyclophilin-cyclosporin A and FKBP-FK506 complexes. *Cell (Cambridge, Mass.)* **66**, 807–815.

Lin, X., Brodeur, S. R., Gish, G., Songyang, Z., Cantley, L. C., Laudano, P., and Pawson, T. (1993). Regulation of c-Src tyrosine kinase activity by the Src SH2 domain. *Oncogene* **5**, 921–923.

Liu, Y.-J., Joshua, D., Williams G., Smith, C., Gordon, J., and MacLennan, I. (1989). Mechanism of antigen-driven selection in germinal centres. *Nature (London)* **342**, 929–931.

Liu, Y.-J., Mason, D. Y., Johnson, G. D., Abbot, S., Gregory, C. D., Hardie, D. L., Gordon, J., and MacLennan, I. C. M. (1991). Germinal center cells express Bcl-2 protein after activation by signals which prevent their entry into apoptosis. *Eur. J. Immunol.* **21**, 1905–1910.

Liu, Z.-G., Smith, S. M., McLaughlin, K. A., Schwartz, L. M., and Osborne, B. A. (1994). Apoptotic signals delivered through the T-cell receptor of a T-cell hybrid require the immediate-early gene *nur77*. *Nature (London)* **367**, 281–284.

Lotteau, V., Teyton, L., Peleraux, A., Nilsson, T., Karlsson, L., Schmid, S. L., Quaranta, V., and Peterson, P. A. (1990). Intracellular transport of class II MHC molecules directed by invariant chain. *Nature (London)* **348**, 600–605.

MacLennan, I. C. M. (1994). Germinal centers. *Annu. Rev. Immunol.* **12**, 117–139.

Manger, B., Weiss, A., Imboden, J., Laing, T., and Stobo, J. D. (1987). The role of protein kinase C in transmembrane signaling by the T cell antigen receptor complex. Effects of stimulation with soluble or immobilized anti-CD3 antibodies. *J. Immunol.* **139**, 2755–2760.

Manolios, N., Letourneur, F., Bonifacino, J. S., and Klausner, R. D. (1991). Pairwise, cooperative, and inhibitory interactions describe the assembly and probable structure of the T-cell antigen receptor. *EMBO J.* **10**, 1643–1651.

Manz, J., Dennis, K., Witte, O., Brinster, R., and Storb, U. (1988). Feedback inhibition of immunoglobulin gene rearrangement by membrane μ but not secreted μ heavy chains. *J. Exp. Med.* **168**, 1363–1381.

Matsushime, H., Roussel, M. F., Ashmun, R. A., and Sherr, C. J. (1991). Colony-stimulating factor 1 regulates novel cyclins during the G1 phase of the cell cycle. *Cell (Cambridge, Mass.)* **65**, 701–713.

Matsuuchi, L., Gold, M. R., Travis, A., Grosschedl, R., DeFranco, A. L., and Kelly, R. B. (1992). The membrane IgM-associated proteins MB-1 and Ig-β are sufficient to promote surface expression of a partially functional B-cell antigen receptor in a non-lymphoid cell line. *Proc. Natl. Acad. Sci. U.S.A.* **89**, 3404–3408.

McCormack, J. E., Pepe, V. H., Kent, R. B., Dean, M., Marshall-Rothstein, A., and Sonenshein, G. E. (1984). Specific regulation of c-*myc* oncogene expression in a murine B-cell lymphoma. *Proc. Natl. Acad. Sci. U.S.A.* **81**, 5546–5550.

Meek, D. W., and Street, A. J. (1992). Nuclear protein phosphorylation and growth control. *Biochem. J.* **287,** 1–15.

Mercep, M., Noguchi, P. D., and Ashwell, J. D. (1989). The cell cycle block and lysis of an activated T cell hybridoma are distinct processes with different Ca^{2+} requirements and sensitivity to cyclosporine A. *J. Immunol.* **142,** 4085–4092.

Meuer, S. C., Hussey, R. C., Cantrell, D. A., Hodgdon, J. C., Schlossman, S. F., Smith, K. A., and Reinherz, E. L. (1984). Triggering of the T3-Ti antigen-receptor complex results in clonal T-cell proliferation through an interleukin 2-dependent autocrine pathway. *Proc. Natl. Acad. Sci. U.S.A.* **81,** 1509–1513.

Miki, H., Miura, K., Matuoka, K., Nakata, T., Hirokawa, N., Prota, S., Kaibuchi, K., Takai, Y., and Takenawa, T. (1994). Association of Ash/Grb-2 with dynamin through the Src homology 3 domain. *J. Biol. Chem.* **269,** 5489–5492.

Milne, D. M., Campbell, D. G., Caudwell, F. B., and Meek, D. W. (1994). Prosphorylation of the tumor suppressor protein p53 by mitogen-activated protein kinases. *J. Biol. Chem.* **269,** 9253–9260.

Mittelstadt, P. R., and DeFranco, A. L. (1993). Induction of early response genes by cross-linking membrane Ig on B lymphocytes. *J. Immunol.* **150,** 4822–4832.

Molina, T. J., Kishihara, K., Siderovski, D. P., van Ewjik, W., Narendran, A., Timms, E., Wakeham, A., Paige, C. J., Hartmann, K.-U., Veillette, A., Davison, D., and Mak, T. W. (1992). Profound block in thymocyte development in mice lacking p56^{lck}. *Nature (London)* **357,** 161–164.

Mond, J. J., Seghal, E., Kung, J., and Finkelman, F. D. (1981). Increased expression of I-region associated antigen (Ia) on B cells after crosslinking of surface immuoglobulin. *J. Immunol.* **127,** 881–888.

Monroe, J. G. (1988). Up-regulation of c-*fos* expression is a component of the mIg signal transduction mechanism but is not indicative of competence for proliferation. *J. Immunol.* **140,** 1454–1460.

Moran, M. F., Polakis, P., McCormick, F., Pawson, T., and Ellis, C. (1991). Protein-tyrosine kinases regulate the phosphorylation, protein interactions, subcellular distribution, and activity of p21^{ras} GTPase-activating protein. *Mol. Cell. Biol.* **11,** 1804–1812.

Mulcahy, L. S., Smith, M. R., and Stacey, D. W. (1985). Requirement for ras proto-oncogene function during serum-stimulated growth of NIH 3T3 cells. *Nature (London)* **313,** 241–248.

Murphy, T. P., Kolber, D. L., and Rothstein, T. L. (1990). Elevated expression of pgp-1, Ly-24, by murine peritoneal B lymphocytes. *Eur. J. Immunol.* **20,** 1137.

Mustelin, T., Coggeshall, K. M., and Altman, A. (1989). Rapid activation of the T-cell tyrosine protein kinase pp56^{lck} by the CD45 phosphotyrosine phosphatase. *Proc. Natl. Acad. Sci. U. S. A.* **86,** 6302–6306.

Mustelin, T., Pessa-Morikawa, T., Autero, M., Gahmberg, C. G., and Burn, P. (1992). Regulation of the p59^{fyn} protien tyrosine kinase by the CD45 phosphotyrosine phosphatase. *Eur. J. Immunol.* **22,** 1173–1178.

Nakanishi, H., and Exton, J. H. (1992). Purification and characterization of the ζ isoform of protein kinase C from bovine kidney. *J. Biol. Chem.* **267,** 16347–16354.

Nakanishi, H., Brewer, K. A., and Exton, J. H. (1993). Activation of the ζ isozyme of protein kinase C by phosphatidylinositol 3,4,5-trisphosphate. *J. Biol. Chem.* **268,** 13–16.

Nel, A. E., Wooten, M. W., Landreth, G. E., Goldschmidt-Clemont, P. J., Stevenson, H. C., Miller, P. J., and Galbraith, R. M. (1986). Translocation of phospholipid/Ca^{2+}-dependent protein kinase in B-lymphocytes activated by phorbol ester or cross-linking membrane immunoglobulin. *Biochem. J.* **233,** 145–149.

Nel, A. E., Hanekom, C., Rheeder, A., Williams, K., Pollack, S., Katz, R., and Landreth, G. (1990). Stimulation of MAP-2 kinase activity in T lymphocytes by anti-CD3 or anti-T1 monoclonal antibody is partially dependent on protein kinase C. *J. Immunol.* **144,** 2683–2689.

Nel, A. E., Taylor, K. E., Kumar, G. P., Gupta, S., Wang, S. C.-T., Williams, K., Liao, O., Swanson, K., and Landreth, G. E. (1994). Activation of a novel serine/threonine kinase that phosphorylates c-Fos upon stimulation of T and B lymphocytes via antigen and cytokine receptors. *J. Immunol.* **152,** 4347–4357.

Nemazee, D. A., and Burki, K. (1989). Clonal deletion of B lymphocytes in a transgenic mouse bearing anti-MHC class I antibody genes. *Nature (London)* **337,** 562–566.

Nevins, J. R. (1992). E2F: A link between the Rb tumor suppressor protein and viral oncoproteins. *Science* **258,** 424–429.

Nishizuka, Y. (1988). The molecular heterogeneity of protein kinase C and its implications for cellular regulation. *Nature (London)* **334,** 661–665.

Noelle, R. J., and Snow, E. C. (1991). T helper cell-dependent B cell activation. *FASEB J.* **5,** 2770–2776.

Noelle, R. J., Daum, J., Bartlett, W. C., McCann, J., and Shepherd, D. M. (1991). Cognate interactions between helper T cells and B cells V. Reconstitution of T helper cell function using purified plasma membranes from activated Th1 and Th2 helper cell and lymphokines. *J. Immunol.* **146,** 1118–1124.

Noelle, R. J., Ledbetter, J. A., and Aruffo, A. (1992a). CD40 and its ligand, an essential ligand-receptor pair for thymuc-dependent B-cell activation. *Immunol. Today* **13,** 431–433.

Noelle, R. J., Roy, M., Shepherd, D. M., Stamenkovic, I., Ledbetter, J. A., and Aruffo, A. (1992b). A 39-kDa protein on activated helper T cells binds to CD40 and transduces the signal for cognate activation of B cells. *Proc. Natl. Acad. Sci. U. S. A.* **89,** 6550–6554.

Northrop, J. P., Ullman, K. S., and Crabtree, G. R. (1993). Characterization of the nuclear and cytoplasmic components of the lymphoid-specific nuclear factor of activated T cells (NF-AT) complex. *J. Biol. Chem.* **268,** 2917–2923.

Nossal, G. J. V. (1994). Negative selection of lymphocytes. *Cell (Cambridge, Mass.)* **76,** 229–239.

Notarangelo, L. D., Duse, M., and Ugazio, A. G. (1992). Immunodeficiency with hyper-IgM (HIM). *Immunodefic. Rev.* **3,** 101–122.

Nunez, G., Hockenbery, D., McDonnell, T. J., Sorensen, C. M., and Korsmeyer, S. J. (1991). Bcl-2 maintains B cell memory. *Nature (London)* **353,** 71–73.

Obar, R. A., Collins, C. A., Hammarback, J. A., Shpetner, H. S., and Vallee, R. B. (1990). Molecular cloning of the microtubule-associated mechanochemical enzyme dynamin reveals homology with a new family of GTP-binding proteins. *Nature (London)* **347,** 256–261.

O'Garra, A., Warren, D. J., Holman, M., Popham, A. M., Sanderson, C. J., and Klaus, G. G. B. (1986). Effects of cyclosporine on responses of murine B cells to T cell-derived lymphokines. *J. Immunol.* **137,** 2220–2224.

Ohtsubo, M., and Roberts, J. M. (1993). Cyclin-dependent regulation of G_1 in mammalian fibroblasts. *Science* **259,** 1908–1912.

Okada, M., Nada, S., Yamanashi, Y., Yamamoto, T., and Nakagawa, H. (1991). CSK: A protein-tyrosine kinase involved in regulation of src family kinases. *J. Biol. Chem.* **266,** 24249–24252.

Okada, T., Sakuma, L., Fukui, Y., Hazeki, O., and Ui, M. (1994a). Blockage of chemotactic peptide-induced stimulation of neutrophils by wortmannin as a result of selective inhibition of phosphatidylinositol 3-kinase. *J. Biol. Chem.* **269,** 3563–3567.

Okada, T., Kawano, Y., Sakakibara, T., Hazeki, O., and Ui, M. (1994b). Essential role of phosphatidylinositol 3-kinase in insulin-induced glucose transport and antilipolysis in rat adipocytes. *J. Biol. Chem.* **269,** 3568–3573.

O'Keefe, S. J., Tamura, J., Kincaid, R. L., Tocci, M. J., and O'Neil, E. A. (1992). FK-506- and CsA-sensitive activation of the interleukin-2 promoter by calcineurin. *Nature (London)* **357,** 692–694.

Oliver, K., Noelle, R. J., Uhr, J. W., Krammer, P. H., and Vitetta, E. S. (1985). B-cell growth factor (B-cell growth factor I or B-cell-stimulating factor, provisional 1) is a differentiation factor for resting B cells and may not induce growth. *Proc. Natl. Acad. Sci. U. S. A.* **82,** 2465–2467.

Olivier, J. P., Raabe, T., Henkemeyer, M., Dickson, B., Mbamalu, G., Margolis, B., Schlessinger, J., Hafen, E., and Pawson, T. (1993). A *Drosophila* SH2-SH3 adaptor protein implicated in coupling the sevenless tyrosine kinase to an activator of Ras guanine nucleotide exchange, Sos. *Cell (Cambridge, Mass.)* **73,** 179–191.

Oltvai, Z. N., Milliman, C. L., and Korsmeyer, S. J. (1993). Bcl-2 heterodimerizes in vivo with a conserved homolog, Bax, that accelerates programmed cell death. *Cell (Cambridge, Mass.)* **74,** 609–619.

Ostergaard, H. L., Shackelford, D. A., Hurley, T. R., Johnson, P., Hyman, R., Sefton, B. M., and Trowbridge, I. S. (1989). Expression of CD45 alters phosphorylation of the lck-encoded tyrosine protein kinase in murine lymphoma T-cell lines. *Proc. Natl. Acad. Sci. U. S. A.* **86,** 8959–8963.

Padeh, S., Levitzki, A., Gazit, A., Mills, G. B., and Roifman, C. (1991). Activation of phospholipase C in human B cells is dependent on tyrosine phosphorylation. *J. Clin. Invest.* **87,** 1114–1118.

Padovan, E., Casorati, G., Deelabonoa, P., Meyer, S., Brockhaus, M., and Lanzavecchia, A. (1993). Expression of two T cell receptor α chains: Dual receptor T cells. *Science* **262,** 422–424.

Page, D. M., and DeFranco, A. L. (1990). Antigen receptor-induced cell cycle arrest in WEHI-231 B lymphoma cells depends on the duration of signaling before the G_1 phase restriction point. *Mol. Cell. Biol.* **10,** 3003–3012.

Page, D. M., Kane, L. P., Allison, J. P., and Hedrick, S. M. (1993). Two signals are required for negative selection of CD4⁺CD8⁺ thymocytes. *J. Immunol.* **151,** 1868–1880.

Parry, S. L., Hasbold, J., Holman, M., and Klaus, G. G. B. (1994). Hypercrosslinking surface IgM or IgD receptors on mature B cells induces apoptosis that is reversed by costimulation with IL-4 and anti-CD40. *J. Immunol.* **152,** 2821–2829.

Paul, W. E., and Ohara, J. (1987). B-cell stimulatory factor-1/interleukin 4. *Annu. Rev. Immunol.* **5,** 429–459.

Paul, W. E., and Seder, R. A. (1994). Lymphocyte responses and cytokines. *Cell (Cambridge, Mass.)* **76,** 241–251.

Pawson, T., and Schlessinger, J. (1993). SH2 and SH3 domains. *Curr. Biol.* **3,** 434–442.

Pecanha, L. M. T., Snapper, C. M., Finkelman, F. D., and Mond, J. J. (1991). Dextran-conjugated anti-Ig antibodies as a model for T cell-independent type 2 antigen-mediated stimulation of Ig secretion in vitro. I. Lymphokine dependence. *J. Immunol.* **146,** 833–839.

Pelech, S. L., and Sanghera, J. S. (1992). Mitogen-activated protein kinases: Versatile transducers for cell signaling. *Trends Biochem. Sci.* **17,** 233–238.

Pelicci, G., Lanfrancone, L., Grignani, F., McGlade, J., Cavallo, F., Forni, G., Nicoletti, I., Pawson, T., and Pelicci, P. G. (1992). A novel transforming protein (SHC) with an SH2 domain is implicated in mitogenic signal transduction. *Cell (Cambridge, Mass.)* **70,** 93–104.

Perlmutter, R. M. (1994). In preparation.

Perlmutter, R. M., Levin, S. D., Appleby, M. W., Anderson, S. J., and Alberola-Ila, J. (1993). Regulation of lymphocyte function by protein phosphorylation. *Annu. Rev. Immunol.* **11,** 451–499.

Pezzutto, A., Dorken, B., Rabinovitch, P. S., Ledbetter, J. A., Moldenhauer, G., and Clark, E. A. (1987). CD19 monoclonal antibody HD37 inhibits anti-immunoglobulin-induced B cell activation and proliferation. *J. Immunol.* **138,** 2793–2799.

Pezzutto, A., Rabinovitch, P. S., Dorken, B., Moldenhauer, G., and Clark, E. A. (1988). Role of the CD22 human B cell antigen in B cell triggering by anti-immunoglobulin. *J. Immunol.* **140,** 1791–1795.

Pines, J. (1993). Cyclin-dependent kinases: Clear as crystal? *Curr. Biol.* **3**, 544–547.

Pines, J. (1994). Arresting developments in cell-cycle control. *Trends Biochem. Sci.* **19**, 143–145.

Pingel, J. T., and Thomas, M. L. (1989). Evidence that the leukocyte-common antigen is required for antigen-induced T lymphocyte proliferation. *Cell (Cambridge, Mass.)* **58**, 1055–1065.

Pleiman, C. M., Hertz, W. M., and Cambier, J. C. (1994). Activation of phosphatidylinositol-3' kinase by Src-family kinase SH3 binding to the p85 subunit. *Science* **263**, 1609–1612.

Pognonec, P., Boulukos, K. E., Gesquiere, J. C., Stehelin, D., and Ghysdael, J. (1988). Mitogenic stimulation of thymocytes results in the calcium-dependent phosphorylation of c-ets-1 proteins. *EMBO J.* **7**, 977.

Prakash, S., Robb, R. J., Stout, R. D., and Parker, D. C. (1985). Induction of high affinity IL-2 receptors on B cells responding to anti-Ig and T cell-derived helper factors. *J. Immunol.* **135**, 117–122.

Prasad, K. V. S., Kapeller, R., Janssen, O., Repke, H., Duke-Cohan, J. S., Cantley, L. C., and Rudd, C. E. (1993). Phosphatidylinositol (PI) 3-kinase and PI 4-kinase binding to the CD4-p56^{lck} complex: The p56^{lck} SH3 domain binds to PI 3-kinase but not to PI 4-kinase. *Mol. Cell. Biol.* **13**, 7708–7717.

Qian, D., Griswold-Prenner, I., Rosner, M. R., and Fitch, F. W. (1993). Multiple components of the T cell antigen receptor complex become tyrosine-phosphorylated upon activation. *J. Biol. Chem.* **268**, 4448–4493.

Raulet, D. (1989). The structure, function, and molecular genetics of the gamma/delta T cell receptor. *Annu. Rev. Immunol.* **7**, 175–207.

Ravichandran, K. S., Lee, K. K., Songyang, Z., Cantely, L. C., Burn, P., and Burakoff, S. J. (1993). Interaction of Shc with the ζ chain of the T cell receptor upon T cell activation. *Science* **262**, 902–905.

Reed, J. C., Tsujimoto, Y., Alpers, J. D., Croce, C. M., and Nowell, P. C. (1987). Regulation of *bcl*-2 proto-oncogene expression during normal human lymphocyte proliferation. *Science* **236**, 1295–1299.

Reif, K., Gout, I., Waterfield, M. D., and Cantrell, D. A. (1993). Divergent regulation of phosphatidylinositol 3-kinase p85α and p85β isoforms upon T cell activation. *J. Biol. Chem.* **268**, 10780–10788.

Reif, K., Buday, L., Downward, J., and Cantrell, D. A. (1994). SH3 domains of the adapter molecule Grb2 comlex with two proteins in T cells: The guanine nucleotide exchange protein Sos and a 75-kDa protein that is a substrate for T cell antigen receptor-activated tyrosine kinases. *J. Biol. Chem.* **269**, 14081–14087.

Remillard, B., Petrillo, R., Maslinski, W., Tsudo, M., Strom, T. B., Cantley, L., and Varticovski, L. (1991). Interleukin-2 receptor regulates activation of phosphatidylinositol 3-kinase. *J. Biol. Chem.* **266**, 14167–14170.

Reth, M. (1989). Antigen receptor tail clue. *Nature (London)* **338**, 383.

Reth, M. (1992). Antigen receptors on B lymphocytes. *Annu. Rev. Immunol.* **10**, 97–121.

Reth, M., Petrac, E., Wiese, P., Lobel, L., and Alt, F. W. (1987). Activation of Vκ gene rearrangement in pre-B cells follows the expression of membrane-bound immunoglobulin heavy chains. *EMBO J.* **6**, 3299–3305.

Richards, J., Gold, M. R., DeFranco, A. L., and Matsuuchi, L. (1994). In preparation.

Ridley, A. J., and Hall, A. (1992). The small GTP-binding protein rho regulates the assembly of focal adhesions and actin stress fibers in response to growth factors. *Cell (Cambridge, Mass.)* **70**, 389–399.

Ridley, A. J., Paterson, H. F., Johnston, C. L., Diekmann, D., and Hall, A. (1992). The small GTP-binding protein rac regulates growth factor-induced membrane ruffling. *Cell (Cambridge, Mass.)* **70**, 401–410.

Rocken, M., Muller, K. M., Saurat, J.-H., Muller, I., Loius, J. A., Cerottini, J.-C., and Hauser, C. (1992). Central role for TCR/CD3 ligation in the differentiation of CD4⁺ T cells toward a Th1 or Th2 functional phenotype. *J. Immunol.* **148,** 47–54.

Rodriguez-Viciana, P., Warne, P. H., Dhand, R., Vanhaesebroeck, B., Gout, I., Fry, M. J., Waterfield, M. D., and Downward, J. Phosphatidylinositol-3-OH kinase as a direct target of Ras. *Nature (London)* **370,** 527–532.

Romeo, C., Amiot, M., and Seed, B. (1992). Sequence requirements for induction of cytolysis by the T cell antigen/Fc receptor ζ chain. *Cell (Cambridge, Mass.)* **68,** 889–897.

Rooney, J. W., Dubois, P. M., and Sibley, C. H. (1991). Crosslinking of surface IgM activates NF-κB in B lymphocyte. *Eur. J. Immunol.* **21,** 2993–3008.

Rossomando, A. J., Wu, J., Weber, M. J., and Sturgill, T. W. (1992). MAP kinase activator from insulin-stimulated skeletal muscle is a protein threonine/tyrosine kinase. *Proc. Natl. Acad. Sci. U.S.A.* **89,** 5221–5225.

Roussel, M. F., Cleveland, J. L., Shurtleff, S. A., and Sherr, C. J. (1991). Myc rescue of a mutant CSF-1 receptor impaired in mitogenic signaling. *Nature (London)* **328,** 445–449.

Roussel, M. F., Davis, J. N., Cleveland, J. L., Ghysdael, J., and Hiebert, S. W. (1994). Dual control of myc expression through a single DNA binding site targeted by ets family proteins and E2F-1. *Oncogene* **9,** 405–415.

Rozakis-Adcock, M., McGlade, J., Mbamalu, G., Pelicci, G., Daly, R., Li, W., Batzer, A., Thomas, S., Brugge, J., Pelicci, P. G., Schlessinger, J., and Pawson, T. (1992). Assocation of the Shc and Grb2/Sem5 SH2-containing proteins is implicated in activation of the Ras pathway by tyrosine kinases. *Nature (London)* **360,** 689–692.

Rozakis-Adcock, M., Fernley, R., Wade, J., Pawson, T., and Bowtell, D. (1993). The SH2 and SH3 domains of mammalian Grb2 couple the EGF receptor to the Ras activator mSOS1. *Nature (London)* **363,** 83–85.

Rubin, G. W. (1991). Signal transduction and the fate of the R7 protoreceptor in *Drosophila. Trends Genet.* **7,** 372–377.

Rubin, C. E., Trevillyan, J. M., Dasgupta, J. V., Wong, L. L., and Schlossman, S. F. (1988). The CD4 antigen is complexed in detergent lysates to a protein tyrosine kinase (pp58) from human T lymphocytes. *Proc. Natl. Acad. Sci. U. S. A.* **85,** 5190.

Russell, D. M., Dembic, Z., Morahan, G., Miller, J. F. A. P., Burki, K., and Nemazee, D. (1991). Peripheral deletion of self-reactive B cells. *Nature (London)* **354,** 308–311.

Russell, J. H., White, C. L., Loh, D. Y., and Meleedy-Rey, P (1991). Receptor-stimulated death pathway is opened by antigen in mature T cells. *Proc. Natl. Acad. Sci. U. S. A.* **88,** 2151–2155.

Rustgi, A. K., Dyson, N., and Bernards, R. (1991). Amino-terminal domains of c-*myc* and N-*myc* proteins mediate binding to the retinoblastoma gene product. *Nature (London)* **352,** 541–544.

Sadowski, H. B., Shuai, K., Darnell, J. E., and Gilman, M. Z. (1993). A common nuclear signal transduction pathway activated by growth factor and cytokine receptors. *Science* **261,** 1739–1744.

Samelson, L. E., Phillips, A. F., Luong, E. T., and Klausner, R. D. (1990). Association of the fyn protein-tyrosine kinase with the T cell antigen receptor. *Proc. Natl. Acad. Sci. U. S. A.* **87,** 4358–4362.

Sanchez, M., Misulovin, Z., Burkhardt, A., Mahajan, S., Costa, T., Franke, R., Bolen, J. B., and Nussenzweig, M. (1993). Signal transduction by immunoglobulin is mediated through Igα and Igβ. *J. Exp. Med.* **178,** 1049–1055.

Saxton, T. M., van Oostveen, I., Bowtell, D., Aebersold, R., and Gold, M. R. (1994). B cell antigen receptor cross-linking induces phosphorylation of the Ras activators SHC and mSOS1 as well as assembly of complexes containing SHC, GRB-2, mSOS1, and a 145-kDa tyrosine-phosphorylated protein. *J. Immunol.* **153,** 623–636.

Schatz, D. G., Oettinger, M. A., and Schlissel, M. S. (1992). V(D)J recombination: Molecular biology and regulation. *Annu. Rev. Immunol.* **10,** 359–383.

Scheuerman, R. H., Racila, E., Tucker, T., Yefenof, Y., Street, N. E., Vitetta, E. S., Picker, L. J., and Uhr, J. W. (1994). Lyn tyrosine kinase signals cell cycle arrest but not apoptosis in B-lineage lymphoma cells. *Proc. Natl. Acad. Sci. U. S. A.* **91,** 4048–4052.

Schmid, S. L., and Jackson, M. R. (1994). Making class II presentable. *Nature (London)* **369,** 103–104.

Schreiber, S. L. (1991). Chemistry and biology of the immunophilins and their immunosuppressive ligands. *Science* **251,** 283–287.

Schreiber, S. L., and Crabtree, G. R. (1992). The mechanism and action of cyclosporin A and FK506. *Immunol. Today* **13,** 136–142.

Schwartz, R. H. (1992). Costimulation of T lymphocytes: The role of CD28, CTLA-4, and B7/BB-1 in IL-2 production and immunotherapy. *Cell (Cambridge, Mass.)* **71,** 1065–1068.

Sentman, C. L., Shutter, J. R., Hockenbery, D., Kanagawa, O., and Korsmeyer, S. J. (1991). Bcl-2 inhibits multiple forms of apoptosis, but not negative selection in thymocytes. *Cell (Cambridge, Mass.)* **67,** 879–888.

Serunian, L. A., Haber, M. T., Fukui, T., Kim, J. W., Rhee, S. G., Lowenstein, J. M., and Cantley, L. C. (1989). Polyphosphoinositides produced by phosphatidylinositol 3-kinase are poor substrates for phospholipase C from rat liver and bovine brain. *J. Biol. Chem.* **264,** 17809–17815.

Settleman, J., Albright, C. F., Foster, L. C., and Weinberg, R. A.(1992a). Association between GTPase activators for Rho and Ras families. *Nature (London)* **359,** 153–154.

Settleman, J., Narashimhan, V., Foster, L. C., and Weinberg, R. A. (1992b). Molecular cloning of cDNAs encoding the GAP-associated protein p190: Implications for a signaling pathway from Ras to the nucleus. *Cell (Cambridge, Mass.)* **69,** 539.

Seyfert, V. L., Sukhatme, V. P., and Monroe, J. G. (1989). Differential expression of a zinc finger-encoding gene in response to positive versus negative signaling through receptor immunoglobulin in murine B lymphocytes. *Mol. Cell. Biol.* **9,** 2083–2088.

Seyfert, V. L., McMahon, S., Glenn, W., Cao, X., Sukhatme, V. P., and Monroe, J. G., (1990). *Egr-1* expression in surface Ig-mediated B cell activation: Kinetics and association with protein kinase C activation. *J. Immunol.* **145,** 3647–3653.

Sgroi, D., Varki, A., Braesch-Andersen, S., and Stamenkovic, I. (1993). CD22, a B cell-specific immunoglobulin superfamily member, is a sialic acid-binding lectin. *J. Biol. Chem.* **268,** 7011–7018.

Sherr, C. J. (1993). Mammalian G_1 cyclins. *Cell (Cambridge, Mass.)* **73,** 1059–1065.

Shi, Y., Sahai, B. M., and Green, D. R. (1989). Cyclosporin A inhibits activation-induced cell death in T cell hybridomas and in thymocytes. *Nature (London)* **339,** 625.

Shiroo, M., Goff, L., Biffen, M., Shivnan, E., and Alexander, D. (1992). CD45 tyrosine phosphatase-activated $p59^{fyn}$ couples the T cell antigen receptor to pathways of diacylglycerol production, protein kinase C activation and calcium influx. *EMBO J.* **11,** 4887–4897.

Siegel, J. N., Klausner, R. D., Rapp, U. R., and Samelson, L. E. (1990). T cell antigen receptor engagement stimulates c-*raf* phosphorylation and induces c-*raf*-associated kinase activity via a protein kinase C-dependent pathway. *J. Biol. Chem.* **265,** 18472–18480.

Sigal, N. H., and Dumont, F. J. (1992). Cyclosporin A, FK-506, and rapamycin: Pharmacologic probes of lymphocyte signal transduction. *Annu. Rev. Immunol.* **10,** 519–560.

Songyang, Z., Shoelson, S. E., McGlade, J., Olivier, P., Pawson, T., Bustelo, X. R., Barbacid, M., Sabe, H., Hanafusa, H., Yi, T., Ren, R., Baltimore, D., Ratnofsky, S., Feldman, R. A., and Cantley, L. C. (1994). Specific motifs recognized by the SH2 domains of Csk, 3BP2, fps/fes, GRB-2, HCP, SHC, Syk, and Vav. *Mol. Cell. Biol.* **14,** 2777–2785.

Stein, D., Wu, J., Fuqua, S. A. W., Roonprapunt, C., Yajnik, V., D'Eustachio, P., Moskow, J. J., Buchberg, A. M., Osborne, C. K., and Margolis, B. (1994). The SH2 domain protein

GRB-7 is co-amplified, overexpressed and in a tight complex with HER2 in breast cancer. *EMBO J.* **13,** 1331–1340.

Stein, P. L., Lee, H.-M., Rich, S., and Soriano, P. (1992). pp59^{fyn} mutant mice display differential signaling in thymocytes and peripheral T cells. *Cell (Cambridge, Mass.)* **70,** 741–750.

Stern, J. B., and Smith, K. A. (1986). Interluekin-2 induction of T-cell G_1 progression and c-*myb* expression. *Science* **233,** 203.

Sternberg, P. W., and Horvitz, H. R. (1991). Signal transduction during *C. elegans* vulval induction. *Trends Genet.* **7,** 366–371.

Stover, D. R., Charbonneau, H., Tonks, N. K., and Walsh, K. A. (1991). Protein-tyrosine phosphatase CD45 is phosphorylated transiently on tyrosine upon activation of Jurkat T cells. *Proc. Natl. Acad. Sci. U. S. A.* **88,** 7704–7707.

Strasser, A., Harris, A. W., and Cory, S. (1991). *bcl*-2 transgene inhibits T cell death and perturbs thymic self-censorship. *Cell (Cambridge, Mass.)* **67,** 889–899.

Strasser, A., Harris, A. W., Corcoran, L. M., and Cory, S. (1994). Bcl-2 expression promotes B- but not T-lymphoid development in *scid* mice. *Nature (London)* **368,** 457–460.

Straus, D. B., and Weiss, A. (1992). Genetic evidence for the involvement of the lck tyrosine kinase in signal transduction through the T cell antigen receptor. *Cell (Cambridge, Mass.)* **70,** 585–593.

Straus, D. B., and Weiss, A. (1993). The CD3 chains of the T cell antigen receptor associates with the ZAP-70 tyrosine kinase and are tyrosine phosphorylated following stimulation. *J. Exp. Med.* **178,** 1523–1530.

Su, B., Jacinto, E., Hibi, M., Kallunki, T., Karin, M., and Ben-Neriah, Y. (1994). JNK is involved in signal integration during costimulation of T lymphocytes. *Cell (Cambridge, Mass.)* **70,** 727–736.

Suda, T., Takahashi, T., Golstein, P., and Nagata, S. (1993). Molecular cloning and expression of the Fas ligand, a novel member of the tumor necrosis family. *Cell (Cambridge, Mass.)* **75,** 1169–1178.

Swain, S. L., Bradley, L. M., Croft, M., Tonkonogy, S., Atkins, G., Weinberg, A. D., Hedrick, S. M., Dutton, R. W., and Huston, G. (1991). Helper T-cell subsets: Phenotype, function, and the role of lymphokines in regulating their development. *Immunol. Rev.* **1234,** 115–144.

Taddie, J. A., Hurley, T. R., Hardwick, B. S., and Sefton, B. M. (1994). Activation of B- and T-cells by the cytoplasmic domains of the B-cell antigen receptor proteins Ig-α and Ig-β. *J. Biol. Chem.* **269,** 13529–13535.

Takata, M., Sabe, H., Hata, A., Inazu, T., Homma, Y., Nukuda, T., Yamamura, H., and Kurosaki, T. (1994). Tyrosine kinases Lyn and Syk regulate B cell receptor-coupled Ca^{2+} mobilization through distinct pathways. *EMBO J.* **13,** 1341–1349.

Terada, N., Lucas, J. J., and Gelfand, E. W. (1991). Differential regulation of the tumor suppressor molecules, retinoblastoma susceptibility gene product (Rb) and p53, during cell cycle progression of normal human T cells. *J. Immunol.* **147,** 698–704.

Thompson, P. A., Gutkind, J. S., Robbins, K. C., Ledbetter, J. A., and Bolen, J. B. (1992). Identification of distinct populations of PI-3 kinase activity following T-cell activation. *Oncogene* **7,** 719–725.

Tiegs, S. L., Russell, D. M., and Nemazee, D. (1993). Receptor editing in self-reactive bone B marrow cells. *J. Exp. Med.* **177,** 1009–1020.

Tilzey, J. F., Chiles, T. C., and Rothstein, T. L. (1991). *Jun-β* gene expression mediated by the surface immunoglobulin receptor of primary B lymphocytes. *Biochem. Biophys. Res. Commun.* **175,** 77–83.

Timson-Gauen, L. K., Kong, A.-N. T., Samelson, L. E., and Shaw, A. S. (1992). The p59^{fyn} tyrosine kinase associates with multiple T cell receptor subunits through its amino-terminal domain. *Mol. Cell. Biol.* **12,** 5438–5446.

Torda, A., Franklin, R. F., Patel, H., Gardner, A. M., Johnson, G. L., and Gelfand, E. W. (1994). Cross-linking of surface IgM stimulates the Ras-Raf-1/MEK/MAPK cascade in human B lymphocytes. *J. Biol. Chem.* **269,** 7538–7543.

Tsubata, T., and Reth, M. (1990). The products of pre-B cell-specific genes ($\lambda 5$ and V_{pre-B}) and the immunoglobulin μ chain form a complex that is transported onto the cell surface. *J. Exp. Med.* **172,** 973–976.

Tsukada, S., Saffran, D. C., Rawlings, D. J., Parolini, O., Allen, R. C., Klisak, I., Sparkes, R. S., Kubagawa, H., Mohandas, T., Quan, S., Belmont, J. W., Cooper, M. D., Conley, M. E., and Witte, O. N. (1993). Deficient expression of a B cell cytoplasmic tyrosine kinase in human X-linked agammaglobulinemia. *Cell (Cambridge, Mass.)* **72,** 279–290.

Tsygankov, A. Y., Broker, B. M., Fargnoli, J., Ledbetter, J. A., and Bolen, J. B. (1992). Activation of tyrosine kinase p60[fyn] following T cell antigen receptor cross-linking. *J. Biol. Chem.* **267,** 18259–18262.

Turka, L. A., Schatz, D. G., Oettinger, M. A., Chun, J. J., Gorka, C., Lee, K., McCormack, W. T., and Thompson, C. B. (1991). Thymocyte expression of RAG-1 and RAG-2: Termination by T cell receptor cross-linking. *Science* **253,** 778–781.

Turka, L. A., Kanner, S. B., Schieven, G. L., Thompson, C. B., and Ledbetter, J. A. (1992). CD45 modulates T cell receptor/CD3-induced activation of human thymocytes via regulation of tyrosine phosphorylation. *Eur. J. Immunol.* **22,** 551–557.

Tuveson, D. A., Carter, R. H., Soltoff, S. P., and Fearon, D. T. (1993). CD19 of B cells as a surrogate kinase insert region to bind phosphatidylinositol 3-kinase. *Science* **260,** 986–989.

Ullman, K. S., Northrop, J. P., Verwij, C. L., and Crabtree, G. R. (1990). Transmission of signals from the T lymphocyte antigen receptor to the genes responsible for cell proliferation and immune function: The missing link. *Annu. Rev. Immunol.* **8,** 421–452.

van Noesel, C. J., Lancaster, A. C., and van Lier, R. A. (1993a). Dual antigen recognition by B cells. *Immunol. Today* **14,** 8–11.

van Noesel, C. J. M., Lankester, A. C., van Schijndel, G. M. W., and van Lier, R. A. W. (1993b). The CR2/CD19 complex on human B cells contains the *src*-family kinase *Lyn*. *Int. Immunol.* **5,** 699–705.

Vaux, D. L., Haecker, G., and Strasser, A. (1994). An evolutionary perspective on apoptosis. *Cell (Cambridge, Mass.)* **76,** 777–779.

Veillette, A., Bookman, M. A., Horak, E. M., and Bolen, J. B. (1988). The CD4 and CD8 T cell surface antigens are associated with the internal membrane tyrosine-protein kinase p56[lck]. *Cell (Cambridge, Mass.)* **55,** 301.

Venkitaraman, A. R., Williams, G. T., Dariavach, P., and Neuberger, M. S. (1991). The B-cell antigen receptor of the five immunoglobulin classes. *Nature (London)* **352,** 777–782.

Venkitaraman, L., Francis, D. A., Wang, V., Liu, J., Rothstein, T. L., and Sen, R. (1994). Cyclosporin A-sensitive induction of NF-AT in murine B cells. *Immunity* **1,** 189–196.

Vetrie, D., Vorechovsky, I., Sideras, P., Holland, J., Davies, A., Flinter, F., Hammarström, L., Kinnon, C., Levinsky, R., Bobrow, M., Smith, C. I. E., and Bentley, D. R. (1993). The gene involved in X-linked agammaglobulinemia is a member of the *src* family of protein-tyrosine kinases. *Nature (London)* **361,** 226–233.

Volarevic, S., Burns, C. M., Sussman, J. J., and Ashwell, J. D. (1990). Intimate association of Thy-1 and the T cell antigen receptor with the CD45 tyrosine phosphatase. *Proc. Natl. Acad. Sci. U.S.A.* **87,** 7085–7089.

von Boehmer, H. (1994). Positive selection of lymphocytes. *Cell (Cambridge, Mass.)* **76,** 219–228.

von Willebrand, M., Baier, G., Couture, C., Burn, P., and Mustelin, T. (1994). Activation of phosphatidylinositol-3-kinase in Jurkat T cells depends on the presence of the p56[lck] tyrosine kinase. *Eur. J. Immunol.* **24,** 234–238.

Wange, R. L., Kong, A. N., and Samelson, L. E. (1992). A tyrosine-phosphorylated 70kda protein binds a photoaffinity analogue of ATP and associates with both the ζ chain and CD3 components of the activated T cell antigen receptor. *J. Biol. Chem.* **267**, 11685–11688.

Wange, R. L., Malek, S. N., Desiderio, S., and Samelson, L. E. (1993). Tandem SH2 domains of ZAP-70 bind to T cell antigen receptor ζ and CD3ε from activated Jurkat T cells. *J. Biol. Chem.* **268**, 19797–19801.

Ward, S. G., Ley, S. C., MacPhee, C., and Cantrell, D. A. (1992). Regulation of D-3 phosphoinositides during T cell activation via the T cell antigen receptor/CD3 comples and CD2 antigens. *Eur. J. Immunol.* **22**, 45–49.

Ward, S., Westwick, J., Hall, N., and Sansom, D. (1993). Ligation of CD28 receptor by B7 induces formation of D-3 phosphoinositides in T lymphocytes independently of T cell receptor/CD3 activation. *Eur. J. Immunol.* **23**, 2572–2577.

Wasylyk, B., Wasylyk, C., Flores, P., Begue, A., Leprince, D., and Stehelin, D. (1990). The c-*ets* proto-oncogenes encode transcription factors that cooperate with c-Fos and c-Jun for transcriptional activation. *Nature (London)* **346**, 191–193.

Watts, J. D., Sanghera, J. S., Pelech, S. L., and Aebersold, R. (1993). Phosphorylation of serine 59 of p56^{lck} in activated T cells. *J. Biol. Chem.* **268**, 23275–23282.

Weiss, A., and Imboden, J. B. (1987). Cell surface molecules and early events involved in human T lymphocyte activation. *Adv. Immunol.* **41**, 1–38.

Weiss, A., and Littman, D. R. (1994). Signal transduction by lymphocyte antigen receptors. *Cell (Cambridge, Mass.)* **76**, 263–274.

Weiss, A., Koretzky, G., Schatzman, R., and Kadlacek, T. (1991). Stimulation of the T cell antigen receptor induces tyrosine phosphorylation of phospholipase C-γ1. *Proc. Natl. Acad. Sci. U.S.A.* **88**, 5484–5488.

Weiss, U., Zoebelein, R., and Rajewsky, K. (1992). Accumulation of somatic mutants in the B cell compartment after primary immunization with a T cell-dependent antigen. *Eur. J. Immunol.* **22**, 511–517.

Wennström, S., Hawkins, P., Cooke, F., Hara, K., Yonezawa, K., Kasuga, M., Jackson, T., Claesson-Welch, L., and Stephens, L. (1994). Activation of phosphoinositide 3-kinase is required for PDGF-stimulated membrane ruffling. *Curr. Biol.* **4**, 385–393.

Wong, G., Muller, O., Clark, R., Conroy, L., Moran, M. F., Polakis, P., and McCormick, F. (1992). Molecular cloning and nucleic acid binding properties of the GAP-associated tyrosine phosphoprotein p62. *Cell (Cambridge, Mass.)* **69**, 551–558.

Woodrow, M., Clipstone, N. A., and Cantrell, D. (1993). p21^{ras} and calcineurin synergize to regulate the nuclear factor of activated T cells. *J. Exp. Med.* **178**, 1517–1522.

Woronicz, J. D., Calnan, B., Ngo, V., and Winoto, A. (1994). Requirement for the orphan steroid receptor Nur77 in apoptosis of T-cell hybridomas. *Nature (London)* **367**, 277–281.

Yamada, T., Taniguchi, T., Yang, C., Yasue, S., Saito, H., and Yamamura, H. (1993). Association with B-cell antigen receptor with protein-tyrosine kinase p72^{syk} and activation by engagement of membrane IgM. *Eur. J. Biochem.* **213**, 455–459.

Yamanashi, Y., Okada, M., Semba, T., Yamori, T., Umemori, H., Tsunasawa, S., Toyo-shima, K., Kitamura, D., Watanabe, T., and Yamamoto, T. (1993). Identification of HS1 protein as a major substrate of protein-tyrosine kinase(s) upon B-cell antigen receptor-mediated signaling. *Proc. Natl. Acad. Sci. U. S. A.* **90**, 3631–3635.

Yamanashi, Y., Fukui, Y., Wongsasant, B., Kinoshita, Y., Ichimori, Y., Toyoshima, K., and Yamamoto, T. (1992). Activation of Src-like protien-tyrosine kinase Lyn and its association with phosphatidylinositol 3-kinase upon B-cell antigen receptor-mediated signaling. *Proc. Natl. Acad. Sci. U. S. A.* **89**, 1118–1122.

Yao, X.-R., and Scott, D. W. (1993). Antisense oligodeoxynucleotides to the blk tyrosine kinase prevent anti-m-chain-mediated growth inhibition and apoptosis in a B-cell lymphoma. *Proc. Natl. Acad. Sci. U. S. A.* **90**, 7946–7950.

Neuroendocrine Epithelial Cell System in Respiratory Organs of Air-Breathing and Teleost Fishes

Giacomo Zaccone[1], Salvatore Fasulo[2], and Luigi Ainis[1]

[1] Department of Animal Biology and Marine Ecology, Faculty of Science, University of Messina, I-98166 Messina, Italy
[2] Department of Animal Biology, Faculty of Science, University of Catania, I-95124 Catania, Italy

This chapter describes the distributional patterns of the neuroendocrine cells in the respiratory surface of fishes and their bioactive secretions, which are compared with similar elements in higher vertebrates.

The neuroendocrine cells in the airways of fishes differentiate as solitary and clustered cells, but the clusters are not converted into neuroepithelial bodies that are found in terrestrial vertebrates.

The lungfish *Protopterus* has innervated neuroendocrine cells in the pneumatic duct region. In *Polypterus* and *Amia* the lungs have neuroendocrine cells that are apparently not innervated. Two types of neuroendocrine cells are located in the gill of teleost fishes. These cells are different by their location, their structure, and immunocytochemistry.

In mammals pulmonary neuroendocrine cells are known to be capable of perceiving changes in the oxygen levels of the airways by releasing secretory substances in response to hypoxic conditions. Other proposed suggestions deserve consideration, such as the regulation of the growth and proliferation of the lung in mammals and osmoregulatory function in fish gill area. The latter supposed function may be supported by the exciting discovery of calcium-binding proteins in the NE cell system of the gill, and the elucidation of their physiological role presents a considerable challenge.

KEY WORDS: Light microscopy, Electron microscopy, Immunocytochemistry, Neuroepithelial bodies, Neuroendocrine cells, Secretory substances, Chemoreceptors, Hypoxia, Pulmonary and Osmoregulatory function.

I. Introduction

Dipnoan fish possess paired ventral esophageal outgrowths that remain connected throughout life with the proximal part of the alimentary tract via a pneumatic duct. These air-filled esophageal pouches serve in gas exchange and are considered to be homologous with the lungs of tetrapods. The pulmonary arteries depart from the left and right aortic roots, which are formed by the branchial efferents, and the pulmonary vein drains in the left auricle. Comparable lungs are present among members of the Polypteridae, another group of primitive fish. Most considerations of the systematics of lung fishes have been based on the structural characteristics of the skeleton, although soft tissue (lungs and vascular system) suggests a closer similarity to tetrapods. Therefore the lungs of dipnoan fish (Dipnoi) serve as a model system for investigating the evolutionary history and the development of the neuroendocrine (NE) epithelial system in the respiratory tract of higher vertebrates.

We have studied the morphology and immunohistochemistry of the NE cells in the airway epithelium of members of the Dipnoi and Polypteridae and compared their distribution to that in the gills of teleostean and holostean fishes in combined immunohistochemical and ultrastructural investigations.

The NE cells were found to be involved in the production of amines and/or polypeptide hormones. Their ability to produce neurochemically active substances and their cell polarization, with respect to the airway lumen, are indicative of a paracrine function even during the phylogenetic discontinuity from gill to obligatory pulmonary respiratory function. The chemoreceptive and neurosecretory functions of the NE cells remain to be studied. Some of these cells probably have completely different functional implications in air-breathing fishes and teleosts. This might be because the gills represent a multifunctional organ with respect to the lung of tetrapods.

This introduction is based on the works and reviews of Burggren and Johansen (1986), Burggren *et al.* (1985), Grigg (1965), Hughes (1966, 1973, 1978), Hughes and Pohunkova (1980), Hughes and Weibel (1976, 1978), Kimura *et al.* (1987), Klika and Lelek (1967), Maina (1987), and Sacca and Burggren (1982), to whom the reader is referred.

As far as dipnoan fish are concerned, the Australian *Neoceratodus* is unique for its obligatory water breathing, whereas both the Africa *Protopterus* and the South American *Lepidosiren* are obligatory air-breathing animals, the former periodically spending prolonged periods of time (months or years) out of water. Thus *Neoceratodus* species possess well-developed gills and a single lung with a pulmonary venous drainage connected to the systemic venous circulation prior to the heart at the level of the sinus venosus (Fox, 1965) or left ductus cuvieri (Satchell, 1976), a condition considered more primitive. As compared to the circulation of *Protopterus* and *Lepidosiren,* the oxygenated pulmonary blood and deoxygenated systemic blood are kept separate during passage to the heart. The lung wall consists of, starting from the outer to inner aspects, a layer of elastic tissue, a thin layer of smooth muscle, a layer of nonelastic connective tissue, and a layer of thin epithelium covering the capillaries.

In *Protopterus,* the alveolar surface is covered by some cells characterized by microvilli on their free surface, whereas others are devoid of such structures (Maina, 1987). The undifferentiated alveolar pneumocytes observed in these species appear to characterize the lung of members of the Dipnoi. This is in agreement with the results of De Groodt *et al.* (1960), who for the first time determined that the alveoli in *Protopterus* lung are lined with epithelial cells.

In *Lepidosiren,* as in most lower vertebrates, the respiratory epithelium contains only one type of cell, which consists of a large cell body containing lamellated bodies. These cells extend processes that cover the capillary surfaces, over the endothelial cells with an interposed thin interstitial layer derived from the thicker connective tissue of the septa. Thus, there are three layers that form the air–blood barrier, just as in tetrapod lungs, mammals included, but their total thickness exceeds 10 times that in fish.

Of all the actinopterygian fishes, only members of the Polypteridae are generally considered to possess lungs of a corresponding type, although they are rather more primitive than those of the Dipnoi (Hughes and Pohunkova, 1980). In *Polypterus senegalensis,* the inner lung surface is mainly covered by low cuboidal to flat epithelial cells (from which microvilli appear to be absent) that constitute a smooth-surfaced respiratory epithelium regularly alternated with ciliated tracts with a mainly longitudinal orientation to the lung axis. There are also occasional groups of ciliated cells, scattered throughout the lung, that appear to harbor similar cell types as the tracts (Hughes and Pohunkova, 1980).

In the smooth areas of the lumenal surface the flat epithelial cells have cell bodies generally sunk a little below the main surface of the lung, and elongated processes that overlay the capillaries. These processes, together with the interstitial layer (often reduced on basement membrane of epithelium only) and endothelium of the capillaries, constitute the air–blood barrier. The cytoplasm of these epithelial cells contains few oganelles and numerous pinocytotic vesicles, suggesting a transport activity.

Apart from ciliated cells, cells with spherical protrusions (considered to be cells of the mucous type) and cells with irregular microvilli that contain lamellated bodies are seen in the ciliated strands and islets (Hughes and Pohunkova, 1980). The ciliated and mucous cells may have a function analogous to similar cells that line the airways in more advanced lungs in the production and removal of mucus and other secretions.

The mucosal NE cells are discussed below.

II. Paraneuronal Endocrine Cells in the Airways of Higher Vertebrates

Before dealing with the ultrastructural and immunohistochemical characteristics of NE cells in the lung and the gill epithelia of fish, it seems appropriate to shed light on the corresponding structures of the NE cell system in mammalian lung. The morphological aspects and ultrastructural and immunohistochemical features have been the subject of previous reviews (Scheuermann, 1987; Fujita *et al.*, 1988; Sorokin and Hoyt, 1989). In brief, we report here on those aspects and advances concerning both immunohistochemistry and cell biology studies on the lung neuroendocrine cells in mammals.

A. Structure, Distribution, and Function of Endocrine Cells and Neuroepithelial Bodies

The epithelium of the air tract of many vertebrates reveals the presence of basal granulated cells (NE cells) with putative chemoreceptor or neurohumoral paracrine function. These cells found in the lungs of mammals and other vertebrate species are arranged within the respiratory mucosa either singly or in groups to form neuroepithelial bodies (NEBs) (Lauweryns and Peuskens, 1969, 1972; Lauweryns *et al.*, 1970; Gould *et al.*, 1983). First reported as *helle zellen* by Fröhlich (1949), their occurrence was later confirmed by Feyrter (1954), who regarded them as cell components of the "diffuse endocrine epithelial system." Studies during the 40 years since their discovery have been focused on the morphological aspects and the cytological nature, ontogeny, and function of the NE cells, using histological techniques. Both solitary NE cells and NEBs share several common cytochemical and ultrastructural characteristics with other NE cells: argyrophilic by silver impregnation, fluorescence by formaldehyde-induced cytochemistry, and the presence of dense-cored secretory granules by electron microscopy (Lauweryns and Cockelaere, 1973; Lauweryns *et al.*, 1977; Hage, 1972a,b). In addition, both cell types are members of the amine precursor uptake and decarboxylation (APUD) system, and therefore have the potential to produce amines or peptides (Pearse, 1969; Pearse and Polak, 1978). Experimental evidence has confirmed the production of serotonin by both NEBs (Lauweryns *et al.*, 1973) and solitary NE cells (Dey *et al.*, 1981). The role and functions of this specialized cell system widely distributed within the lung remain elusive. On the basis of a series of experiments using a neonatal rabbit model, Lauweryns *et al.* suggested that NEBs may function as hypoxia-sensitive airway chemoreceptors (Lauweryns and Cockelaere, 1973; Lauweryns *et al.*, 1972). Because of their content of bombesin or related peptides, pulmonary NE cells have been suggested to have a role in the growth and proliferation of lung cells, especially during fetal life, and possibly also in lung neoplasia (Gould *et al.*, 1983). In fact, some studies have demonstrated that NE cells and NEBs develop early during morphogenesis (Cutz *et al.*, 1984); their amine and/or peptide contents peak at the time of birth and decline postnatally (Track and Cutz, 1982; Cutz *et al.*, 1984). A role in lung neoplasia is also suggested for NE cells (Gould *et al.*, 1983).

On the basis of ultrastructural similarity, some investigators have suggested that these cells may be the precursors of pulmonary carcinoids and oat cell carcinomas (Bensch *et al.*, 1968; Gould and Chejfec, 1978; McDowell *et al.*, 1976). Despite these advances their precise cytological identities, physiological functions, and pathological roles remain controversial or poorly defined (Lauweryns and Peuskens, 1969; Johnson and Georgieff, 1989). Neuroepithelial bodies located in the human lung (Lauweryns and Peuskens, 1972) were also observed in metatheria (Haller, 1992), monotremes (both platypus and echidna; c.d. Haller, personal communication), birds (Cook and King, 1969), chelonians (Scheuermann *et al.*, 1983; Scheuermann, 1987), and amphibians (Goniakowska-Witalinska, 1993). The NE cells of the mammalian lung were more recently reviewed by Scheuermann (1987), Fujita *et al.* (1988), and Sorokin and Hoyt (1989). The fine structure of NE cells forming NEBs corresponds to that of the solitary cells. Neuroepithelial bodies were found to occur in the bronchiolar epithelium and their surfaces are in direct contact with the airway lumen. In the rabbit lung NEBs are revealed by light microscopy using argyrophilic (Grimelius) staining (Cutz *et al.*, 1978). All the levels of the bronchiolar tree contained NEBs randomly distributed in large airways, but in terminal bronchioles the majority were close to or lining the bifurcations. Cytoplasmic argyrophilia is also revealed in NE cells and NEBs of the lung tissues of a large number of vertebrate species (Scheuermann, 1987). Transmission electron microscopy (TEM) observations reveal that NEBs are clearly differentiated by adjacent bronchial epithelium. Most are ovoid with delimited lateral borders. Their apex extended to the bronchial lumen. Their basis, which is enclosed by bronchial basement membrane, protruded into the submucosa. Plasma membranes of NEBs formed complex interdigitations or desmosomes. Numerous nerve endings were seen between the cells. Fine structural investigations revealed the presence of dense-cored vesicles (DCVs) within the cytoplasm of the NE cells. They vary in size and number and are seen to accumulate in the basal portion of the cell. Few are in the apical region. The numerous round or oval DCVs are 120–140 nm in diameter and have a homogeneous electron-dense center. Other cell organelles include varied numbers of small mitochondria, rough and smooth endoplasmic reticulum, microtubules, and supranuclear Golgi apparatus. In some instances plasma membranes form short microvilli. Most of the bronchial epithelial cells adjacent to NEBs are nonciliated (Clara-like) and protrude above the NEB surface. Ciliated cells are present in the usual numbers but appear scanty in peripheral bronchioles.

Electron-microscopy images indicate exocytotic release of basal granules from the solitary NE cells of the mouse trachea (Taira and Shibasaki, 1978). The size of DCVs may be species dependent, as reported for the NE cells of mammalian and amphibian species (Goniakowska-Witalinska et al., 1993a). But fixation and staining methods may result in some shrinkage of the specific granules. The NEBs of cell vertebrates receive an extensive innervation, in contrast to the solitary NE cells, a number of which are free of nerves. Innervation studies show the presence of unmyelinated axons penetrating the subepithelial basement membrane to form synapses with the NE cells. Neuroendothelial bodies show dual innervation by afferent and efferent pathways, which was demonstrated by different nerve profiles. In the rat and mouse, nerve terminals contain numerous mitochondria and synaptic vesicles (Wasano, 1977). Differences among various species were reviewed by Scheuermann (1987). A synaptic association of a nerve fiber with the endocrine cell is demonstrated (Hage et al., 1977; Taira, 1985) but EM images do not clearly show the efferent or afferent (sensory) nature of the nerve terminals.

B. Secretory Substances

Pulmonary NE cells contain secretory products similar to those found in other endocrine paraneurons of the body. An overview of the distribution of specific substances in the NE cells of the airways of various vertebrate species has been provided by Polak et al. (1993).

1. Amines

Both solitary and aggregated types of pulmonary NE cells are known to contain serotonin. This monoamine was found using the fluorescence histochemical method of Falck et al. (1962). With this technique the NE cells emit a characteristic yellow fluorescence due to serotonin (Lauweryns *et al.*, 1972, 1973; Wasano, 1977). Glyoxylic acid-induced fluorescence (GIF) has been another fluorescence method for the demonstration of biogenic amines and the serotonin fluorescence has been found in the neuroepithelial bodies of frog and toad (Rogers and Haller, 1978). Lung NE cells were not so easily detected by fluorescence microscopy, owing to low contents of endogenous amine. Cutz *et al.* (1974, 1975) demonstrated intracellular amines in NE cells by the formaldehyde-induced fluorescence (FIF) method, but only after *in vivo* or *in vitro* administration of amino acid precursors (L-dihydroxyphenylalanine, L-5-hydroxytryptophan). These fluorescent cells were regarded as Kultschitzky type (K type) cells, sharing some features in common with cells of the amine precursor uptake and decarboxylation (APUD) endocrine system (Pearse, 1969). These cells were found to occur *in vivo* or after *in vitro* treatment of other specimens with amine precursor (Ericson *et al.*, 1972; Hage, 1973, 1974; Lauweryns *et al.*, 1977; Dey *et al.*, 1981).

Serotonin was more recently located by immunohistochemistry in lung NE cells (Wharton *et al.*, 1981; Takahashi and Yui, 1983; Lauweryns *et al.*, 1986) and seems to coexist with a variety of peptides, such as calcitonin gene-related peptide (CGRP; Keith and Ekman, 1988), gastrin-releasing peptide (GRP) or somatostatin (Dayer *et al.*, 1985), helodermin-like peptide (Luts *et al.*, 1991), calcitonin (Haller, 1992), and cholecystokinin-like peptide (Wang and Cutz, 1993).

Serotonin appeared to be a good histological marker for NE cells in the airway epithelium of lower vertebrates (Zaccone *et al.*, 1989c, 1992a; Scheuermann *et al.*, 1988), but its coexistence with peptides has not been fully investigated.

2. Regulatory Peptides

A rich variety of regulatory peptides is known to be contained in lung NE cells. It has long been argued that these cells produce peptides and biogenic amines that occur in a number of endocrine and NE cell types (Hökfelt *et al.*, 1980; Changeux, 1986; Heym and Kummer, 1988). However, the precise role and relationship between amines and peptides in NE cells are, at present, unknown.

Immunohistochemical studies have localized immunoreactivities for several peptides, including bombesin (Wharton *et al.*, 1978), gastrin-releasing peptide (GRP) (Tsutsumi *et al.*, 1983), leucine-enkephalin (Leu-enkephalin) (Cutz *et al.*, 1981), and calcitonin (Becker *et al.*, 1980; Gosney and Sissons, 1985) within single bronchopulmonary NE cells or neuroepithelial bodies.

The presence of CGRP was also reported more recently within the NE cells of the lung by several groups (Fig. 1) (Uddman *et al.*, 1985; Cadieux

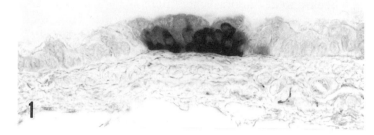

FIG. 1   Neuroendocrine cells in the rat lung stained by the ABC method with antiserum to the calcitonin gene-related peptide (CGRP). Staining is seen in a group of cells clustered in the bronchiolar epithelium. Magnification: ×560. (Micrograph courtesy of Prof. J. Polak, Department of Histochemistry, Royal Postgraduate Medical School, University of London, United Kingdom.)

et al., 1986; Martling *et al.*, 1988; Johnson and Wobken, 1987; Keith and Ekman, 1988). Coproduction of calcitonin and CGRP was detected by Shimosegawa and Said (1991) in the single and grouped NE cells in the respiratory epithelium of the rat. This coexpression of calcitonin and CGRP is of interest in view of the accepted concept of the calcitonin gene as encoding nucleotide sequences for both calcitonin and CGRP. The biological functions of the two peptides remain to be defined.

A number of the above-mentioned regulatory peptides are found to occur in the NE cells of human lung (Wharton *et al.*, 1978; Cutz *et al.*, 1981; Becker *et al.*, 1980) but not in the lungs of other animals (Lauweryns and Van Ranst, 1987; Polak and Bloom, 1982; Cadieux *et al.*, 1986).

It seems that cholecystokinin (CCK) is a peptide marker widely distributed in the mammalian lung. It was demonstrated by Wang and Cutz (1993) by light microscopic immunohistochemistry and immunoelectron microscopy in the pulmonary NE cells of humans and in a wide number of mammalian species. This demonstrates that the expression of this peptide in lung tissue is well conserved among species. Multiple molecular forms of CCK were identified in the endocrine cells and nerves of the diffuse NE system, where they probably act as both hormones and neurotransmitters (Dockray, 1983).

Cutz *et al.* (1991) and Wang *et al.* (1993) demonstrated expression of CCK mRNA in human infant and rat lungs, using methods of RNAse protection, Northern blot, and *in situ* hybridization. They found a possible regulation of the CCK gene in the development of the lung, with a maximal concentration of CCK mRNA after birth. By analogy with bombesin function (see above) a role has been suggested for CCK regarding their involvement in lung development and/or neonatal adaptation (Cutz *et al.*, 1985; Sunday *et al.*, 1990). Both bombesin and C-terminal tetradecapeptide of

GRP are growth factors for normal human bronchial epithelial cells (Willey *et al.*, 1984), although the growth factor-like properties of CCK are still to be documented.

The pattern of peptide coexistence and its functional implication in corresponding NE cells of different species is not known; neither are species differences in relation to their immunoreactivity to different peptides.

Identification of the numerous regulatory peptides and neuromodulators in pulmonary NE cells and their coproduction within a single cell, as well as the acquisition of genes sustaining the regulation of lung function, is more complex than initially realized, and they are recognized by expression of genes. The role played by the various peptides and their interactions between the various mediators remains to be determined.

3. Other Substances

Cholinesterase activity was found in the cytoplasm of NE cells in the fetal lung of the rabbit (Lauweryns and Cockelaere, 1973; Sonstengard *et al.*, 1982). The APUD ability (Pearse, 1969) of these cells was demonstrated by the neuroepithelial bodies of the embryonic rat lung, differentiating *in vitro*, which display acetylcholinesterase-containing granules (Scheuermann, 1987).

Chromogranin A-like immunoreactivity has been demonstrated in the cell granules of pulmonary NE cells (Lauweryns *et al.*, 1987). This acid carrier protein is known to occur universally in endocrine cells now categorized as paraneurons (Fujita and Kobayashi, 1973).

ATP and other adenine nucleotides are contained in the secretory granules of every endocrine paraneuronal cell as well as in the larger and smaller granules/vesicles of every neuron. But their presence in the bronchopulmonary NE cells remains to be proved.

Neuron-specific enolase (NSE), a brain-specific isozyme of the glycolytic enzyme enolase, is characterized by its consistent occurrence in the cytoplasm of mature neurons. Immunoreactivity for NSE has been found in almost all endocrine paraneurons of both sensory and endocrine nature (Iwanaga *et al.*, 1989). Neuron-specific enolase immunoreactivity has been found in the NE cells of bronchopulmonary paraneurons (Wharton *et al.*, 1981; Dayer *et al.*, 1983). The functional significance of NSE in the endocrine paraneurons remains to be elucidated.

Seldeslagh and Lauweryns (1993a,b) demonstrated immunoreactivity for endothelins (ETs) 1, 2, and 3 and for the precursor, big-ET-1, in single NE cells and NEBs in the lung of newborn cat, rat, hamster, and mouse. Endothelins coexist with both serotonin and CGRP (Seldeslagh and Lauweryns, 1993a). Endothelin isoforms are also colocalized with sarafotoxin

TABLE I

Serotonin and Peptide Markers of Neuroendocrine Cells in Airways of the Various Vertebrates

Marker	Vertebrate	Ref.
Serotonin	Mammals	Cutz *et al.* (1974, 1975)
		Lauweryns *et al.* (1973, 1977, 1986)
		Wasano (1977)
		Dey *et al.* (1981)
		McDowell *et al.* (1994a,b)
		Dey and Hoffpanir (1986)
	Birds	Lopez *et al.* (1983)
	Reptiles	Pastor *et al.* (1987)
		Adriaensen *et al.* (1991)
		Scheuermann *et al.* (1983, 1988)
	Amphibians	Cutz *et al.* (1986)
		Rogers and Haller (1978)
		Scheuermann *et al.* (1989)
		Goniakowska-Witalinska *et al.* (1992a)
	Air-breathing fish	Zaccone *et al.* (1989b,c)
		Adriaensen *et al.* (1990)
	Teleost fish	Zaccone *et al.* (1992a)
		Bailly *et al.* (1992)
Bombesin	Mammals	Wharton *et al.* (1978)
		Dey and Hoffpanir (1986)
	Birds	Lopez *et al.* (1993)
	Amphibians	Cutz *et al.* (1986)
Gastrin-releasing peptide	Mammals	Tsutsumi *et al.* (1983)
Calcitonin	Mammals	Becker *et al.* (1980)
		Cutz *et al.* (1981)
		Gosney and Sissons (1985)
		McDowell *et al.* (1994a,b)
Calcitonin gene-related peptide	Mammals	Uddman *et al.* (1985)
		Cadieux *et al.* (1986)
		Martling *et al.* (1988)
		Lauweryns and Van Ranst (1987)
		Johnson and Wobken (1987)
		Keith and Ekman (1988)
		McDowell *et al.* (1994b)
Enkephalin	Mammals	Cutz *et al.* (1981)
	Reptiles	Adriaensen *et al.* (1991)
	Amphibians	Goniakowska-Witalinska *et al.* (1992a)
	Teleost fish	Zaccone *et al.* (1992a)
	Lamprey	Ainis *et al.* (1990)
Cholecystokinin	Mammals	Wang and Cutz (1993)
Helodermin-like peptide	Mammals	Luts *et al.* (1991)
Chromogranin	Mammals	Lauweryns *et al.* (1987)
Neuron-specific enolase	Mammals	Wharton *et al.* (1981)

(Continued)

TABLE I   (*Continued*)

Marker	Vertebrate	Ref.
		Dayer *et al.* (1983)
Endothelin	Mammals	Seldeslagh and Lauweryns (1993a,b)
	Teleost fish	Zaccone *et al.* (1994)
Calcium-binding proteins		
Calbindin	Teleost fish	Zaccone *et al.* (1992b)
Calmodulin and S-100 protein	Teleost fish	Fasulo *et al.* (unpublished data)
Neuropeptide Y	Teleost fish	Wendelaar-Bonga (1993)

(Seldeslagh and Lauweryns, 1993b), a potent snake venom peptide that has coronary constrictor activity (Lee *et al.*, 1986).

C. Pulmonary Neuroepithelial Bodies in Mammals

Advances in the knowledge of the functions of pulmonary NE cells and neuroepithelial bodies have not been notable in comparison with their immunohistochemical detection. It is the small number and diffuse distribution of the cells that discourage direct cellular studies. Pioneers in this field of study were Cutz and associates, who have developed *in vitro* models for the study of NE cells and neuroepithelial bodies by using isolated enriched cell preparations derived from fetal rabbit lungs (Cutz *et al.*, 1985). Authors demonstrated that lung NE cells can be isolated and maintained in short-term culture. Low concentrations of fetal bovine serum enhanced the *in vitro* maintenance of NE cells, whereas nerve growth factor had no such effect. Fetal NE cells incorporated [^{3}H]thymidine *in vitro*, suggesting that the origin, cell kinetics, and differentiation of these cells may be similar to that of gut endocrine cells. The methodology aspects of this model provide an insight into the role (autocrine/paracrine) of the pulmonary NE cell system.

A variety of morphological, biochemical, and molecular approaches have been used by Cutz *et al.* (1993) to further validate their previous *in vitro* model. A detailed morphological characterization of cultured NEB cells has been carried out by the use of high-resolution light microscopy, TEM and scanning electron microscopy (SEM), high-performance liquid chromatography (HPLC), and molecular analysis of mRNA encoding 5-HT-synthesizing enzymes tryptophan 5-hydroxylase (TPH) and aromatic-L-amino-acid decarboxylase (AADC). Expression of TPH mRNA and AADC mRNA signal levels appeared higher in cultures in comparison with whole tissue (Figs. 2 and 3).

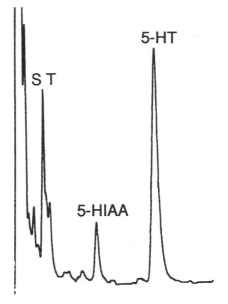

FIG. 2 Chromatogram of 3-day control culture of NEB cells using HPLC with electrochemical detector. A large peak of serotonin (5-HT) is shown. A small peak of 5-HIAA, a metabolite of 5-HT, is also present. ST, Standard. (From Cutz *et al.*, 1993, *Anat. Rec.* **236.** Copyright © 1993 permission of Wiley-Liss, a division of John Wiley and Sons, Inc.)

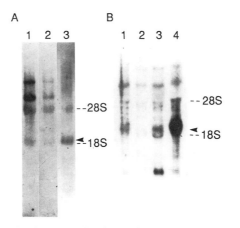

FIG. 3 Northern blots of 3-day control culture of NEB cells analyzed for TPH (A) and AADC (B) mRNA. Each lane was loaded with 15 μg of total RNA. (A) TPH mRNA marked by arrowhead; lane 1, rabbit fetal gut; lane 2, lung (control); lane 3, culture of NEB cells (fraction P50). (B) AADC mRNA marked by arrowhead; lane 1, rabbit fetal brain (positive control); lane 2, rabbit fetal liver (negative control); lane 3, rabbit fetal lung; lane 4, cell culture of NEB (fraction P50). (From Cutz *et al.*, 1993, *Anat. Rec.*, **236.** Copyright © 1993 permission of Wiley-Liss, a division of John Wiley and Sons, Inc.)

Additional experiments (Cutz *et al.*, 1993; Youngson *et al.*, 1993) have also been done to define the role of pulmonary NE cells as positive airway chemoreceptors. It has been reported that during hypoxia an increased exocytosis of dense-cored vesicles accompanied by a concomitant decrease in intracellular content of 5-HT in the NEB cultures was observed (Fig. 4). This suggests that the amount of 5-HT release is strictly correlated with the degree of hypoxia. An increased exocytosis of dense-cored vesicles comparable to the hypoxia experiments is also observed in an ultrastructural analysis of NEB cells exposed to Ca^{2+} ionophore (Fig. 5).

Other *in vitro* aspects of cell biology of pulmonary NE cells concern the paracrine effects of bombesin/GRP and other putative growth factors on these cells. These aspects have been illustrated by Speirs *et al.* (1993). It has been suggested that GRP would act in a paracrine rather than an autocrine manner. An autoradiographic technique combined with cytokeratin staining (Speirs *et al.*, 1993) demonstrated that GRP receptor sites are observed on fibroblasts, but not on the pulmonary NE cells. This leads to the hypothesis that GRP would act indirectly through the fibroblasts rather than directly on the NE cells. Epidermal growth factor (EGF), a pleiotrophic polypeptide that has been implicated to play a role in normal growth and development, including the lung (Stahlman *et al.*, 1989), had no effect on NE cell cultures (Speirs *et al.*, 1993).

Cell biology could be a useful approach to study the functions of NE cells in the airways of lower vertebrates (see below). The lung/gill of these animals, however, is a multifunctional organ that complicates experimental studies on the function of the NE cell system in the above-described vertebrates.

D. Evolutionary Considerations

Neuroendocrine cells in the lungs of air-breathing fish and the gills of teleosts are solitary. These cells may also be noninnervated or associated with nerve endings. Noninnervated NE cells, both single or clustered, have also been reported in the external gills of the neotenous salamander, *Ambystoma tigrinum* (Goniakowska-Witalinska *et al.*, 1993a). In the lung of the semiaquatic salamander *Hynobius nebulosus* Matsumura (1985) reported both solitary NE cells and NEBs, and Scheuermann *et al.* (1989) described single and grouped NE cells in the lung of the entirely aquatic *Ambystoma mexicanum*. In the terrestrial tiger salamander (*A. tigrinum*) the uptake of oxygen proceeds mainly via the lung. This salamander has NEBs with open-type cells and a most complicated structure, with an innervation similar to that of the mammalian NEBs (Goniakowska-Witalinska *et al.*, 1992a). These reports comparing NE cells in the lung

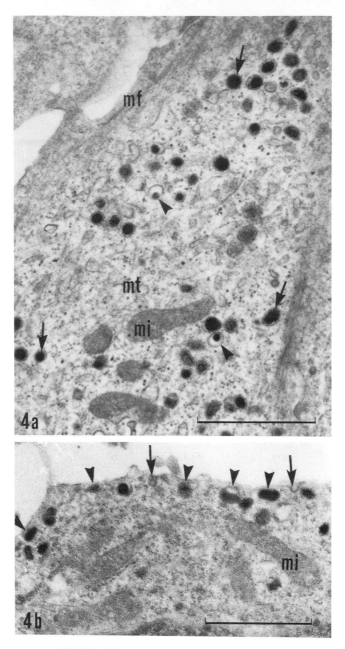

FIG. 4 Ultrastructure of NEB cells exposed *in vitro* to hypoxia compared with normoxic control. (a) Cytoplasm of control NEB cell maintained in normoxic environment with numerous DCVs, showing dense core (arrows) and a few partially filled DCVs (arrowheads). Bundles of microfilaments (mf), scattered microtubules (mt), and mitochondria (mi) are also present. The cytoplasm close to the plasma membrane contains no DCVs. Magnitification: ×31,000. (From Cutz *et al.*, 1993.) (b) Different exocytotic profiles in NEB cell exposed to

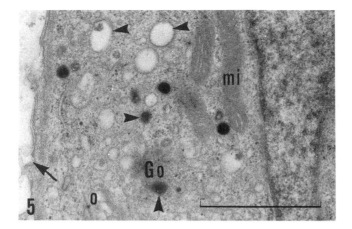

FIG. 5 Portion of NEB cell cytoplasm from a culture incubated with Ca^{2+} ionophore (5 μg/ml), showing exocytotic vesicles (arrow) and empty vesicles (arrowheads). Go, Golgi zone; mi, mitochondria. Magnification: ×32,000. (From Cutz *et al.*, 1993, *Anat. Rec.*, **236.** Copyright © 1993 permission of Wiley-Liss, a division of John Wiley and Sons, Inc.)

of amphibian reveal some correlation between the mode of life of animal and the structure of NEBs.

Interestingly, NEBs appear to be well developed in the lungs of terrestrial amphibia, whereas these structures seem not to be a characteristic of the lungs of aquatic amphibians (largely urodeles) and of the gills of fish. For a review of comparative studies on the NE cells in the lungs of amphibians, the reader is referred to the work of Goniakowska-Witalinska (1993). Our knowledge of the NE cells in reptiles is only fragmentary, except for the studies of Scheuermann *et al.* (1983) about NEBs in turtles. No more recent studies are available on the occurrence of NE cells in the avian airways except for those recorded by Cook and King (1969) and King *et al.* (1974).

Two kinds of endocrine cells in the gills of fish express Met-5-enkephalin and Leu-5-enkephalin, in addition to serotonin; these enkephalins do not coexist in the same cell. Serotonin is the main NE marker of both solitary cells and NEBs in the mammalian and amphibian lung. Enkephalins also appear in the NE system of the red-eared turtle (Adriansen *et al.*, 1991).

---

severe hypoxia. Several DCVs (arrowheads) are in contact with the plasma membrane and a few exocytotic vesicles. Magnification: ×28,000. (From Cutz *et al.*, 1993, *Anat. Rec.* **236.** Copyright © 1993 permission by Wiley-Liss, a division of John Wiley and Sons, Inc.)

These bioactive compounds are not fully ascertained in the NE cells of mammalian lung, whereas bombesin (GRP) is reported as a powerful marker of these cells in view of its involvement in the contraction of the airway smooth muscles and possibly also blood vessels.

It seems that the combinations of peptides coexisting in NE cells have changed to acquire particular physiological significance in the cells of species living in different environments. Therefore NE cells that appear physiologically similar but that contain different peptides may be fulfilling additional functions linked to the specific requirements of their target tissues. This may be true for the peptides of the gill area in fish and aquatic amphibians. It is the major site of both gas exchange and ion transport in comparison with the normal (terrestrial) pattern of obligatory pulmonary respiration in higher vertebrates.

III. Diffuse Neuroendocrine System of the Lung in Air-Breathing Fish

Since the introduction of the concept of diffuse neuroendocrine system (DNES) by Feyrter (1954), the APUD (amine precursor uptake and decarboxylase) system by Pearse (1968, 1969), or the paraneuronal system (Fujita *et al.*, 1988), particular attention has been devoted to the study of the anatomical characteristics and functional properties of such a regulatory system within the respiratory organs of mammals and, to some degree, of other tetrapods. Except for the monoamines found in the NE-like cells in the pulmonary mucosa of *Polypterus* (Scheuermann and De Groodt-Lasseel, 1982), almost no interest has been paid to the lungs of Dipnoi and the respiratory organs of ancient fish, in spite of their presumable importance for the understanding of the phylogenic aspects connected with the adaptation of the respiratory system during evolution.

Because serotonin represents the main monoamine of mammalian bronchopulmonary paraneurons and of amphibian NEBs (Fujita *et al.*, 1988), it may be expected to be contained in the corresponding pulmonary cells of *Polypterus* and *Protopterus*. Both immunofluorescence and peroxidase–anti-peroxidase (PAP) methods show serotonin-immunopositive NE cells scattered throughout the epithelial layer in all the specimens examined. For the most part, such cells are pyramidal in shape and of the open type, contacting both the basal membrane and the air lumen; only in the pulmonary epithelium of *Protopterus* do serotonin cells seem to be of the closed type, that is, not contacting the lumen (Zaccone *et al.*, 1989b,c). All the above-described NE cells are considered similar to those found within the walls of lung airways in mammals and submammalian tetrapods.

However, none of the studies on *Protopterus, Lepidosiren,* and *Neocerat-odus* have demonstrated the presence of endocrine cells in the respiratory epithelium (L. Goniakowska-Witalinska and C. J. Haller, personal communication).

Ultrastructural observations revealed the occurrence of NE cells, in small islets containing ciliated and goblet cells as well as several pneumocytes, in the lung of *Polypterus* (Zaccone *et al.,* 1989b), and in the ciliated epithelium of the pneumatic duct in *Protopterus* (Adriaensen *et al.,* 1990) (Figs. 6 and 7). These single NE cells are characterized by the occurrence of DCVs in the basal part of the cells, with an average diameter of 80–165 nm in *Polypterus* and of 30–150 nm in *Protopterus* (Adriaensen *et al.,* 1990). They have a well-developed Golgi apparatus, rough endoplasmic reticulum, and numerous mitochondria and microfilaments. Contact of DCVs and cell membrane in the basal part of the cell, where the secretion to the extracellular space occurs, is often seen. Their free surface is narrow and enriched with microvilli, the apical cytoplasm containing numerous electron-lucent vesicles. In *Polypterus,* intraepithelial nerve endings contacting the NE cells were not detected (Zaccone *et al.,* 1989c), whereas profiles of afferent-type nerve terminals were observed in the vicinity of NE cells in *Protopterus* (Adriaensen *et al.,* 1990). The presence of NE cells was observed in both the lung and gill of the bowfin *Amia calva* (Goniakowska-Witalinska *et al.,* 1992b, 1993b). Briefly, the NE cells in these epithelia are solitary or in groups of two to three cells and are surrounded and covered by epithelial cells. The NE cells are not inner-

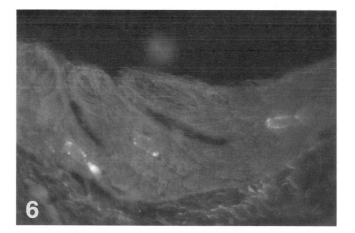

FIG. 6   Formaldehyde-induced fluorescent neuroendocrine cells in the ciliated epithelium ot the pneumatic duct of *Protopterus aethiopicus* lung. Magnification: ×425. (From Adriaensen *et al.,* 1990.)

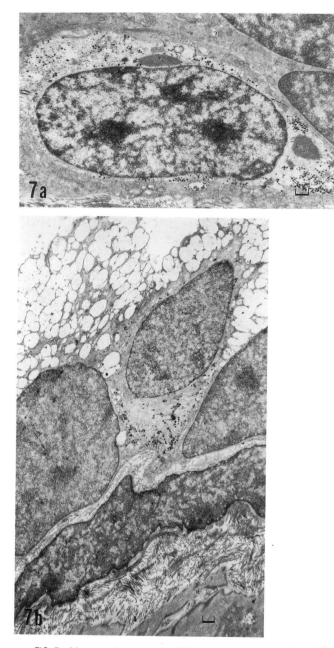

FIG. 7   Neuroendocrine cells (NE) in the basal part of the ciliated epithelium of the lung of
P. aethiopicus, with numerous electron-dense secretory granules. Magnification: (a) ×3400;
(b) ×3200. (From Adriaensen *et al.*, 1990.)

vated. The cytoplasm of NE cells contains a large nucleus with patches of condensed chromatin, numerous elongated mitochondria, free ribosomes, a well-developed Golgi apparatus and rough endoplasmic reticulum (RER), as well as characteristic DCVs with electron-dark interiors surrounded by clear space. The DCVs are dispersed throughout the cytoplasm (Fig. 8). Dense-cored vesicles range from 300 to 360 nm in diameter, whereas the DCVs in the lung are 110 to 1150 nm in diameter and polymorphic in shape.

The NE cells with small DCVs are characteristic of the lung epithelium of amphibia and other vertebrates (Goniakowska-Witalinska and Cutz, 1990), whereas NE cells with large DCVs occurring in gill epithelium were found exclusively in *A. calva* and the neotenous tiger salamander *A. tigrinum* (Goniakowska-Witalinska *et al.*, 1993a).

Immunohistochemical observations have revealed the occurrence of serotonin, neuropeptides (Leu-enkephalin and Met-enkephalin), and NSE in the NE cells of both the gills and the lung.

The exact function of NE cells and of the serotonin contained inside them remains obscure. It is well known that serotonin exerts a local vasoconstrictive action in the lung of mammals, in a paracrine way, induced by hypoxia (Fujita *et al.*, 1988). The frequent coexistence of serotonin and peptidic hormones points to a role in the synthesis, storage, and release of regulatory peptides, in other words a trophic function, during development and differentiation and in adaptation of the respiratory system at birth. Regulatory peptides may be discharged into intercellular spaces, fulfilling neurotransmitter and/or neuromodulator functions, or local hormones may be released into adjacent blood capillaries. The pattern of development of the NE system differs between higher and lower vertebrates. In mammals the NE system prevails during fetal life and reaches a peak during the neonatal period, reducing subsequently. In amphibians and fish, on the contrary, it is also well developed in the adult. It is noteworthy that serotonin had not yet been colocalized with any peptidic messenger in the aquatic vertebrates. In fish gill, NE cell regulatory peptides, and serotonin do not coexist in the same cell (see below).

IV. Diffuse Neuroendocrine System of the Gills in Teleost Fish

This section is mainly concerned with a description of our more recent immunohistochemical findings on the endocrine paraneurons in teleost gills. The localization of paraneuronal cells in fish gills initiated by Laurent, Dunel-Erb and Bailly, using histofluorescence procedures and more recently immunohistochemical methods to detect serotonin, represents im-

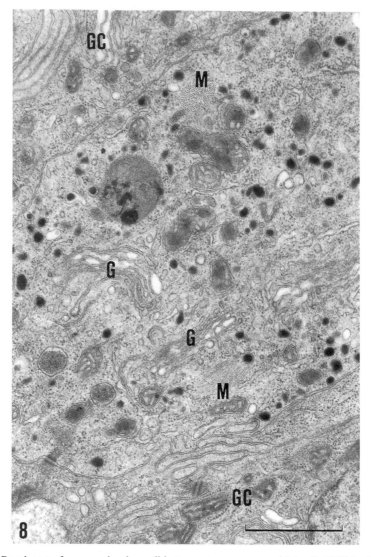

FIG. 8   Basal part of neuroendocrine cell between two mucous goblet cells (GC) in the air
bladder of the bowfin, *Amia calva*. Small, dense-cored vesicles are dispersed in the cytoplasm, but they are mainly grouped in the basal part of cell. Golgi (G) apparatus and
cross e/o longitudinally sectioned microfilaments (M) are also seen. Magnification: ×25,500.
(Micrograph courtesy of Prof. L. Goniakowska-Witalinska, Department of Comparative
Anatomy, Jagiellonian University, Cracow, Poland.)

portant progress in fish gill endocrine studies in assessing the role of NE cells of the gill as paracrine cells. Our contribution is now focused on the immunohistochemical characterization of neuropeptides in open-type NE cells, which are reported here for the first time. Because the gills provide physiological evidence for an anatomical dichotomy of respiratory and osmoregulatory functions, the localization of serotonin and regulatory peptides in the NE cells represents one of the most fundamental approaches to understand the phylogenic aspects of gill functioning. The reader will find an excellent description of gill morphology in the reviews by Laurent (1984, 1989).

The gills are as sophisticated as the lungs of tetrapods and consist of a complex arrangement of several epithelia vascularized by different circulatory circuits. The gill shows a complex organization fitting two functions: respiration and osmoregulation. They are important physiological processes and modulation of the epithelial functioning is under the control of hormones and neural factors.

Ultrastructural (Dunel-Erb et al., 1982) and immunohistochemical evidence has led to the discovery of receptosecretory cells (Bailly et al., 1992; Zaccone et al., 1992b) within the gill filament and lamellar epithelia that qualify as paraneurons according to the terminology of Fujita et al. (1988). Therefore the gills may be the site of entero- and exteroceptors presumably sensing characteristics of the internal and external milieu.

A first indication in this direction has been the presence of serotonin, reported by some authors at the ultrastructural level, in cells of the branchial epithelium, considered to be of the NE type, in selachian, chondrostean (sturgeon), and teleostean species, thus suggesting the probable involvement of serotonin in the modulation of fish gill function (Dunel-Erb et al., 1982; Bailly et al., 1992). However, further studies are required to detect whether some regulatory peptides are also present in these cells. In view of the relevance of these aspects for understanding the mechanisms implicated in the passage from the branchial to the pulmonary respiration, we contributed to the elucidation of the distributional pattern of NE cells in the branchial epithelium. That is why we tried to immunohistochemically characterize these elements by localizing a panel of substances already known as typical markers for NE cells of mammalian airways. Preliminary results are also available regarding the localization of NE cells using both ultrastructural and argyrophilic methods. Besides serotonin, immunoreactivity for human calcitonin, amphibian bombesin, the synthetic Met-5- and Leu-5-enkephalins, and other specific markers such as NSE and chromogranin has been tested. Of all the antibodies tested, positive reactions were noted with serotonin in NE cells of the closed type, and with the two enkephalins in NE cells of the open type. In serial sections we have verified the colocalization of the enkephalins (Met- and Leu-) in the same

NE cells, whereas serotonin always occurs alone, in contrast to the mammalian lung, where it is present together with the enkephalins (Zaccone *et al.*, 1992a). Neuron-specific enolase was found to occur in both the two cell types.

From a morphological point of view, in all the species examined the NE cells have been checked solely or in small groups, with both more numerous in trout, eel, *Ictalurus,* and *Heteropneustes fossilis* compared to other species [the blenny (*Blennius sanguinolentus*) and the lungfish *Protopterus annectens*]. The cells often are ovoidal in shape and distributed throughout the filament epithelium, with a predominance of cell clusters toward the filament ends (Dunel-Erb *et al.*, 1982; Zaccone *et al.*, 1992a) (Figs. 9 and 10). Neuroendocrine cells were also located in the fish gill filament, using indolamine and serotonin immunocytochemistry (Bailly *et al.*, 1992) (Fig. 11), and neuropeptide Y (NPY) immunostaining (Wendelaar-Bonga, 1993). Neuropeptide Y is a potent constrictor of mammalian blood vessels. Its release is in general associated with a catecholamine. The occurrence of NPY in gill NE cells make a significant contribution in certain vascular areas of the gill by strengthening the effect

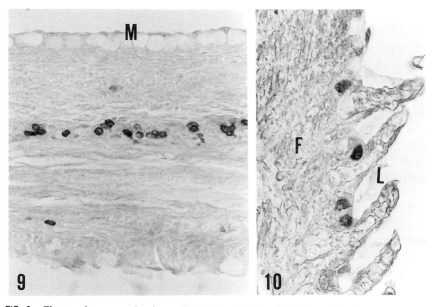

FIG. 9 Clustered neuroendocrine cells in the distal end thicker epithelium of trout gill filament, immunostained for serotonin. Note the concentration of the immunoreactive cells in the basal epithelium. M, mucous cell. Magnitification: ×340.

FIG. 10 Met-5-enkephalin-immunoreactive neuroendocrine cells at the basis of trout gill lamellae. F, gill filament; L, gill lamella. Magnification: ×340.

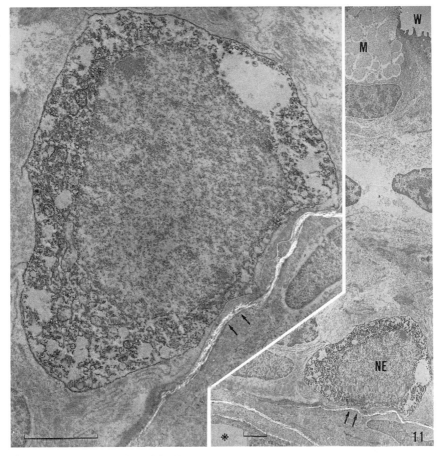

FIG. 11   Longitudinal section of the distal half of the gill filament of rainbow trout (*Oncorhynchus mykiss*) immunostained for serotonin, showing the presence of a neuroendocrine cell (NE) in the basal epithelium facing toward an efferent lamellar artery (asterisk). Cell cytoplasm is filled with numerous intensely labeled granular vesicles (inset). Arrows indicate the basement membrane. M, Mucous cell; W, external environment. Bar: 1 μm. (Modified from Bailly *et al.*, 1992, *Anat. Rec.*, **233**. Copyright © 1992. Reprinted by permission of John Wiley & Sons, Inc.)

produced by sympathetic nerve stimulation. The gills are also the prime source of calcitonin gene-related peptide (CGRP) (Fouchereau-Peron *et al.*, 1990; Arlot-Bonnemains *et al.*, 1991). However, we were not able to demonstrate its presence in NE cells by immunohistochemistry (G. Zaccone, S. Fasulo and A. Mauceri, unpublished data). This may be due to differences in CGRP molecules among vertebrate species, resulting in a lack of cross-reaction with mammalian antibodies.

Cytokeratin immunoreactivities should also be found in NE cells of the gills of the catfish *H. fossilis,* using AE1 and KL1 antibodies (Ainis *et al.,* 1993). Simple-type keratins (not synthesized by epidermal keratinocytes) were found by immunohistochemistry in the NE Merkel cells of mammalian and fish skin (Moll *et al.,* 1984; Saurat *et al.,* 1984; Markl *et al.,* 1989; Wollina, 1990), and regarded as good markers of epithelial NE cells of the gills for future investigations. Current investigations conducted at our laboratories demonstrate the occurrence of S-100 protein, calmodulin, and parvalbumin in the gills of *H. fossilis* (S. Fasulo, A. Mauceri, G. Tagliafierro, M. B. Ricca, P. L. Cascio, and L. Ainis, unpublished observation) (Figs. 12 and 13). They are calcium-binding proteins and markers of the DNES of fish gills. They were found to occur in NE cells, except for parvalbumin, which was located in the surface pavement cell layers. Their function may be correlated with a regulation of intracellular calcium levels, like calbindin, and thus with participation in the calcium-modulated cellular processes. Immunoreactivity for S-100 protein has been extensively observed in sensory and endocrine organs of mammals (Nakajima *et al.,* 1984).

Their role in gills can only be hypothesized, in view of the lack of comparative, and functional data, which remain a challenging task for the coming years. Indeed, the gills are multifunctional organs heavily implicated not only in respiratory exchanges, but also in regulation of the acid–base balance, osmoregulation (mediating ionic absorption and excretion), and (at least in freshwater fish) in selective absorption of Ca^{2+}. In this respect it is worth mentioning the regulation of the branchial osmoregulatory function by 5-HT-dependent mechanisms or the powerful action of this amine on the transepithelial movements of water and electrolytes (Berridge and Schlue, 1978; Donowitz *et al.,* 1980; Zimmermann and Binder, 1984; Ahlman *et al.,* 1984). In addition, evidence (Zaccone *et al.,* 1992b) has shown the presence of calbindin DK28 in NE cells identified in both gill filament and lamellar epithelia (Fig. 14). An as yet unidentified role for calbindin relates to Ca^{2+} signaling coupled to calbindin expression. We speculate that calcium signals could be transmitted to chloride cells—a target organ for the uptake of calcium (G. Zaccone, S.

FIG. 12   Neuroendocrine cells in the gill filament of the catfish *H. fossilis,* immunostained for S-100 protein. Magnification: ×340.

FIG. 13   Neuroendocrine cells in the distal end of a gill filament (*H. fossilis*) immunostained for calmodulin. Magnification: ×340.

FIG. 14   Neuroendocrine cells in the gill filament and lamellar epithelium of the catfish *H. fossilis,* immunostained with an antibody to calbindin DK28. Magnification: ×340.

FIG. 15   Endothelin immunoreactivity in neuroendocrine cells of the conger eel (*Conger conger*) gill filament epithelium. Magnification: ×340.

FIG. 16   Endothelin-immunopositive neuroendocrine cells in the gill filament and lamellar epithelium of the catfish *H. fossilis.* Magnification: ×340.

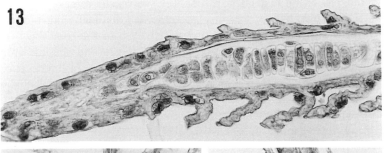

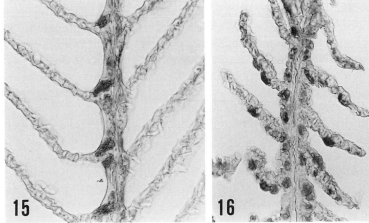

Fasulo, L. Ainis, unpublished observations). The above-described calcium-binding proteins, that is, calmodulin, calbindin, and parvalbumin, are also seen to be expressed ectopically and therefore are found in the surface epithelial cell layers in the skin and gills of some teleost species (Zaccone et al., 1989a, 1992b). This is also in agreement with findings by Wollina (1993), who found calmodulin immunopositivity in the surface skin epithelium of the salamander A. mexicanum and immunoreactivity in the deeper compartments of the skin of terrestrial vertebrates, thus speaking in favor of the involvement of calmodulin-reactive cells in osmoregulatory processes (Zaccone et al., 1989a; Wollina, 1993).

With regard to their morphology and content of bioactive substances, a paracrine function should be attributed to these NE cells (Bailly et al., 1992; Zaccone et al., 1992b). Probably the two types of NE cells that are recognized in the branchial epithelium have different regulatory functions, as suggested by Zaccone et al. (1992b) and Dunel-Erb (1994). Those of the closed type might exert a control on oxygen levels of water flow. In fact, the NE cells located in the filament epithelium of the leading edge are on the side of "inhaled water" and store serotonin (Dunel-Erb, 1994). It is known that the vasoconstrictive effect of the serotonin released from mammalian lung NE cells, in response to hypoxia, displaces the blood flow toward the well-oxygenated pulmonary districts. In the basal part of NE cells, where serotonin granules are secreted, nerve endings are present, and efferent lamellar arteries are found close to these. In trout, in which hypoxia has been investigated, a release of serotonin from NE cells in the surrounding tissues has been reported. It has also been suggested that it exerts direct action on lamellar arteries by vasoconstriction (Dunel-Erb, 1994).

Neuropeptides found in open-type NE cells were correlated with a paracrine function on the surrounding cells, that is, the chloride cells of the gill epithelium. These cells are scattered along the filament epithelium and proliferate when the water ionic composition is manipulated.

Enkephalin-containing NE cells can perceive stimuli from the external environment and so might exert a chemoreceptive function by local modulation of the branchial osmoregulatory functions and by paracrine secretion of substances such as serotonin and neuropeptides that are produced by two different types of cell (Zaccone et al., 1992a).

Both ultrastructural and immunohistochemical methods have been used to differentiate the nerves associated with NE cells in the gills of fishes. Bailly et al. (1992), using the electron microscope, found two main morphological types of nerve profiles, which were located beneath or close to NE cells in the gills of trout. Both have already been described as facing presynaptic accumulations of exocytotic granular vesicles in the NE cells (Dunel-Erb et al., 1982). Only the endings of the second type belong to serotonin filament neurons. They contact the basal poles of NE cells in the gills. In addition to serotonergic innervation, NE cells are innervated by a catecholaminergic sympathetic nervous system with varicose nerve fibers located beneath the NE cells and around the blood vessels (Dunel-Erb and Bailly, 1986; Dunel-Erb et al., 1989; Dunel-Erb, 1994). According to Laurent, Dunel-Erb, and Bailly the nerve functioning is influenced by serotonin, other than indolamines. A double branchial respiratory and osmoregulatory function by serotonin-dependent mechanisms has been postulated. A hypoxic signal from NE cells should be monitored by these cells, which release diffusing neuroactive agents such as serotonin in a paracrine manner (Bailly et al., 1992). Also, the ion regulation may be affected by serotonin, which stimulates adenylate cyclase activity in the ionocytes of the gill. Our present studies in cooperation with S. E. Wendelaar-Bonga are being focused on the interrelationship between the adenylate cyclase activity within the ionocytes and the NE cells acting as paracrine cells in ion transport. Immunohistochemistry is being used to differentiate between the two cell types, using histochemical antibodies both to adenylate cyclase activity and neuroactive substances.

Neuroendocrine cells of fish gill were detected by endothelin immunohistochemistry (Zaccone et al., 1994). They were found in the gill filament and lamellar epithelium of several species of teleost (Figs. 15 and 16), including the bowfin. These findings represent the first study to demonstrate cell localization of endothelin in the NE cells of the airway of lower vertebrates. These also suggest an old evolutionary origin and biological significance of endothelins, which until now have been located only in mammalian (NEBs (see above).

Neuroendocrine cells have also been detected in the gill epithelium of *Lampetra japonica* (Ainis et al., 1990). They are differentiated from other epithelial cells by a positive immunoreaction for Leu-enkephalin, while reacting negatively to serotonin and bombesin. Gill innervation is poor in lampreys; in addition, the vascular circuits are rather innervated in this group. It appears that localization of NE cells serving as receptors remains difficult and awaits further investigations.

Serotonin-containing cells in the peribranchial epithelium covering fish gill filaments in the lancelet, *Branchiostoma floridae*, were also reported by Holland and Holland (1993). These cells might be homologous to those found in fish gills.

V. Accessory Respiratory Organs

The organs for air breathing have evolved independently among many fish species. As a result, they have elaborated a variety of anatomical, physiological, biochemical, and behavioral strategies for gas exchange.

Like most aquatic organisms, fish possess multiple respiratory sites, either external (e.g., the general body surface or external gills) or internal gills or air sacs.

Indian tropical air-breathing fishes show the presence of air-breathing organs (AROs) that are different from the lung/swimbladder found among members of the Polypteridae. In many of them, the AROs are modified gills (Hughes and Munshi, 1973a,b; Munshi, 1962). Also, the lungs of members of the Dipnoi and Polypteridae are of different origin embryologically than are the air sacs found among members of the holostei and primitive teleostei. Only the former structures are homologous with the lungs of the tetrapods, and therefore used for air–gas exchange. Accessory respiratory organs are structures utilized by fish coping with hypoxic fresh water. Unfortunately there is little information about the presence of widespread epithelial cells with NE function in the above-described structures evolved for air-breathing purposes.

The cells illustrated by Hughes and Munshi (1973a) look like solitary chemosensory cells of the skin (M. Whitear, personal communication), which have also been reported in the gill epithelia of various teleosts, usually on the pharyngeal bars. A chemoreceptor-like function is postulated. Distal to the nucleus vesicles typically appear that may be electron lucent or have electron-dense contents, but that do not have a "core"—a characteristic feature of airway cells that have a chemoreceptor or paracrine function.

Immunohistochemical investigations on the occurrence of NE cells in the respiratory organs of the Indian catfish *H. fossilis* and *Anabas testudineus* are under way in our laboratory.

Some preliminary investigations conducted on *Heteropneustes* showed that more immunoreactive NE cells are present in the gills than in the air sac epithelium. This could be because this catfish is clearly more adapted to an aquatic existence (Hughes *et al.*, 1992). By contrast, some air-breathing fish (e.g., *Amia*) have NE cells in their respiratory organs (Goniakowska-Witalinska *et al.*, 1993b).

VI. Concluding Remarks

To date the ultrastructural and immunohistochemical features as well as the bioactive components of the endocrine paraneurons of the lung in air-breathing species and of the gill in teleost fish are not yet fully investigated. The function of these paraneurons, as well as their role during the transition from the gill to pulmonary respiration, is far from clear. Unlike the NE cells of mammalian lung, the NE cells of *Polypterus* lung (Zaccone *et al.*, 1989b,c) appears devoid of innervation, whereas in teleost species their connections with both central and branchial nervous systems suggest a possible chemoreceptor role (Bailly *et al.,* 1992).

The shape of NE cells, whether solitary or grouped into clusters, with an apical portion reaching the surface epithelium as evidenced in *Polypterus,* and the open-type cells of teleost gill epithelium (Zaccone *et al.,* 1992b), are indicative of a receptor function. Furthermore, in the gill area most of these cells are located nearest to chloride cells and might influence the physiological state of these cells in a paracrine way. Calbindin DK28, calmodulin, and S-100 protein are considered novel markers of NE cells of teleost gills; these may be functionally related to the calcium-modulated cellular processes in the gills (Zaccone *et al.,* 1992b; Fasulo *et al.,* 1993). The function of endothelin in the NE cells of fish gills is not known.

Most studies on dipnoan lung focus on gross anatomy, morphometry, and physiology of gas exchange, but have failed to demonstrate pulmonary endocrine paraneurons in its respiratory epithelium. Thus, further investigations are needed to clarify the functional relationship between the endocrine paraneurons and the respiratory organs in fish.

The presence of NE elements in the gill and the swimbladder/lung of fish might indicate a common ancestry with the tetrapods, but NE elements may differ in their distribution patterns and functional significance, fulfilling various functions besides the respiration processes. Their presence is restricted to the pneumatic duct region and adjacent parts of the common anterior chamber of some species, but the functional and the developmental significance of these locations is not fully understood.

In *Polypterus* both the lung and the pneumatic duct region are the histological sites of NE cells, as demonstrated by serotonin immunohistochemistry. Evidence is emerging that NE cells in the lung and gills of fish contain a combination of biologically active substances, but it is difficult to formulate any physiological generalizations, because the combination is different from that encountered in the NE cells of mammalian vertebrates, where neuropeptides coexist with amines.

Although the immunoreactivity to regulatory peptides differs among species, it seems that some neuropeptides are consistently associated with specific pathways. Enkephalins were, for example, first located in NE cells of gill tissues and have never been identified unequivocally in the pulmonary cells of mammals.

The presence of NE cells in both the lung and gills of phylogenically ancient fish (i.e., *A. calva*) is reiterated, as its helps to understand the evolution of respiratory processes in vertebrates.

No accounts are available on the endocrine nature of air-breathing organs in those fish showing structural adaptations of the gill apparatus for gas exchange.

We can conclude that the paraneuronal cells found in the lung and gill of air-breathing fish and teleosts share secretory granules of similar structure and peptides present in the endocrine cells of the regulatory system of the vertebrate body. It is possible that bioactive substances in the airways of fish are diversified to some extent to fulfill specific roles, owing to the more complex functions of respiratory organs in these animals in comparison with those occurring in the lungs of terrestrial vertebrates.

Acknowledgments

Professor D. W. Scheuermann gave useful comments on a draft manuscript, and Dr. S. Dunel-Erb kindly made a careful revision before the finalization of the paper. Thanks are also due to those colleagues who made micrographs of NE cells available during the preparation of the manuscript.

References

Adriaensen, D., Scheuermann, D. W., Timmermans, J.-P., and De Groodt-Lasseel, M. H. A. (1990). Neuroepithelial endocrine cells in the lung of the lungfish *Protopterus aethiopicus:* An electron and fluorescence-microscopical investigation. *Acta Anat.* **139**, 70–77.

Adriaensen, D., Scheuermann, D. W., and De Groodt-Lasseel, M. H. A. (1991). Coexistence of calcitonin gene-related peptide, enkephalin and serotonin in the neuroepithelial endocrine system of the lung of the red-eared turtle. *Verh. Anat. Ges.* **85**, (*Anat. Anz., Suppl.* **170**), 437–438.

Ahlman, H, Grönstad, K., Nilsson, O., and Dahlström, A. (1984). Biochemical and morphological studies on the secretion of 5-HT into the gut lumen of the rat. *Biog. Amines* **1**, 63–73.

Ainis, L., Tagliafierro, G., Licata, A., Lo Cascio, P., Mauceri, A., and Zaccone, G. (1990). Leu-enkephalin-like immunoreactivity in the neuroepithelial cells of lamprey gills. *Proc. Unione Zool. Ital. (UZI) Meet.,* 387.

Ainis, L., Tagliafierro, G., Mauceri, A., Licata, A., Ricca, M. B., and Fasulo, S. (1993). Cytokeratin patterns of the skin and gill epithelia in the freshwater Indian catfish *Heteropneustes fossilis. Eur. J. Histochem.* **37**, Suppl., 9.

Arlot-Bonnemains, Y., Fouchereau-Peron, M., Jullienne, A., Milhaud, G., and Moukhtar, M. S. (1991). Binding sites of calcitonin gene-related peptides (CGRP) to trout tissues. *Neuropeptides (Edinburgh)* **20**, 181–186.

Bailly, Y., Dunel-Erb, S., and Laurent, P. (1992). The neuroepithelial cells of the fish gill filaments: Indolamine-immunocytochemistry and innervation. *Anat. Rec.* **233**, 143–161.

Becker, K. L., Monaghan, K. G., and Silva, O. L. (1980). Immunocytochemical localization of calcitonin in Kulchitsky cells of human lung. *Arch. Pathol. Lab. Med.* **104**, 196–198.

Bensch, K. G., Corrin, B., Pariente, R., and Spencer, H. (1968). Oat-cell carcinoma of the lung: Its origin and relationship to bronchial carcinoid. *Cancer (Philadelphia)* **22**, 1163–1172.

Berridge, M. J., and Schlue, W. R. (1978). Ion-selective electrode studies on the effects of 5-hydroxytryptamine on the intracellular level of potassium in an insect gland. *J. Exp. Biol.* **78**, 203–216.

Burggren, W. W., and Johansen, K. (1986). Circulation and respiration in lungfishes. *J. Morphol., Suppl.* **1**, 217–236.

Burggren, W. W., Johansen, K., and McMahon, B. (1985). Respiration in phyletically ancient fishes. *In* "Evolutionary Biology of Primitive Fishes" (R. E. Foreman, A. Gorbman, J. M. Dodd, and R. Olson, eds.), pp. 217–252. Plenum, New York.

Cadieux, A., Springall, D. R., Mulderry, P. K., Rodrigo, J., Ghatei, M. A., Terenghi, G., Bloom, S. R., and Polak, J. M. (1986). Occurrence, distribution and ontogeny of CGRP immunoreactivity in the rat lower respiratory tract: Effect of capsaicin treatment and surgical denervations. *Neuroscience* **19**, 605–607.

Changeux, J. P. (1986). Coexistence of neuronal messengers and molecular selection. *Prog. Brain Res.* **68**, 373–403.

Cook, R. D., and King, A. S. (1969). A neurite-receptor complex in the avian lung: Electron microscopical observation. *Experientia* **25**, 1162–1164.

Cutz, E., Chan, W., Wong, V., and Conen, P. E. (1974). Endocrine cells in rat fetal lungs. *Lab. Invest.* **30**, 458–464.

Cutz, E., Chan, W., Wong, V., and Conen, P. E. (1975). Ultrastructure and fluorescence histochemistry of endocrine (APUD-type) cells in tracheal mucosa of human and various animal species. *Cell Tissue Res.* **158**, 425–437.

Cutz, E., Chan, W., and Sonstegard, K. S. (1978). Identification of neuro-epithelial bodies in rabbit fetal lungs by scanning electron microscopy: A correlative light, transmission and scanning electron microscopic study. *Anat. Rec.* **192**, 459–466.

Cutz, E., Chan, W., and Track, N. S. (1981). Bombesin, calcitonin and Leu-enkephalin immunoreactivity in endocrine cells of human lung. *Experientia* **37**, 765–767.

Cutz, E., Gillan, J. E., and Track, N. S. (1984). Pulmonary endocrine cells in the developing human lung and during neonatal adaptation. *In* "The Endocrine Lung in Health and Disease" (K. L. Becker and A. F. Gazdar, eds.), pp. 210–231. Saunders, Philadelphia.

Cutz, E., Yeger, H., Wong, V., Bienkowski, E., and Chan, W. (1985). *In vitro* characteristics of pulmonary neuroendocrine cells isolated from rabbit fetal lung. I. Effects of culture media and nerve growth factor. *Lab. Invest.* **53**, 672–683.

Cutz, E., Goniakowska-Witalinska, L., and Chan, W. (1986). An immunohistochemical study of regulatory peptides in lungs of amphibians. *Cell Tissue Res.* **244**, 227–233.

Cutz, E., Wang, D., and Perrin, D. G. (1991). Bombesin/gastrin releasing peptide and cholecystokinin gene expression in lungs of sudden infant death syndrome victims. *Lab. Invest.* **64**, 119A.

Cutz, E., Speirs, V., Yeger, H., Newman, C., Wang, D., and Perrin, D. G. (1993). Cell biology of pulmonary neuroepithelial bodies—Validation of an *in vitro* model. I. Effects of hypoxia and Ca^{2+}-ionophore on serotonin content and exocytosis of dense core vesicles. *Anat. Rec.* **236**, 41–52.

Dayer, A. M., Kapanci, Y., Rademakers, A., Marangos, P. J., De Mey, J., and Will, J. A. (1983). Neuron-specific enolase and serotonin distribution in the fetal rhesus monkey lung by immunocytochemistry. *Fed. Proc., Fed. Am. Soc. Exp. Biol.* **42**, 798.

Dayer, A. M., De Mey, Y., and Will, J. A. (1985). Localization of somatostatin-, bombesin-, and serotonin-like immunoreactivity in the lung of the fetal rhesus monkey. *Cell Tissue Res.* **239**, 621–625.

De Groodt, M., Lagasse, A., and Sebruyns, M. (1960). Elektronen-mikroskopische morphologie der Lungenalveolen des *Protopterus* und *Amblystoma*. *Int. Conf. Electron Microsc., Proc. 4th, 1958*, pp. 418–421.

Dey, R. D., and Hoffpanir, J. M. (1986). Ultrastructural colocalization of the bioactive mediators 5-hydroxytryptamine and bombesin in endocrine cells of human fetal airways. *Cell Tissue Res.* **246**, 119–124.

Dey, R. D., Echt, R., and Dinerstein, R. J. (1981). Morphology, histochemistry and distribution of serotonin-containing cells in tracheal epithelium of adult rabbit. *Anat. Rec.* **199**, 23–31.

Dockray, G. J. (1983). Cholecystokinin. *In* "Brain Peptides" (D. T. Krieger, M. J. Brownstein, and J. B. Martin, eds.), pp. 851–869. Wiley, New York.

Donowitz, M., Charney, A. N., and Asarkof, N. (1980). Effect of serotonin on active electrolyte transport in rabbit ileum, gall bladder and colon. *Am. J. Physiol.* **239**, G463–G472.

Dunel-Erb, S. (1994). The neuroepithelial cells of the fish gill filament. *In* "Advances in Fish Research" (B. R. Singh, ed.). Narendra, Delhi.

Dunel-Erb, S., and Bailly, Y. (1986). The sphincter of the efferent filament artery in teleost gills. II. Sympathetic innervation. *J. Morphol.* **187**, 239–246.

Dunel-Erb, S., Bailly, Y., and Laurent, P. (1982). Neuroepithelial cells in fish gill primary lamellae. *J. Appl. Physiol.* **53**, 1342–1353.

Dunel-Erb, S., Bailly, Y., and Laurent, P. (1989). Neurons controlling the gill vasculature in five species of teleosts. *Cell Tissue Res.* **255**, 567–573.

Ericson, L. E., Håkanson, R., Larson, B., Owmann, C., and Sundler, F. (1972). Fluorescence and electron microscopy of amine-storing enterochromaffin-like cells in tracheal epithelium of mouse. *Z. Zellforsch. Mikrosk. Anat.* **124**, 532–545.

Falck, B., Hillarp, N. A., Thieme, G., Torp, A. (1962). Fluorescence of catecholamines and related compounds condenseded with formaldehyde. *J. Histochem. Cytochem.* **10**, 348–354.

Feyrter, F. (1954). Zur pathologie des argyrophilen Helle-Zellen-Organes in bronchialbaum des menschen. *Virchows Arch. Pathol. Anat. Physiol.* **325**, 723–732.

Fouchereau-Peron, M., Arlot-Bonnemains, Y., Taboulet, J., Milhaud, G., and Moukhtar, M. S. (1990). Distribution of calcitonin gene-related peptide and calcitonin-like immunoreactivity in trout. *Regul. Pept.* **27**, 171–179.

Fox, H. (1965). Early development of the head and pharynx of *Neoceratodus* with a consideration of its phylogeny. *J. Zool.* **146**, 470–554.

Fröhlich, F. (1949). Die "Helle-Zelle" der Bronchialschleimhaut und ihre beziehungen zum Problem der Chemoreceptoren. *Frankf. Z. Pathol.* **60**, 517–559.

Fujita, T., and Kobayashi, S. (1973). The cells and hormones of the GEP endocrine system—current studies. *In* "Gastroentero-pancreatic Endocrine System: A Cell-biological Approach" (T. Fujita, ed.), pp. 1–16. Igaku-Shoin, Tokyo.

Fujita, T., Kanno, T., and Kobayashi, S. (1988). "The Paraneuron." Springer-Verlag, Tokyo.

Goniakowska-Witalinska, L. (1993). The lungs of Amphibia. *In* "The Respiratory System of Non-mammalian Vertebrates" (L. M. Pastor, ed.), Vol. 2. Murcia University Press.

Goniakowska-Witalinska, L., and Cutz, E. (1990). Ultrastructure of neuroendocrine cells in the lungs of three anuran species. *J. Morphol.* **203**, 1–9.

Goniakowska-Witalinska, L., Lauweryns, J. M., Zaccone, G., Fasulo, S., and Tagliafierro, G. (1992a). Ultrastructure and immunocytochemistry of the neuroepithelial bodies in the lung of the tiger salamander, *Ambystoma tigrinum* (Urodela, Amphibia). *Anat. Rec.* **234**, 419–431.

Goniakowska-Witalinska, L., Zaccone, G., Fasulo, S., Youson, J., Licata, A., and Contini, A. (1992b). Neuroepithelial neuroendocrine cells in the lung and the gill of the bowfin, *Amia calva. Proc. Conf. Eur. Comp. Endocrinol., 16th,* University of Padova, p. 205.

Goniakowska-Witalinska, L., Zaccone, G., and Fasulo, S. (1993a). Immunocytochemistry and ultrastructure of the solitary neuroepithelial cells in the gills of the neotenic tiger salamander *Ambystoma tigrinum* (Urodela, Amphibia). *Eur. Arch. Biol.* **104**, 45–50.

Goniakowska-Witalinska, L., Zaccone, G., Fasulo, S., and Youson, J. (1993b). The neuroendocrine cells in the gills of the bowfin *Amia calva.* The ultrastructural and immunocytochemical study. Submitted for publication.

Gosney, J. R., and Sissons, M. C. J. (1985). Widespread distribution of bronchopulmonary endocrine cells immunoreactive for calcitonin in the lung of the normal adult rat. *Thorax* **40**, 194–198.

Gould, V. E. (1983). The endocrine lung. *Lab. Invest.* **48**, 507–509.

Gould, V. E., and Chejfec, G. (1978). Ultrastructural and biochemical analysis of "undifferentiated" pulmonary carcinomas. *Hum. Pathol.* **9**, 377–384.

Gould, V. E., Linnoila, R. I., Memoli, V. A., and Warren, W. H. (1983). Biology of disease neuroendocrine components of the bronchopulmonary tract: Hyperplasias, dysplasias and neoplasms. *Lab. Invest.* **49**, 519–537.

Grigg, G. C. (1965). Studies on the Queensland lungfish, *Neoceratodus forsteri* (Krefft). *Aust. J. Zool.* **13**, 243–253.

Hage, E. (1972a). Electron microscopic identification of endocrine cells in the bronchial epithelium of human foetuses. *Acta Pathol. Microbiol. Scand., Sect. A* **80A**, 143–144.

Hage, E. (1972b). Endocrine cells in the bronchial mucosa of human foetuses. *Acta Pathol. Microbiol. Scand., Sect. A* **80A**, 225–234.

Hage, E. (1973). Electron microscopic identification of several types of endocrine cells in the bronchial epithelium of human foetuses. *Z. Zellforsch. Mikrosk. Anat.* **141**, 401–412.

Hage, E. (1974). Histochemistry and fine structure of endocrine cells in fetal lungs of rabbit, mouse, and guinea-pig. *Cell Tissue Res.* **149**, 513–524.

Hage, E., Hage, J., and Juel, G. (1977). Endocrine-like cells of the pulmonary epithelium of the human adult lung. *Cell Tissue Res.* **178**, 39–48.

Haller, C. J. (1992). Evidence for the coexistence of serotonin and calcitonin gene-related peptide at the subcellular level in neuroepithelial bodies in the lung of a marsupial, *Isoodon macrourus*. *Cell Tissue Res.* **270**, 199–203.

Heym, C., and Kummer, W. (1988). Regulatory peptides in paraganglia. *Prog. Histochem. Cytochem.* **18**, 1–92.

Hökfelt, T., Johansson, O., Ljungdahl, A., Lundberg, J. M., and Schultzberg, M. (1980). Peptidergic neurones. *Nature (London)* **284**, 515–521.

Holland, N. D., and Holland, L. Z. (1993). Serotonin-containing cells in the nervous system and other tissues during ontogeny of a lancelet, *Branchiostoma floridae*. *Acta Zool.* **74**, 195–204.

Hughes, G. M. (1966). Evolution between air and water. *Ciba Found. Symp. Dev. Lung*, pp. 64–80.

Hughes, G. M. (1973). Ultrastructure of the lung of *Neoceratodus* and *Lepidosiren* in relation to the lung of the other vertebrates. *Folia Morph. Prague* **2**, 155–161.

Hughes, G. M. (1978). A morphological and ultrastructural comparison of some vertebrate lungs. *In* "XIXth Congressus Morphologicus Symposia" (E. Klika, ed.), pp. 393–403. Charles University Press, Prague.

Hughes, G. M., and Munshi, J. S. D. (1973a). Surface area of the respiratory organs of the climbing perch, *Anabas testudineus*. *J. Zool.* **170**, 227–243.

Hughes, G. M., and Munshi, J. S. D. (1973b). Nature of air breathing organs of the Indian fishes *Channa, Amphipnous, Clarias* and *Saccobranchus* as shown by electron microscopy. *J. Zool.* **170**, 245–270.

Hughes, G. M., and Pohunkova, H. (1980). Scanning and transmission electron microscopy of the lungs of *Polypterus senegalensis*. *Folia Morphol. (Prague)* **28**(1), 110–112.

Hughes, G. M., and Weibel, E. R. (1976). Morphometry of fish lungs. *In* "Respiration of Amphibious Vertebrates" (G. M. Hughes, ed.), pp. 213–232. Academic Press, London and New York.

Hughes, G. M., and Weibel, E. R. (1978). Visualization of layers lining the lung of the south american lungfish (*Lepidosiren paradoxa*) and a comparison with the frog and rat. *Tissue Cell* **10**, 343–353.

Hughes, G. M., Roy, P. K., and Munshi, J. S. D. (1992). Morphometric estimation of oxygen diffusing capacity for the air-sac in the catfish *Heteropneustes fossilis*. *J. Zool.* **227**, 193–209.

Iwanaga, T., Takahashi, Y., and Fujita, T. (1989). Immunohistochemistry of neuron-specific and glia-specific proteins. *Arch. Histol. Cytol.* **52**, Suppl., 13–24.

Johnson, D. E., and Georgieff, M. K. (1989). Pulmonary neuroendocrine cells: Their secretory products and their potential roles in health and chronic lung disease in infancy. *Am. Rev. Respir. Dis.* **140**, 1807–1812.

Johnson, D. E., and Wobken, J. D. (1987). Calcitonin gene-related peptide immunoreactivity in airway epithelial cells of the human fetus and infant. *Cell Tissue Res.* **250**, 579–583.

Keith, I. M., and Ekman, R. (1988). Calcitonin gene-related peptide in hamster lung and its coexistence with serotonin: A chemical and immunocytochemical study. *Regul. Pept.* **22**, 315–323.

Kimura, A., Gomi, T., Kikuchi, Y., and Hashimoto, T. (1987). Anatomical studies of the lung of air-breathing fish. I. Gross anatomical and light microscopic observations on the lungs of the african lungfish (*Protopterus aethiopicus*). *J. Med. Soc. Toho Univ.* **34**, 7–15.

King, A. S., McLelland, J., Cook, R. D., King, D. Z., and Walsh, C. (1974). The ultrastructure of afferent nerve endings in the avian lung. *Respir. Physiol.* **22**, 21–40.

Klika, E., and Lelek, A. (1967). A contribution to the study of the lungs of *Protopterus annectens* and *Polypterus senegalensis*. *Folia Morphol.* (*Prague*) **2**, 168–175.

Laurent, P. (1984). Gill internal morphology. *In* "Fish Physiology" (W. S. Hoar and D. J. Randall, eds.), Vol. 10A, pp. 73–183. Academic Press, New York.

Laurent, P. (1989). Gill structure and function: Fish. *In* "Comparative Pulmonary Physiology: Current Concepts" (S. C. Wood, ed.), pp. 69–120. Dekker, New York.

Lauweryns, J. M., and Cockelaere, M. (1973). Hypoxia-sensitive neuro-epithelial bodies: Intrapulmonary secretory neuroreceptors, modulated by the CNS. *Z. Zellforsch. Mikrosk. Anat.* **145**, 521–540.

Lauweryns, J. M., and Peuskens, J. C. (1969). Argyrophil (kinin and amine producing?) cells in human infant airway epithelium. *Life Sci.* **8**, 577–585.

Lauweryns, J. M., and Peuskens, J. C. (1972). Neuro-epithelial bodies (neuroreceptor or secretory organs?) in human infant bronchial and bronchiolar epithelium. *Anat. Rec.* **172**, 471–481.

Lauweryns, J. M., and Van Ranst, L. (1987). Calcitonin gene-related peptide immunoreactivity in rat lung: Light and electron microscopic study. *Thorax* **42**, 183–189.

Lauweryns, J. M., Peuskens, J. C., and Cockelaere, M. (1970). Argyrophil, fluorescent and granulated (peptide and amine producing?) AFG cells in human infant bronchial epithelium. Light and electron microscopic studies. *Life Sci.* **9**, 1417–1429.

Lauweryns, J. M., Cockelaere, M., and Theunynck, P. (1972). Neuroepithelial bodies in the respiratory mucosa of various mammals. A light optical, histochemical and ultrastructural investigation. *Z. Zellforsch. Mikrosk. Anat.* **135**, 569–592.

Lauweryns, J. M., Cockelaere, M., and Theunynck, P. (1973). Serotonin producing neuroepithelial bodies in rabbit respiratory mucosa. *Science* **180**, 410–413.

Lauweryns, J. M., Cockelaere, M., Deleersnyder, M., and Liebens, M. (1977). Intrapulmonary neuro-epithelial bodies in newborn rabbits. Influence of hypoxia, hyperoxia, hypercapnia, nicotine, reserpine, L-DOPA and 5-HT. *Cell Tissue Res.* **182**, 425–440.

Lauweryns, J. M., Van Ranst, L., and Verhofstad, A. A. J. (1986). Ultrastructural localization of serotonin in the intrapulmonary neuroepithelial bodies of neonatal rabbits by use of immunoelectron microscopy. *Cell Tissue Res.* **243**, 455–459.

Lauweryns, J. M., Van Ranst, L., Lloyd, R. V., and O'Connor, D. T. (1987). Chromogranin in bronchopulmonary neuroendocrine cells. Immunocytochemical detection in human, monkey and pig respiratory mucosa. *J. Histochem. Cytochem.* **35,** 113–118.

Lee, S. Y., Lee, C. Y., Chen, Y. M., and Kochva, E. (1986). Coronary vasospasm as the primary cause of death due to the venom of the burrowing asp, *Atractaspis engaddensis. Toxicon* **24,** 285.

Lopez, J., Barrenechea, M. A., Burrel, M. A., and Sesma, P. (1993). Immunocytochemical study of the lung of domestic fowl and pigeon: Endocrine cells and nerves. *Cell Tissue Res.* **273,** 89–95.

Luts, A., Uddman, R., Absood, A., Häkanson, R., and Sundler, F. (1991). Chemical coding of endocrine cells of the airways: Presence of helodermin-like peptides. *Cell Tissue Res.* **265,** 425–433.

Maina, J. N. (1987). The morphology of the lung of the african lungfish, *Protopterus aethiopicus:* A scanning electron-microscopic study. *Cell Tissue Res.* **250,** 191–196.

Markl, J., Winter, S., and Franke, W. W. (1989). The catalog and the expression complexity of cytokeratins in a teleost fish, the rainbow trout. *Eur. J. Biol.* **50,** 1–6.

Martling, C. R., Saria, A., Fisher, J. A., Hökfelt, T., and Lundberg, J. M. (1988). Calcitonin gene-related peptide and the lung: Neuronal coexistence with substance P release by capsaicin and vasodilatory effect. *Regul. Pept.* **20,** 125–139.

Matsumura, H. (1985). Electron microscopic studies of the lung of the salamander, *Hynobius nebulosus.* III. A scanning and transmission electron microscopic observations on neuroepithelial bodies. *Okajimas Folia Anat. Jpn.* **62,** 187–204.

McDowell, E. M., Barrett, L. A., and Trump, B. F. (1976). Observations on small granule cells in adult human bronchial epithelium and in carcinoid and oat cell tumors. *Lab. Invest.* **34,** 202–206.

McDowell, E. M., Sorokin, S. P., and Hoyt, R. F., Jr. (1994a). Ontogeny of endocrine cells in the respiratory system of Syrian golden hamster. I. Larynx and trachea. *Cell Tissue Res.* **275,** 143–156.

McDowell, E. M., Sorokin, S. P., and Hoyt, R. F., Jr. (1994b). Ontogeny of endocrine cells in the respiratory system of Syrian golden hamster. II. Intrapulmonary airways and alveoli *Cell Tissue Res.* **275,** 157–167.

Moll, R., Moll, I., and Franke, W. W. (1984). Identification of Merkel cells in human skin by specific keratin antibodies: Changes of cell density and distribution in fetal and adult plantar epidermis. *Differentiation (Berlin)* **28,** 136–154.

Munshi, J. S. D. (1962). The accessory respiratory organs of *Heteropneustes fossilis* Bloch. *Proc.—R. Soc. Edinburgh, Sect. B: Biol.* **68,** 128–146.

Nakajima, T., Kameya, T., Watanabe, S., Hirota, T., Shimosato, Y., and Isobe, T. (1984). S-100 protein distribution in normal and neoplastic tissues. *In* "Advances in Immunohistochemistry" (R. A. Delellis, ed.), pp. 141–158. Masson, Chicago.

Pastor, L. M., Ballesta, J., Perez-Tomas, R., Marin, J. A., Hernandez, F., and Madrid, J. F. (1987). Immunocytochemical localization of serotonin in the reptilian lung. *Cell Tissue Res.* **248,** 713–715.

Pearse, A. G. E. (1968). Common cytochemical and ultrastructural characteristics of cells producing polypeptide hormones (APUD series) and their relevance to thyroid and ultimobranchial cells and calcitonin. *Proc. R. Soc. London, Ser. B* **170,** 71–80.

Pearse, A. G. E. (1969). The cytochemistry and ultrastructure of polypeptide hormone producing cells of the APUD series and the embryologic, physiologic and pathologic implications of the concept. *J. Histochem. Cytochem.* **17,** 303–313.

Pearse, A. G. E., and Polak, J. M. (1978). The diffuse neuroendocrine system and the APUD concept. *In* "Gut Hormones" (S. R. Bloom, ed.), pp. 33–39. Churchill-Livingstone, Edingburgh, London, and New York.

Polak, J. M., and Bloom, S. R. (1982). Distribution of regulatory peptides in the respiratory tract of man and mammals. *In* "Systemic Role of Regulatory Peptides" (S. R. Bloom, J. M. Polak, and E. Lindenlaub, eds.), pp. 241–269. Schattauer Verlag, Stuttgart, and New York.

Polak, J. M., Becker, K. L., Cutz, E., Gail, D. B., Goniakowska-Witalinska, L., Gosney, J. R., Lauweryns, J. M., Linnoila, I., McDowell, E. M., Miller, Y. E., Scheuermann, D. W., Springall, D. R., Sunday, M. E., and Zaccone, G. (1993). Lung endocrine cell markers, peptides and amines. *Anat. Rec.* **236,** 169–171.

Rogers, D. C., and Haller, C. J. (1978). Innervation and cytochemistry of the neuroepithelial bodies of the ciliated epithelium of the toad lung (*Bufo marinus*). *Cell Tissue Res.* **195,** 395–410.

Sacca, R., and Burggren, W. (1982). Oxygen uptake in air and water in the air-breathing reedfish *Calamoichthys calabaricus:* Role of skin, gills and lungs. *J. Exp. Biol.* **97,** 179–186.

Satchell, G. H. (1976). The circulatory system of air-breathing fish. *In* "Respiration of Amphibious Vertebrates" (G. M. Hughes, ed.), pp. 105–124. Academic Press, New York.

Saurat, J. H., Didierjean, L., Skalli, O., Siegenthaler, G., and Gabbiani, G. (1984). The intermediate filament proteins of rabbit normal epidermal Merkel cells are cytokeratins. *J. Invest. Dermatol.* **83,** 431–435.

Scheuermann, D. W. (1987). Morphology and cytochemistry of the endocrine epithelial system in the lung. *Int. Rev. Cytol.* **106,** 35–88.

Scheuermann, D. W., and De Groodt-Lasseel, M. H. A. (1982). Monoamine-containing granulated cells in the Polypterus lung. *Verh. Anat. Ges.* **76,** 301–302.

Scheuermann, D. W., De Groodt-Lasseel, M. H. A., Stilman, C., and Meister, M.-L. (1983). A correlative light-, fluorescence-, and electron-microscopic study of neuroepithelial bodies in the lung of the red-eared turtle, *Pseudemys scripta elegans. Cell Tissue Res.* **234,** 249–269.

Scheuermann, D. W., Stilman, C., and De Groodt-Lasseel, M. H. A. (1988). Microspectrofluorimetric analysis of the formaldehyde-induced fluorophores of 5-hydroxytryptamine and dopamine in intrapulmonary neuroepithelial bodies after administration of L-5-hydroxytryptophan and L-DOPA. *Histochemistry* **88,** 219–225.

Scheuermann, D. W., Adriaensen, D., and Timmermans, J. P. (1989). Neuroepithelial endocrine cells in the lung of *Ambystoma mexicanum. Anat. Rec.* **225,** 139–149.

Seldeslagh, K., and Lauweryns, J. (1993a). Endothelin in normal lung tissue of newborn mammals: Immunocytochemical distribution and co-localization with serotonin and calcitonin gene-related peptide. *J. Histochem. Cytochem.* **41,** 1495–1502.

Seldeslagh, K., and Lauweryns, J. (1993b). Safarotoxin expression in the bronchopulmonary tract: Immunohistochemical occurrence and co-localization with endothelins. *Histochemistry* **100,** 257–263.

Shimosagawa, T., and Said, S. I. (1991). Co-occcurrence of immunoreactive calcitonin and calcitonin gene-related peptide in neuroendocrine cells of rat lungs. *Cell Tissue Res.* **264,** 555–561.

Sonstengard, K. S., Mailmann, R. B., Cheek, J. M., Tomlin, T. E., and Di Augustine, R. P. (1982). Morphological and cytochemical characterization of the neuroepithelial bodies in fetal rabbit lung. I. Studies of isolated neuroepithelial bodies. *Exp. Lung Res.* **3,** 349–377.

Sorokin, S. P., and Hoyt, R. F. (1989). Neuroepithelial bodies and solitary small-granule cells. *In* "Lung Cell Biology" (D. Massaro, ed.), pp. 191–334. Dekker, New York and Basel.

Speirs, V., Bienkowski, E., Wong, V., and Cutz, E. (1993). Paracrine effects of bombesin/gastrin-releasing peptide and growth factors on pulmonary neuroendocrine cells in vitro. *Anat. Rec.* **236,** 53–61.

Stahlman, M. T., Orth, D. N., and Gray, M. E. (1989). Immunocytochemical localization of epidermal growth factor in the developing human respiratory system and in acute and chronic lung disease in the neonate. *Lab. Invest.* **60,** 539–547.

Sunday, M. E., Hua, J., Dai, H. B., and Nusrat, A., and Torday, J. S. (1990). Bombesin increases fetal lung growth and maturation in utero and in organ culture. *Am. J. Respir. Cell. Mol. Biol.* **3,** 199–205.

Taira, K. (1985). Endocrine-like cells in the laryngeal mucosa of adult rabbit demonstrated by electron microscopy and by the Grimelius silver-impregnation method. *Biomed. Res.* **6,** 377–385.

Taira, K., and Shibasaki, S. (1978). A fine structure study of nonciliated cells in the mouse tracheal epithelium with special reference to the relation of "brush cells" and "endocrine cells." *Arch. Histol. Jpn.* **41,** 351–366.

Takahashi, S., and Yui, R. (1983). Gastrin-releasing peptide (GRP) and serotonin in the human fetal lung: An immunohistochemical study. *Biomed. Res.* **4,** 315–320.

Track, N. S., and Cutz, E. (1982). Bombesin-like immunoreactivity in developing human lung. *Life Sci.* **30,** 1553–1556.

Tsutsumi, Y., Osamura, R. Y., Watanabe, K., and Yanaihara, N. (1983). Simultaneous immunohistochemical localization of gastrin-releasing peptide (GRP) and calcitonin (CT) in human bronchial endocrine type cells. *Virchows Arch. A: Pathol. Anat. Histol.* **400,** 163–171.

Uddman, R., Luts, A., and Sundler, F. (1985). Occurrence and distribution of calcitonin gene-related peptide in the mammalian respiratory tract and middle ear. *Cell Tissue Res.* **241,** 551–555.

Wang, D., Cutz, E., Wang, Y. Y., and Perrin, D. G. (1993). Cholecystokinin gene expression in neuroendocrine cells of mammalian lungs. *Am. J. Respir. Cell. Mol. Biol.* (submitted for publication).

Wang, Y. Y., and Cutz, E. (1993). Localization of cholecystokinin-like peptide in neuroendocrine cells of mammalian lungs: A light and electron microscopic immunohistochemical study. *Anat. Rec.* **236,** 198–205.

Wasano, K. (1977). Neuro-epithelial bodies in the lung of the rat and the mouse. *Arch. Histol. Jpn.* **40,** 207–219.

Wendelaar-Bonga, S. E. (1993). Endocrinology. *In* "The Physiology of Fishes" (D. H. Evans, ed.), pp. 469–502. CRC Press, Boca Raton, FL.

Wharton, J., Polak, J. M., Bloom, S. R., Ghatei, M. A., Solcia, E., Brown, M. R., and Pearse, A. G. E. (1978). Bombesin-like immunoreactivity in the lung. *Nature (London)* **273,** 769–770.

Wharton, J., Polak, J. M., Cole, G. A., Marangos, P. J., and Pearse, A. G. E. (1981). Neuron-specific enolase as an immunocytochemical marker for the diffuse neuroendocrine system in human foetal lung. *J. Histochem. Cytochem.* **29,** 1359–1364.

Willey, J. C., Lechner, J. F., and Harris, C. C. (1984). Bombesin and C-terminal tetradeca-peptide of gastrin-releasing peptide are growth factors for normal human brochial epithelial cells. *Exp. Cell Res.* **153,** 245–248.

Wollina, U. (1990). The relationship of epidermal hyper-plasia and Merkel cell hyperplasia. *Dermatologica* **181,** 73.

Wollina, U. (1993). Diversity of epithelial skin tumours: Thoughts and comments on some basic principles. *Recent Results Cancer Res.* **128,** 153–178.

Youngson, C., Nurse, C., Yeger, H., and Cutz, E. (1993). Oxygen sensing in airway chemoreceptors. *Nature (London)* **365,** 153–155.

Zaccone, G., Fasulo, S., Ainis, L., and Contini, A. (1989a). Localization of calmodulin positive immunoreactivity in the surface epidermis of the brown trout, *Salmo trutta*. *Histochemistry* **91,** 13–16.

Zaccone, G., Tagliafierro, G., Goniakowska-Witalinska, L., Fasulo, S., Ainis, L., and

Mauceri, A. (1989b). Serotonin-like immunoreactive cells in the pulmonary epithelium of ancient fish species. *Histochemistry* **92,** 61–63.

Zaccone, G., Goniakowska-Witalinska, L., Lauweryns, J. M., Fasulo, S., and Tagliafierro, G. (1989c). Fine structure and serotonin immunohistochemistry of the neuroendocrine cells in the lungs of the bichirs *Polypterus delhezi* and *P. ornatipinnis*. *Bas. Appl. Histochem.* **33,** 277–287.

Zaccone, G., Lauweryns, J. M., Fasulo, S., Tagliafierro, G., Ainis, L., and Licata, A. (1992a). Immunocytochemical localization of serotonin and neuropeptides in the neuroendocrine paraneurons of teleost and lungfish gills. *Acta Zool.* (*Stockholm*) **73**(3), 177–183.

Zaccone, G., Wendelaar Bonga, S., Flik, G., Fasulo, S., Licata, A., Lo Cascio, P., Mauceri, A., and Lauriano, E. R. (1992b). Localization of calbindin D28k-like immunoreactivity in fish gill: A light microscopic and immunoelectron histochemical study. *Regul. Pept.* **41,** 195–208.

Zaccone, G., Mauceri, A., Fasulo, S., Ainis, L., Lo Cascio, P., and Ricca, M. B. (1994). Endothelin in neuroendocrine cells of fish gill. *Neuropeptides* (*Edinburgh*) (in press).

Zimmermann, T. W., and Binder, H. J. (1984). Serotonin-induced alteration in colonic transport in the rat. *Gastroenterology* **86,** 310–317.

Structure–Function Relationships in Gap Junctions

Hartwig Wolburg* and Astrid Rohlmann†

* Institute of Pathology, University of Tübingen, D-72076 Tübingen, Germany, and † Department of Anatomy, Developmental Neurobiology Unit, University of Göttingen, D-37075 Göttingen, Germany

Gap junctions are metabolic and electrotonic pathways between cells and provide direct cooperation within and between cellular nets. They are among the cellular structures most frequently investigated. This chapter primarily addresses aspects of the assembly of the gap junction channel, considering the insertion of the protein into the membrane, the importance of phosphorylation of the gap junction proteins for coupling modulation, and the formation of whole channels from two hemichannels. Interactions of gap junctions with the subplasmalemmal cytoplasm on the one side and with tight junctions on the other side are closely considered. Furthermore, reviewing the significance and alterations of gap junctions during development and oncogenesis, respectively, including the role of adhesion molecules, takes up a major part of the chapter. Finally, the literature on gap junctions in the central nervous system, especially between astrocytes in the brain cortex and horizontal cells in the retina, is summarized and new aspects on their structure–function relationship included.

KEY WORDS: Gap junction, Connexon, Connexin, Phosphorylation of connexins, Tight junction, Freeze-fracturing, Central nervous system, Gap junctions in development, Gap junctions in oncogenesis, Astrocytes, Horizontal cells, Brain, Retina, Heart, Liver, Lens.

I. Introduction

Gap junctions (GJs) are specialized membrane areas of many cells that are believed to serve as sites of intercellular communication. Tissues and organs are not only the sum of their cells but also incorporate cell regulation circuits, and their physiological efficiency is largely determined by their capability to communicate in a network. There are several modalities as to how cells can cooperate: by the release of hormones, growth factors, or other messengers that are bound by specific receptors, enabling the cell to react to the type of signal. More locally, cells communicate through cell surface glycoproteins or adhesion molecules such as cadherins, integrins, or members of the immunoglobulin superfamily that exchange information between cells and between cells and the extracellular matrix. These molecular interactions are concerned with migration, sorting, and grouping of cells during development, initially giving rise to cell clusters, multicellular structures, and finally to organs (Bell, 1978; Jessell, 1988; Takeichi, 1990, 1991; Hynes, 1992; Singer, 1992; Gumbiner, 1993). However, the phenomenon of cell adhesion is by no means confined to developmental stages. Association of cells at least in epithelial tissues requires the existence of specialized junctional complexes, the adhesion junctions or zonulae or maculae adhaerentes (see, e.g., Behrens *et al.*, 1989; Schoenenberger and Matlin, 1991; Chen and Öbrink, 1991).

Network cooperation is most rapidly achieved by formation of channels between cells and their processes. However, in contrast to the large variety of ionic channels distributed all over the cellular surface, intercellular channels have the problem that each cell must supply one-half of the channel and that both hemichannels must be closed so as to hinder unspecific exchange with the extracellular space. Finally, hemichannels must fit each other exactly in order to form a complete channel bridging the intercellular space. The probability of this molecular finding process would seem to be low. One possibility of enhancement would be a clustering of many hemichannels in a certain membrane area and confrontation of this area with the corresponding one of an adjacent cell. This would imply probably complicated adhesion mechanisms preceding channel formation (Kanno *et al.*, 1984; Edelman, 1986, 1988; Jongen *et al.*, 1991).

The detection of GJs dates back about 30 years. Dewey and Barr (1962) were the first to describe an electron microscopically detected structure, called a *nexus,* in smooth muscle cells in the jejunum. Robertson (1963) described hexagonally packed subunits in electrical synapses between Mauthner cells in the medulla oblongata of the goldfish brain, and Revel and Karnovsky (1967) found in heart and liver an intercellular space that measured 1–2 nm in width and allowed tracer molecules to penetrate, thus confirming a difference from the well-established zonula occludens or tight junction (Revel *et al.*, 1967).

During the following years a tremendous number of studies described the fine structure of GJs, and, most importantly, it became clear that sites of low electrical resistance between excitable cells correspond to GJs. These junctions were structurally identical to those of many cell types throughout organisms and were called electrical synapses or electrotonic junctions. Another exciting discovery concerning the significance of GJs was that cells injected with fluorescent dyes transferred the dye to adjacent cells (Loewenstein and Kanno, 1964; Bennett *et al.*, 1967; Payton *et al.*, 1969; Rose, 1971). A proof of this transfer via GJs has been presented by a combination of physiological and morphological methods (Loewenstein, 1972; MacVicar and Dudek, 1981; Schmalbruch and Jahnsen, 1981). Early in GJ research, tumor cells were found to be coupled in terms of dye coupling and low electrical resistance, but failed to show intercellular communication (Loewenstein, 1967; Johnson and Sheridan, 1971). Later, numerous investigations referred to the general problem of how to describe the relationship between dye transfer and electrical communication and coupling in neoplastic cells. Using neutral and negatively charged tracer molecules of different molecular weights, Simpson *et al.* (1977) estimated the largest size of molecules transferred from one salivary gland cell of *Chironomus* to the coupled partner cell to be about 1158 Da. This molecular mass corresponds to an estimated channel diameter of about 1.4 nm.

In addition to the transfer of artificial or synthetic tracers, physiological substances have been investigated. Amino acids, sugars, and nucleotides (Johnson and Sheridan, 1971; Ricske *et al.*, 1975; Pitts and Sims, 1977; Pederson *et al.*, 1980), and second messengers such as cAMP (Imanaga, 1974; Tsien and Weingart, 1976) and Ca^{2+} (Sanderson *et al.*, 1990; Cornell-Bell *et al.*, 1990; Charles *et al.*, 1992; Enkvist and McCarthy, 1992), were found to be transferred preferentially through GJs, and this metabolic coupling of a cellular network has become a basis for the increasing success in understanding the complicated interrelationship in cellular systems.

The huge mass of data on GJs has been reviewed many times (see, e.g., Bennett, 1973; McNutt and Weinstein, 1973; Staehelin, 1974; Griepp and Revel, 1977; Bennett and Goodenough, 1978; Sotelo and Korn, 1978; Gilula, 1979; Korn and Faber, 1979; Peracchia, 1980; Loewenstein, 1981; Bennett *et al.*, 1981, 1991; Larsen, 1977, 1983; Meda *et al.*, 1984; Peracchia and Girsch, 1985; Manjunath and Page, 1985; Caveney, 1985; Hertzberg, 1985; Sheridan and Atkinson, 1985; Spray and Bennett, 1985; Revel *et al.*, 1985; Neyton and Trautmann, 1986; Pitts and Finbow, 1986; Evans, 1988; Faber and Korn, 1989; Guthrie and Gilula, 1989; Kistler and Bullivant, 1989; Dermietzel *et al.*, 1990; Malewicz *et al.*, 1990; Musil and Goodenough, 1990; Spray and Burt, 1990; Schuster, 1990; Kolb and Somogyi, 1991; Goodenough and Musil, 1993; Dermietzel and Spray, 1993; Beyer, 1993). However, the last comprehensive review summarizing the results on the structural correlates of GJ permeation dates from 1983 (Larsen, 1983). The present chapter collects more recent data on the relationship between structure of GJs on the one hand and the functional states as far as they have been deduced from structural evaluation on the other. Furthermore, more recent aspects of hemichannel insertion and their completion as intercellular channels are discussed, including the cytoskeleton–GJ interactions that might be involved in the dynamics of gap junction proteins in membrane lipid layers.

II. Structure and Assembly of Gap Junction Channel

A. Building Blocks of a Gap Junction: The Connexins

Gap junctions are evolutionally conserved structures with their different connexin types showing about 59% sequence homology (Bennett *et al.,* 1991). Beyer *et al.* (1987) presented a topological model of connexin 32 and connexin 43, with positions of identical amino acids indicated in relation to the junctional membrane.

The members of the connexin family share topological criteria, such as four transmembrane regions and two extracellular loops. The cytoplasmic domains represent regions specific for the different connexin types (for a review of molecular cloning, topological structure, and gene structure of connexins, see Bennett *et al.,* 1991; Beyer *et al.,* 1987, 1989; Beyer, 1993). Although 12 connexins have been cloned (Beyer, 1993), for example, chicken connexin 56 and sheep lens fiber MP70 (Rup *et al.,* 1993) or connexin 43 and MP70 (Paul *et al.,* 1991; Beyer, 1993), many homologies and relations between connexins are still unclear. A protein called ductin has been proposed to be a constituent of GJs (Finbow and Pitts, 1993). Ductin seems to be identical with the 16-kDa protein coisolated with GJs from bovine brain by Dermietzel *et al.* (1989b). The 16-kDa protein appears to be a constituent of the proton channel of a vacuolar H^+-ATPase, which is widely distributed in the animal kingdom (including the invertebrates) (Bohrmann, 1993). Generally, the identification of GJ proteins in invertebrates, in particular in arthropods, is only at its beginning (Finbow *et al.,* 1984; Pitts and Finbow, 1986; Berdan and Gilula, 1988; for literature on the diversity of arthropod GJ proteins, see Bohrmann, 1993; Ryerse, 1993; for connexon architecture of arthropod GJs, see Sikerwar *et al.,* 1991).

Besides the considerable degree of homology of different connexins in vertebrates, sequences of one connexin type show similarities among different species: for example, connexin 43 of *Xenopus,* chick, human, mouse, and cow, or connexin 32 of rat, compared to connexin 30 of the frog (Beyer, 1993; Hennemann *et al.,* 1992a). Moreover, antibodies to the 27-kDa rat liver GJ polypeptide were found to bind to liver sections from mammals, fish, and avian species (Hertzberg and Skibbens, 1984). Comparison of the developing avian cardiovascular system with the mammalian cardiovascular system revealed the presence of connexin 43 in both systems, but these tissues exhibited different electrophysiological properties. This discrepancy may be due to the predominance of connexin 42 in avian cardiac tissue and the fact that connexin 43 expression is restricted to specific vascular myocytes (Minkoff *et al.*, 1993). Hence, although one connexin type occurs in comparable organs or tissues of various species, the coexistence with other connexins and their relative amount may determine GJ channel properties.

Genes for different connexins might be located on one chromosome, for example, genes encoding connexins 37 and 40 were found on human chromosome 1 (Willecke *et al.*, 1990), and genes encoding connexins 26 and 46 were found on human chromosome 13 (Willecke *et al.*, 1990) and rat chromosome 14 (Haefliger *et al.*, 1992), respectively. Similarly, connexin 31.1 and 30.3 genes were located on mouse chromosome 4 (Hennemann *et al.*, 1992b) and connexin 37 and 31.1 genes were found on rat chromosome 4 (Haefliger *et al.*, 1992). Close proximity of different connexin genes may suggest similarities in expression patterns and functions, as proposed for connexin 31.1 and connexin 30.3 (Hennemann *et al.*, 1992b), or for connexin 37 and connexin 40 (Willecke *et al.*, 1991; Hennemann *et al.*, 1992c). However, no correlation between colocalization and coexpression has been shown for human connexins 26 and 46 (Willecke *et al.*, 1990). Haefliger *et al.* (1992) concluded that distribution of connexins could overlap but were not completely coincident.

Thus, (1) similarities among different connexin types, (2) the similarity of a connexin type in one species with those in other species, and (3) the overlapping of colocalized connexin genes and their expression patterns could be criteria that together would characterize the different connexin types.

Because different connexins may be expressed in a single cell, a greater variety of GJ channels is possible. For example, five connexins have been found in mouse keratinocytes (Hennemann *et al.*, 1992b) and connexins 37 and 43 are coexpressed in endothelial cells (Reed *et al.*, 1993; for further examples, see Dermietzel and Spray, 1993). The capacity of one cell to express multiple connexins may lead to a diversity in expression patterns of GJ channels.

Two connexins may also occur in the same junctional cluster (heterotypic junction), for example, connexins 26 and 32 in hepatocytes (Nicholson *et al.*, 1987; Traub *et al.*, 1989; Zhang and Nicholson, 1989; Kuraoka *et al.*, 1993). After microinjection of mRNAs in single *Xenopus* oocytes, connexins 32 and 43 were found to form heterologous channels (Swenson *et al.*, 1989; Werner *et al.*, 1991) or connexins 43 and 26 in epidermal cells (Risek *et al.*, 1994). Hence, at the cellular level, the diversity of GJs may be enhanced by expression of different connexins and formation of heterologous channels between cellular compartments. It is still unknown whether different connexins can form heterooligomeric connexons (heterohexamers). However, in a developmental study of epidermal GJs in the rat, Risek *et al.* (1994) provided evidence for the presence of homooligomeric connexons in heterotypic junctions.

Besides leading to a greater variety of GJ channels, the connexin diversity of one cell may limit communication between cells. Cells expressing *Xenopus* connexin 38 may couple to cells expressing rat connexin 43 but not to neighboring cells that express connexin 32 (Swenson *et al.*, 1989; Werner *et al.*, 1989). Theoretically, this manner of coupling would allow communication to certain neighboring cells and prohibit coupling to others (Haefliger *et al.*, 1992). Another study exemplifies two blood vessel wall cell types expressing connexin 43. Each cell type possesses in addition another connexin type (endothelial cell connexin 37; smooth muscle cell connexin 40), which endows the specific cell type with individual interaction possibilities (Reed *et al.*, 1993). Future studies will be needed to clarify dynamics of coupling between cells that exhibit different connexins.

B. Assembly of Connexon

Gap junctions are plaques of channels bridging the intercellular cleft between closely apposed plasma membranes of two cellular compartments. Each channel is composed of two connexons (Caspar *et al.*, 1977; Unwin and Zampighi, 1980; Unwin and Ennis, 1984). Connexons, in turn, are complexes of six identical integral membrane protein subunits, each consisting of one connexin molecule (Caspar *et al.*, 1977; Unwin and Zampighi, 1980; Beyer *et al.*, 1987).

Studies have revealed that there are at least two differences between the GJ proteins and other integral membrane proteins. First, they have a much more rapid turnover (see Section IV,A), and second, they seem to be fully assembled in the trans-Golgi network (Musil and Goodenough, 1993), unlike other integral plasma membrane proteins, which are already fully assembled in the endoplasmic reticulum (ER; Gething *et al.*, 1986; Hurtley and Helenius, 1989; Rose and Doms, 1988). This post-ER assembly pathway may therefore occur late in connexin protein transport, because a high connexin content in the Golgi apparatus could favor oligomerization. Connexins seem to accumulate in the Golgi apparatus at least *in vitro* (Musil *et al.*, 1990; Berthoud *et al.*, 1992), and (possibly) *in vivo* (Hendrix *et al.*, 1992). In addition, connexins are supposed to assemble only after insertion of the proteins into the membrane (Rahman *et al.*, 1993). Late assembly of GJ proteins might be reasonable in order for the cell to avoid GJ formation intracellularly, for example, between membranes of organelles (Musil and Goodenough, 1993). After oligomerization of connexons (diameter of 9 nm) in the trans-Golgi network, they could be rapidly transported in vesicles to their insertion sites. Because connexons are believed to occur even in GJ incompetent cells, such as S180 sarcoma cells and L929 fibroblasts, neither adhesion molecule-mediated intercellular contact nor phosphorylation to higher molecular mass connexin forms would be required to build connexons (Musil and Goodenough, 1993). Young *et al.* (1987) managed to incorporate GJs into membranes but failed to determine whether the incorporated 27-kDa protein existed as single oligomers (hemichannels) or as bipartite oligomers (complete channels). Electrophysiological data, however, indicate that unpaired connexons in membranes might exist (Paul *et al.*, 1991; DeVries and Schwartz, 1992). Using specific experimental conditions of high salt, high pH, reducing agent, and detergents for treatment of liver GJ plaques, connexons could be isolated (Stauffer *et al.*, 1991). Rotational power spectra of these connexons revealed a clear dominance of the sixfold symmetry and a dimension of about 8 nm (Stauffer *et al.*,1991), as expected from the three-dimensional maps of Unwin and Zampighi (1980). With the aid of atomic force microscopy the structure of isolated hepatic GJs was analyzed. Connexons appeared to be smaller, 4–6 nm in diameter (Hoh *et al.*, 1991), compared to the results of Stauffer *et al.*, (1991) and Unwin and Zampighi (1980). Center-to-center spacing has been determined as being 9.1 nm and connexons seem to protrude 0.4–0.5 nm from the surface of the plaque (Hoh *et al.*, 1991; for comparison of different membrane cross-sections of GJs depending on irradiation conditions, see Sosinsky *et al.*, 1988).

The native structure of crystalline, hexagonally packed connexon assemblies of arthropods showed a similar center-to-center distance of about 8.5 nm compared to those of mouse and rat liver junctions (Sikerwar *et al.*, 1991). Three-dimensional surface representations revealed connexons as annular oligomers protruding 3.0–4.5 nm into the cytoplasm. The channel was found to be 4–4.5 nm wide in the extracellular region (Sikerwar *et al.*, 1991).

Studies have further analyzed connexon assembly and connexon structure. Up to now, it has not been clear whether connexons may be inserted into membranes as hemichannels or, if so, whether they function as water-permeable channels connecting the cytoplasmic compartment with the extracellular space (Paul *et al.*, 1991). Musil and Goodenough (1993) found that the channels are, with a high probability, closed.

C. Formation of Intercellular Channels from Two Hemichannels

The presence of GJs not only implies the insertion of connexins into the membrane but also the formation of whole channels, each consisting of two hexamer connexin molecules. Cells that form GJs *in vivo* lose their GJ integrity after isolation procedures although hemichannels remain preserved in the membrane. However, the hemichannels in isolated cells cannot be identified by freeze-fracturing because the typical arrangement of connexons including the pits that are left in the E-face disappear by lateral dispersal following dissociation (for a comprehensive discussion of interpretative problems of freeze-fracture replicas of GJs in dissociated cells, see Preus *et al.*, 1981).

Spray *et al.* (1986) and Young *et al.* (1987) investigated gating properties of isolated liver GJs after reconstituting the GJs in artificial lipid bilayers. Antibodies raised against the major 27-kDa protein blocked the junctional conductance. From immunocytochemical (Young *et al.*, 1987) as well as electrophysiological evidence (Hertzberg *et al.*, 1985) it has been suggested that antibodies recognized epitopes on the cytoplasmic side of the connexon. The antibody-induced reduction of GJ conductance by at least 90% prompted Spray *et al.* (1986) to assume that the current had passed through complete channels rather than through hemichannels. An antibody-induced reduction in conductance by more than 50% can be explained only if cytoplasmic domains can be recognized by the antibody regardless of the orientation of the bilayer in the recording pipette. Young *et al.*(1987) were unable to determine whether the incorporated GJ proteins existed as single oligomers or as complete channels. In the experiments designed by these authors, the rate by which antibodies reduced channel conductance varied between 30 and 90%. However, in some but not all experiments antibody-binding sites were found to exist on both sides of the artificial membrane.

Connexons have been demonstrated to require stabilization by lipids (Malewicz *et al.*, 1990), and cholesterol has been shown to increase GJ assembly in Novikoff hepatoma cells (Meyer *et al.*, 1990). This and other studies dealing with the influence of lipids on GJ communication (Aylsworth *et al.*, 1986, 1987; Chang *et al.*, 1988) suggest the importance of the lipid microenvironment of both hemichannels for the functional integrity of the complete GJ channel.

In addition, the binding forces between the hemichannels have been shown to be established by hydrogen rather than by covalent bonds. John and Revel (1991) have presented evidence that the two extracellular loops of each connexin molecule are linked by intramolecular disulfide bonds. These bonds are crucial for the spatial stabilization of the peptide chain and, as a result, for the correct alignment of the whole channel across the gap. Intermolecular disulfide bonds could not be demonstrated, suggesting that the integrity of the complete dodecamer is maintained by noncovalent hydrogen bonds (Manjunath et al., 1984a; Manjunath and Page, 1985). This suggestion was confirmed by experiments using hypertonic solutions such as 0.5 M disaccharides or 8 M urea. The technique developed by Barr et al. (1965) and demonstrated morphologically by Goodenough and Gilula (1974), Hirokawa and Heuser (1982), Peracchia and Peracchia (1985), and Manjunath and Page (1985) allows separation of the membranes of the coupled cells in the GJ area. Although living, the cells lose their GJ arrays under electron microscopy investigation when separated by isolation procedures. The gap and tight junctional structures, however, remain intact after short-term treatment with hypertonic solutions, which pulls apart the junctional membranes. Peracchia and Peracchia (1985) used this method to visualize the true external surface of gap junctional membranes and demonstrated extracellular filaments bridging the connexons. Milks et al. (1988) and Goodenough et al. (1988) used split gap junctions to demonstrate specific binding sites of antibodies directed against different sequences of the connexin polypeptide. Whereas gold-conjugated antibodies directed against cytoplasmic domains of the connexin molecule labeled both intact and split gap junctions, an antibody directed against an amino acid sequence of an extracellular loop of the polypeptide labeled only split gap junctions. In a study on the isolation and purification of liver GJ channels, Stauffer et al. (1991) found that it was not possible to obtain dodecamers, because they dissociated into single hexamers. However, if the purified connexons had been incubated in the presence of 4% polyethylene glycol 2000 at 4° C or in the presence of 10% polyethylene glycol 2000 at 15°C, the connexons formed filaments or crystalline sheets, respectively. This shows that connexons can grow by end-to-end association and also, under modified conditions, by side-to-side association. As has been shown by Kistler and Bullivant (1988a) and Manjunath and Page (1986), connexins of lens fibers and heart muscle cells, respectively, can be conserved as paired entities despite solubilization procedures. Mazet et al. (1992) solubilized connexin 32 GJs from rat liver and reconstituted them into proteoliposomes. Data indicate incorporation of whole channels rather than hemichannels (connexons). The different results following identical biochemical treatments are difficult to explain. The existence of heterologous GJs (junctions between different types of cells equipped with different connexins) requires recognition areas in the extracellular domains of connexin loops. The stability of the amino acid sequences in these domains has been demonstrated several times (Beyer et al., 1987; Kistler and Bullivant, 1988b; Beyer, 1993). Despite this property of all connexins investigated so far and the noncovalent nature of the intermolecular bonds (John and Revel, 1991), the strength of the intermolecular bonds between two hemichannels seems to differ considerably from tissue to tissue. In any case, extracellular domains seem to be necessary for the formation of the GJ channel (Meyer *et al.*, 1992). However, no type of connexin is attracted strongly enough within the "gap" to avoid the cleavage of the junction into two hemichannels during the freeze-fracture process.

The primary questions as to why connexons of isolated cells do not maintain the clustering of connexons (and, vice versa, how connexon clusters are raised during GJ formation) are insufficiently answered by considering the interactions between two hemichannels of opposite membranes. The interactions (probably hydrogen bonds) are prerequisites to stabilize the single channel that couples cellular profiles but do not explain why thousands of paired connexons form a GJ. We are just beginning to understand this crucial event, and as yet few data are available. In general, all considerations on the structure–function relations in GJs regarding the interconnexonal spacing must start with the awareness of being dependent on preparation procedures. Raviola *et al.* (1980) and Miller and Goodenough (1985) have claimed that glutaraldehyde fixation of GJs in different tissues causes a dramatic condensation of connexon packing. However, there are several factors that relativize the disadvantage of being constrained to fixation of the GJs. In a given experimental approach, the fixation procedure is always the same and can, therefore, not account for the differences in connexon densities. If the aggregating influence of glutaraldehyde on a dispersed GJ is extrapolated, the dispersing effect should actually be even greater than observed in the replica. However, in a study on isolated hepatic GJs using atomic force microscopy, Hoh *et al.* (1991) were not able to discern any changes in morphology resulting from glutaraldehyde fixation. An additional aspect, concerning the significance of the interconnexonal spacing, has been provided by Risek *et al.* (1994). They tried to correlate the presence of aggregates of tightly and loosely packed connexons in one GJ to that of heterotypic junctions (different connexins in the hemichannel population of one junctional plaque). Different connexins would then exhibit different interconnexonal spacing.

Using biochemical methods, Manjunath and Page (1986) have analyzed GJs from heart and liver for the presence of intersubunit and interconnexon disulfide linkages by solubilization in deoxycholate and cleavage of disulfide bonds with 2-mercaptoethanol. They found that in heart GJs—in contrast to liver GJs—both intersubunit (intraconnexon) and interconnexon disulfide bonds are present. They pointed out that myocardial cells suffer permanently from shearing forces in the direction of the membrane plane. Thus, contraction demands a particular high degree of stabilization and anchorage of GJ components in the membrane, which seems to be less necessary in the liver. Correspondingly, not only do the types of connexin in heart and liver differ but the fine structure of the cytoplasmic surfaces differs as well (see Section III,A). Braun *et al.* (1984) postulated on the basis of a statistical–mechanical analysis of freeze-fracture replicas from liver GJs that aggregates of connexons are not maintained by an attractive force between the connexons in the plane of one given membrane, but by the mimimization of the repulsive force between apposed membranes. Linkage of connexon proteins across the extracellular gap requires overcoming of electrostatic forces. The energy for overcoming the repulsion must be minimized by minimizing the membrane area of close apposition. Because the escape of single connexons from the GJ plaque increases the area of close apposition providing the driving force for the cohesion, this model predicts the free lateral movement of connexons in the membrane and thus an independence of cytoskeletal attachment to the connexons. Indeed, Hirokawa and Heuser (1982) and Shibata *et al.* (1985) visualized the cytoplasmic surface of liver cell GJs by means of deep etching and found these surfaces but not the nonjunctional surfaces to be devoid of attached cytoskeletal elements. On the other hand, there are studies demonstrating alterations of GJ structures evoked by cytoskeleton inhibitors. The question arises as to whether or not GJ–cytoskeleton interactions are activated during dynamic alterations of GJs.

III. Gap Junctions and Their Molecular Environment

A. Interactions between Connexons and Cytoskeleton/Cytoplasm

The well-known observation that in freeze-fracture replicas GJs are associated consistently with the P-face, at least in vertebrate tissues, suggests that connexons may be anchored by certain mechanisms in the submembranous cytoplasm. The P-face association is independent of whether or not cytoplasmic components are present at the cytoplasmic surface (for an overview of freeze-fracture techniques in GJ research, see Rash and Yasumura, 1992). To our knowledge, only one exception in vertebrate GJs has been described (Lo and Reese, 1993). Gap junctions in mouse lens fibers show some particles associated with the E-face. Some invertebrates such as crayfish or insects show E-face-associated GJs (Peracchia and Dulhunty, 1976; Lane and Swales, 1980; Hanna *et al.*, 1984; Bosch, 1989). In contrast, in the earthworm *Eisenia foetida*, Hama (1987) described GJ particles adhering to either the P- or the E-face, and in the squid *Loligo*

pealei, the GJs are morphologically indistinguishable from GJs in vertebrates (Ginzberg *et al.,* 1985).

Most data on the interaction between GJ membrane and the cytoplasm
were obtained from cellular systems in vertebrates. Rassat *et al.* (1981)
described an increase in number and size of GJs in liver cells after treatment with the microtubule inhibitor vinblastine sulfate. They were not able
to determine whether these alterations resulted from *de novo* synthesis of
GJ proteins or from aggregation of GJ precursor molecules. In epithelial
cells of the prostate, Tadvalkar and Pinto da Silva (1983) also observed
the rapid assembly of GJs by administration of colchicine and cytochalasin
B, and this assembly proceeded even in the presence of a metabolic
inhibitor (dinitrophenol) or an inhibitor of protein synthesis (cycloheximide). The authors proposed that preexisting connexon precursors are
under the positional control of cytoskeletal elements. When disrupting
these elements the connexons migrate laterally and assemble into GJ
plaques. It seems likely that only in this final destination site (the GJ
plaque) does the recognition between apposed hemichannels of both cells
occur, because the nearest apposition of partner cells in the "gap" area
occurs only in those areas where, for example, in freeze-fracture replicas
connexon clusters are visualized. Outside the GJ proper the extracellular
cleft is wider, hindering the completion of whole channels. These results
suggest that outside the GJ plaque a pool of disintegrated connexons or
even connexon subunits might be available but arrested in the membrane
by cytoskeletal anchorage. However, differences between cell types such
as liver, cardiac, lens, or retinal cells prohibit the formulation of a general
principle on the role of cytoskeletal attachment in GJ regulation. In cardiac
GJs, Shibata and Yamamoto (1986) described cytoplasmic surfaces free
of cytoskeletal filaments. However, in contrast to the smooth surface in
liver GJs observed by Hirokawa and Heuser (1982), Shibata *et al.* (1985),
and Kuraoka *et al.* (1993), cardiac GJ membranes show a granular sub-
structure that has been found to be sensitive to Ca^{2+}-free, high K^+ buffer.
These results are consistent with those published previously by Manjunath
et al. (1984b). They found a protein subunit in cardiac but not in liver GJs
that had to be isolated with a serine protease inhibitor, phenylmethylsulfo-
nyl fluoride. Proteolyzed cardiac GJs and liver GJs without protease inhibitor showed the same ultrastructure. In contrast, and unexpected by the
authors, Hatae *et al.* (1993) found no difference in lens fiber GJs in the deep
etching morphology of the cytoplasmic surface between endoproteinase-
treated and nontreated specimens. The surface particles remained even
after the epitope of the antibody against MP70 (the GJ protein in lens
fibers) was lost. This suggests that particles at the cytoplasmic surface of
a GJ plaque indicate the coexistence of several types of connexins within
the same GJ. In cardiac GJs, Kardami *et al.* (1991) found biochemical

and ultrastructural evidence for the association of basic fibroblast growth factor (bFGF) with cardiac GJs. The authors suggest that the antibody directed against the amino-terminal domain of bFGF recognized epitopes located beneath the GJ membrane that may be constituents of the fuzzy coat found in cardiac GJs. Also, in astrocytic GJs (which consist of connexin 43), Yamamoto *et al.* (1991) demonstrated bFGF immunoreactivity at the electron microscope level. Because connexin 43 has been identified as a substrate for protein kinase C (Saez *et al.*, 1990a), the GJs of the connexin 43 type could be regulated via the phosphorylation of bFGF.

In addition, in a freeze-fracture study of cardiac GJs, Green and Severs (1984a) found that furrows at the P-face that had no counterparts as ridges at the E-face were in direct association with GJs and appeared only at intervals of several minutes after cellular death, when reorganization could be expected. The furrows were suggested to reflect attachment areas between cytoskeletal elements and the connexons during or after connexon shifts in the membrane. As has been demonstrated by Page *et al.* (1983), perfusion with Ca^{2+}-loaded solution evoked, in cardiac GJs, a dispersion of a few connexons outside the GJ, and this dispersion indeed justifies the assumption of large-field reorganization of GJ membranes in which cytoskeletal elements are involved. However, within the GJ the connexon density is altered less dramatically within the first several minutes postmortem, as would be expected if the filaments were activated to reorganize GJ areas and leave furrows between or around them. The connexon configuration in cardiac GJs discussed by Green and Severs (1984b) changes from a hexagonal package (by freezing at "0 min" after animal death) to a less ordered arrangement (by freezing at 20–40 min after animal death; see Fig. 1). Page *et al.* (1983) reported on a decrease in the connexon density in cardiac GJs following perfusion with Ca^{2+}-loaded solution. The nearest neighbor distance increased from 6.89 to 7.41 nm. In contrast, Dahl and Isenberg (1980) and Délèze and Hervé (1983) found a decrease in the center-to-center spacing from 9.4 to 9.1 nm and from 10.23 to 9.25 nm following uncoupling of cardiac cells by dihydroouabain and heptanol, respectively, and De Mazière and Scheuermann (1990) described an increase in the connexon density from 9400 to 11,600/μm^2 after hypoxia of the isolated rat heart that corresponded to a decrease of the center-to-center spacing from 9.99 to 9.60 nm. Bernardini *et al.* (1984) demonstrated a 14% decrease in particle spacing in GJs of pancreatic cells following treatment with heptanol. Both uncoupling procedures are believed to be followed by an increase in the intracellular Ca^{2+} concentration. Calmodulin has been suggested to be involved in the Ca^{2+} uncoupling mechanism in that it occupies binding sites located at the cytoplasmic surface between the amino and carboxy terminals of connexins (Peracchia and Bernardini, 1984; Peracchia and Girsch, 1985).

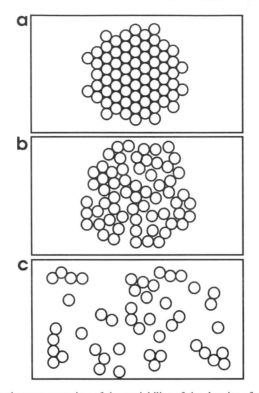

FIG. 1   Diagrammatic representation of the variability of the density of connexons that can
be visualized in freeze-fracture replicas as ~8-nm particles. Each circle represents one
connexon. Magnification: ~x600,000 : 1 (a) Hexagonal pattern of connexons corresponding
to a connexon density of 12,000–14,000/μm^2. These GJs are not frequently found but believed
by several authors to reflect the uncoupled state (see Section III,A). (b) Slight derangement
of the crystalline connexon pattern that is connected to an increase of the interconnexonal
spacing by one or a few nanometers is termed in several studies as "low connexon density,"
and is interpreted in several studies as correlating to the coupling state (see Section III,A).
Also in the outer horizontal cells of the goldfish retina, this connexon density (in the magnitude of 8000/μm^2 and termed as "high") appears to represent the coupling state (see Fig.
3 and Section V,B). (c) The connexon dispersion found in horizontal cells under conditions
of light adaptation or dopamine administration is at its maximum when comparing it to
dispersion rates in other systems [see (b) and Section III,A]. In the retina, this dispersion
seems to correlate with uncoupling, but possibly this mode of uncoupling is pH gated (see
Figs. 5 and 6 and Section V,B).

These conformational alterations should go along with a decrease in the
interconnexonal distance. This assumption is in agreement with many
past studies describing an increase in the connexon density including a
slight decrease of the channel extension parallel to the membrane plane
following an uncoupling procedure (Peracchia and Dulhunty, 1976; Peracchia, 1977; Chuang-Tseng *et al.,* 1982; Unwin and Ennis, 1983; Chuang

et al., 1985; Peracchia and Peracchia, 1980a,b). However, most data on coupling/uncoupling-related connexon spacing describe subtle reorganizations of connexon arrays on the order of 1 nm. In view of critical methodological considerations (Appleyard *et al.,* 1985; Sosinsky *et al.,* 1988; Rash and Yasumura, 1992), it may be asked whether distances of less than 1 nm can reliably be measured in freeze-fracture replicas under different physiological conditions. Indeed, there are some other studies that deny a consistent change in connexon package as a result of experimental alteration of junctional conductance (see, for example, Hanna *et al.,* 1984; Berdan and Caveney, 1985; Miller and Goodenough, 1985). On the other hand, and in sharp contrast to the findings in all other systems investigated so far, physiological and reversible alterations of connexon densities in horizontal cells of the fish retina are considerably larger. Here, the averaged interconnexon spacing varies from 20 to 40 nm (Kurz-Isler *et al.,* 1992; see Fig. 1).

In the GJs of the outer horizontal cells of the goldfish retina, Kurz-Isler and Wolburg (1986, 1988) described a rapid alteration in connexon densities after changing the illumination conditions. In the dark, they found a density of 7000–8000 connexons/μm^2; after red light adaptation the connexon density decreased to 3000/μm^2 In addition, during dark adaptation most GJs are small in size and undercoated by an electron-dense material. Frequently, this material was associated with microtubules and microfilaments (Kurz-Isler and Wolburg, 1988). The association of osmiophilic material beneath densified GJs during long-lasting dark adaptation speaks in favor of a functional interaction between cytoskeleton and connexons. In short-term experiments (a few minutes of red light exposure) the authors found, in ultrathin sections, a patchlike pattern of electron dense material that resembled the pattern of connexon clusters seen in corresponding freeze-fracture replicas. They suggested that the connexons within a cluster are more tightly connected to the cytoskeleton. However, another assumption cannot be excluded. The membrane areas between the connexon clusters are on the same order of magnitude as both the clusters and the electron-dense plaques. Thus, which part of the GJ—the connexon clusters or the membrane between—may be connected with cytoskeleton-associated plaque material is, at least in the short-term experiments, not possible to distinguish. Schmitz and Wolburg (1991) have shown that the connexon density of the outer horizontal cell GJs in the goldfish retina explanted *in vitro* is sensitive to the pH value, as has also been shown in the calf lens by Peracchia and Peracchia (1980b). At pH 6.5, it was on the order of 8000–9000 connexons/μm^2, at pH 7.1 it was about 6000–7000 connexons/μm^2, and at pH 7.5 it was about 4000 connexons/μm^2. The highest connexon density was found parallel to the appearance of membrane-associated plaque material. This material disappeared following an increase in the pH. Regardless of the physiological coupling state, represented by high or low connexon densities in horizontal cells, the results of Kurz-Isler and Wolburg (1988) and Schmitz and Wolburg (1991) consistently show the coincidence of high connexon density with cytoplasm/cytoskeleton–GJ association (see also Section V,B). A high connexon density seems to require the presence of cytoplasmic material with high osmium-binding capacity. On the other hand, it might be interesting that Vaughan and Lasater (1990a) have not been able to electrically uncouple fish horizontal cells by exposing them to F-actin-disrupting cytochalasin D. If coupling in this cellular system occurs at a high connexon density as has been suggested by Kurz-Isler and Wolburg (1988) and Kurz-Isler *et al.* (1992), F-actin is not responsible for the maintenance of this connexon density. Unfortunately, there are no immunocytochemical or freeze-fracture data available to demonstrate the effect of cytochalasin D on the fate and composition of the cytoplasmic plaques and the connexon positioning during and after microfilament disruption. In the lens of primates, but not of other vertebrate species, GJs were found to be associated with actin filament bundles, which are suggested to provide structural stability for the lens (Lo *et al.*, 1994).

The literature on GJ–cytoskeleton interactions cited above has not been able to demonstrate a general principle of the regulation of GJ formation and shape and the control of interconnexonal spacing. Hence, in the next section, we consider a further possibility with which to regulate connexon positioning in the membrane and describe relations between GJs and tight junctions.

B. Relations between Gap and Tight Junctions

Gap junctions and tight junctions are frequently codistributed in many cell types (see, e.g., Friend and Gilula, 1972; Decker and Friend, 1974; Elias and Friend, 1976; Simionescu *et al.*, 1976; Larsen, 1977; Van Deurs and Koehler, 1979; Meda *et al.*, 1979; Raviola *et al.*, 1980; Azarnia *et al.*, 1981; Rassat *et al.*, 1981; Bernardini *et al.*, 1984; Wolburg *et al.*, 1983; Van den Hoef *et al.*, 1984b; In't Veld *et al.*, 1985; Lane *et al.*, 1986; Mack *et al.*, 1987; Mancel and Hirsch, 1989; Risek *et al.*, 1994). However, in most studies the morphological interrelationship between the two junctions has not been regarded with particular emphasis. Only a few reviews consider this topic, which nevertheless deserves some interest. Friend and Gilula (1972) described GJs and tight junctions in a number of epithelia such as exocrine pancreas, liver, adrenal cortex, epididymis, small intestine, and duodenum. Whereas in the pancreas epithelium GJs are completely enclosed by tight junctional strands, the two junctions are rarely intermingled in the liver. In the small intestine there are no GJs between tight junctions, and in the epididymis the elaborate tight junctional network is also free of GJs. In the adrenal cortex, the authors found it difficult to ascertain whether linear arrays adjacent to large GJ plaques represented linear GJs or beaded tight junctional ridges.

Peracchia (1977) demonstrated that GJs and tight junctions are separated in the plane of the lateral membrane in the mucous cells of the rat stomach. In contrast, epithelial cells of the rat stomach show tight junctional strands intermixed with GJs (Bernardini *et al., 1984*).

In the choroid plexus of the rat, Van Deurs and Koehler (1979) described an elaborate network of tight junctional strands, although GJs were not mentioned. In the transitional zone between ependymal cells (GJs only) and choroid plexus cells (tight junctions only) of the same species, Mack *et al.* (1987) found tight junctions with intercalated GJs. In a developmental study on the chick choroid plexus, Dermietzel *et al.* (1977) found both types of junctions. The first junctional structures do not develop before embryonic day 5, at which time particles arise and form strands and clusters. The authors do not exclude the possibility that a common pool of junctional precursors for both kinds of junctions may exist.

In the mouse oocyte–follicle cell complex, Van den Hoef *et al.* 1984b) found, besides unidentified rhombic particle arrays, heterologous GJs between oocytes and the surrounding follicle cells. After defolliculation, the oocytes acquired linear arrays resembling tight junctions that were occasionally associated with small GJs. These may not have been genuine tight junctions but particles of GJs rearranged after defolliculation.

In dissociated and reaggregated B-cells of the rat endocrine pancreas, In't Veld *et al.* (1985) described GJs and tight junctions that form irregularly shaped aggregates of strands and connexon clusters. This coexistence of both types of contacts under *in situ* conditions was denied by In't Veld *et al.* (1984) but demonstrated by Meda *et al.* (1979). These authors described linear arrays and suggested that they represent an unusual form of GJ or a step in the formation of tight junctions or even one in the interconversion between GJs and tight junctions. Decker and Friend (1974), in a study on the neurulating amphibian embryo, proposed that the tight junctions provide a framework for early GJs. In another study on chick embryo shank skin, Elias and Friend (1976) observed a retinoic acid-induced GJ proliferation that preceded the generation of tight junctions. As discussed by Dermietzel *et al.* (1977) and Larsen (1977), a common pool of membrane particles was thought to be shared by the two types of intercellular contacts. Alternatively, in a theoretical consideration on junction formation, Weinbaum (1980) assumed a close arrangement of particles with different biochemical and electrical properties. First, randomly dispersed particles that are uniformly charged, or second, strands with electric dipole properties, or third, arrays of units with quadrupole interactions could represent (on the basis of negatively charged sialic groups at the surface of the membrane) three different configurations that would agree with existing interpositions of different types of junctions. More directly, Simionescu *et al.* (1976) showed, in endothelial cells of arteries and veins and in addition to typical tight junctions and GJs, so-called "special junctions" that consisted of a close apposition of connexons to tight junctional strands. In a diagrammatic proposal, the authors located occluding particles at the edge of a GJ cluster, possibly intending a role of tight junctions for the stabilization of connexons in the membrane plane.

These early speculations on the close interrelationship between GJs and tight junctions are of particular interest considering more recent results. Furuse *et al.* (1993) identified the first integral membrane protein of tight junctions, which they called occludin. This 55.9-kDa protein was sequenced and the hydrophilicity plot revealed four putative transmembrane domains with two extracellular loops. Thus, although the protein does not show any sequence similarity to connexins, the molecular structure may suggest a functional similarity as regards the occlusion of the extracellular space. In contrast to the connexins, which are arranged in a circular manner leaving a central channel between cellular processes, the occludin molecules may be arranged in a linear fashion. The function of insulated discontinuous tight junctions is not at all clear, because a paracellular barrier is not established. The strange similarity of some GJs or tight junctions and their close interposition in many cell types suggest an association of tight junctions with GJ particles according to the proposals of Weinbaum (1980). However, this association certainly is not of general importance for the junction formation because in many cells only one or the other type of contact is expressed.

IV. Modulation of Gap Junctions

A. Synthesis and Regulation of Connexins

In spite of about 30 years of work on GJs, the regulation of their formation, persistence, and removal is still enigmatic. Currently two regulatory steps in GJ formation are under investigation: first, the turnover rates of GJ proteins as predominantly studied in liver and, second, modifications of the connexins before insertion of these modified complexes into membranes. In general, important proteins with key functions in regulatory circuits (e.g., enzymes in limiting steps of metabolic pathways) tend to have short half-lives (Goldberg and John, 1976). Early studies of Gurd and Evans (1973) using a double-label isotope technique found a slow turnover for the "sarcosine-resistant fraction" of rat liver membranes containing GJs. However, a high content of collagen with low turnover rates could have contaminated this fraction (Fallon and Goodenough, 1981). [^{35}S]Methionine pulse injection studies (Yancey et al., 1980) revealed a peak incorporation of M_r 10,000 proteins into isolated liver gap junction after 3 hr. The half-life of GJ protein in rat liver was then determined to be approximately 19 hr (Yancey et al., 1981). Because of reutilization from the radioactive amino acids, however, the true half-life time is probably overestimated by a factor of 2. Comparing the turnover of liver GJ proteins to that of other proteins in liver plasma membranes with half-life times of 3.6 days (Yancet et al., 1981), the GJ protein has a 4.5-fold more rapid turnover. Hence, short half-life times of GJ protein implies that a block in their synthesis results in a rapid disappearance of the junctions and may therefore provide an alternative mechanism of regulating intercellular communication besides uncoupling by closure of channels (Yancey et al., 1981). In vivo studies with [^{14}C]bicarbonate labeling revealed the shortest half-life times for mouse liver GJs known so far. Fallon and Goodenough (1981) found a strikingly rapid turnover of 5 hr. Several in vitro studies of connexin biosynthesis [e.g., the cardiomyocyte connexin 43 half-life of 1–2 hr (Laird et al., 1991)] demonstrated short half-life times of 1–4 hr (Beyer, 1993).

On the one hand, therefore, it seems likely that intercellular communication by GJs may be controlled at the level of protein synthesis and degradation rather than by changes in reversible molecular conformations at the level of individual connexons (Fallon and Goodenough, 1981).

On the other hand, rapid turnover rates of the GJ proteins depend on processing steps during the protein translocation to the cellular membrane. In rat cardiac myocyte cultures, Puranam et al. (1993) used the ionophore monensin as a metabolic blocker to study biosynthesis and translocation of connexin 43 GJ protein. They found that 40- and 41-kDa forms of connexin 43 accumulate in swollen Golgi vesicles and concomitantly found a reduction in the number of plaques. These results suggested that connexin 43 is initially phosphorylated in the endoplasmic reticulum or at cis-medial Golgi from the 40-kDa to the 41-kDa form during its pathway to the membrane. The more extensive phosphorylation to 42- and 44-kDa forms occurs at locations distal from the block at the trans-Golgi cisternae.

Components of the extracellular matrix, for example, proteoglycans and glycosaminoglycans, also influence the expression of GJs at multiple levels (Spray et al., 1988). Continuing work may identify more matrix molecules involved in signaling processes that regulate GJ communication (Rosenberg et al., 1993). Other regulatory molecules include hormones

(Cole and Garfield, 1986; for review, see Larsen, 1985; Spray and Bennett, 1985; Spray *et al.*, 1985) and cAMP derivatives (Kessler *et al.*, 1984). Protein kinase actions on junctions have been shown in a number of systems (Flagg-Newton *et al.*, 1981; Saez *et al.*, 1986, 1990b; Chanson *et al.*, 1988; Takeda *et al.*, 1989; Murray and Gainer, 1989; Enkvist and McCarthy, 1992; for kinase actions on junctional conductance, see Bennett *et al.*, 1991).

Accumulating evidence, from studies using different cell lines, suggests that connexins may build protein pools prior to membrane insertion (Zidell and Loch-Caruso, 1990; Hendrix *et al.*, 1992). This would explain why transcription and translation inhibitors in some cases had no significant effect on the establishment of functional GJ communication (Cox *et al.*, 1976; Epstein *et al.*, 1977). Processes controlling the levels of connexins 43, 32, and 26, however, have been shown in regenerating liver (Kren *et al.*, 1993; for review, see Musil and Goodenough, 1990). Posttranscriptional mechanisms, of which mRNA stability may represent one critical factor, are suggested to regulate GJ expression dynamically in this system (Kren *et al.*, 1993).

In pancreatic B cells, In't Veld *et al.* (1985) described an increase in connexons in response to dibutyryl cyclic AMP or to phosphodiesterase inhibitor. However, the authors did not prove if the augmented number of GJs resulted in a higher level of cell–cell communication. In human adrenal cortical tumor cells (Murray and Taylor, 1988) and in rat liver parenchymal cells (De Mazière and Scheuermann, 1988), the GJ areas are increased after administration of dibutyryl cyclic AMP. In a series of experiments, Flagg-Newton *et al.* (1981), Flagg-Newton and Loewenstein (1981), and Azarnia *et al.* (1981) investigated the influence of cAMP on the intercellular permeability and GJ morphology in a mouse cancer cell line. These studies suggested the importance of cAMP-mediated phosphorylation in GJ modulation. Saez *et al.* (1986) directly demonstrated an effect of cAMP on phosphorylation of the 27-kDa GJ protein in liver cells that correlated with an increase in the junctional conductance. In rat cardiac myocytes *in vitro* (Oyamada *et al.*, 1994) and in cultured rat or bovine lens epithelial cells (Jiang *et al.*, 1993), connexin 43 and connexin 46, respectively, were described to be phosphorylated with culturing time. The phosphorylation was suggested to support the intercellular coupling and to be controled by intracellular calcium concentration (Crow *et al.*, 1994). Activation of protein kinase C has been found to phosphorylate connexin 32 in hepatocytes (Takeda *et al.*, 1989). However, this study did not address the coupling behavior of these cells. In another study, Oh *et al.* (1993) investigated intercellular communication-deficient (GJIC⁻) mutant clones of a rat liver epithelial cell line, using photobleaching and scrape-loading dye transfer techniques. They suggested that the absence of a hyperphosphorylated form of connexin 43 should be responsible for the inability of mutant cells to communicate. Because distribution and localization of connexin 43 did not differ in GJIC$^-$ and GJIC$^+$ cells the loss of phosphorylation apparently did not impair the insertion of gap junctional protein into the membranes. However, the occurrence and phosphorylation state of the main gap junction protein of liver cells, connexin 32 (Kumar and Gilula, 1986; Kuraoka et al., 1993), in this communication-deficient mutant cell line of hepatocytes were not investigated. On the contrary, there is some support for the inverse hypothesis of a blocked GJ communication following tyrosine phosphorylation. According to a study on communication-competent cardiomyocytes, tyrosine and threonine residues are not phosphorylated in connexin 43 protein (Laird et al., 1991). Evans et al. (1993) described a loss of recognition of the 60-kDa membrane protein (MP60) in the center of the developing mouse lens by a monoclonal antibody. In view of the study of Voorter and Kistler (1989), according to which protein kinase A phosphorylates GJ proteins only in the lens cortex but not in the lens nucleus, the finding of Evans et al. (1993) may suggest a cleavage of the GJ protein in the lens nucleus associated with a loss of phosphorylation sites and recognition by the antibody. Absence of phosphorylation sites would establish a constitutive metabolic coupling in the center of the lens, which is required to overcome reduction of nutrient supply.

Another set of molecules involved in the regulation of GJs includes growth factors. Epidermal growth factor (EGF)-induced disruption of GJs indicated phosphorylation of serine residues (Lau et al., 1992). In particular, the multifunctional polypeptide bFGF has been immunohistochemically shown to be associated with connexin 43 GJs of rat astrocytes (Yamamoto et al., 1991). Biochemical studies revealed that bFGF-like peptides are either an integral part of, or exist in close association with, connexin 43 in rat cardiac GJs (Kardami et al., 1991). Considering the hypothesis that phosphorylation is involved in the regulation of assembly/disassembly of GJs it is noteworthy that bFGF is able to activate protein kinase C, for which connexin 43 has been identified as a substrate (Saez et al., 1990a). The phosphate content of connexin 43 could, therefore, be modulated via bFGF induction, which in turn could mediate the GJ communication (Yamamoto et al., 1991).

B. Dynamics of Gap Junctions in Development,
   Oncogenesis, and Other Diseases

In the previous sections we describe the lateral shift of connexons within the membrane plane, the modulation of the connexin molecule through phosphorylation processes, and the formation of two separate hemichannels to complete intercellular GJ channels. In the following sections we focus briefly on developmental aspects of GJ-mediated communication, which include the formation and degradation of gap junctions as well as their behavior in neoplastic cells.

1. Gap Junctions during Development

The role of GJs in embryonic development has been reviewed several times (see, e.g., Bennett *et al.*, 1981; Caveney, 1985; Guthrie and Gilula, 1989; Warner, 1992; Goodenough and Musil, 1993). From the very early developmental stages on, GJs are involved in the transfer of morphogenic signals (Edelman, 1986, 1988). One of the best-known examples for the significance of GJ mediated spread of such signals is the development of the freshwater coelenterate *Hydra* (Green, 1988). An established head of *Hydra* appears to release a head inhibitor diffusing down the body column and passing from cell to cell through GJs. The disruption of this GJ communication by transplanting grafts from just below the head of a donor animal that had been previously loaded with GJ-specific antibodies resulted in a reduction in the ability of an existing head to inhibit the formation of an additional head (Fraser *et al.*, 1987). Similar principles of GJ involvement in pattern formation were studied in the early amphibian embryo (Warner *et al.*, 1984), chicken embryo (Dealy *et al.*, 1994), and the leech embryo (Wolszon *et al.*, 1994). Zeng (1987) studied the effect of antibodies directed against mouse liver GJ protein 26 KDa on GJ formation in early *Xenopus* embryo and found GJs of very small size in comparison to controls. The assumed attenuation of connexon assembly by anti-connexin antibodies has also been suggested by Meyer *et al.* (1992), who investigated dye transfer and junctional morphology in dissociated and reaggregated Novikoff hepatoma cells following treatment with Fab fragments of antibodies against the extracellular domains of connexins. The authors were not able to observe many GJ in freeze-fracture replicas, suggesting the extracellular domains of the connexin molecule to be essential for junction formation.

A novel field of GJ research was opened by the discovery that adhesion molecules play an important role in the regulation and formation of GJ communication. Kanno *et al.* (1984) reported on a monoclonal antibody recognizing uvomorulin (E-cadherin), which interfered with the Ca^{2+}-dependent cell-to-cell adhesion and inhibited the Lucifer Yellow transfer between teratocarcinoma stem cells. Another indication for the involvement of cadherins in GJ function was provided by Mege *et al.* (1988), who transfected liver cell adhesion molecule (L-CAM) cDNA into L-CAM-deficient mouse sarcoma cells that were incompetent for GJ communication. After transfection, functional GJs appeared. Correspondingly, Musil *et al.* (1990) showed that transfection with L-CAM cDNA may restore communicating junctions in S180 and L929 cell lines. Before transfection these cells were deficient in known cell–cell adhesion molecules and expressed incompletely phosphorylated connexin 43. Jongen *et al.* (1991) found a good correlation between Ca^{2+}-dependent GJ communication and E-cadherin expression in mouse epithelial cells, suggesting that Ca^{2+}-dependent E-cadherin is involved in the regulation (assembly and/or function) of connexin 43. In Novikoff hepatoma cells which also contain connexin 43, another cadherin, N-cadherin or A-CAM, was described to be involved in GJ regulation (Meyer *et al.*, 1992). Cells treated with Fab fragments of antibodies against A-CAM were not able to form GJs. In contrast, anti-connexin Fab fragments inhibited the assembly of adherens junctions. These results show close interactions between adhesion molecules and GJ formation during development. A study using an immortalized clonal cell line from a rat ovarian epithelium (Stein *et al.*, 1993) complemented these results by correlating the loss of desmosome-mediated adhesiveness with the GJ communication.

It may be concluded that adhesion molecule-mediated contacts, which might include synergistic processing steps such as phosphorylation and cell–cell contact via CAMs and extracellular domains of connexins, must precede GJ formation. Most studies refer to Ca^{2+}-dependent cadherins as the competent adhesion molecules, but N-CAM, a Ca^{2+}-independent member of the immunoglobulin superfamily, has also been shown to be involved in GJ regulation (Keane *et al.*, 1988). Moreover, Churchill *et al.* (1993) detected intercellular currents carried by GJs and dye coupling in the absence of Ca^{2+} in hemocytes of the cockroach. The role of Ca^{2+}-independent adhesion molecules in GJ function could, however, not be ruled out.

2. Gap Junctions in Oncogenesis

The interruption of the GJ-mediated communication pathway seems to be one step in the malignant transformation of cells. This was already hypothesized as early as 1967 by Loewenstein and reviewed by Janssen-Timmen *et al.*, 1986 and Klaunig and Ruch (1990). Conversely, the proliferation of neoplastic cells has been shown to be inhibited by supporting the GJ coupling between these cells (Charles *et al.*, 1992; Zhu *et al.*, 1992; Naus *et al.*, 1993). The invasive behavior belongs to the peculiarities of neoplastic cells. Because invasive properties have been shown to be acquired after the loss of E-cadherin-mediated cell–cell adhesion (Behrens *et al.*, 1989; Chen and Öbrink, 1991), and E-cadherin is essential for GJ communication acquired during development (see above), a direct relationship between invasiveness of neoplastic cells and loss of GJ communication seems to be obvious.

There is a large body of evidence that the activation of protein kinase C mediated by tumor-producing phorbol esters correlates with a change in the state of phosphorylation of connexins and a reduction in GJ communication (for an overview, see Berthoud *et al.*, 1992). However, as pointed out already by Chanson *et al.* (1988), phorbol esters have been convincingly demonstrated to decrease junctional permeability only in established cell lines and, for example, not in acinar cell pairs of the rat exocrine pancreas. Incorporation of phosphate into tyrosine residues of connexin 43 is correlated with incompetence for communication between cells via GJs (Crow *et al.*, 1990; Filson *et al.*, 1990; Swenson *et al.*, 1990). In accordance with this, Goldberg and Lau (1993) showed that transformation of fibroblasts with pp60$^{v\text{-}src}$ oncogene led to a massive phosphorylation of tyrosine residues, whereas in nontransformed, still communication-competent cells, no phosphotyrosine was detectable. However, the amount of connexin 43 mRNA and protein of the transformed cells was even elevated, and not diminished, as could have been expected for these communication-incompetent cells. In clone 9 cells, a rat liver epithelial cell line, anti-connexin 43 immunoreactivity disappeared rapidly after phorbol ester treatment, suggesting a retrieval of connexons from the membrane (Berthoud *et al.*, 1992). In contrast, Hülser and Brümmer (1982) previously showed a gradual decrease in Lucifer Yellow spread in cultured spheroids of BICR/M1R-K tumor cells that was, nevertheless, connected with the maintenance of GJ plaques in freeze-fracture replicas. Naus *et al.* (1991) reported on a reduction in the level of connexin 43 mRNA in C6 glioma cells compared to astrocytes. Treatment of astrocytes with phorbol esters blocks GJ communication and the spread of Ca^{2+} waves (Enkvist and McCarthy, 1992), which are thought to indicate the movement of Ca^{2+} and other second messengers such as inositol triphosphate through the astrocytic syncytium (Cornell-Bell *et al.*, 1990; Jensen and Chiu, 1993; Kim *et al.*, 1994). If C6 glioma cells were transfected with connexin 43 cDNA the proliferation rate of these transfectants was heavily reduced, their dye coupling increased, and GJs between them formed (Charles *et al.*, 1992; Zhu *et al.*, 1992; Naus *et al.*, 1993). In addition, the proliferation rate of C6 cells was reduced by culturing them in medium conditioned by connexin 43 cDNA-transfected cells (Zhu *et al.*, 1992). This suggests the release of diffusible growth-controlling signal(s) by communication-competent cells. Gap junction coupling may therefore exert influences on the synthesis and/or secretion of growth inhibitors.

A study on connexin expression during oncogenesis in rat liver demonstrates the complexity of the relationship between connexin expression and development of neoplasms (Neveu *et al.*, 1994). Neoplasms generated by diethylnitrosamine displayed deficiencies in connexins 32 and 26. However, Northern blotting failed to demonstrate reduction in connexin mRNAs, suggesting that some tumors downregulate connexin immunoreativity independent of message expression. Several posttranslational mechanisms may be responsible for the equilibrium between synthesis, insertion into membranes, assembly, and turnover rates of connexins during carcinogenesis, all of which remain to be elucidated in detail.

3. Degradation of Gap Junctions

During development, several mechanisms may serve as regulators of GJ communication. As already mentioned, uncoupling procedures may (among others) include phosphorylation of connexins, Ca^{2+}-dependent calmodulin-binding, or changes in connexon density in the membrane plane. Moreover, another important mechanism seems to be the degradation of GJs, which occasionally has been described as a dispersal of GJ clusters (in the metamorphosing insect *Manduca sexta;* Lane and Swales, 1980), but frequently as the occurrence of annulated GJs. These were described in such different tissues, for example, as ovarian cells (Burghardt and Anderson, 1979; Larsen *et al.,* 1979; Van den Hoef *et al.,* 1984a; Koike *et al.,* 1993), cardiac muscle (Spray *et al.,* 1985; Mazet *et al.,* 1985), neuroepithelium of the rat embryo (Schuster, et al., 1990), hamster cheek pouch mucosa (White *et al.,* 1984), chicken otocyst sensory epithelium (Ginzberg and Gilula, 1979), and fish retina (Vaughan and Lasater, 1990b) The motive force for the formation of annulated GJs seems to be an actin-dependent contractile endocytotic mechanism (Larsen *et al.,* 1979) that finally terminates in lysosomal degradation (Larsen and Tung, 1978; Vaughan and Lasater, 1990b). On the other hand, an extensive association of actin filament bundles with GJs must not in any case be involved in the removal of junctional domains but rather in the stabilization of membrane structure, as was shown in the primate lens by Lo *et al.* (1994; see also Section III,A).

There is little information about the linkage between these morphological and enzymological aspects of degradation on the one hand, and more molecular aspects on the other. In any case, several studies indicate that phosphorylation of amino acids is involved in the degradation of GJs and provide evidence for the idea that the majority of the protein seems to be degraded while in its most highly phosphorylated state (Crow *et al.,* 1990; Filson *et al.,* 1990; Swenson *et al.,* 1990; Goldberg and Lau, 1993).

In addition, phosphorylation by protein kinase A and protein kinase C apparently has different effects on the proteolysis of connexin 32 *in vitro* (Elvira *et al.,* 1993). Data implicate that phosphorylation is not restricted to the gating mechanism of the GJ channels (Elvira *et al.,* 1993). Phosphorylation mechanisms involved in blocking communication between cells via a possible degradation mechanism, however, still need further investigation.

4. Gap Junctions Involved in Diseases

The first studies have been published on the involvement of GJs in the etiology of disorders. Chagas' disease is caused by an infection through *Trypanosoma cruzi* and represents the main cause of cardiomyopathies in Latin America. Cocultures of neonatal rat cardiomyocytes and *T. cruzi* revealed a loss of GJs in heart cells that correlated with a loss of connexin 43 immunostaining (Campos de Carvalho *et al.*, 1993). Hence, disorders in cell–cell interactions via GJs may contribute to some forms of cardiac arrhythmias, based on altered patterns of propagation caused by a less intensive coupling of cells. Abnormal electrical conductions in heart diseases also occurred, for example, in Wolff–Parkinson–White syndrome. Here, a malformation built an accessory conduction pathway between atrium and ventricle of the heart and caused ventricular preexcitation. This phenomenon could be correlated to a connexin variant and its unusual distribution, analyzed at the ultrastructural level (Severs *et al.*, 1993). In tissue of patients with advanced ischemic heart disease a different GJ organization was found at the border of healed infarct areas, compared to normal tissue. Furthermore, the GJ surface area per unit myocyte volume and GJ content per cell were reduced (Severs *et al.*, 1993). In summary, alterations in GJs seem to play an important role in electrical rhythm disturbances associated with heart disease.

In the central nervous system, communication between neurons has been supposed to be involved in the genesis of epilepsy. However, the strength of dye-coupling patterns could not be related to the degree of epileptogenicity (Cepeda *et al.*, 1993). In the peripheral nervous system, Charcot–Marie–Tooth disease is subdivided in forms with a primary pathological defect in degeneration of the myelin or in degeneration of the axons. In CMTX (X-linked Charcot–Marie–Tooth) disease myelin of peripheral nerves degenerates. Studies have led to the idea that connexin 32 may be involved in this inherited defect. Therefore, genes of affected patients of eight families were analyzed and different mutations of the connexin 32 gene were found. In normal tissue, connexin 32 appeared at the nodes of Ranvier and at the Schmidt–Lanterman incisures in myelinated peripheral nerves. In CMTX disease, which is apparently limited to peripheral nerves, mutations in the connexin 32 gene and subsequent malfunction could explain the disorders in nerve structure and functions (Bergoffen *et al.*, 1993).

This line of research on possible involvement of altered GJ coupling will certainly provide further cues to their physiological functions.

V. Gap Junctions in Central Nervous System

A. Brainstem and Cortex

1. Distribution of Connexins

Gap junction proteins were found to be expressed by specific cell populations in the brain (Dermietzel and Spray, 1993). In addition, the expression of different connexins changed in developing brain (Dermietzel et al., 1989a). During neurogenesis, connexin 43 and connexin 26 dominated in undifferentiated neuroepithelial cells, whereas connexins 43 and 32 were expressed mainly in the adult brain (Dermietzel et al., 1989a). Connexin 32 staining in adult brain was prominent in some layers of cerebral cortex, basal ganglia, thalamus, cerebellum, and brainstem and confined to neurons and oligodendrocytes. Double immunostaining with nEN (an antibody against the neuron-specific enolase) or myelin basic protein (MBP) as neuronal and oligodendrocytic markers, respectively, showed that the majority of anti-connexin 32-stained subcortical structures (basal ganglia, thalamus, and various nuclei of brainstem) may arise from coupled oligodendrocytes (Dermietzel et al., 1989a). However, only weak electrical coupling has been shown between oligodendrocytes in culture (Kettenmann and Ransom, 1988). It was suggested that oligodendrocytes are seldom coupled to each other, but that coupling is abundant between astrocytes and oligodendrocytes (Mugnaini, 1986). Because the coupling behavior of connexin 43 (present in astrocytes) is voltage insensitive and that of connexin 32 (present in oligodendrocytes) is voltage dependent, Dermietzel et al. (1991) hypothesized that this could be the basis for a rectifying effect in the signal transduction process between astrocytes and oligodendrocytes. However, Ransom and Kettenmann (1990) were not able to support this hypothesis electrophysiologically.

Another study used the monoclonal antibody 92B, against the carboxy-terminus region of connexin 32, to stain neurons throughout the central nervous system (CNS) (Yamamoto et al., 1990b). Brainstem and spinal α motoneurons exhibited the most prominent staining. Staining consisted of puncta, with a diameter of about 1.0 μm and a length of 5 μm, in these cells (Yamamoto et al., 1990b). This was in contrast to punctate staining (with a diameter of about 0.5 μm or less) of other brain regions. Staining of facial nucleus motoneurons was most frequent on the soma, but extended into the proximal dendrites, where it was reduced in size. At the electron microscope level, facial motoneurons exhibited a dense staining of restricted regions of the endoplasmic reticulum (ER) closely apposed to the plasma membrane of the motoneuron soma, called subsurface cisterns, and various other intracellular structures. Subsurface cisterns have been hypothesized to be involved in ionic (e.g., Ca^{2+}) communication between cell exterior and the ER. Yamamoto *et al.* (1990b) suggested GJ channels across ER membranes in certain classes of cisterns. This would, in turn, enable ionic fluxes between cisternal clefts, and/or between the extracellular space and the cisternal lumen, the latter because of stained appositional membranes. These authors suggested an intracellular communication system within motoneurons, in addition to intercellular fluxes through GJs. These systems could be mediated by channels, which were marked by connexin 32 antibodies.

2. Gap Junctions in Astrocytes

Astrocytes form large networks coupled by GJs (Brightman and Reese, 1969; Massa and Mugnaini, 1982; Landis and Reese, 1982; Yamamoto *et al.*, 1990b; Binmöller and Müller, 1992). Their coupling behavior as well as the formation of GJs is maintained in cell culture (Massa and Mugnaini, 1985; Fischer and Kettenmann, 1985; Anders, 1988; Kettenmann and Ransom, 1988; Ransom and Kettenmann, 1990; Dermietzel *et al.*, 1991; Charles *et al.*, 1993; Kim *et al.*, 1994). Sontheimer *et al.* (1990) found cell coupling, as assessed by Lucifer yellow spread and electrophysiological recording, to be restricted to A2B5⁻ astrocytes cultured from rat optic nerve. A2B5⁻ astrocytes are termed A1 astrocytes and are believed to form the superficial and perivascular glia limitans (Raff, 1989). A2B5⁺ (A2) astrocytes from rat optic nerve were not found to be coupled under culture conditions (Sontheimer *et al.*, 1990). Functionally, the astrocytes are widely believed to be involved in the spatial buffering of the extracellular space (Gardner-Medwin, 1983; Odette and Newman, 1988; Jensen and Chiu, 1993), and GJs are postulated to redistribute K^+ rapidly within the astrocytic net to transport it to the circulation. However, each astrocyte is believed to be connected with a blood vessel (Reichenbach, 1989). From this, the distance between a site of neuronal activity and increased K^+ concentration and the next capillary may be too short to need a GJ syncytium. Therefore, other functions of astrocytic GJs such as the spread of Ca^{2+} and inositol triphosphate (Cornell-Bell *et al.*, 1990; Jensen and Chiu, 1993; Kim *et al.*, 1994) may be of even greater significance than that of the spatial buffering of the extracellular K^+. The astrocytic network dynamically reacts to neuronal stimuli, and in turn, astrocytes are known to influence neurons in different ways (for astrocyte–neuron interactions, see Smith, 1992).

Connexin 43 staining was shown to be confined to astrocyte/astrocyte coupling (Dermietzel et al., 1991). This protein is coexpressed with connexin 26 only in ependyma and leptomeninges (Dermietzel and Spray, 1993). On postnatal day 1, connexin 43-immunoreactive stellate processes were found to be densely stained in rat brain, but the adult staining pattern was reached by postnatal day 10 (Yamamoto et al., 1992). Binmöller and Müller (1992) were not able to demonstrate Lucifer Yellow spread in astrocytes of the rat visual cortex slice before postnatal day 11. In addition to the expression of connexin 43, developmental cues appear to be needed for GJ formation (see Sections IV,A and IV,B,1). The emergence of connexin 43 immunoreactivity differed around brainstem motoneurons. Staining was found earlier next to oculomotor and trigeminal motoneurons and afterward it appeared around hypoglossal or facial motoneurons (Yamamoto et al., 1992). Connexin 43 staining first appeared in radial glial cells and was then mainly found in astrocytic processes, in which staining changed from whole process labeling to an exclusively punctate pattern. This developmental sequence differed among brain regions and even within one region among different nuclei, for example, for the brainstem nuclei (Yamamoto et al., 1992). The overall medullary staining level including facial motor nucleus was moderate for connexin 43 in adult rat brain, compared to other regions (Yamamoto et al., 1990c). The time required to achieve the adult connexin 43 staining pattern parallels several important maturation steps in the rat brain, for example, formation of synaptic contacts, glial ensheathment of neuronal elements, and generation of electrical activity. The expression of connexin 43 in astrocytes of the adult brain was proposed to depend on neurons being regulated according to local neuronal needs (Yamamoto et al., 1990c; Vukelic et al., 1991; Nagy et al., 1992; Lee et al., 1994).

Direct evidence for an induction of astroglial GJs by neuronal stimuli has been shown by Rohlmann et al. (1993). Following facial nerve transection in adult rats, immunostaining of astrocytic GJs changed in the corresponding, ipsilateral motor nucleus of the brainstem. Enhanced immunostaining was confined to astrocytes surrounding the lesioned motoneurons, whereas on the control side staining remained faint, as in intact rats. These data showed that the astrocytic network can deviate from a moderately stained gap junction (GJ) pattern of the facial nucleus (Yamamoto et al., 1990c) by enhancing the GJ-coupling capacity as a consequence of axotomy in neighboring motoneurons. As yet, it is not known whether the number of GJs is augmented or whether the enhanced immunoreactivity parallels an increase in connexin 43 synthesis, which would increase the potential to form GJs. The amount of connexin 43 mRNA is already high in the facial nucleus of intact adult rats (Micevych and Abelson, 1991) and the half-life time of connexin 43 has not yet been determined in astrocytes, although connexin 43 of cardiomyocytes has a rapid turnover (see Section IV,A).

Astrocytic coupling has also been modified by intrathalamic kainic acid injection. The following neuronal degeneration caused alterations (increase after 5 hr and disappearance after 24 hr) of connexin 43 immunoreactivity at the lesion site (Vukelic et al., 1991; Hossain et al., 1994a). In these studies, the blood–brain barrier was not left intact, and neurons degenerated. In contrast, Rohlmann et al. (1993) set the motor nerve lesion distant from the motoneurons in the peripheral nervous system (PNS), and neurons were allowed to regenerate. In this case, retrograde transneuronal stimuli were found to induce astrocytic coupling after a short latency (Rohlmann et al., 1994a). Enhanced astrocytic GJ staining occurred in facial nucleus only 45 min to 1.5 hr after facial nerve cut. This rapid modification in the astrocytic coupling potential preceded chromatolysis of the lesioned motoneurons themselves (Neiss et al., 1993) and antedated the increase in ornithine decarboxylase activity occurring after 5–8 hr (Tetzlaff and Kreutzberg, 1985). The short latency of alterations in the coupling capacity of the astrocytic network may therefore play a role in triggering appropriate programs in motoneurons that could lead to axonal regeneration (Rohlmann et al., 1994a).

Homogenates of the brain revealed two forms of connexin 43: one dephosphorylated form migrating as a 41-kDa species, and a second, phosphorylated form of about 43 kDa (Nagy et al., 1992). Interestingly, the ratio of the two forms was found to be relatively uniform in various brain regions: 0.71 (43-1 to 41-kDa form). Even where immunoreactivity differed, for example, between striatum and globus pallidus, the ratio was equal (Nagy et al., 1992). Although phosphorylation is thought to be involved in different processes of protein synthesis and degradation (see Section IV,A and B), the constant ratio between dephosphorylated and phosphorylated forms deserves further investigation. Hossain et al. (1994b) found this ratio to increase during rat brain development. They suggested that the brain contains a phosphatase that may be involved in the dephosphorylation of connexin 43 to connexin 41 and therefore in the regulation of intercellular communication. Furthermore, this phosphatase may be responsible for the observation that in adult brain the level of connexin 41 is lower when brain metabolism is rapidly inactivated, indicating the postmortem conversion of connexin 43 to connexin 41 (Hossain et al., 1994b). In vitro studies of Batter et al. (1992) on cultured hypothalamic and striatal astrocytes suggested differences in their GJ expression patterns regulated on the mRNA level, but, in agreement with Nagy et al. (1992), changes in the relative levels of phosphorylated connexin 43 forms were not found. Nagy *et al.* (1992) found a positive correlation between densley stained brain regions and the amount of connexin 43 biochemically detected in Western blots. Because the density of immunostaining seems to represent the amount of connexin 43 protein, it has been suggested that this is an indicator of the number of GJs (Nagy *et al.*, 1992).

At the light microscopy level, connexin 43 immunostaining was reported to be intense in layer 1 of the cerebral cortex, whereas all other layers were only moderately stained (Yamamoto *et al.*, 1990c). Layers 3 and 4 included patches of more heavily stained areas. Staining consisted of small puncta and numerous annular profiles (see Section IV,B,3). Staining patterns of different cortical areas seemed to be identical; only parts of the piriform cortex were more intensely stained. Intracellularly located granular structures, 20–30 nm in diameter, were found frequently near or even attached to the inner junctional membrane and were suggested to be ribosomes (Yamamoto *et al.*, 1990a). Sometimes connexin 43 staining occurred on cytoplasmic membranes that were continuous with junctional membranes (Yamamoto *et al.*, 1990a). Astrocytic end feet around blood vessels were marked by a honeycomb meshwork composed of puncta with a larger diameter than puncta in the neuropil, which were about 1–2 μm in diameter. Blood vessels analyzed at the EM level were ensheathed by astrocytic processes or lamellae (Yamamoto *et al.*, 1990c). Gap junctions are straight-line membrane appositions. Immunohistochemistry using antibodies against connexin 43 (Fig. 2A) revealed stained GJs but also staining of neighboring cytoplasm. This might be due to a diffusion of diaminobenzidine (DAB) reaction product, as has been shown by Yamamoto *et al.* (1990c) and Nagy *et al.* (1992).

Horseradish peroxidase (HRP) filling of one astrocyte showed GJs between filled processes and lamellae of the cell (Fig. 2B and C), a way of coupling that has been named autocellular coupling (see below). Surface and volume analysis using the point-counting method revealed that about 86% of the cell surface of astrocytes in the visual cortex of adult rats was built up of lamellae. The remaining 14% was predominantly composed of astrocytic processes, in tissue spaces excluding the cell soma (A. Rohlmann *et al.*, unpublished observations). Hence, it seems to be reasonable that GJs were mainly found on plasma membranes of lamellae. Using the stereological disector-counting method, the number of GJs was quantified on electron micrographs (Rohlmann *et al.*, 1994b). For this purpose, ultrathin sections were tilted in the electron microscope, 10° at a time, to actually count all GJs of the examined angle, including the diagonally cut GJs. From their data Rohlmann *et al.* (1994b) estimated about 890 × 10^6 GJs/mm³ of neuropil tissue. Considering 29,700 astrocytes in the same volume of tissue (Gabbott and Stewart, 1987), 1 astrocyte would possess about 30,000 GJs in the visual cortex of adult rats. Electron microscopy

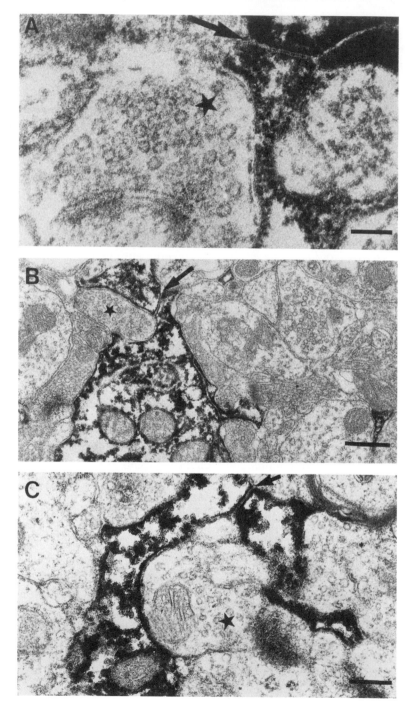

analysis revealed an average diameter of unstained GJs about 0.17 μm, indicating a small to medium size of GJs in the cortex (Rohlmann and Wolff, 1994). For comparison, Yamamoto *et al.* (1990a,c) estimated a width of 0.3–1 μm for puncta representing immunohistochemically stained GJs in light microscopy studies. This may be an overestimation because of staining procedures and possible diffusion of diaminobenzidine reaction product (Nagy *et al.*, 1992; Yamamoto *et al.*, 1990c).

All data above suggest a tremendous coupling capacity between astrocytes. Not all of the GJs, however, serve intercellular coupling (heterocoupling). Rohlmann and Wolff (1994) demonstrated GJs connecting different processes (including heterocellular coupling) as well as lamellae of the same process (autocellular coupling). More detailed morphometric analysis revealed that GJs are distributed over the whole surface of astrocytes irrespective of whether different cells or compartments of one cell are coupled by them. It was concluded that astrocytic GJ coupling must apparently be considered to be coupling of subcellular compartments rather than of cells. The function of autocoupling of a given astrocyte is enigmatic. Connexins normally form hexamers with 12 extracellular loops locking 12 other extracellular loops at the corresponding hemichannel of the adjacent partner cell (or process), forming a channel between cells (or processes). At these sites, the extracellular space may be occluded. As was investigated *in vitro* by Stauffer *et al.* (1991) and described in Section II,C in more detail, connexons can (under certain conditions) grow by end to end or by side-to-side association. Thus, there is no absolute certainty about the topology of connexin oligomerization in the membrane. Possibly, under certain conditions, connexins may not (only) produce intercellular channels but likewise may seal the intercellular cleft, serving as some sort of tight junction (although without the morphology and chemical composition of tight junctions; see Section III,B).

3. Gap Junctions in Neurons

What about the coupling of neurons? In the mammalian CNS GJs have been found in different brain regions, for example, hypothalamus (Andrew *et al.*, 1981; Cobbett and Hatton, 1984), inferior olivary nucleus (Sotelo *et al.*, 1974; Benardo and Foster, 1986), hippocampus (Schmalbruch and

---

FIG. 2 Stained GJs (marked by arrows) next to presynaptic elements (asterisks) in rat brain. (A) Connexin-43 (Cx-43) staining of an astrocytic GJ and neighboring cytoplasm. Scale bar: 0.08 μm. (B) Horseradish peroxidase (HRP)-filled cellular compartments of an astrocyte. Scale bar: 0.4 μm (Provided by Prof. J. R. Wolff, Göttingen, Germany.) (C) HRP-filled cellular compartments of an astrocyte. Scale bar: 0.2 μm. (Provided by Prof. J. R. Wolff, Göttingen, Germany.)

Jahnsen, 1981; Andrew *et al.,* 1982; Kosaka, 1983; Katsumaru *et al.,* 1988; Shiosaka *et al.,* 1989; Yamamoto *et al.,* 1989), main olfactory bulb (for references, see Reyher *et al.,* 1991), the retina (see Section V,B), and the cerebellar cortex (Sotelo and Llinas, 1972). For a review on ultrastructural studies of vertebrate electrical synapses see Leitch (1992). In addition, occurrence of GJs in the cortex of higher vertebrates has been known for some time (Sloper, 1972; Sloper and Powell, 1978; Smith and Moskovitz, 1979; Peters, 1980) and these have been shown to exhibit significant levels of cells containing connexin 32 mRNA (Micevych and Abelson, 1991). According to the different layers the amount of connexin 32 mRNA varied from ''no cells labeled'' in layer 6 to ''very high number of cells labeled'' in layers 2 and 3, which could include stained neurons and oligodendrocytes. However, connexin 32 mRNA was detected in discrete cell populations, such as layer 2 of the neocortex, that appeared to be neurons (Micevych and Abelson, 1991). Peinado *et al.* (1993) studied neurons in living slices of rat neocortex by the addition of the intracellular tracer neurobiotin at different postnatal stages. At early stages, coupling between neurons, primarily through their dendrites, was high, whereas injections at later stages resulted in little or no dye coupling. It was concluded that a transient local coupling pattern could provide neurons to exchange information and enable them to take part in a temporary coordination during circuit formation. Thus, GJ-related information exchange between neighboring cells could possibly guide synaptic establishment. *In vivo* injection of Lucifer Yellow into the rat neocortex (Connors *et al.,* 1983) already showed that neuronal coupling was especially extensive in immature rat neocortex, and declined with further maturation, with a subsequent enhanced establishment of chemical synapses. From the data above, one can draw the conclusion that GJ coupling between neurons of the adult cortex may have little effect on the spread of information or synchrony between cells. However, Dermietzel and Spray (1993) discussed possible functions of neuronal coupling, for example, enlargement of receptive field sizes in visual systems or second messenger exchange, the latter possibly even with low coupling strength. This view is supported by Christensen (1983) who analyzed lamprey neurons. Christensen found a few GJs to be more effective in depolarizing the soma as compared to the effect of a higher number of chemical synapses.

B. Retina

The retina as a part of the central nervous system is, owing to its easy accessibility, particularly suitable for multimethodological approaches in neurobiology. For many years it has been the subject of numerous morphological and physiological investigations, many of which have studied the processing of spatiotemporal information in the neuronal network. The electrotonic coupling via GJs of cells in the retina including photoreceptors, horizontal cells, bipolar cells, amacrine cells, ganglion cells (at least in the cat), and glial cells is well known (Dowling and Boycott, 1966; Raviola and Gilula, 1973, 1975; Uga and Smelser, 1973; Hayes, 1976, 1977; Burns and Tyler, 1990; for an overview, see Vaney, 1994; Cook and Becker, 1994). Remarkably, the identity of connexins in retina GJs is not published as yet. However, connexin 32 and connexin 43 were found to be distributed in a distinct manner in retinal neurons and glial cells (R. Weiler and R. Dermietzel, personal communication). Hitherto, the main interest was focused on the spread of electrotonic signals and dyes such as Lucifer Yellow or biocytin (Naka and Rushton, 1967; Kaneko, 1971; Piccolino et al., 1982; Teranishi et al., 1984; Kaneko and Stuart, 1984; Kujiraoka and Saito, 1986; Marc et al., 1988; Goddard et al., 1991; Cuenca et al., 1993; Vaney, 1993; Teranishi and Negishi, 1994). Compared with the number of physiological studies on electrical synapses, morphological studies on GJs in the retina are less numerous (see, e.g., Lasansky, 1976; Raviola, 1976; Raviola and Gilula, 1973, 1975; Raviola et al., 1980; Cooper and McLaughlin, 1981; Zimmerman, 1983; Witkovsky et al., 1983; Owen, 1985; Tonosaki et al., 1985). Among all retina cells coupled by GJs, the outer horizontal cells were most frequently studied and their GJ morphology correlated to physiological parameters.

Horizontal cells (HCs) are second order neurons in the outer part of the inner nuclear layer that are involved in the generation of the antagonistic surround of the receptive fields of bipolar and ganglion cells. In various species, some subtypes possess axons with enlarged axon terminals extending into the inner nuclear layer; these axon terminals form their own network coupled by GJs (Kouyama and Watanabe, 1986; Marshak and Dowling, 1987; Kurz-Isler et al., 1992; Vaney, 1993, 1994). Horizontal cells generate a different type of light-induced response (S-potential), the amplitude of which increases with the area of illumination. The spread of S-potentials is mediated by GJs (Naka and Rushton, 1967) and can be decreased by dopamine (Teranishi et al., 1984). In the light, dopamine is released from the interplexiform cells and binds to D1 and D2 receptors of HCs, which were shown to be involved in HC uncoupling in the turtle retina (Piccolino et al., 1984, 1989). In the dark, γ-aminobutyric acid (GABA) is released from the HCs and binds to receptors of the interplexiform cell, thus keeping the rate of dopamine release low. γ-Aminobutyric acid antagonists have an uncoupling effect on HC GJs (Piccolino et al., 1982) by allowing an increase in dopamine release (Negishi et al., 1983; Piccolino et al., 1987). The decreasing effect of dopamine on the GJ conductance is achieved through a cyclic AMP-dependent protein kinase, which in turn is believed to phosphorylate the GJ protein(s) (Lasater and Dowling, 1985; Lasater, 1987; Laufer *et al.*, 1989)

In freeze-fracture studies, the connexon density of HC GJs in fish retina was found to be dependent on illumination conditions (Wolburg and Kurz-Isler, 1985; Kurz-Isler and Wolburg, 1986, 1988; Figs. 3–6) as well as on the presence of dopamine (Baldridge *et al.*, 1987, 1989; Weiler *et al.*, 1988; Kurz-Isler *et al.*, 1992). In the dark, many small GJs on the order of 0.1 μm^2 revealed a high connexon density (about 8000/μm^2); in the light, the GJs became larger (up to 3 μm^2) and the connexon density decreased (to about 3000/μm^2; Kurz-Isler and Wolburg, 1988; Figs. 5 and 6). It should be pointed out that the connexon density that we term as "high" is low in comparison with what in other systems was described as crystalline or densely packed (Figs. 1, 3, and 4; for references, see Section III,A); inversely, our "low" density of connexons as a short-term and

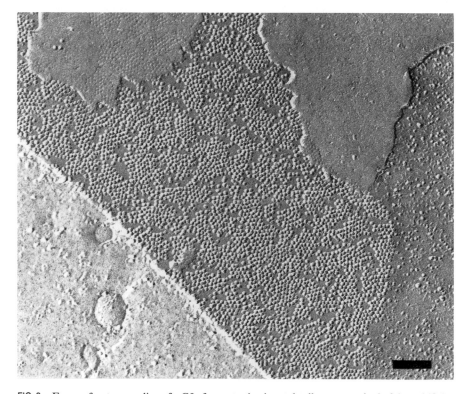

FIG. 3   Freeze-fracture replica of a GJ of an outer horizontal cell axon terminal of the goldfish retina, 60 min after dark adaptation. The connexons are densely packed. Compare to Fig. 1b. Scale bar: 0.1 μm. (Provided by Dr. T. Voigt, Tübingen, Germany.)

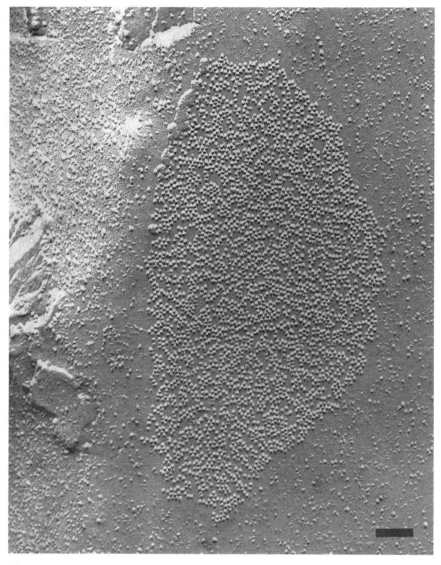

FIG. 4   Freeze-fracture replica of a GJ of an outer horizontal cell axon terminal of the goldfish retina, 30 min after dark adaptation. The connexons are densely packed, and the number of nonconnexonal intramembranous particles directly beneath the GJ area is decreased. Scale bar: 0.1 μm. (Provided by Dr. T. Voigt, Tübingen, Germany.)

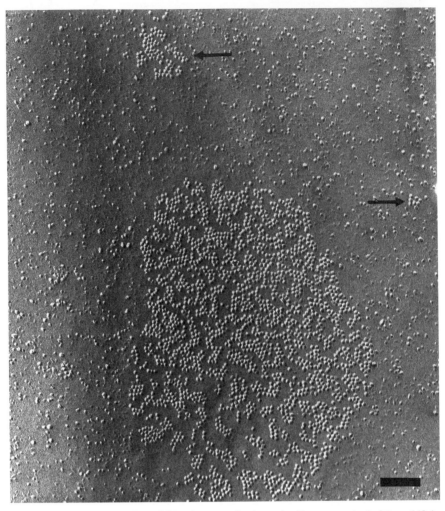

FIG. 5   Freeze-fracture replica of GJs of an outer horizontal cell axon terminal of the goldfish retina, 15 min after light adaptation. Although many connexons lie closely packed together the space between connexon clusters within the whole GJ area is increased when compared to the dark-adapted state. The GJ area proper is devoid of any other nonconnexonal intramembranous particles. Arrows point to smaller GJs. Scale bar: 0.1 μm. (Provided by Dr. T. Voigt, Tübingen, Germany.)

reversible physiological phenomenon (Fig. 6) was not observed in other systems except under developmental conditions (e.g., Ryerse *et al.*, 1984; Ginzberg *et al.*, 1985; Koike *et al.*, 1993). The distance between single small GJs appearing a few minutes after exposure to darkness seemed to be too large to be attritbuted to pure dissociation of a formerly large,

FIG. 6 Freeze-fracture replica of a GJ of an outer horizontal cell axon terminal of the goldfish retina, 60 min after light adaptation. The connexon density is very low. Nevertheless, some connexons form small densely packed clusters or linear formations. Compare to Fig. 1c. Scale bar: 0.1 μm. (Provided by Dr. T. Voigt, Tübingen, Germany.)

light-adapted GJ. In contrast, the dramatic reduction in size of most GJs in the dark-adapted state cannot be explained solely by simple aggregation of the formerly dispersed connexons via lateral shift. An adequate explanation may be the *de novo* formation of small GJ plaques from a pool of formerly unidentified single hemichannels.

In the dark, the HC receptive field size is larger than in the light. Therefore, Kurz-Isler and Wolburg (1986, 1988) concluded from their observations that the GJs revealing high connexon densities (in the dark; Figs. 3 and 4) may correspond to the coupled state, and the low connexon density (in the light; Figs. 5 and 6) to the uncoupled state of HCs. This assumption was supported by the consistent finding that dopamine, which is known to uncouple HC GJs, decreases the density of connexons (Baldridge *et al.*, 1987; Kurz-Isler *et al.*, 1992); following administration of 6-OH-dopamine, which depletes the retina of dopamine by loss of dopaminergic neurons, the connexons are densely packed (Weiler *et al.*, 1988; Baldridge *et al.*, 1989). On the other hand, under conditions of decreased intracellular pH, the connexon density is increased (Schmitz and Wolburg, 1991), although HCs have been described as uncoupled (Laufer *et al.*, 1989). At this point, two lines of consideration will be followed:

1. Hampson *et al.* (1994) reported that in the rabbit retina dopamine uncouples the HCs at pH 7.2. However, dopamine did not exert any effect on the number of Lucifer Yellow-coupled cells at pH 7.4. The uncoupling effect of low extracellular pH was not reversed by dopaminergic antagonists, suggesting that dopaminergic uncoupling is pH gated. If we could transfer the results of Hampson *et al.* (1994) from the rabbit retina to the fish retina, they appear to be suitable to explain the contradictory effects of both uncoupling agents, pH and dopamine, on the connexon density. Lowering the (intra- or extracellular) pH results in a stereotypical increase of the connexon density; however, dopamine seems to decrease the connexon density only under conditions of a pH value between 7.2 and 7.4. Therefore, Schmitz and Wolburg (1991) possibly were not able (1) to neutralize the effect of pH values 7.1, 6.8, and 6.5 with dopamine, and (2) to observe any effect of haloperidol at pH 7.5. This would mean that below a certain pH threshold the uncoupling of HC GJs is indicated by a high connexon density; only within a small pH window does dopamine have an effect on coupling that is characterized by a decreased connexon density.

2. The high connexon density seems to be associated with the presence of electron-dense material independent of whether the GJs were observed following acidification or dark adaption (Kurz-Isler and Wolburg, 1988; Schmitz and Wolburg, 1991). Hidaka *et al.* (1989) observed similar densifications beneath GJs in cultured HCs of the catfish retina; unfortunately, the authors were not able to correlate their appearance to the presence

or absence of electrotonic coupling in a given cell pair. Vaughan and Lasater (1990a) observed the distribution of F-actin by means of immuno-cytochemistry of the bass retina. They found actin label where GJs were formed. However, following disruption of the microfilament network by cytochalasin D, the electrotonic coupling between HCs was maintained. This contradicts the assumption that actin directly influences the coupling of HC GJs.

VI. Conclusions

During the last years, the methodological approach in GJ research has become more and more complex. A remarkable feature of GJ research is the fact that the progress in the investigation of GJs is different in different systems. In the retina, an enormous quantity of physiological data on electrotonic coupling contrasts with the complete lack of data on connexins. As well, in invertebrates, many studies described the morphology and physiology of coupled cells, but the identification of GJ proteins in these animal groups is just at its beginning. In contrast, in fields in which the study of the biochemistry and molecular biology of GJs has mostly proceeded (e.g., liver GJs), the functional understanding of which molecules are transferred from cell to cell and how this transfer may be regulated is poorly developed. Nevertheless, the molecular regulation of connexin synthesis; processing by phosphorylation before, during, and after the insertion of the connexons into the membrane; and the study of gating phenomena by means of biochemical and electrophysiological methods have been addressed increasingly. New tools have been used to investigate the initiation of GJ formation during normal development as well as during neoplastic transformation. The expression of adhesion molecules as a prerequisite of GJ formation can rarely be overestimated and many new findings can be expected. It is evident that in more recent studies morphological aspects have been less well recognized than in earlier years. Nevertheless, phenomena such as autocoupling of astrocytes, or the interconnexonal spacing of GJs, may still require further morphological inspections to reveal relationships between GJ structure and function.

Acknowledgments

The authors would like to thank Mrs. Cerilla Maelicke for correcting the language, U. Kniesel for the help with Fig. 1, and Prof. J. R. Wolff for providing Fig. 2B and C. A.R. is supported by the Graduate College "Dynamics and Organization of Neuronal Networks" project. Research by H.W. is financially supported by the Deutsche Forschungsgemeinschaft.

References

Anders, J. J. (1988). Lactic acid inhibition of gap junctional intercellular communication in vitro astrocytes as measured by fluorescence recovery after laser photobleaching. *Glia* **1**, 371–379.

Andrew, R. D., MacVikar, B. A., Dudek, F. E., and Hatton, G. H. (1981). Dye transfer through gap junctions between neuroendocrine cells of rat hypothalamus. *Science* **211**, 1187–1189.

Andrew, R. D., Taylor, C. P., Snow, R. W., and Dudek, F. E. (1982). Coupling in rat hippocampal slices. *Brain Res. Bull.* **8**, 211–222.

Appleyard, S. T., Witkowski, J. A., Ripley, B. D., Shotton, D. M., and Dubowitz, V. (1985). A novel procedure for pattern analysis of features present on freeze-fractured plasma membranes. *J. Cell Sci.* **74**, 105–117.

Aylsworth, C. F., Trosko, J. E., and Welsch, C. W. (1986). Influence of lipids on gap junction mediated intercellular between Chinese hamster cells in vitro. *Cancer Res.* **46**, 4527–4533.

Aylsworth, C. F., Welsch, C. W., Kabara, J. J., and Trosko, J. E. (1987). Effects of fatty acids on gap junctional communication: Possible role in tumor promotion by dietary fat. *Lipids* **22**, 445–454.

Azarnia, R., Dahl, G., and Loewenstein, W. R. (1981). Cell junction and cyclic AMP. III. Promotion of junctional membrane permeability and junctional membrane particles in a junction-deficient cell type. *J. Membr. Biol.* **63**, 133–146.

Baldridge, W. H., Ball, A. K., and Miller, R. G. (1987). Dopaminergic regulation of horizontal cell gap junction particle density in goldfish retina. *J. Comp. Neurol.* **265**, 428–436.

Baldridge, W. H., Ball, A. K., and Miller, R. G. (1989). Gap Junction particle density of horizontal cells in goldfish retinas lesioned with 6-OHDA. *J. Comp. Neurol.* **287**, 238–246.

Barr, L., Dewey, M. M., and Berger, W. (1965). Propagation of action potentials and the structure of the nexus in cardiac muscle. *J. Gen. Physiol.* **48**, 797–823.

Batter, D. K., Corpina, R. A., Roy, C., Spray, D. C., Hertzberg, E. L., and Kessler, J. A. (1992). Heterogeneity in gap junction expression in astrocytes cultured from different brain regions. *Glia* **6**, 213–221.

Behrens, J., Mareel, M. M., Van Roy, F. M., and Birchmeier, W. (1989). Dissecting tumor cell invasion: Epithelial cells acquire invasive properties after the loss of uvomorulin-mediated cell-cell adhesion. *J. Cell Biol.* **108**, 2435–2447.

Bell, G. I. (1978). Models for the specific adhesion of cells to cells. *Science* **200**, 618–627.

Benardo, L. S., and Foster, R. E. (1986). Oscillatory behavior in inferior olive neurons: Mechanism, modulation, cell aggregates. *Brain Res. Bull.* **17**, 773–784.

Bennett, M. V. L. (1973). Function of electrotonic junctions in embryonic and adult tissues. *Fed. Proc., Fed. Am. Soc. Exp. Biol.* **32**, 65–75.

Bennett, M. V. L., and Goodenough, D. A. (1978). Gap junctions, electrotonic coupling, and intercellular communication. *Neurosci. Res. Program Bull.* **16**, 371–486.

Bennett, M. V. L., Nakajima, Y., and Pappas, G. D. (1967). Physiology and ultrastructure of electrotonic junctions. III. Giant electromotor neurons of *Malapterus electricus*. *J. Neurophysiol.* **30**, 161–179.

Bennett, M. V. L., Spray, D. C., and Harris, A. L. (1981). Gap junctions and development. *Trends Neurosci.* **4**, 159–163.

Bennett, M. V. L., Barrio, L. C., Bargiello, T. A., Spray, D. C., Hertzberg, E. L., and Saze, J. C. (1991). Gap junctions: New tools, new answers, new questions. *Neuron* **6**, 305–320.

Berdan, R. C., and Caveney, S. (1985). Gap junction ultrastructure in three states of conductance. *Cell Tissue Res.* **239**, 111–122.

Berdan, R. C., and Gilula, N. B. (1988). The arthropod gap junction and pseudo-gap junction: Isolation and preliminary biochemical analysis. *Cell Tissue Res.* **251**, 257–274.

Bergoffen, J., Scherer, S. S., Wang, S., Oronzi Scott, M., Bone, L. J., Paul, D. L., Chen, K., Lensch, M. W., Chance, P. F., and Fischbeck, K. H. (1993). Connexin mutations in X-linked Charcot-Marie-Tooth disease. *Science* **262**, 2039–2042.

Bernardini, G., Peracchia, C., and Peracchia, L. L. (1984). Reversible effects of heptanol on gap junction structure and cell-to-cell electrical coupling. *Eur. J. Cell Biol.* **34**, 307–312.

Berthoud, V. M., Ledbetter, M. L. S., Hertzberg, E. L., and Saez, J. C. (1992). Connexin43 in MDCK cells: Regulation by a tumor-promoting phorbol ester and Ca^{2+}. *Eur. J. Cell Biol.* **57**, 40–50.

Beyer, E. C. (1993). Gap junctions. *Int. Rev. Cytol.* **137**, 1–37.

Beyer, E. C., Paul, D. L., and Goodenough, D. A. (1987). Connexin 43: A protein from rat heart homologous to a gap junction protein from liver. *J. Cell Biol.* **49**, 2621–2629.

Beyer, E. C., Kistler, J., Paul, D. L., and Goodenough, D. A. (1989). Antisera directed against connexin 43 peptides react with a 43-kD protein localized to gap junctions in myocardium and other tissues. *J. Cell Biol.* **108**, 595–605.

Binmöller, F.-J., and Müller, C. M. (1992). Postnatal development of dye-coupling among astrocytes in rat visual cortex. *Glia* **6**, 127–137.

Bohrmann, J. (1993). Antisera against a channel-forming 16 kDa protein inhibit dye-coupling and bind to cell membranes in *Drosophila* ovarian follicles. *J. Cell Sci.* **105**, 513–518.

Bosch, E. (1989). Comparative study of neuronal and glial gap-junctions in crayfish nerve cords. *J. Comp. Neurol.* **285**, 399–411.

Braun, J., Abney, J. R., and Owicki, J. C. (1984). How a gap junction maintains its structure. *Nature (London)* **310**, 316–318.

Brightman, M. W., and Reese, T. S. (1969). Junctions between intimately apposed cell membranes in the vertebrate brain. *J. Cell Biol.* **40**, 648–677.

Burghardt, R. C., and Anderson, E. (1979). Hormonal modulation of ovarian interstitial cell with particular reference to gap junctions. *J. Cell Biol.* **81**, 104–114.

Burns, M. S., and Tyler, N. K. (1990). Interglial cell gap junctions increase in urethane-induced photoreceptor degeneration in rats. *Invest. Ophthalmol. Vis. Sci.* **31**, 1690–1701.

Campos de Carvalho, A. C., Tanowitz, H. B., Wittner, M., Dermietzel, R., and Spray, D. C. (1993). Trypanosome infection decreases intercellular communication between cardiac myocytes. *Prog. Cell Res.* **3**, 193–197.

Caspar, D. L. D., Goodenough, D. A., Makowski, L., and Phillips, W. C. (1977). Gap junction structures. I. Correlated electron microscopy and X-ray diffraction. *J. Cell Biol.* **74**, 605–628.

Cavency, S. (1985). The role of gap junctions in development. *Annu. Rev. Physiol.* **47**, 319–336.

Cepeda, C., Walsh, J. P., Peacock, W., Buchwald, N. A., and Levine, M. S. (1993). Dye-coupling in human neocortical tissue resected from children with intractable epilepsy. *Cereb. Cortex* **3**, 95–107.

Chang, C. C., Jones, C., Trosko, J. E., Peterson, A. R., and Sevanian, A. (1988). Effect of cholesterol epoxides on the inhibition of intercellular communication and on mutation induction in Chinese hamster V79 cells. *Mutat. Res.* **206**, 471–478.

Charles, A. C., Naus, C. C. G., Zhu, D., Kidder, G. M., Dirksen, E. R., and Sanderson, M. J. (1992). Intercellular calcium signalling via gap junctions in glioma cells. *J. Cell Biol.* **118**, 195–201.

Chanson, M., Bruzzone, R., Spray, D. C., Regazzi, R., and Meda, P. (1988). Cell uncoupling and protein kinase C: correlation in a cell line but not in a differentiated tissue. *Am. J. Physiol.* **255**, C699–C704.

Charles, A. C., Dirksen, E. R., Merrill, J. E., and Sanderson, M. J.(1993). Mechanisms of intercellular calcium signaling in glial cells studied with dantrolene and thapsigargin. *Glia* **7,** 134–145.

Chen, W., and Öbrink, B. (1991). Cell-cell contacts mediated by E-cadherin (uvomorulin) restrict invasive behavior of L-cells. *J. Cell Biol.* **114,** 319–327.

Christensen, B. N. (1983). Distribution of electrotonic synapses on identified lamprey neurons: A comparison of a model prediction with an electron microscopic analysis. *J. Neurophysiol.* **49,** 705–716.

Chuang, H. H., Chuang-Tseng, M. P., Wu, W. L., Sandri, C., and Akert, K. (1985). Coupling of gap junctions by induction of impulse conductivity in cultured epithelium of newt embryo *(Cynops orientalis). Cell Tissue Res.* **241,** 25–29.

Chuang-Tseng, M. P., Chuang, H. H., Sandri, C., and Akert, K. (1982). Gap junctions and impulse propagation in embryonic epithelium of amphibia. A freeze-etching study. *Cell Tissue Res.* **225,** 249–258.

Churchill, D., Coodin, S., Shivers, R. R., and Caveney, S. (1993). Rapid de novo formation of gap junctions between insect hemocytes in vitro: A freeze-fracture, dye-transfer and patch-clamp study. *J. Cell Sci.* **104,** 763–772.

Cobbett, P., and Hatton, G. I. (1984). Dye-coupling in hyphothalamic slices: Dependence on in vivo hydration state and osmolarity of incubation medium. *J. Neurosci.* **4,** 3034–3038.

Cole, W. C., and Garfield, R. C. (1986). Evidence for physiological regulation of myometrial gap junction permeability. *Am. J. Physiol.* **251,** C411–C420.

Connors, B. W., Benardo, L. S., and Prince, D. A. (1983). Coupling between neurons of the developing rat neocortex. *J. Neurosci.* **3,** 773–782.

Cook, J. E., and Becker, D. L. (1994). Gap junctions in the vertebrate retina. *Micr. Res. Techn.* In Press.

Cooper, N. G. F., and McLaughlin, B. J. (1981). Gap junctions in the outer plexiform layer of the chick retina: Thin section and freeze-fracture studies. *J. Neurocytol.* **10,** 515–529.

Cornell-Bell, A. H., Finkbeiner, S. M., Cooper, M. S., and Smith, S. J. (1990). Glutamate induces calcium waves in cultured astrocytes: Long-range glial signaling. *Science* **247,** 470–473.

Cox, R. P., Krauss, M. R., Balis, M. E., and Dancis, J. (1976). Studies on cell communication with enucleated human fibroblasts. *J. Cell Biol.* **71,** 693–703.

Crow, D. S., Beyer, E. C., Paul, D., Kobe, S. S., and Lau, A. F. (1990). Phosphorylation of connexin43 gap junction protein in uninfected and RSV-transformed mammalian fibroblasts. *Mol. Cell. Biol.* **10,** 1754–1763.

Crow, J. M., Atkinson, M. M., and Johnson, R. G. (1994). Micromolar levels of intracellular calcium reduce gap junctional permeability in lens cultures. *Invest. Ophthalmol. Vis. Sci.* **35,** 3332–3341.

Cuenca, N., Fernández, E., Garciá, M., and De Juan, J. (1993). Dendrites of rod dominant ON-bipolar cells are coupled by gap junctions in carp retina. *Neurosci. Lett.* **162,** 34–38.

Dahl, G., and Isenberg, G. (1980). Decoupling of heart muscle cells: Correlation with increased cytoplasmic calcium activity and with changes of nexus ultrastructure. *J. Membr. Biol.* **53,** 63–73.

Dealy, C. N., Beyer, E. C., and Kosher, R. A. (1994). Expression patterns of mRNAs for the gap junction proteins connexin 43 and connexin 42 suggest their involvement in chick limb morphogenesis and specification of the arterial vasculature. *Dev. Dyn.* **199,** 156–167.

Decker, R. S., and Friend, D. S. (1974). Assembly of gap junctions during amphibian neurulation. *J. Cell Biol.* **62,** 32–47.

Délèze, J., and Hervé, J. C. (1983). Effect of several uncouplers of cell-to-cell communication on gap junction morphology in mammalian heart. *J. Membr. Biol.* **74,** 203–215.

De Mazière, M. G. L., and Scheuermann, D. W. (1988). Morphometrical analysis of the gap junctional area in parenchymal cells of the rat liver after administration of dibutyryl cAMP and aminophylline. *Cell Tissue Res.* **252**, 611–618.

De Mazière, M. G. L., and Scheuermann, D. W. (1990). Structural changes in cardiac gap junctions after hypoxia and reoxygenation: A quantitative freeze-fracture analysis. *Cell Tissue Res.* **261**, 183–194.

Dermietzel, R., and Spray, D. C. (1993). Gap junctions in the brain: Where, what type, how many and why? *Trends Neurosci.* **16**, 186–192.

Dermietzel, R., Meller, K., Tetzlaff, W., and Waelsch, M. (1977). In vivo and in vitro formation of the junctional complex in choroid epithelium. A freeze-etching study. *Cell Tissue Res.* **181**, 427–441.

Dermietzel, R., Traub, O., Hwang, T. K., Beyer, E., Bennett, M. V. L., Spray, D. C., and Willecke, K. (1989a). Differential expression of three gap junction proteins in developing and mature brain tissues. *Proc. Natl. Acad. Sci. U.S.A.* **86**, 10148–10152.

Dermietzel, R., Völker, M., Hwang, T.-K., Berzborn, R. J., and Meyer, H. E. (1989b). A 16 kDA protein co-isolating with gap junctions from brain tissue belonging to the class of proteolipids of the vacuolar H^+-ATPases. *FEBS Lett.* **253**, 1–5.

Dermietzel, R., Hertzberg, E. L., Kessler, J. A., and Spray, D. C. (1991). Gap junctions between cultured astrocytes. Immunocytochemical, molecular, and electrophysiological analysis. *J. Neurosci.* **11**, 1421–1432.

DeVries, S. H., and Schwartz, E. A. (1992). Hemi-gap-junction channels in solitary horizontal cells of the catfish retina. *J. Physiol. (London)* **445**, 201–230.

Dewey, M. M., and Barr, L. (1962). Intercellular connection between smooth muscle cells: The nexus. *Science* **137**, 670–672.

Dowling, J. E., and Boycott, B. B. (1966). Organization of the primate retina: Electron microscopy. *Proc. R. Soc. London* **166**, 80–111.

Edelman, G. M. (1986). Cell adhesion molecules in the regulation of animal form and tissue pattern. *Annu. Rev. Cell Biol.* **2**, 81–116.

Edelman, G. M. (1988). Morphoregulatory molecules. *Biochemistry* **17**, 3533–3543.

Elias, P. M., and Friend, D. S. (1976). Vitamin-A-induced mucous metaplasia. An in vitro system for modulating tight and gap junction differentiation. *J. Cell Biol.* **68**, 173–188.

Elvira, M., Diez, J. A., Wang, K. K. W., and Villalobo, A. (1993). Phosphorylation of connexin-32 by protein kinase C prevents its proteolysis by μ-calpain and m-calpain. *J. Biol. Chem.* **268**, 14294–14300.

Enkvist, M. O., and McCarthy, K. D. (1992). Activation of protein kinase C blocks astroglial gap junction communication and inhibits the spread of calcium waves. *J. Neurochem.* **59**, 519–526.

Epstein, M. L., Sheridan, J. D., and Johnson, R. G. (1977). Formation of low-resistance junctions in vitro in the absence of protein synthesis and ATP production. *Exp. Cell Res.* **104**, 25–30.

Evans, W. H. (1988). Gap junctions: Towards a molecular structure. *BioEssays* **8**, 3–6.

Evans, C. W., Eastwood, S., Rains, J., Gruijters, W. T. M., Bullivant, S., and Kistler, J. (1993). Gap junction formation during development of the mouse lens. *Europ. J. Cell Biol.* **60**, 243–249.

Faber, D. S., and Korn, H. (1989). Electrical field effects: Their relevance in central neural networks. *Physiol. Rev.* **69**, 821–863.

Fallon, R. F., and Goodenough, D. A. (1981). Five-hour half-life of mouse liver gap-junction protein. *J. Cell Biol.* **90**, 521–526.

Filson, A. J., Azarnia, R., Beyer, E. C., Loewenstein, W. R., and Brugge, J. S. (1990). Tyrosine phosphorylation of a gap junction protein correlates with inhibition of cell-to-cell communication. *Cell Growth Differ.* **1**, 661–668.

Finbow, M. E., and Pitts, J. D. (1993). Is the gap junction channel—the *connexon*—made of connexin or ductin? *J. Cell Sci.* **106**, 463–472.

Finbow, M. E., Buultjens, T. E. J., Lane, N. J., Shuttleworth, J., and Pitts, J. D. (1984). Isolation and characterization of arthropod gap junctions. *EMBO J.* **3**, 2271–2278.

Fischer, G., and Kettenmann, H. (1985). Cultured astrocytes form a syncytium after maturation. *Exp. Cell Res.* **159**, 273–279.

Flagg-Newton, J. L., and Loewenstein, W. R. (1981). Cell junction and cyclic AMP: II. Modulations of junctional membrane permeability, dependent on serum and cell density. *J. Membr. Biol.* **63**, 123–131.

Flagg-Newton, J. L., Dahl, G., and Loewenstein, W. R. (1981). Cell junction and cyclic AMP: I. Upregulation of junctional membrane permeability and junctional membrane particles by administration of cyclic nucleotide of phosphodiesterase inhibitor. *J. Membr. Biol.* **63**, 105–121.

Fraser, S. E., Green, C. R., Bode, H. R., and Gilula, N. B. (1987). Selective disruption of gap junctional communication interferes with a patterning process in *Hydra*. *Science* **237**, 49–55.

Friend, D. S., and Gilula, N. B. (1972). Variations in tight and gap junctions in mammalian tissues. *J. Cell Biol.* **53**, 758–776.

Furuse, M., Hirase, T., Itoh, M., Nagafuchi, A. Yonemura, S., Tsukita, S., and Tsukita, S. (1993). Occludin: A novel integral membrane protein localizing at tight junctions. *J. Cell Biol.* **123**, 1777–1788, 1993.

Gabbott, P. L. A., and Stewart, M. G. (1987). Distribution of neurons and glia in the visual cortex (area 17) of the adult albino rat: A quantitative description. *Neuroscience* **21**, 833–845.

Gardner-Medwin, A. R. (1983). Analysis of potassium dynamics in mammalian brain tissue. *J. Physiol.* **335**, 393–426.

Gething, M.-J., McCammon, K., and Sambrook, J. (1986). Expression of wild type and mutant forms of influenza hemagglutinin: The role of folding in intracellular transport. *Cell (Cambridge, Mass.)* **46**, 939–964.

Gilula, N. B. (1979). Electrotonic junctions. *In* "The Neurosciences: Fourth Study Program" (F. O. Schmitt and F. G. Worden, eds.), pp. 359–366. MIT press, Cambridge, MA.

Ginzberg, R. D., and Gilula, N. B. (1979). Modulation of cell junctions during differentiation of the chicken otocyst sensory epithelium. *Dev. Biol.* **68**, 1–21.

Ginzberg, R. D., Morales, E. A., Spray, D. C., and Bennett, M. V. L. (1985). Cell junctions in early embryos of squid (*Loligo pealei*). *Cell Tissue Res.* **239**, 477–484.

Goddard, J. C., Behrens, U. D., Wagner, H. J., and Djamgoz, M. B. A. (1991). Biocytin: Intracellular staining, dye-coupling and immunocytochemistry in carp retina. *NeuroReport* **2**, 755–758.

Goldberg, A. L., and John, A. C. S. (1976). Intracellular protein degradation in mammalian and bacterial cells: Part 2. *Annu. Rev. Biochem.* **45**, 747–803.

Goldberg, G. S., and Lau, A. F. (1993). Dynamics of connexin43 phosphorylation in pp60$^{v\text{-}src}$-transformed cells. *Biochem. J.* **295**, 735–742.

Goodenough, D. A., and Gilula, N. B. (1974). The splitting of hepatocyte gap junctions and zonulae occludentes with hypertonic disaccharides. *J. Cell Biol.* **61**, 575–590.

Goodenough, D. A., and Musil, L. S. (1993). Gap junctions and tissue business: problems and strategies for developing specific reagents. *J. Cell Sci., Suppl.* **17**, 133–138.

Goodenough, D. A., Paul, D. L., and Jesaitis, L. (1988). Topological distribution of two connexin 32 antigenic sites in intact and split rodent hepatocyte gap junctions. *J. Cell Biol.* **107**, 1817–1824.

Green, C. R. (1988). Evidence mounts for the role of gap junctions during development. *BioEssays* **8,** 7–10.

Green, C. R., and Severs, N. J. (1984a). Connexon rearrangement in cardiac gap junctions: Evidence for cytoskeletal control? *Cell Tissue Res.* **237,** 185–186.

Green, C. R., and Severs, N. J. (1984b). Gap junction connexon configuration in rapidly frozen myocardium and isolated intercalated disks. *J. Cell Biol.* **99,** 453–463.

Griepp, E. B., and Revel, J.-P. (1977). Gap junctions in development. *In* "Intercellular Communication" (W. C. De Mello, ed.), pp. 1–32. Plenum, New York and London.

Gumbiner, B. M. (1993). Proteins associated with the cytoplasmic surface of adhesion molecules. *Neuron* **11,** 551–564.

Gurd, J. W., and Evans, W. H. (1973). Relative rates of degradation of mouse liver surface membrane proteins. *Eur. J. Biochem.* **36,** 273–279.

Guthrie, S. C., and Gilula, N. B. (1989). Gap junctional communication and development. *Trends Neurosci.* **12,** 12–16.

Haefliger, J.-A., Bruzzone R, Jenkins, N. A., Gilbert, D. J., Copeland, N. G., and Paul, D. L. (1992). Four novel members of the connexin family of gap junction proteins. *J. Biol. Chem.* **267,** 2057–2064.

Hama, F. (1987). Fine-structural and morphometrical studies on the segmental septa of the giant fibers in the earthworm, *Eisenia foetida. Cell Tissue Res.* **249,** 565–575.

Hampson, E. C. G. M., Weiler, R., and Vaney, D. I. (1994). pH-gated dopaminergic modulation of horizontal cell gap junctions in mammalian retina. *Proc. R. Soc. London, Ser. B* **255,** 67–72.

Hanna, R. B., Pappas, G. D., and Bennett, M. V. L. (1984). The fine structure of identified electrotonic synapses following increased coupling resistance. *Cell Tissue Res.* **235,** 243–249.

Hatae, T., Iida, H., Kuraoka, A., and Shibata, Y. (1993). Cytoplasmic surface ultrastructures of gap junctions in bovine lens fibers. *Invest. Ophthalmol. Visual Sci.* **34,** 2164–2173.

Hayes, B. P. (1976). The distribution of intercellular gap junctions in the developing retina and pigment epithelium of *Xenopus laevis. Anat. Embryol.* **150,** 99–111.

Hayes, B. P. (1977). Intercellular gap junctions in the developing retina and pigment epithelium of the chick. *Anat. Embryol.* **151,** 325–334.

Hendrix, E. M., Mao, S. J., Everson, W., and Larsen, W. J. (1992). Myometrial connexin 43 trafficking and gap junction assembly at term and in preterm labor. *Mol. Reprod. Dev.* **33,** 27–38.

Hennemann, H., Schwarz, H.-J., and Willecke, K. (1992a). Characterization of gap junction genes expressed in F9 embryonic carcinoma cells: Molecular cloning of mouse connexin 31 and 45 cDNAs. *Eur. J. Cell Biol.* **57,** 51–58.

Hennemann, H., Dahl, E., White, J. B., Schwarz, H-J., Lalley, P. A., Chang, S., Nicholson B. J., and Willecke K. (1992b). Two gap junction genes, connexin 31.1 and 30.3, are closely linked on mouse chromosome 4 and preferentially expressed in skin. *J. Biol. Chem.* **267,** 17225–17233.

Hennemann, H., Suchyna, T., Lichtenberg-Frate, H., Jungbluth, S., Dahl, E., Schwarz, H.-J., Nicholson, B. J., and Willecke, K.(1992c). Molecular cloning and functional expression of mouse connexin40, a second gap junction gene preferentially expressed in lung. *J. Cell Biol.* **117,** 1299–1310.

Hertzberg, E. L. (1985). Antibody probes in the study of gap junctional communication. *Annu. Rev. Physiol.* **47,** 305–318.

Hertzberg, E. L., and Skibbens, R. V. (1984). A protein homologous to the 27,000 dalton liver gap junction protein is present in a wide variety of species and tissues. *Cell (Cambridge, Mass.)* **39,** 61–69.

Hertzberg, E. L., Spray, D. C., and Bennett, M. V. L. (1985). Reduction of gap junctional conductance by microinjection of antibodies agianst the 27-kDa liver gap junction polypeptide. *Proc. Natl. Acad. Sci. U.S.A.* **82**, 2412–2416.

Hidaka, S., Shingai, R., Dowling, J. E., and Naka, K. (1989). Junctions form between catfish horizontal cells in culture. *Brain Res.* **498**, 53–63.

Hirokawa, N., and Heuser, J. (1982). The inside and outside of gap-junction membranes visualized by deep etching, *Cell (Cambridge, Mass.)* **30**, 395–406.

Hoh, J. H., Lal, R., John, S. A., Revel J.-P., and Arnsdorf, M. F. (1991). Atomic force microscopy and dissection of gap junctions. *Science* **253**, 1405–1408.

Hossain, M. Z., Sawchuk, M. A., Murphy, L. J., Hertzberg, E. L., and Nagy, J. I. (1994a). Kainic acid induced alterations in antibody recognition of connexin43 and loss of astrocytic gap junctions in rat brain. *Glia* **10**, 250–265.

Hossain, M. Z., Murphy, L. J., Hertzberg, E. L., and Nagy, J. I. (1994b). Phosphorylated forms of connexin 43 pedominate in rat brain: Demonstration by rapid inactivation of brain metabolism. *J. Neurochem.* **62**, 2394–2403.

Hülser, D. F., and Brümmer, F. (1982). Closing and opening of gap junction pores between two- and three-dimensionally cultured tumor cells. *Biophys. Struct. Mech.* **9**, 83–88.

Hurtley, S. M., and Helenius, A. (1989). Protein oligomerization in the endoplasmic reticulum. *Annu. Rev. Cell Biol.* **5**, 277–307.

Hynes, R. O. (1992). Integrins: Versatility, modulation, and signaling in cell adhesion. *Cell (Cambridge, Mass.)* **69**, 11–25.

Imanaga, J. (1974). Cell to cell diffusion of Procion yellow in sheep and calf Purkinje fibres. *J. Membr. Biol.* **16**, 381–388.

In't Veld, P. A., Pipeleers, D. G., and Gepts, W. (1984). Evidence against the presence of tight junctions in normal endocrine pancreas. *Diabetes* **33**, 101–104.

In't Veld, P. A., Schuit, F., and Pipeleers, D. G. (1985). Gap junctions between pancreatic B-cells are modulated by cyclic AMP. *Eur. J. Cell Biol.* **36**, 269–276.

Janssen-Timmen, U., Traub, O., Dermietzel, R., Rabes, H. M., and Willecke, K. (1986). Reduced number of gap junctions in rat hepatomas detected by monoclonal antibody. *Carcinogenesis (London)* **7**, 1475–1481.

Jensen, A. M., and Chiu, S.-Y. (1993). Astrocyte networks. *In* "Astrocytes, Pharmacology and Function" (S. Murphy, ed.), pp. 309–330. Academic Press, San Diego.

Jessell, T. M. (1988). Adhesion molecules and the hierarchy of neural development. *Neuron* **1**, 3–13.

Jiang, J. X., Paul, D. L., and Goodenough, D. A. (1993). Posttranslational phosphorylation of lens fiber connexin 46: a slow occurrence. *Invest. Ophthalmol. Vis Sci.* **34**, 3558–3565.

John, S. A., and Revel, J.-P. (1991). Connexon integrity is maintained by non-covalent bonds: Intramolecular disulfide bonds link the extracellular domains in rat connexin 43. *Biochem. Biophys. Res. Commun.* **178**, 1312–1318.

Johnson, R. G., and Sheridan, J. D. (1971). Junctions between cancer cells in culture. Ultrastructure and permeability. *Science* **174**, 717–719.

Jongen, W. M. F., Fitzgerald, D. J., Asamoto, M., Piccoli, C., Slaga, T. J., Gros, D., Takeichi, M., and Yamasaki, H. (1991). Regulation of connexin 43-mediated gap junctional intercellular communication by Ca^{2+} in mouse epidermal cells is controlled by E-cadherin. *J. Cell Biol.* **114**, 545–555.

Kaneko, A. (1971). Electrical connexions between horizontal cells in the dogfish retina. *J. Physiol. (London)* **213**, 95–105.

Kaneko, A., and Stuart, A. E. (1984). Coupling between horizontal cells in the carp retina revealed by diffusion of Lucifer Yellow. *Neurosci. Lett.* **47**, 1–7.

Kanno, Y., Sasaki, Y., Shiba, Y., Yoshida-Noro, C., and Takeichi, M. (1984). Monoclonal antibody ECCD-1 inhibits intercellular communication in teratocarcinoma PCC3 cells. *Exp. Cell Res.* **152**, 270–274.

Kardami, E., Stoski, R. M., Doble, B. W., Yamamoto, T., Hertzberg, E. L., and Nagy, J. I. (1991). Biochemical and ultrastructural evidence for the association of basic fibroblast growth factor with cardiac gap junctions. *J. Biol. Chem.* **266,** 19551–19557.

Katsumaru, H., Kosaka, T., Heizmann, C. W., and Hama, K. (1988). Gap junctions on GABAergic neurons containing the calcium-binding protein parvalbumin in the rat hippocampus (CA1 region). *Exp. Brain Res.* **72,** 363–370.

Keane, R. W., Mehta, P. P., Rose, B., Honig, L. S., Loewenstein W. R., and Rutishaur, U. (1988). Neural differentiation, NCAM-mediated adhesion, and gap junctional communication in neuroectoderm. A study in vitro. *J. Cell Biol.* **106,** 1307–1319.

Kessler, J. A., Spray, D. C., Saez, J. C., and Bennett, M. V. L. (1984). Modulation of synaptic phenotype: insulin and cAMP independently initiate formation of electrotonic synapse in cultured sympathetic neurons. *Proc. Natl. Acad. Sci. U.S.A.* **81,** 6235–6239.

Kettenmann, H., and Ransom, B. R. (1988). Electrical coupling between astrocytes and between oligodendrocytes studied in mammalian cell cultures. *Glia* **1,** 64–73.

Kim, W. T., Rioult, M. G., and Cornell-Bell, A. H. (1994). Glutamate-induced calcium signaling in astrocytes. *Glia* **11,** 173–184.

Kistler, J., and Bullivant, S. (1988a). Dissociation of lens fibre gap junctions releases MP 70. *J. Cell Sci.* **91,** 415–422.

Kistler, J., and Bullivant, S. (1988b). The gap junction proteins: Vive la différence! *BioEssays* **9,** 167–168.

Kistler, J., and Bullivant, S. (1989). Structural and molecular biology of the eye lens membranes. *Crit. Rev. Biochem. Mol. Biol.* **24,** 151–181.

Klaunig, J. E., and Ruch, R. J. (1990). Biology of disease: Role of inhibition of intercellular communication in carcinogenesis. *Lab. Invest.* **62,** 135–146.

Koike, K., Watanabe, H., Hiroi, M., and Tonosaki, A. (1993). Gap junction of stratum-granulosum cells of mouse follicles. Immunohistochemistry and electron microscopy. *J. Electron Micr.* **42,** 94–106.

Kolb, H. A., and Somogyi, R. (1991). Biochemical and biophysical analysis of cell-to-cell channels and regulation of gap junctional permeability. *Rev. Physiol. Biochem. Pharamcol.* **118,** 1–47.

Korn, H., and Faber, D. S. (1979). Electrical interactions between vertebrate neurons: Field effects and electrotonic coupling. *In* "The Neurosciences. Fourt Study Program" (F. O. Schmitt and F. G. Worden, eds.), pp. 333–358. MIT Press, Cambridge, MA.

Kosaka, T. (1983). Neuronal gap junctions in the polymorph layer of the rat dentate gyrus. *Brain Res.* **277,** 347–351.

Kouyama, N., and Watanabe, K. (1986). Gap junctional contacts of luminosity-type horizontal cells in the carp retina: A novel pathway of signal conduction from the cell body to the axon terminal. *J. Comp. Neurol.* **249,** 404–410.

Kren, B. T., Kumar, N. M., Wang, S., Gilula, N. B., and Steer, C. J. (1993). Differential regulation of multiple gap junction transcripts and proteins during rat liver regeneration. *J. Cell Biol.* **123,** 707–718.

Kujiraoka, T., and Saito, T. (1986). Electrical coupling between bipolar cells in the carp retina. *Proc. Natl. Acad. Sci. U.S.A.* **83,** 4063–4066.

Kumar, N. M., and Gilula, N. B. (1986). Cloning and characterization of human and rat liver cDNAs coding for a gap junction. *J. Cell Biol.* **103,** 767–776.

Kuraoka, A., Iida, H., Hatae, T., Shibata, Y., Itoh, M., and Kurita, T. (1993). Localization of gap junction proteins, connexin-32 and connexin-26, in rat and guinea pig liver as revealed by quick-freeze, deep-etch immunoelectron microscopy. *J. Histochem. Cytochem.* **41,** 971–980.

Kurz-Isler, G., and Wolburg, H. (1986). Gap junctions between horizontal cells in the cyprinid fish alter rapidly their structure during light and dark adaptation. *Neurosci. Lett.* **67,** 7–12.

Kurz-Isler, G., and Wolburg, H. (1988). Light-dependent dynamics of gap junctions between horizontal cells in the retina of the crucian carp. *Cell Tissue Res.* **251,** 641–649.

Kurz-Isler, G., Voigt, T., and Wolburg, H. (1992). Modulation of connexon densities in gap junctions of horizontal cell perikarya and axon terminals in fish retina: Effects of light/dark cycles, interruption of the optic nerve and application of dopamine. *Cell Tissue Res.* **268,** 267–275.

Laird, D. W., Puranam K. L., and Revel, J. P. (1991). Turnover and phosphorylation dynamics of connexin 43 gap junction protein in cultured cardiac myocytes. *Biochem. J.* **273,** 67–72.

Landis, D. M. D., and Reese, T. S. (1982). Regional organization of astrocytic membranes in cerebellar cortex. *Neuroscience* **7,** 937–950.

Lane, N. J., and Swales, L. S. (1980). Dispersal of gap junctional particles, not internalization, during the in vivo disappearance of gap junctions. *Cell (Cambridge, Mass.)* **19,** 579–586.

Lane, N. J., Dallai, R., Burighel, P., and Martinucci, G. B. (1986). Tight and gap junctions in the intestinal tract of tunicates (Urochordata): A freeze-fracture study. *J. Cell Sci.* **84,** 1–18.

Larsen, W. J. (1977). Structural diversity of gap junctions. A review. *Tissue Cell* **9,** 373–394.

Larsen, W. J. (1983). Biological implication of gap junction structure, distribution and composition: A review. *Tissue Cell* **15,** 645–671.

Larsen, W. J. (1985). Relating the population dynamics of gap junctions to cellular function. *In* "Gap Junctions" (M. V. L. Bennett and D. C. Spray, eds.), pp. 289–306. Cold Spring Harbor Lab., Cold Spring Harbor, NY.

Larsen, W. J., and Tung, H.-N. (1978). Origin and fate of cytoplasmic gap junctional vesicles in rabbit granulosa cells. *Tissue Cell* **10,** 585–598.

Larsen, W. J., Tung, H.-N., Murray, S. A., and Swenson, C. A. (1979). Evidence for the participation of actin microfilaments and bristle coats in the internalization of gap junction membrane. *J. Cell Biol.* **83,** 576–587.

Lasansky, A. (1976). Interactions between horizontal cells of the salamander retina. *Invest. Ophthalmol.* **15,** 909–916.

Lasater, E. M. (1987). Retinal horizontal cell gap junctional conductance is modulated by dopamine through a cyclic AMP-dependent protein kinase. *Proc. Natl. Acad. Sci. U.S.A.* **84,** 7319–7323.

Lasater, E. M., and Dowling, J. E. (1985). Dopamine decreases conductance of the electrical junctions between cultured retinal horizontal cells. *Proc. Natl. Acad. Sci. U.S.A.* **82,** 3025–3029.

Lau, A. F., Kanemitsu, M. Y., Kurata, W. E., Danesh, S., and Boynton, A. L. (1992). Epidermal growth factor disrupts gap-junctional communication and induces phosphorylation of connexin 43 on serine. *Mol. Biol. Cell.* **3,** 865–874.

Laufer, M., Salas, R., Medina, R., and Drujan, B. (1989). Cyclic adenosine monophosphate as a second messenger in horizontal cell uncoupling in the telost retina. *J. Neurosci. Res.* **24,** 299–310.

Lee, S. H., Kim, W. T., Cornell-Bell, A. H., and Sontheimer, H. (1994). Astrocytes exhibit regional specificity in gap-junction coupling. *Glia* **11,** 315–325.

Leitch, B. (1992). Ultrastructure of electrical synapses: Review. *Electron Microsc. Rev.* **5,** 311–339.

Lo, W.-K., and Reese, T. S. (1993). Multiple structural types of gap junctions in mouse lens. *J. Cell Sci.* **106,** 227–235.

Lo, W.-K., Mills, A., and Kuck, J. F. R. (1994). Actin filament bundles are associated with fiber gap junctions in the primate lens. *Exp. Eye Res.* **58,** 189–196.

Loewenstein, W. R. (1967). Intercellular communication and tissue growth. *J. Cell Biol.* **33,** 225–234.

Loewenstein, W. R. (1972). Cellular communication through membrane junctions. *Arch. Intern. Med.* **129,** 299–305.

Loewenstein, W. R. (1981). Junctional intercellular communication: The cell-to-cell membrane channel. *Physiol. Rev.* **61,** 829–913.

Loewenstein, W. R., and Kanno, Y. (1964). Studies on an epithelial (gland) cell junction. I. Modifications of surface membrane permeability. *J. Cell Biol.* **22,** 565–586.

Mack, A., Neuhaus, J., and Wolburg, H. (1987). Relationship between orthogonal arrays of particles and tight junctions as demonstrated in cells of the ventricular wall of the rat brain. *Cell Tissue Res.* **248,** 619–625.

MacVicar, B. A., and Dudek, F. E. (1981). Electrotonic coupling between pyramidal cells: A direct demonstration in rat hippocampal slices. *Science* **213,** 782–785.

Malewicz, B., Kumar, V. V., Johnson, R. G., and Baumann, W. J. (1990). Lipids in gap junction assembly and function. *Lipids* **25,** 419–427.

Mancel, E., and Hirsch, M. (1989). Development of tight junctions in the human ciliary epithelium. *Exp. Eye Res.* **48,** 87–98.

Manjunath, C. K., and Page, E. (1985). Cell biology and protein composition of cardiac gap junctions. *Am. J. Physiol.* **248,** H783–H791.

Manjunath, C. K., and Page, E. (1986). Rat heart gap junctions as disulfide-bonded connexon multimers: Their depolymerization and solubilization in deoxycholate. *J. Membr. Biol.* **90,** 43–45.

Manjunath, C. K., Goings, G. E., and Page, E. (1984a). Detergent sensitivity and splitting of isolated liver gap junctions. *J. Membr. Biol.* **78,** 147–155.

Manjunath, C. K., Goings, G. E., and Page, E. (1984b). Cytoplasmic surface and intramembrane components of rat heart gap junctional membrane. *Am. J. Physiol.* **246,** H865–H875.

Marc, R. E., Liu, W.-L.S., and Muller, J. F. (1988). Gap junctions in the inner plexiform layer of the goldfish retina. *Vision Res.* **28,** 9–24.

Marshak, D. W., and Dowling, J. E. (1987). Synapses of cone horizontal cell axons in goldfish retina. *J. Comp. Neurol.* **256,** 430–443.

Massa, P. T., and Mugnaini, E. (1982). Cell junctions and intramembrane particles of astrocytes and oligodendrocytes: A freeze-fracture study. *Neuroscience* **7,** 523–538.

Massa, P. T., and Mugnaini, E. (1985). Cell-cell junctional interactions and characteristic plasma membrane features of cultured rat glial cells. *Neuroscience* **14,** 695–709.

Mazet, F., Wittenberg, B. A., and Spray, D. C. (1985). Fate of intercellular junctions in isolated adult cardiac rat cells. *Circ. Res.* **56,** 195–204.

Mazet, J.-L., Jarry, T., Gros, D., and Mazet, F. (1992). Voltage dependence of liver gap-junction channels reconstituted into liposomes and incorporated into planar bilayers. *Eur. J. Biochem.* **210,** 249–256.

McNutt, N. S., and Weinstein, R. S. (1970). The ultrastructure of the nexus. A correlated thin-section and freeze-cleave study. *J. Cell Biol.* **47,** 666–688.

Meda, P., Perrelet, A., and Orci, L. (1979). Increase of gap junctions between pancreatic B-cells during stimulation of insulin secretion. *J. Cell Biol.* **82,** 441–448.

Meda, P., Perrelet, A., and Orci, L. (1984). Gap junctions and cell-to-cell coupling in endocrine glands. *Mod. Cell Biol.* **3,** 131–196.

Mege, R.-M., Matsuzaki, F., Gallin, W. J., Goldberg, J. I., Cunningham, B. A., and Edelman, G. M. (1988). Construction of epithelioid sheets by transfection of mouse sarcoma cells with cDNAs for chicken cell adhesion molecules. *Proc. Natl. Acad. Sci. U.S.A.* **85,** 7274–7278.

Meyer, R. A., Malewicz, B., Baumann, W. J., and Johnson, R. G. (1990). Increased gap junction assembly between cultured cells upon cholesterol supplementation. *J. Cell Sci.* **96,** 231–238.

Meyer, R. A., Laird, D. W., Revel, J. P., and Johnson, R. G. (1992). Inhibition of gap junction and adherens junction assembly by connexin and A-CAM antibodies. *J. Cell Biol.* **119,** 179–189.

Micevych, P. E., and Abelson, L. (1991). Distribution of mRNAs coding for liver and heart gap junction proteins in the rat central nervous system. *J. Comp. Neurol.* **305,** 96–118.

Milks, L. C., Kumar, N. M., Houghton, R., Unwin, N., and Gilula, N. B. (1988). Topology of the 32-kd liver gap junction protein determined by site-directed antibody localizations. *EMBO J.* **7,** 2967–2976.

Miller, T. M., and Goodenough, D. A. (1985). Gap junction structures after experimental alteration of junctional channel conductance. *J. Cell Biol.* **101,** 1741–1748.

Minkoff, R., Rundus, V. R., Parker, S. B., Beyer, E. C., and Hertzberg, E. L. (1993). Connexin expression in the developing avian cardiovascular system. *Circ. Res.* **73,** 71–78.

Mugnaini, E. (1986). Cell junctions of astrocytes, ependyma and related cells in the mammalian central nervous system, with emphasis on the hypothesis of a generalized syncytium of supporting cells. *In* "Astrocytes" (S. Fedoroff and A. Vernadakis, eds.), Vol. 1, pp. 329–371. Academic Press, London.

Murray, A. W., and Gainer, H. S. C. (1989). Regulation of gap junctional communication by protein kinases. *In* "Cell Interactions and Gap Junctions" (N. Sperelakis and W. Cole, eds.), Vol. 1, pp. 97–106. CRC Press, Boca Raton, FL.

Murray, S. A. and Taylor, F. (1988). Dibutyryl cyclic AMP modulation of gap junctions in SW-13 human adrenal cortical tumor cells. *Am. J. Anat.* **181,** 141–148.

Musil, L. S., and Goodenough, D. A. (1990). Gap junctional intercellular communication and the regulation of connexin expression and function. *Curr. Opin. Cell Biol.* **2,** 875–880.

Musil, L. S., and Goodenough, D. A. (1993). Multisubunit assembly of an integral plasma membrane channel protein, gap junction connexin 43, occurs after exit from the ER. *Cell (Cambridge, Mass.)* **74,** 1065–1078.

Musil, L. S., Cunningham, B. A., Edelman, G. M., and Goodenough, D. A. (1990). Differential phosphorylation of the gap junction protein connexin-43 in junctional communication-competent and communication-deficient cell lines. *J. Cell Biol.* **111,** 2077–2088.

Nagy, J. I., Yamamoto, T., Sawchuk, M. A., Nance, D. M., and Hertzberg, E. L. (1992). Quantitative immunohistochemical and biochemical correlates of connexin 43 localization in rat brain. *Glia* **5,** 1–9.

Naka, K.-I., and Rushton, W. A. H. (1967). The generation and spread of S-potentials in fish (Cyprinidae). *J. Physiol.* **192,** 437–461.

Naus, C. C. G., Bechberger, J. F., Caveney, S., and Wilson, J. X. (1991). Expression of gap junction genes in astrocytes and C6 glioma cells. *Neurosci. Lett.* **126,** 33–36.

Naus, C. C. G., Hearn, S., Zhu, D., Nicholson, B. J., and Shivers, R. R. (1993). Ultrastructural analysis of gap junctions in C6 glioma cells transfected with connexin 43 cDNA. *Exp. Cell Res.* **206,** 72–84.

Negishi, K., Teranishi, T., and Kato, S. (1983). A GABA antagonist, bicuculline, exerts its uncoupling action on external horizontal cells through dopamine cells in carp retina. *Neurosci. Lett.* **37,** 261–266.

Neiss, W. F., Schulte, E., Guntinas-Lichius, O., Gunkel, A., and Stennert, E. (1993). Quantification of chromatolysis in motoneurons of the Wistar rat. *Ann. Anat. Suppl.* **175,** 6–7.

Neveu, M. J., Hully, J. R., Babcock, K. L., Hertzberg, E. L., Nicholson, B. J., Paul, D. L., and Pitot, H. C. (1994). Multiple mechanisms are responsible for altered expression of gap junction genes during oncogenesis in rat liver. *J. Cell Sci.* **107,** 83–95.

Neyton, J., and Trautmann, A. (1986). Physiological modulation of gap junction permeability. *J. Exp. Biol.* **124,** 93–114.

Nicholson, B., Dermietzel, R., Teplow, D., Traub, O., Willecke, K., and Revel, J.-P. (1987). Two homologous protein components of hepatic gap junctions. *Nature (London)* **329,** 732–734.

Odette, L. L., and Newman, E. A. (1988). Model of potassium dynamics in the central nervous system. *Glia* **1,** 198–210.

Oh, S. Y., Dupont, E., Madhukar, B. V., Briand, J.-P., Chang, C.-C., Beyer, E., and Trosko, J. E. (1993). Characterization of gap junctional communication-deficient mutants of a rat liver epithelial cell line. *Eur. J. Cell Biol.* **60,** 250–255.

Owen, W. G. (1985). Chemical and electrical synapses between photoreceptors in the retina of the turtle, *Chelydra serpentina. J. Comp. Neurol.* **240,** 423–433.

Page, E., Karrison, T., and Upshaw-Earley, J. (1983). Freeze-fractured cardiac gap junctions: Structural analysis by three methods. *Am. J. Physiol.* **244,** H525–H539.

Oyamada, M., Kimura, H., Oyamada, Y., Miyamoto, A., Ohshika, H., and Mori, M. (1994). The expression, phosphorylation, and localization of connexin 43 and gap-junctional intercellular communication during the establishment of a synchronized contraction of cultured neonatal rat cardiac myocytes. *Exp. Cell Res.* **212,** 351–358.

Paul, D. L., Ebihara, L., Takemoto, L. J., Swenson, K. I., and Goodenough, D. A. (1991). Connexin46, a novel lens gap junction protein, induces voltage-gated currents in nonjunctional plasma membrane of *Xenopus* oocytes. *J. Cell Biol.* **115,** 1077–1089.

Payton, B. W., Bennett, M. V. L., and Pappas, G. D. (1969). Permeability and structure of junctional membranes at an electrotonic synapse. *Science* **166,** 1641–1643.

Pederson, D., Sheridan, J. D., and Johnson, R. (1980). The development of metabolite transfer between reaggregating Novikoff hepatoma cells. *Exp. Cell Res.* **127,** 159–178.

Peinado, A., Yuste, R., and Katz, L. C. (1993). Extensive dye coupling between rat neocortical neurons during the period of circuit formation. *Neuron* **10,** 103–114.

Peracchia, C. (1977). Gap junctions. Structural changes after uncoupling procedures. *J. Cell Biol.* **72,** 628–641.

Peracchia, C. (1980). Structural correlates of gap junction permeation. *Int. Rev. Cytol.* **66,** 81–146.

Peracchia, C., and Bernardini, G. (1984). Gap junction structure and cell-to-cell coupling regulation: Is there a calmodulin involvement? *Fed. Proc. Fed. Am. Soc. Exp. Biol.* **43,** 2681–2691.

Peracchia, C., and Dulhunty, A. F. (1976). Low resistance junctions in crayfish. Structural changes with functional uncoupling. *J. Cell Biol.* **70,** 419–439.

Peracchia, C., and Girsch, S. J. (1985). Functional modulation of cell coupling: Evidence for a calmodulin-driven channel gate. *Am. J. Physiol.* **248,** H765–H782.

Peracchia, C., and Peracchia, L. L. (1980a). Gap junction dynamics. Reversible effects of divalent cations. *J. Cell Biol.* **87,** 708–718.

Peracchia, C., and Peracchia, L. L. (1980b). Gap junction dynamics. Reversible effects of hydrogen ions. *J. Cell Biol.* **87,** 719–727.

Peracchia, C., and Peracchia, L. L. (1985). Bridges linking gap junction particles extracellularly. A freeze-etching rotary-shadowing study of split junctions. *Eur. J. Cell Biol.* **36,** 286–293.

Peters, A. (1980). Neurobiology—General principles related to epilepsy. Morphological correlates of epilepsy: Cells in the cerebral cortex. *In* "Antiepileptic Drugs: Mechanisms of Action" (G. H. Glaser, J. K. Penry, and D. M. Woodburgy, eds.), pp. 21–48. Raven Press, New York.

Piccolino, M., Neyton, J., Witkovsky, P., and Gerschenfeld, H. M. (1982). Gamma-aminobutyric acid antagonists decrease junctional communication between horizontal cells of the retina. *Proc. Natl. Acad. Sci. U.S.A.* **79,** 3671–3675.

Piccolino, M., Neyton, J., and Gerschenfeld, H. M. (1984). Decrease of the gap-junction permeability induced by dopamine and cyclic 3'-5' adenosine-monophosphate in horizontal cells of the turtle retina. *J. Neurosci.* **4,** 2477–2488.

Piccolino, M., Witkovsky, P., and Trimarchi, C. (1987). Dopaminergic mechanisms underlying the reduction of electrical coupling between horizontal cells of the turtle retina induced by d-amphetamine, bibuculline, and veratridine. *J. Neurosci.* **7,** 2273–2284.

Piccolino, M., Demontis, G., Witkovsky, P., Strettoi, E., Cappagli, G. C., Porceddu, M. L., De Montis, M. G., Pepitoni, S., Biggio, G., Meller, E., and Bohmaker, K. (1989). Involvement of D1 and D2 dopamine receptors in the control of horizontal cell electrical coupling in the turtle retina. *Eur. J. Neurosci.* **1,** 247–257.

Pitts, J. D., and Finbow, M. E. (1986). The gap junction. *J. Cell Sci., Suppl.* **4,** 239–266.

Pitts, J. D., and Sims, J. W. (1977). Permeability of junctions between animal cells. Intercellular transfer of nucleotides but not of macromolecules. *Exp. Cell Res.* **104,** 153–163.

Preus, D., Johnson, R., and Sheridan, J. (1981). Gap junctions between Novikoff hepatoma cells following dissocation and recovery in the absence of cell contact. *J. Ultrastruct. Res.* **77,** 248–262.

Puranam, K. L., Laird, D. W., and Revel, J. P. (1993). Trapping an intermediate form of connexin 43 in the Golgi. *Exp. Cell Res.* **206,** 85–92.

Raff, M. C. (1989). Glial cell diversification in the rat optic nerve. *Science* **243,** 1450–1455.

Rahman, S., Carlile, G., and Evans, W. H. (1993). Assembly of hepatic gap junctions. Topography and distribution of connexin 32 in intracellular and plasma membranes determined using sequence-specific antibodies. *J. Biol. Chem.* **268,** 1260–1265.

Ransom, B. R., and Kettenmann, H. (1990). Electrical coupling, without dye coupling, between mammalian astrocytes and oligodendrocytes in cell culture. *Glia* **3,** 258–266.

Rash, J. E., and Yasumura, T. (1992). Improved structural detail in freeze-fracture replicas: High-angle shadowing of gap junctions cooled below −170°C and protected by liquid nitrogen-cooled shrouds. *Microsc. Res. Tech.* **20,** 187–204.

Rassat, J., Robenek, H., and Themann, H. (1981). Ultrastructural changes in mouse hepatocytes exposed to vinblastine sulfate with special reference to the intercellular junctions. *Eur. J. Cell Biol.* **24,** 203–210.

Raviola, E. (1976). Intercellular junctions in the outer plexiform layer of the retina. *Invest. Ophthalmol. Visual Sci.* **15,** 881–895.

Raviola, E., and Gilula, N. B. (1973). Gap junctions between photoreceptor cells in the vertebrate retina. *Proc. Natl. Acad. Sci. U.S.A.* **70,** 1677–1681.

Raviola, E., and Gilula, N. B. (1975). Intramembrane organization of specialized contacts in the outer plexiform layer of the retina. *J. Cell Biol.* **65,** 192–222.

Raviola, E., Goodenough, D. A., and Raviola, G. (1980). Structure of rapidly frozen gap junctions. *J. Cell Biol.* **87,** 273–279.

Reed, K. E., Westphale, E. M., Larson, D. M., Wang, H.-Z., Veenstra, R. D., and Beyer, E. C. (1993). Molecular cloning and functional expression of human connexin 37, an endothelial cell gap junction protein. *J. Clin. Invest.* **91,** 997–1004.

Reichenbach, A. (1989). Attempt to classify glial cells by means of their process specialization using the rabbit retinal Müller cell as an example of cytotopographic specialization of glial cells. *Glia* **2,** 250–259.

Revel, J.-P., and Karnovsky, M. J. (1967). Hexagonal array of subunits in intercellular junctions of the mouse heart and liver. *J. Cell Biol.* **33,** C7–C12.

Revel, J.-P., Olson, W., and Karnovsky, M. J. (1967). A twenty-angström gap junction with a hexagonal array of subunits in smooth muscle. *J. Cell Biol.* **35,** 112A.

Revel, J.-P., Nicholson, B. J., and Yancey, S. B. (1985). Chemistry of gap junctions. *Annu. Rev. Physiol.* **47**, 263–280.

Reyher, C. K. H., Lübke, J., Larsen, W. J., Hendrix, G. M., Shipley, M. T., and Baumgarten, H. G. (1991). Olfactory bulb granule cell aggregates: Morphological evidence for interperikaryal electrotonic coupling via gap junctions. *J. Neurosci.* **11**, 1485–1495.

Rieske, E., Schubert, P., and Kreutzberg, G. W. (1975). Transfer of radioactive material between electrically coupled neurons of the leech central nervous system. *Brain Res.* **84**, 365–382.

Risek, B., Klier, F. G., and Gilula, N. B. (1994). Developmental regulation and structural organization of connexins in epidermal gap junctions. *Develop. Biol.* **164**, 183–196.

Robertson, J. D. (1963). The occurrence of a subunit pattern in the unit membranes of club endings in Mauthner cell synapses in goldfish brains. *J. Cell Biol.* **19**, 201–221.

Rohlmann, A., and Wolff, J. R. (1994). A morphometric study of the astrocytic gap junction coupling. Submitted for publication.

Rohlmann, A., Laskawi, R., Hofer, A., Dodo, E., Dermietzel, R., and Wolff, J. R. (1993). Facial nerve lesions lead to increased immunostaining of the astrocytic gap junction protein (connexin 43) in the corresponding facial nucleus of rats. *Neurosci. Lett.* **154**, 206–208.

Rohlmann, A., Laskawi, R., Hofer, A., Dermietzel, R., and Wolff, J. R. (1994a). Astrocytes as rapid sensors of peripheral axotomy in the facial nucleus of rats. *NeuroReport* **5**, 409–412.

Rohlmann, A., Laskawi, R., and Wolff, J. R. (1994b). Structure and dynamic regulation of the astrocytic gap junction coupling pattern. *J. Brain Res.* **35**, 132.

Rose, B. (1971). Intercellular communication and some structural aspects of membrane junctions in a simple cell system. *J. Membr. Biol.* **5**, 1–19.

Rose, J. K., and Doms, R. W. (1988). Regulation of protein export from the endoplasmic reticulum. *Annu. Rev. Cell Biol.* **4**, 257–288.

Rosenberg, E., Spray, D. C., and Reid, L. M. (1993). Matrix regulation of gap junctions and of excitability in cells. *In* "Extracellular Matrix" (M. A. Zern and L. M. Reid, eds.), pp. 449–462. New York and Basel.

Rup, D. M., Veenstra, R. D., Wang, H.-Z., Brink, P. R., and Beyer, E. C. (1993). Chick connexin-56, a novel lens gap junction protein. *J. Biol. Chem.* **268**, 706–712.

Ryerse, J. S. (1993). Structural, immunocytochemical and initial biochemical characterization of NAOH-extracted gap junctions from an insect, *Heliothis virescens. Cell Tissue Res.* **274**, 393–403.

Ryerse, J. S., Nagel, B. A., and Hammel, I. (1984). The role of connexon aggregate fusion in gap junction growth. *J. Submicrosc. Cytol.* **16**, 649–657.

Saez, J. C., Spray, D. C., Nairn, A. C., Hertzberg, E. L., Greengard, P., and Bennett, M. V. L. (1986). c-AMP increases junctional conductance and stimulates phosphorylation of the 27-kDa principal gap junction polypeptide. *Proc. Natl. Acad. Sci. U.S.A.* **83**, 2473–2477.

Saez, J. C., Nairn, A. C., Spray, D. C., Czernik, A. J., and Hertzberg, E. L. (1990a). Protein kinase C-dependent regulation of heart connexin 43. *J. Cell Biol.* **111**, 145a.

Saez, J. C., Nairn, A. C., Czernik, A. J., and Spray, D. C. (1990b). Phosphorylation of connexin-32, a hepatocyte gap-junction protein, by cAMP-dependent protein kinase, protein kinase-C and Ca^{2+} calmodulin-dependent protein kinase-II. *Eur. J. Biochem.* **192**, 263–273.

Sanderson, M. J., Charles, A. C., and Dirksen, E. R. (1990). Mechanical stimulation and intercellular communication increases intracellular Ca^{2+} in epithelial cells. *Cell Regul.* **1**, 585–596.

Schmalbruch, H., and Jahnsen, H. (1981). Gap junctions on CA3 pyramidal cells of guinea pig hippocampus shown by freeze-fracture. *Brain Res.* **217**, 175–178.

Schmitz, Y., and Wolburg, H. (1991). Gap junction morphology of retinal horizontal cells is sensitive to pH alterations in vitro. *Cell Tissue Res.* **263,** 303–310.

Schoenenberger, C.-A., and Matlin, K. S. (1991). Cell polarity and epithelial oncogenesis. *Trends Cell Biol.* **1,** 87–92.

Schuster, T. (1990). "Kommunikation zwischen Zellen. Gap Junctions, Nexus, elektrische Synapsen." Akademie-Verlag, Berlin.

Schuster, T., Bergmann, M., and Wendler, D. (1990). Neuroepithelial cell contacts in 12 day old rat embryo. Normal and annulated gap junctions. *J. Hirn Forsch.* **31,** 405–414.

Severs, N. J., Gourdie, R. G., Harfst, E., Peters, N. S., and Green, C. R. (1993).Intercellular junctions and the application of microscopical techniques: The cardiac gap junction as a case model. *J. Microsc. (Oxford)* **169,** 299–328.

Sheridan, J. D., and Atkinson, M. M. (1985). Physiological roles of permeable junctions: Some possibilities. *Annu. Rev. Physiol.* **47,** 337–354.

Shibata, Y., and Yamamoto, T. (1986). Cytoplasmic surface ultrastructures of cardiac gap junctions as revealed by quick-freeze, deep-etch replicas. *Anat. Rec.* **214,** 107–112.

Shibata, Y., Manjunath, C. K., and Page, E. (1985). Differences between cytoplasmic surfaces of deep-etched heart and liver gap junctions. *Am. J. Physiol.* **249,** H690–H693.

Shiosaka, S., Yamamoto, T., Hertzberg, E. L., and Nagy, J. I. (1989). Gap junction protein in rat hippocampus: Correlative light and electron microscope immunohistochemical localization. *J. Comp. Neurol.* **281,** 282–297.

Sikerwar, S. S., Downing, K. H., and Glaeser, R. M. (1991). Three-dimensional structure of an invertebrate intercellular communicating junction. *J. Struct. Biol.* **106,** 255–263.

Simionescu, M., Simionescu, N., and Palade, G. E. (1976). Segmental differentiations of cell junctions in the vascular endothelium. Arteries and veins. *J. Cell Biol.* **68,** 705–723.

Simpson, J., Rose, B., and Loewenstein, W. R. (1977). Size limit of molecules permeating the junctional membrane channels. *Science* **195,** 294–301.

Singer, S. J. (1992). Intercellular communication and cell-cell adhesion. *Science* **255,** 1671–1677.

Sloper, J. J. (1972). Gap junctions between dendrites in the primate neocortex. *Brain Res.* **44,** 641–646.

Sloper, J. J., and Powell, T. P. S. (1978). Gap junctions between dendrites and somata of neurones in the primate sensorimotor cortex. *Proc. R. Soc. London* **203,** 39–47.

Smith, D. E., and Moskovitz, N. (1979). Ultrastructure of layer IV of the primary auditory cortex of the squirrel monkey. *Neuroscience* **4,** 349–360.

Smith, S. J. (1992). Do astrocytes process neural information? *Prog. Brain Res.* **94,** 119–136.

Sontheimer, H., Minturn, J. E., Black, J. A., Waxman, S. G., and Ransom, B. R. (1990). Specificity of cell-cell coupling in rat optic nerve astrocytes in vitro. *Proc. Natl. Acad. Sci. U.S.A.* **87,** 9833–9837.

Sosinsky, G. E., Jésior, J. C., Caspar, D. L. D., and Goodenough, D. A. (1988). Gap junction structures. VIII. Membrane cross-sections. *Biophys. J.* **53,** 709–722.

Sotelo, C., and Korn, H. (1978). Morphological correlates of electrical and other interactions through low-resistance pathways between neurons of the vertebrate central nervous system. *Int. Rev. Cytol.* **55,** 67–107.

Sotelo, C., and Llinas, R. (1972). Specialized membrane junctions between neurons in the vertebrate cerebellar cortex. *J. Cell Biol.* **53,** 271–289.

Sotelo, C., Llinas, R., and Baker, R. (1974). Structural study of inferior olivary nucleus of the cat: Mophological correlates of electrotonic coupling. *J. Neurophysiol.* **37,** 541–559.

Spray, D. C., and Bennett, M. V. L. (1985). Physiology and pharmacology of gap junctions. *Annu. Rev. Physiol.* **47,** 281–303.

Spray, D. C., and Burt, J. M. (1990). Structure-activity relations of the cardiac gap junctional channel. *Am. J. Physiol.* **258,** C195–C205.

Spray, D. C., White, R. L., Mazet, F., and Bennett, M. V. L. (1985). Regulation of gap junctional conductance. *Am. J. Physiol.* **248,** H753–H764.

Spray, D. C., Saez, J. C., Brosius, D., Bennett, M. V. L., and Hertzberg, E. L. (1986). Isolated liver gap junctions: Gating of transjunctional currents is similar to that in intact pairs of rat hepatocytes. *Proc. Natl. Acad. Sci. U.S.A.* **83,** 5494–5497.

Spray, D. C., Saez, J. C., Burt, J. M., Watanabe, T., Reid, L. M., Hertzberg, E. L., and Bennett, M. V. L. (1988). Gap junctional conductance: Multiple sites of regulation. *In* "Gap Junctions" (E. L. Hertzberg and R. G. Johnson, eds.), pp. 227–244. Alan R. Liss, New York.

Staehelin, L. A. (1974). Structure and function of intercellular junctions. *Int. Rev. Cytol.* **39,** 191–283.

Stauffer, K. A., Kumar, N. M., Gilula, N. B., and Unwin, N. (1991). Isolation and purification of gap junction channels. *J. Cell Biol.* **115,** 141–150.

Stein, L. S., Stein, D. W. J., Echols, J., and Burghardt, R. C. (1993). Concomitant alterations of desmosomes, adhesiveness, and diffusion through gap junction channels in a rat ovarian transformation model system. *Exp. Cell Res.* **207,** 19–32.

Swenson, K. I., Jordan, I. R., Beyer, E. C., and Paul, D. L. (1989). Formation of gap junctions by expression of connexins in *Xenopus* oocyte pairs. *Cell (Cambridge, Mass.)* **57,** 145–155.

Swenson, K. I., Piwnica-Worms, H., McNamee, H., and Paul, D. L. (1990). Tyrosine phosphorylation of the gap junction protein connexin 43 is required for the pp60^{v-src}-induced inhibition of communication. *Cell Regul.* **1,** 989–1002.

Tadvalkar, G., and Pinto da Silva, P. (1983). In vitro rapid assembly of gap junctions is induced by cytoskeleton disruptors. *J. Cell Biol.* **96,** 1279–1287.

Takeda, A., Saheki, S., Shimazu, T., and Takeuchi, N. (1989). Phosphorylation of the 27-kDa gap junction protein by protein kinase C in vitro and in rat hepatocytes. *J. Biochem. (Tokyo)* **106,** 723–727.

Takeichi, M. (1990). Cadherins: A molecular family important in selective cell-cell adhesion. *Annu. Rev. Biochem.* **59,** 237–252.

Takeichi, M. (1991). Cadherin cell adhesion receptors as a morphogenetic regulator. *Science* **251,** 1451–1455.

Teranishi, T., and Negishi, K. (1994). Double-staining of horizontal and amacrine cells by intracellular injection with Lucifer Yellow and biocytin in carp retina. *Neuroscience* **59,** 217–226.

Teranishi, T., Negishi, K., and Kato, S. (1984). Regulatory effect of dopamine on spatial properties of horizontal cells in carp retina. *J. Neurosci.* **4,** 1271–1280.

Tetzlaff, W., and Kreutzberg, G. W. (1985). Ornithine decarboxylase in motoneurons during regeneration. *Exp. Neurol.* **89,** 679–688.

Tonosaki, A., Washioka, H., Nakamura, H., and Negishi, K. (1985). Complementary freeze-fracture replication: An example of its use in the study of horizontal cell gap junctions of the carp retina. *J. Electron. Microsc. Tech.* **2,** 187–192.

Traub, O., Look, J., Dermietzel, R., Brümmer, F., Hülser, D., and Willecke, K. (1989). Comparative characterization of the 21-kD and 26-kD gap junction proteins in murine liver and cultured hepatocytes. *J. Cell Biol.* **108,** 1039–1051.

Tsien, R. W., and Weingart, R. (1976). Inotropic effects of cyclic AMP in calf ventricular muscle studied by a cut end method. *J. Physiol. (London)* **260,** 117–141.

Uga, S., and Smelser, G. K. (1973). Comparative study of the fine structure of retinal Müller cells in various vertebrates. *Invest. Ophthalmol.* **12,** 434–448.

Unwin, P. N. T., and Ennis, P. D. (1983). Calcium-mediated changes in gap junction structure: Evidence from the low angle X-ray pattern. *J. Cell Biol.* **97,** 1459–1466.

Unwin, P. N. T., and Ennis, P. (1984). Two configurations of a channel-forming membrane protein. *Nature (London)* **307**, 609–613.

Unwin, P. N. T., and Zampighi, G. A. (1980). Structure of the junction between communicating cells. *Nature (London)* **283**, 545–549.

Van den Hoef, M. H. F., Dictus, W. J., Hage, W. J., and Bluemink, J. G. (1984a). The ultrastructural organization of gap junctions between follicle cells and the oocyte in *Xenopus laevis*. *Eur. J. Cell Biol.* **33**, 242–247.

Van den Hoef, M. H. F., Bluemink, J. G., and De Laat, S. W. (1984b). Gap junctions and rhombic particle arrays in the freeze-fractured mouse oocyte-follicle cell complex. *Eur. J. Cell Biol.* **35**, 312–316.

Van Deurs, B., and Koehler, J. K. (1979). Tight junctions in the choroid plexus epithelium. A freeze-fracture study including complementary replicas. *J. Cell Biol.* **80**, 662–673.

Vaney, D. I. (1993). The coupling pattern of axon-bearing horizontal cells in the mammalian retina. *Proc. R. Soc. London, Ser. B* **252**, 93–101.

Vaney, D. I. (1994). Patterns of neuronal coupling in the retina. *Prog. Retinal Eye Res.* **13**, 301–355.

Vaughan, D. K., and Lasater, E. M. (1990a). Distribution of F-actin in bipolar and horizontal cells of bass retinas. *Am. J. Physiol.* **259**, C205–C214.

Vaughan, D. K., and Lasater, E. M. (1990b). Renewal of electrotonic synapses in teleost retinal horizontal cells. *J. Comp. Neurol.* **299**, 364–374.

Voorter, C. E. M., and Kistler, J. (1989). cAMP-dependent protein kinase phosphorylates gap junction proteins in lens cortex but not in lens nucleus. *Biochim. Biophys. Acta* **986**, 8–10.

Vukelic, J. I., Yamamoto, T., Hertzberg, E. L., and Nagy, J. I. (1991). Depletion of connexin43-immunoreactivity in astrocytes after kainic acid-induced lesions in rat brain. *Neurosci. Lett.* **130**, 120–124.

Warner, A. E. (1992). Gap junctions in development—a perspective. *Semin. Cell Biol.* **3**, 81–91.

Warner, A. E., Guthrie, S. C., and Gilula, N. B. (1984). Antibodies to gap-junctional protein selectively disrupt junctional communication in the early amphibian embryo. *Nature (London)* **311**, 127–131.

Weiler, R., Kohler, K., Kolbinger, W., Wolburg, H., Kurz-Isler, G., and Wagner, H.-J. (1988). Dopaminergic neuromodulation in the retina of lower vertebrates. *Neurosci. Res., Suppl.* **8**, S183–S196.

Weinbaum, S. (1980). Theory for the formation of intercellular junctions based on intramembranous particle patterns observed in the freeze-fracture technique. *J. Theor. Biol.* **83**, 63–92.

Werner, R., Levine, E., Rabadan-Diehl, D., and Dahl, G. (1989). Formation of hybrid cell-cell channels. *Proc. Natl. Acad. Sci. U.S.A.* **86**, 5380–5384.

Werner, R., Levine, E., Rabadan-Diehl, C., and Dahl, G. (1991). Gating properties of connexin32 cell-cell channels and their mutants expressed in *Xenopus* oocytes. *Proc. R. Soc. London, Ser. B* **243**, 5–11.

White, F. H., Thompson, D. A., and Gohari, K. (1984). Ultrastructural morphometry of gap junctions during differentiation of stratified squamous epithelium. *J. Cell Sci.* **69**, 67–86.

Willecke, K., Jungbluth, S., Dahl, E., Hennemann, H., Heynkes, R., and Grzeschik, K. H. (1990). Six genes of the human connexin gene family coding for gap junctional proteins are assigned to four different human chromosomes. *Eur. J. Cell Biol.* **53**, 275–280.

Willecke, K., Heynkes, R., Dahl, E., Stutenkemper, R., Hennemann, H., Jungbluth, S., Suchyna, T., and Nicholson, B. J. (1991). Mouse connexin37: Cloning and functional expression of a gap junction gene highly expressed in lung. *J. Cell Biol.* **114**, 1049–1057.

Witkovsky, P., Owen, W. G., and Woodworth, M. (1983). Gap junctions among the peri-karya, dendrites, and axon terminals of the luminosity-type horizontal cell of the turtle retina. *J. Comp. Neurol.* **261,** 359–368.

Wolburg, H., and Kurz-Isler, G. (1985). Dynamics of gap junctions between horizontal cells in the goldfish retina. *Exp. Brain Res.* **60,** 397–401.

Wolburg, H., Kästner, R., and Kurz-Isler, G. (1983). Lack of orthogonal particle assemblies and presence of tight junctions in astrocytes of goldfish. A freeze-fracture study. *Cell Tissue Res.* **234,** 389–402.

Wolszon, L. R., Gao, W.-Q., Passani, M. B., and Macagno, E. R. (1994). Growth cone "collapse" *in vivo:* Are inhibitory interactions mediated by gap junctions? *J. Neurosci.* **14,** 999–1010.

Yamamoto, T., Shiosaka, S., Whittaker, M. E., Hertzberg, E. L., and Nagy, J. I. (1989). Gap junction protein in rat hippocampus: Light microscope immunohistochemical localization. *J. Comp. Neurol.* **281,** 269–281.

Yamamoto, T., Ochalski, A., Hertzberg, E. L., and Nagy, J. I. (1990a). LM and EM immunolocalization of the gap junctional protein connexin 43 in rat brain. *Brain Res.* **508,** 313–319.

Yamamoto, T., Hertzberg, E. L., and Nagy, J. I. (1990b). Epitopes of gap junctional proteins localized to neuronal subsurface cisterns. *Brain Res.* **527,** 135–139.

Yamamoto, T., Ochalski, A., Hertzberg, E. L., and Nagy, J. I. (1990c). On the organization of astrocytic gap junctions in rat brain as suggested by LM-immunohistochemistry and EM-immunohistochemistry of connexin-43 expression. *J. Comp. Neurol.* **302,** 853–883.

Yamamoto, T., Kardami, E., and Nagy, J. I. (1991). Basic fibroblast growth factor in rat brain. Localization to glial gap junctions correlates with connexin 43 distribution. *Brain Res.* **554,** 336–343.

Yamamoto, T., Vukelic, J., Hertzberg, E. L., and Nagy, J. I. (1992). Differential anatomical and cellular patterns of connexin43 expression during postnatal development of rat brain. *Dev. Brain Res.* **66,** 165–180.

Yancey, S. B., Dewees, P., and Revel, J.-P. (1980). In vivo incorporation of [35S]methionine into gap junctional proteins. *J. Cell Biol.* **87,** 212.

Yancey, S. B., Nicholson, B. J., and Revel, J.-P. (1981). The dynamic state of liver gap junctions. *J. Supramol. Struct. Cell. Biochem.* **16,** 221–232.

Young, J. D.-E., Cohn, Z. A., and Gilula, N. B. (1987). Functional assembly of gap junctional conductance in lipid bilayers: Demonstration that the major 27kD protein forms the junctional channel. *Cell (Cambridge, Mass.)* **48,** 733–743.

Zeng, M. (1987). Effect of antibodies against gap junction protein on gap function formation in early amphibian embryogenesis. *J. Electron. Microsc. Tech.* **7,** 85–89.

Zhang, J. T., and Nicholson, B. J. (1989). Sequence and tissue distribution of a second protein of hepatic gap junctions, Cx26, as deduced from its cDNA. *J. Cell Biol.* **109,** 3391–3401.

Zhu, D., Kidder, G. M., Caveney, S., and Naus, C. C. G. (1992). Growth retardation in glioma cells cocultured with cells overexpressing a gap junction protein. *Proc. Natl. Acad. Sci. U.S.A.* **89,** 10218–10221.

Zidell, R. H., and Loch-Caruso R. (1990). A simple dye-coupling assay for evaluating gap junctional communication: The importance of transcription and translation on the establishment of dye-coupling. *Cell Biol. Int. Rep.* **14,** 613–628.

Zimmerman, R. P. (1983). Bar synapses and gap junctions in the inner plexiform layer: Synaptic relationships of the interstitial amacrine cell of the retina of the cichlid fish, *Astronotus ocellatus. J. Comp. Neurol.* **218,** 471–479.

Index

A

AA4.1, marker of B cell differentiation, 136, 142
Acetylation, posttranslational modification of microtubules, 12–14, 18
Antibody, *see* Immunoglobulin
Antigen
  presentation, 190
  processing, 190
Antigen recognition activation motif
  consensus sequence, 232–233
  kinase binding, 238–241
  phosphorylation, 233–234, 238, 240
  role in lymphocyte signaling, 184, 188, 231–233
AP-1, transcription factor complex, 244, 253
Apoptosis
  B cell, 142, 183, 197, 208–209, 250–256
  morphological features, 250
  role of
    Bax, 255
    bcl-2, 254–255
    CD40, 209
    Fas ligand, 204
    nur 77, 251, 255–256
    PTK, 213
  T cell, 183, 198–199, 250–256
ARAM, *see* Antigen recognition activation motif
Astrocyte
  connexin staining, 343, 345
  gap junction
    cell coupling, 342–343, 347
    coupling, 347
    induction by neuronal stimuli, 343
    phosphorylation, 344
    quantitation, 345

    size, 345, 347
    staining, 344
Axon, microtubule stability, 17–18

B

B29, marker of B cell differentiation, 141
B220
  marker of B cell differentiation, 137–138, 141–142
  role in B cell development, 156, 161–162
Bax, role in apoptosis, 255
B cell
  activation, 204–208
  apoptosis, 142, 183, 197, 208–209
  CD5 expression, 133
  clonal anergy, 198
  clonal deletion, 197
  cytokine inhibition of progenitor growth, 154
  germinal center cells, 208–209
  lineage commitment, 160–162
  lipopolysaccharide assay, 147
  markers of differentiation
    AA4.1, 136, 142
    B29, 141
    B220, 137–138, 141–142
    BP-1/6C3, 138
    CD10, 139
    CD43, 137–138
    CD45, 137–138
    cell size, 141
    $\gamma 5$, 140–141
    heat-stable antigen, 138, 142
    immunoglobulin proteins, 135–136

375